BEHAVIOURAL
ECOLOGY An Evolutionary Approach

THIRD EDITION

BEHAVIOURAL
ECOLOGY An Evolutionary Approach

EDITED BY

J. R. Krebs FRS
Royal Society Research Professor at the
Edward Grey Institute of Field Ornithology,
Department of Zoology,
University of Oxford; and
E. P. Abraham Fellow of Pembroke College

N. B. Davies
Lecturer in Zoology at the
University of Cambridge; and
Fellow of Pembroke College

DRAWINGS BY
Jan Parr

OXFORD

BLACKWELL SCIENTIFIC PUBLICATIONS

LONDON EDINBURGH BOSTON

MELBOURNE PARIS BERLIN VIENNA

© 1978, 1984, 1991 by
Blackwell Scientific Publications
Editorial offices:
Osney Mead, Oxford OX2 0EL
25 John Street, London WC1N 2BL
23 Ainslie Place, Edinburgh EH3 6AJ
3 Cambridge Center, Cambridge
  Massachusetts 02142, USA
54 University Street, Carlton
  Victoria 3053, Australia

Other editorial offices:
Arnette SA
2, rue Casimir-Delavigne
75006 Paris
France

Blackwell Wissenschaft
Meinekestrasse 4
D-1000 Berlin 15
Germany

Blackwell MZV
Feldgasse 13
A-1238 Wien
Austria

First published 1978
Second edition 1984
Reprinted 1986, 1989
Third edition 1991

Set by Enset (Photosetting),
Midsomer Norton, Bath, Avon
Printed and bound in Great Britain
at The University Press, Cambridge

DISTRIBUTORS

Marston Book Services Ltd
PO Box 87
Oxford OX2 0DT
(*Orders*: Tel: 0865 791155
      Fax: 0865 791927
      Telex: 837515)

USA
Blackwell Scientific Publications, Inc.
3 Cambridge Center
Cambridge, MA 02142
(*Orders*: Tel: 800 759-6102)

Canada
Oxford University Press
70 Wynford Drive
Don Mills
Ontario M3C 1J9
(*Orders*: Tel: 416 441-2941)

Australia
Blackwell Scientific Publications
(Australia) Pty Ltd
54 University Street
Carlton, Victoria 3053
(*Orders*: Tel: 03 347-0300)

British Library
Cataloguing in Publication Data

Behavioural ecology.—3rd. ed.
  1. Animals. Behaviour. Evolution.
  Ecological aspects
  I. Krebs, J.R. (John R.)
  II. Davies, N.B. (Nicholas B.)
  591.51

  ISBN 0-632-02792-9

Library of Congress
Cataloging in Publication Data

Behavioural ecology: an evolutionary
approach/
  edited by J.R. Krebs,
  N.B. Davies.—3rd ed.
      p.   cm.
  Includes bibliographical references and
  index.
  ISBN 0-632-02702-9
  1. Animal behavior.   2. Animal ecology
  3. Behavior evolution.
  I. Krebs, J.R. (John R.)
  II. Davies, N.B. (Nicholas B.), 1952–.
  QL751.B345   1991
  591.5′ 1—dc20

Contents

Part 4: Cooperation and Conflict

List of authors

Monique Borgerhoff Mulder *Department of Anthropology, University of California, Davis, CA 95616, USA*

Jack W. Bradbury *Department of Biology, University of California, La Jolla, San Diego, CA 92093, USA*

Tim Clutton-Brock *Large Animal Research Group, Department of Zoology, 34A Storeys Way, Cambridge CB2 0DT, UK*

Nicholas B. Davies *Department of Zoology, Downing Street, Cambridge CB2 3EJ, UK*

Stephen T. Emlen *Section of Neurobiology and Behavior, Division of Biological Sciences, Cornell University, Seeley G. Mudd Hall, Ithaca, NY 14853-2702, USA*

John A. Endler *Department of Biological Sciences, University of California, Santa Barbara, CA 93106, USA*

Charles Godfray *Department of Biology and Centre for Population Biology, Imperial College at Silwood Park, Ascot, Berkshire SL5 7PY, UK*

Alan Grafen *Department of Plant Sciences, University of Oxford, South Parks Road, Oxford OX1 3RA, UK*

David G. C. Harper *School of Biological Sciences, Biology Building, Falmer, Brighton, Sussex BN1 9QG, UK*

Paul H. Harvey *Department of Zoology, University of Oxford, South Parks Road, Oxford OX1 3PS, UK*

Alejandro Kacelnik *Edward Grey Institute of Ornithology, Department of Zoology, South Parks Road, Oxford OX1 3PS, UK*

John R. Krebs *Edward Grey Institute of Ornithology, Department of Zoology, South Parks Road, Oxford OX1 3PS, UK*

Catherine M. Lessells *Department of Animal and Plant Sciences, University of Sheffield, Sheffield S10 2TN, UK*

Manfred Milinski *Abteilung Verhaltensökologie, Ethologische Station Hasli,
Zoologisches Institut, Universität Bern, 50a Wohlenstrasse, CH-3032,
Hinterkappelen, Switzerland*

Geoffrey A. Parker *Department of Environmental and Evolutionary Biology,
University of Liverpool, Liverpool L69 3BX, UK*

Jon Seger *Department of Biology, University of Utah, Salt Lake City, Utah
84112, USA*

Preface

In editing the 3rd edition of *Behavioural Ecology*, we have followed the tradition established by the two previous versions. Our authors were asked to write up-to-date reviews of the major conceptual issues, using examples to illustrate the principal ideas. The intention is that the book will be used by graduate and upper level undergraduate courses, where students are already familiar with the basic ideas in behavioural ecology.

Four chapters (Chapters 3, 5, 6 and 8) in this edition cover material not dealt with in separate chapters in the 2nd edition, and six of the remaining chapters involve new authors or authors writing on new subjects. Only two chapters (1 and 10) are written by the same author on the same subject as in the 2nd edition and these are both extensively modified to take into account new results and ideas. Several chapters have been dropped from the present edition (comparative method, game theory, group living, territoriality, learning, sex, and the behavioural ecology of plants). To some extent these topics are now covered in other chapters: comparative methods are used in Chapters 2 and 7, learning in behavioural ecology is alluded to in Chapter 4, sexual selection in plants in Chapter 7 and game theory is used in many chapters. The omission of other topics does not imply that we think they are no longer important; it reflects our wish to give the book a new appearance.

What are the major changes in behavioural ecology since the last edition? Without attempting a comprehensive list, and with anticipatory apologies to those who feel slighted by our choice, we pick out four points that are reflected both in this book and in the talks presented at the 1990 International Behavioural Ecology Conference in Uppsala (21–26 August) which we both attended just before writing this preface.

1 *Sexual selection.* There is no doubt that sexual selection is sexy. Both theoretical and empirical work is moving at a greater pace (Chapter 7). Zahavi's handicap model, albeit in modified form, has been rehabilitated. In particular, sexual displays as honest indicators of disease resistance is the fashionable focus of both theoretical and empirical work, the latter offering suggestive hints but as yet no conclusive evidence. Sociologists of science will no doubt write PhD theses on the fact that the fashion in behavioural ecology has swung from energetic costs and benefits of resource acquisition in the oil-crisis-dominated 1970s to sex and disease in the aids-crisis-dominated late 1980s.

2 *Mechanisms.* Behavioural ecology was spawned as a wayward offspring of ethology, focusing on just one of Tinbergen's (1963) 'four questions'

(function) and eschewing the other three (mechanism, ontogeny, evolutionary history). Inexorably behavioural ecologists are being drawn back into the study of mechanisms. Understanding sexual selection, for example, may come to depend increasingly on knowledge of perceptual processes in animals (as suggested by Mike Ryan at Uppsala) and, in the case of 'disease resistance', an unravelling of the links between male hormones, secondary sexual traits and immunological competence, as Marlene Zuk proposed at the Uppsala meeting.

A return to mechanisms does not imply turning the clock back. Behavioural ecologists, approaching mechanisms from a sophisticated and coherent functional framework, will often ask unusual questions and come up with surprising answers. For example, reading a traditional account of mechanisms of regulation of food intake in vertebrates would give no hint that environmental stochasticity or social dominance would have major effects on the equilibrium level of fat reserves, as reported by Jan Ekman in Uppsala. Yet these effects, and therefore the underlying mechanisms that produce them, were predicted by stochastic dynamic optimization models (McNamara & Houston 1990b).

3 *Molecular techniques.* DNA fingerprinting, using multilocus probes, has already overturned behavioural ecologists' view of monogamy in birds: the rich and varied sex lives of so-called monogamous birds revealed by genetic analysis is surely just the first of many surprises in store. For example, in Uppsala, Terry Burke reported that the traditional view of lekking birds, in which most matings are thought to be obtained by very few males, may not be borne out by genetic analysis. Not only would this cause behavioural ecologists to rethink their models of lek evolution, but it would also cause a re-analysis of correlates (plumage, behaviour) of mating success.

The importance of DNA fingerprinting (and other molecular techniques that will soon follow) should encourage behavioural ecologists to devote more time and effort to thinking about and devising new techniques to answer seemingly unanswerable questions.

4 *Links to ecology.* For many years behavioural ecologists have been promising that our understanding of individual mating, foraging, life history and other decisions will lead to insights into population and community ecology. Although the promise is some way from being fulfilled, Mike Rosenzweig showed, at the Uppsala meeting, one possible way forward, building on the ideal free distribution to predict conditions of coexistence of two species of gerbil in the Negev Desert. Given the increasing importance of environmental issues in general, behavioural ecologists should give thought to how their exquisite studies of individual adaptation feed back into ecological processes.

John R. Krebs
Nicholas B. Davies

Acknowledgements

We thank Robert Campbell, Susan Sternberg, Jane Andrew and Edward Wates of Blackwell Scientific Publications for their advice and enthusiasm during the preparation of this book. Authors of various chapters would like to acknowledge the following for their help.

Chapter 1: Julee Greenough, Laurence Hurst and David Haig for comments on the manuscript.

Chapter 3: Tim Caro, Lee Cronk, Robin Dunbar, Patrick Gray, Ray Hames, Kristen Hawkes, Barry Hewlett, Kim Hill, Sarah Hrdy, Austin Hughes, Peter Richerson, Daniela Sieff, Eric Smith, Don Symons and Margo Wilson. Thanks also to John Hartung, Alan Rogers and Nicholas Blurton Jones for permission to adapt material already published or in press.

Chapter 4: Alasdair Houston and Jonathan Newman. J.R. Krebs is supported by AFRC, NERC and The Royal Society. A. Kacelnik was supported by King's College, Cambridge, and by a research grant from the Wellcome Trust.

Chapter 6: Tim Guilford, Victor Rush, Kevin Long, Tracy McLellan and Pat Ross. Discussion with Tim Guilford was particularly valuable in helping to clarify problems in the effects of prey detection and its consequences.

Chapter 7: R.M. Gibson, S. Nee, A. Pomiankowski, A.F. Read and S. Vehrencamp.

Chapter 9: Andrew Balmford, Jack Bradbury, Tim Clutton-Brock, Lisa Petit and Kevin Teather.

Chapter 11: A.F.G. Bourke, H.J. Brockmann, N.F. Carlin, J.M. Carpenter, E.L. Charnov, D.W. Davidson, D.H. Feener, Jr, A. Grafen, N.A. Morgan, D.C. Queller, R.L. Trivers and J.W. Wenzel.

PART 1

NATURAL SELECTION AND LIFE HISTORIES

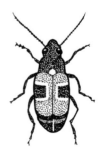

Introduction

Behavioural ecologists often preface their studies with statements such as 'individuals are expected to behave so as to maximize their reproductive success'. They then go on to consider some component which serves this ultimate goal such as foraging behaviour, mate choice, fighting strategy or parental investment. We begin the book with two chapters which take a critical look at this important assumption which underlies our subject.

Since W.D. Hamilton's revolutionary insight that individuals can pass on copies of their genes not only by producing offspring but also by helping non-descendent kin, such as siblings, the behavioural ecologists' slogan has become 'individuals maximize their inclusive fitness'. As Grafen illustrates in Chapter 1, this is one of the most widely misunderstood concepts. Grafen shows how 'Hamilton's rule' for the evolution of altruism is derived and discusses how it can be used to make predictions about altruistic behaviour in populations. He also discusses how behavioural ecologists might be able to justify their use of non-genetic models. Most of us are unaware of the genetic mechanisms underlying the strategies we model or study empirically, yet genetic variation lies at the heart of evolutionary adaptation. Grafen addresses this dilemma and considers the question of whether genetic constraints limit the feasible set of optimal strategies. To take a simple example where they would, consider a single locus for which the optimal genotype is a heterozygote. Because of the laws of Mendelian inheritance, the population can never evolve to a pure culture of optimal individuals.

In Chapter 2 Lessells considers the evolution of life histories, which includes problems such as when to mature, what clutch size to produce, and whether to breed repeatedly (iteroparity) or in one 'big-bang' effort (semelparity). She shows that to predict the optimal clutch size we need to measure two trade-offs, namely that between current and future reproduction and that between number and quality of offspring within a brood. She emphasizes that we cannot simply use natural variation to measure these trade-offs. For example, within a population, individuals which lay larger clutches may be *more* likely to survive to the next breeding season. This does not mean that there is no trade-off between survival and reproduction. The result may arise simply because good quality individuals can both survive better and afford to produce larger clutches. It is clear then that experimental manipulations are needed where, for example, clutches of different sizes are randomly allocated across individuals to measure the effects of reproductive effort on adult survival. Lessells reviews these experiments, which mainly involve laboratory studies of insects and field studies of birds, showing that in some cases the observed clutch size is close to the theoretical prediction based on the premise that selection maximizes lifetime reproductive success. She raises some methodological problems for field-workers. One of the most intriguing is that for an experiment to 'work', in the sense of producing, for example, good measurements of survival costs from raising larger broods, the animals must 'play the game' set by the experimenter (e.g. work harder for larger broods). However, in some cases the animals' response to the manipulation may be a strategic one rather than one forced on them by the experiment.

Lessells concludes her chapter with a brief discussion on how the habitat imposes selective pressures which mould life histories and how individuals may exhibit 'phenotypic plasticity', namely a single genotype shows a range of phenotypes (e.g. clutch size) depending on environmental conditions (e.g. food supply). She reviews the beautiful work by Daan and his co-workers on the kestrel, which shows variability in laying date and clutch size depending on territory quality.

Text books on sociobiology and behavioural ecology often conclude with a chapter on humans. We have included Borgerhoff Mulder's chapter in this opening section not only to mark the recent flourishing of studies which have applied models from behavioural ecology to human foraging and reproductive behaviour but also to emphasize that human studies have raised issues which should feed back into the work on animal behaviour. In the early days of sociobiology the invasion of evolutionary ideas into the complex world of mankind met the same stormy reception which greeted Darwin's work a hundred years earlier. One eminent anthropologist remarked that 'a marriage between biology and the social sciences would be welcome but what the sociobiologists are attempting is a rape!' As Borgerhoff Mulder emphasizes in her critical review, it is a matter for empirical research to discover how well functional models from behavioural ecology can predict human behaviour patterns. For example, Blurton Jones' study of interbirth intervals in Kalahari bushmen reveals a close fit to models based on maximization of lifetime reproductive success. Even when an individual's goal is no longer maximization of biological fitness, as is likely in many modern industrial societies, some aspects of human behaviour may still reflect past selective pressures. For example, in Daly and Wilson's study of intrafamily homicide, careful consideration of alternative hypotheses has shown that a Darwinian account gives a better explanation than many others.

Borgerhoff Mulder shows how models from optimal foraging theory can illuminate aspects of human diet choice, how the polygyny threshold model can make testable prediction about marriage in Kipsigis, and how kinship theory can help to explain patterns of adoption, food sharing and inheritance of wealth. The critical studies she reviews show how much progress has been made since the early naïve acceptance of sociobiological ideas for human studies, and hopefully marks the start of a fruitful interchange between traditional anthropology and behavioural ecology.

1: Modelling in behavioural ecology

Alan Grafen

1.1 Introduction

The chapter begins with a general justification for using non-genetic models to study adaptation. This is extremely important because the genetics of interesting traits is rarely known. The second section discusses Hamilton's model of social interactions (Hamilton, 1964; 1967), explaining it in a new way. The scope of inclusive fitness theory is discussed, and applications of Hamilton's rule are described. Hamilton's model underlies most current research on social interactions, and shows how a model can become so much part of the framework of our thought that we are unaware of it. A model of signalling forms the basis for the third section, in which the much vilified handicap principle of Zahavi (1975, 1977a) is vindicated. The clarification of Zahavi's ideas is upheld as an example of a cardinal virtue of modelling.

1.2 Population genetics underlies behavioural ecology

The starting point for much behavioural ecology is that animals are maximizers of one sort or another—efficient predators or foragers, or elusive prey. The usual ground for believing this is the presumption that natural selection has made them so. If not now then at some time in the past (Dawkins, 1982, pp. 20–24), there existed heritable variation in hunting and foraging techniques, and in ploys to escape predators. Changes in allele frequencies have made animals good at what they do.

The behavioural ecologist, though, does not usually know the genetics underlying the character she studies. While she would be interested to know this genetic system, it is not of primary importance. Her main aim is to uncover the selective forces that shape the character. The behavioural ecologist has to hope in her ignorance that her method will work almost regardless of which particular genetic system underlies the character (Lloyd, 1977). This hope raises two questions. First, is it justified? Second, is the assumption so powerful and plausible that a whole research strategy should be based on it?

1.2.1 The phenotypic gambit

Let us start with a brief caricature, with examples, of an important method in behavioural ecology. It has two elements.

(i) A strategy set This is a list or set of (perhaps all) possible states of the character of interest. Here are three examples of strategy sets. McGregor *et al.* (1981) studied the song of male great tits, and in particular their repertory size. The strategy set they used was simply every different repertory size they observed: integers from one to five. Brockmann *et al.* (1979) studied the nesting of great golden digger wasps. These wasps sometimes acquire a nest by digging, and sometimes by entering an already existing nest. Brockmann and co-workers were interested in the relative frequency of these two ways of acquiring a nest, and so the strategy set was simply all possible proportions of digging rather than entering—numbers between zero and one. In the hawk–dove game devised by Maynard Smith and Price (1973), the strategy set consists of two strategies, called hawk and dove.

(ii) A rule for determining the success of a strategy The success of a strategy is the number of offspring left by an animal adopting it, or alternatively its inclusive fitness (see section 1.3 below). The rule for determining success may involve the frequencies with which strategies are adopted in the population. One way to determine the rule is to observe it, as McGregor *et al.* They counted how many offspring every male fathered in his lifetime, and averaged across all males sharing the same repertory size. Another way is to model the rule, which Brockmann *et al.* did because they needed to know how the success of the strategies changed as their frequencies changed. They used data to estimate parameters in the rule. When the purpose is to investigate theoretically the consequences of a particular form of frequency dependence, then an appropriate rule is simply assumed: in the hawk–dove game, the rule is represented in the pay-off matrix.

The phenotypic gambit is to examine the evolutionary basis of a character as if the very simplest genetic system controlled it: as if there were a haploid locus at which each distinct strategy was represented by a distinct allele, as if the pay-off rule gave the number of offspring for each allele, and as if enough mutation occurred to allow each strategy the opportunity to invade.

The gambit implies that all strategies occurring in the population are equally successful, and that they are at least as successful as any non-occurring strategy would be if it arose in small numbers. The application of the gambit to a given strategy set and pay-off rule is a powerful way of testing the joint hypothesis that the strategy set and the pay-off function have been correctly identified, and that the gambit is true.

In their first model, Brockmann *et al.* rejected this joint hypothesis when two existing strategies turned out not to be equally successful. They adopted a new strategy set in their second model.

The joint hypothesis might be false because the genetic system underlying the character does not produce the same phenotypic effects as the very simplest genetic system, the one assumed in the gambit. The mere fact that the predictions of equal success of existing strategies and the inferiority of unplayed strategies are rejected does not reveal which element in the joint hypothesis is false. The research strategy implied by the phenotypic gambit is to treat such rejection as evidence that the pay-off function or strategy set is wrong, and not that the genetic system is causing the discrepancy (see also Chapter 4).

1.2.2 Is it a winning gambit?

Taken literally, the gambit is usually unjustified: few species studied by behavioural ecologists are haploid. But will the genetic system that does underlie the character produce the same phenotypic effects as the genetic system the gambit assumes?

Two points are important here. First, an example is known in which the gambit would be misleading. In some human populations affected by malaria, there are three distinct phenotypes corresponding to the three possible genotypes at a diploid locus with two alleles (Allison, 1954). One type almost invariably dies before reproducing, of sickle cell anaemia. The other two types differ in their resistance to malaria. The coexistence of these three genotypes with markedly different fitnesses would be very puzzling to a behavioural ecologist applying the phenotypic gambit. The mechanics of Mendelian segregation prevent the whole population from sharing the optimal phenotype, because it is produced by the heterozygous genotype. Here, as undoubtedly elsewhere, it is essential to know the underlying genetics in order to understand the distribution of phenotypes observed in the population.

The second point is that such cases are probably rare. Only certain features of genetic systems, such as over-dominance in the sickle cell case, can sustain dramatic differences in fitness, and these features are not known to be common. Maynard Smith (1982) has analysed how well different genetic systems support the simplification represented by the gambit, and he concludes that by and large they do so very well. The sorts of character studied by behavioural ecologists are likely to be controlled by many loci, and this reduces the scope for the maintenance of large fitness differences.

Genetic systems are themselves subject to evolution. In its simplest form, this is the creation of a new allele by mutation, but more substantial changes could occur. In the sickle cell case, a (functional) gene duplication of the locus would allow one locus to fixate for each allele. Every individual in the population could then have the 'intermediate' genotype that confers malarial protection without the sickle cell anaemia. That this has not happened for sickle cell may be because this intermediate genotype would be disadvantageous where malaria is not a major selective force. The existence of fitness differences between genotypes at equilibrium creates selection for evolution of the genetic system itself.

The behavioural ecologist hopes that genetic systems that do not support the gambit are rare or transient. If the discrepancies produced by genetic systems are smaller than the accuracy of the data, then field-workers can safely ignore them. We know this might not be so, and we should be anxious to find out whether this hope is justified. The soundness of behavioural ecologists' methods depends on arguments concerning population genetics, but our methods are actually designed to avoid doing genetics.

We have seen that the gambit cannot be made with perfect safety. It is a leap of faith. But what would behavioural ecology be like if we refused to use it in our research? It would be very different. Detailed studies in which the precise nature of a character is examined as an adaptation would have to be accompanied by a study in which the genetic mechanism underlying the character was uncovered so precisely that an explicit genetic model could be constructed. There would be no decimal places without genetics. This would reduce drastically the range of characters we can study. Genetically simple and well-studied characters are usually straightforwardly disadvantageous mutants maintained by judicious artificial selection in strains that have spent tens of generations in the laboratory—and so are rarely of evolutionary interest. A behavioural ecological study would have to be very large if genetics were included, and it would be impossible to complete a study on elephants, say, within the lifetime of a scientist.

If the gambit is generally justified, therefore, the genetics is an almost irrelevant complication in understanding the selective forces that shape a character. The gambit makes truly phenotypic explanations possible, and the effort expended in discovering the genetics would be wasted. Better to allocate that effort to studying in an evolutionary ways characters of evolutionary interest, and in a genetic way characters of genetic interest.

These are the reasons why the gambit is so attractive, whether it is justified or not. The advantages seem to me to justify continuing to employ the gambit, always providing we remember that we may be wrong. Theoretical work by Thomas (1985a,b,c,d) and other studies reviewed by Hines (1987) tackle the problem of when the evolutionarily stable strategy approach applied to phenotypes gives the same answer as a more genetical approach would. They lend strong support to the phenotypic gambit.

1.3 Inclusive fitness and Hamilton's rule

It is all very well to say that animals are maximizers, but what do they maximize? Is it number of offspring? This section displays and discusses one of the most important models in modern evolutionary theory, Hamilton's social interactions model (Hamilton, 1964), made more elegant by Hamilton (1970). It is the basis for the powerful principle that animals act as if maximizing their inclusive fitness.

Inclusive fitness is a fundamental concept of evolutionary biology. A widespread misconception is that the point of inclusive fitness is to help us understand interactions between relatives. The model will make clear that the scope of inclusive fitness covers all interactions in which the genotype of one individual affects the fitness of conspecifics. The special role of relatives is a powerful result of the theory, not a restricting assumption. Indeed, the most important case on which inclusive fitness theory sheds light is where interactants are unrelated. Here, they should behave so as to maximize their own number of offspring, and have no regard for the effect of their actions on the number of offspring of the other individual. This will be the case in many interactions, perhaps the majority.

1.3.1 Hamilton's model of social interactions

Inclusive fitness is based on Hamilton's model of social interactions. We begin by illustrating in Fig. 1.1 a simpler, conventional model used in population genetics. Each individual in the population is represented by a square with a rectangular addition. The combined area represents the individual's number of offspring. The idea is that the square is a standard unit of fitness, while the rectangular addition represents the effect of the individual's own genotype on its number of offspring. The rectangle will always be drawn outside the square, but, for convenience of illustration and for generality, let us agree that its area may count positively or negatively, depending on whether the individual has an advantageous or disadvantageous genotype. The sign of the numerical value attached to the rectangle will specify where necessary whether an area counts positively or negatively.

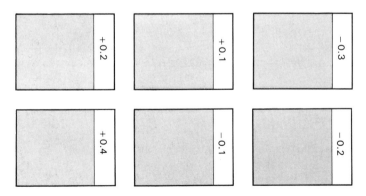

Fig. 1.1 The figure shows in schematic form the numbers of offspring of six individuals. The square represents a standard unit of one offspring, and the rectangular sections represent differences from one caused by the individual's own genotype. These differences can be positive or negative. The six individuals therefore have 1.2, 1.1, 0.7, 1.4, 0.9 and 0.8 offspring.

Consider any locus, say the G locus, and focus on one allele G and amalgamate all other alleles at the locus under the name g. The first use for our model will be to compute the frequency of G in the next generation, on the assumption of Mendelian segregation. Each individual will have a genotype at the G locus, which for diploids will be either GG, Gg or gg. The number of G alleles in offspring is the combined area of all GG individuals and half the area of Gg individuals. The number of G alleles in offspring is the combined area of all gg individuals and the remaining half of the area of Gg individuals. The frequency of G among the offspring is just the number of G alleles divided by the sum of the number of G and g alleles.

This elementary calculation has an important consequence. G will increase in frequency if it is associated with larger number of offspring. But we have not specified which allele G is, nor which locus it is at, and neither have we made any detailed assumptions about how the fitness effects arise. Therefore, every allele will increase in frequency if it is associated with a larger number of offspring. If every allele at every locus is under selection of this sort, then it is reasonable to say that the organism is under selection to maximize its number of offspring. This kind of statement about selection on the organism transcends the picky details of genetics and justifies the application of 'selection thinking' by organismal biologists.

This crucial organismal conclusion from a genetic model for non-social traits attracted Hamilton to try to devise a parallel model for traits in which one individual's genotype was allowed to affect the fitness of others. The model is altered by introducing two new kinds of areas. An elementary social interaction is represented by a triangle attached to an actor and a circle attached to a recipient. The triangle represents the action's effect on the actor's number of offspring, and the circle represents the effect on the recipient's number of offspring. A dotted line can be drawn between corresponding triangles and circles. It is possible to have more than one recipient, so one triangle may connect to more than one circle. Figure 1.2 therefore represents Hamilton's model of social interactions.

The next step is to notice that number of offspring will do some things in the same way as in the simpler model, but fails to provide the organismal conclusion. The areas of GG, Gg and gg individuals can still be appropriately combined to compute the frequency of G among the offspring. (For this purpose, the areas of the triangles and circles are simply added to the individual's number of offspring; all that matters is how many offspring an individual has, not why.) It is still true that an allele G will be selected if it is associated with greater number of offspring, in just the same way as before. The trouble is that number of offspring is no longer under the exclusive control of the individual. So we cannot say that organisms are selected to maximize the number of offspring, because one component of their offspring is not under their genetic control, but someone else's. Also, an individual controls the number of offspring other individuals have, and if social behaviour can be selected, then this effect on others must also be

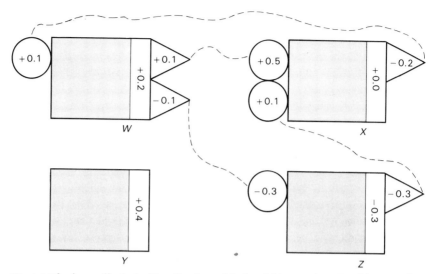

Fig. 1.2 The figure illustrates Hamilton's model of social interactions. In addition to the square and rectangle of Fig 1.1, there are four social actions represented here, each requiring a triangle and a circle. The triangle represents the effect on the actor, while the circle connected by a dashed line represents the effect of the action on the recipient. Each effect can be positive or negative. The letters *W*, *X*, *Y* and *Z* name the individuals. The topmost line, for example, shows *X* paying a cost of 0.2 offspring with a net gain of 0.1 for *W*. In this arrangement, the areas representing offspring are drawn in contact with the parent of the offspring. *W* individual has 1.3 offspring, *X* has 1.4 offspring, *Y* has 1.4, while *Z* has 0.1.

included in any measure that selection might cause organisms to maximize. In other words, we need a measure based on control, not one based on results.

Hamilton produced a different accounting system for computing the frequency of *G* among the offspring. His measure is illustrated in Fig. 1.3. It includes the square (the standard unit), the rectangle (effect of own genotype through non-social traits), the triangles (effect of own genotype through social traits) and the circles linked to those triangles (effect of own genotype on the number of offspring of others). Only a fraction of these circles counts towards the new measure of fitness, and that fraction is the relatedness to the recipient. In recognition of the fact that the effects on others are included, Hamilton called this new measure 'inclusive fitness'. Also important is the fact that it *excludes* the effects of others on the individual's own offspring (the circles that were directly attached to an individual in Fig. 1.2).

Now Hamilton proved an important fact about inclusive fitness. We can calculate the combined inclusive fitness of *GG* individuals, and half of the inclusive fitness of *Gg* individuals, and call this the summed inclusive fitness of *G*. We can calculate similarly the summed inclusive fitness of *g*. If we then divide the summed inclusive fitness of *G* by the total summed inclusive fitnesses of *G* and *g*, we might hope to get a frequency of *G* among the offspring. We do not. But the answer we get has a very important property. It is in the same direction from the frequency of G among the parents as the frequency of G among the offspring. It therefore gets the direction of gene frequency range right, even if the magnitude is wrong. Thus, a gene will increase in frequency if it is associated with higher inclusive fitness and decrease if it is associated with a lower inclusive fitness.

In return for this weakening so far as computing gene frequencies is concerned, we obtain an enormous strengthening on the organismal side. All the effects of an individual's genotype are now included in its inclusive fitness. G increases in frequency if it is associated with an increased inclusive fitness. Remember that G can be any allele at any locus. We may conclude that all alleles are selected to increase the inclusive fitness of the individual bearing them, and so we may reach another organismal conclusion that transcends the details of genetics: organisms are under selection to maximize their inclusive fitness. This was Hamilton's goal, and it is a result of fundamental importance in the study of social behaviour.

Two implications can be noted here. First, it is common to specify a

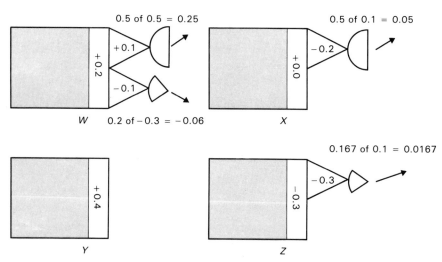

Fig. 1.3 This figure shows the areas of Fig. 1.2 rearranged. An area is now attached to the individual whose genotype caused to exist the offspring represented by the area. Each individual has lost its own circles, but has gained the circles it gave to others. Figure 1.2 showed accounting by results; here we see accounting by control. To calculate inclusive fitness, we need to add up the areas attached to an individual but diminishing the area of a circle. Only a fraction of the circles is counted, and that fraction is the relatedness of the actor to the recipient. Assume that the relatednesses are as follows: W to X, 0.5; W to Z, 0.2; X to W, 0.5; Z to X, 0.167. The inclusive fitness of W is therefore 1 (square, standard unit)+0.2 (rectangle, non-social effect of own genotype)+0.1−0.1 (the effects on his own number of offspring of his social actions)+0.5×0.5−0.3×0.2 (the effects of his social actions on others (0.5, −0.3) weighted by the appropriate relatednesses (0.5, 0.2), coming to 1.39 in total. The other inclusive fitnesses are 0.85 for X, 1.4 for Y and 0.4167 for Z.

social action by saying that it loses the actor one offspring and gains its full sibling three offspring. Inclusive fitness immediately allows us to calculate that this action would be favoured by selection. The loss of one to the actor's inclusive fitness is outweighed by the gain of a half times three. The second implication is the powerful idea of a rate of exchange between own offspring and the offspring of others. An individual acts as if it valued each other individual's offspring as worth a fraction of one of its own—and that fraction is the relatedness between the individuals.

The organismal level of the conclusions reached by inclusive fitness theory is important first of all because it achieves a radical simplification if genetics can be passed over. It is also important because in social traits, even more than in non-social traits, virtually nothing is ever known of the genetics of evolutionarily interesting characters. We observe organisms and interesting morphology and behaviour, but we rarely observe interesting genes. Practical applications therefore require a principle at the level of the organism.

In making use of inclusive fitness, it is often helpful to go back to the model of social interactions on which it is based. The model has been examined in some detail because a full understanding of inclusive fitness is important for students of behavioural ecology. No proof of Hamilton's central result has been given, but in the next section a logically parallel result is proved about what has come to be known as 'Hamilton's rule'.

1.3.2 Relatedness defined and Hamilton's rule deduced

The notion of relatedness was left unexamined in the previous section, with the hope that the reader had some loose notion that identical twins have a relatedness of one, full sibs have a relatedness of one half, and cousins have a relatedness of one eighth. It turns out that relatedness is quite a subtle notion, and it is convenient here to explain the meaning of relatedness at the same time as giving a convincing if informal proof of Hamilton's rule.

The idea of inclusive fitness is that helping relatives is a bit like helping yourself, because they share your genes. This approach will now be made more precise, by finding a definition of relatedness with the 'design requirement' that the relatedness of a potential actor A to the potential recipient R measures the extent to which A helping R is like A helping itself. Relatedness is usually introduced in connection with common ancestry: full sibs share both parents, half sibs share one parent, and cousins share two grandparents. But for the moment we are interested only in measuring genetic similarity, which can be caused by common ancestry but can also be caused by other processes.

There are many senses of genetic similarity, and correspondingly many ways to measure it. For example, we could concentrate on only one locus, and assign a similarity of 1 if two individuals shared both alleles at that locus, 0.5 if they shared only one allele, and 0 if they shared no allele. Or we could ask what fraction of all alleles at all loci are shared between two individuals. But neither of these suggestions satisfies our design requirements.

The measure of genetic similarity that does the trick is illustrated in Fig. 1.4. A line indicates possible gene frequencies, numbers from zero to one. Three points are marked on it, whose relative positions define relatedness. We concentrate on just one allele at one locus. The first point, μ, is the frequency of that allele in the population whose gene frequencies we are tracking. The second point, A, is the frequency of that allele in the potential actor, and the third point, R, is the frequency of that allele in the potential recipient.

We now trace the consequence for the spread of the special allele of altruism performed by A to R. In two special cases illustrated in Fig. 1.4, our design requirement specifies the relatedness right away. If R and μ coincide, then the recipient is the same as the population average (Fig. 1.4b). When A helps R, he adds alleles in the existing proportion, and so does not help to change the population gene frequency at all. From the point of view of the allele G that controls the action, A may as well throw his help away, because that does not change the population gene frequency either. The relatedness should therefore be zero in the case where R and μ coincide. On the other hand, when A and R coincide, helping R has just the same consequence for the changing gene frequency as if A helped himself to the

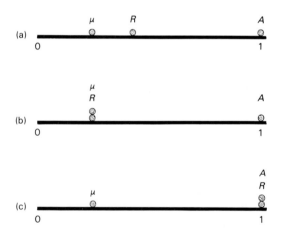

Fig. 1.4 This figure illustrates the meaning of relatedness. The line between zero and one represents the gene frequency of a particular allele at a particular locus. A represents the actor's gene frequency, μ represents the population average gene frequency, and R represents the average position of the recipient. Relatedness is the fraction of distance from μ to A at which R lies—in case (a) about one quarter. In case (b), the recipient has the same gene frequency as the population average and so A may as well throw his help away on a random recipient because relatedness is zero. In case (c), R has the same gene frequency as A, and so A helping R is as good as A helping himself. The relatedness in this case is therefore one.

same extent (Fig. 1.4c). When *A* and *R* coincide, therefore, the relatedness must be one.

These two special cases suggest that relatedness be defined more generally as 'the fraction of the distance from μ to *A* at which *R* lies'. This definition corresponds to the regression definition of relatedness first put forward by Hamilton (1970), and which is now gaining popularity in modern theory. Regression relatednesses are not always the same as Sewall Wright's correlation coefficients of relatedness (Wright, 1969) or as Hamilton's life-for-life coefficients (Hamilton, 1972). Table 1.1 shows the life-for-life and regression coefficients for some relationships under haplodiploidy. All three values are the same under out-breeding diploidy with no selection in an infinite population.

The terms *correlation* and *regression* arise because the definitions can be formulated to look exactly like statistical formulae for correlation and regression coefficients (see Hamilton, 1975). If we code a 'score' for each individual in the population as the fraction of alleles at the *G* locus that are *G*, then Wright's correlation coefficient is simply the correlation between the scores of potential altruists and potential recipients. It is symmetric, and can range from −1 to 1. The regression coefficient is the slope in a regression of the potential recipients' scores on the potential altruists'. It can in principle take any value and, as Table 1.1 shows, need not be symmetric.

With the regression definition of relatedness it is now very easy to provide a convincing if informal proof of Hamilton's rule. For the sake of definiteness, suppose *R* is one quarter of the way from μ to *A*, perhaps because *A* and *R* are half-sibs. Imagine that *A* has resources that he can convert into four offspring for *R*. These are shown at *R* in Fig. 1.5. These offspring will have the same effect on the gene frequency as the combination of one offspring at *A* and three offspring at μ. This is so because the groups are the same size and because their gene frequencies are the same. This shows that *A* helping *R* to have four offspring has the same effect as if he were able to have one offspring himself, and produce three with the population gene frequency.

In deciding in what direction the population gene frequency will change, we can ignore offspring at μ and sum up the effects of the other offspring produced by all the individuals in the population. But the offspring at μ are not altogether irrelevant. The more offspring there are at μ, the more the next generation will resemble the current generation. Hence offspring at μ slow down the magnitude of the change without affecting its direction. Hamilton (1964) called this the 'diluting effect'.

We can say that *A*'s creation of four children for *R* has the same effect on the direction of gene frequency change as producing one offspring for itself. This is simply a verbal formulation of Hamilton's rule. It is easy to see that the same argument applies for any relatedness, not just for one quarter.

The regression definition of relatedness therefore makes Hamilton's rule work, whether the gene in question is rare or common. What, though,

Table 1.1 The relatednesses under haplodiploidy between some categories of relatives under two definitions of relatedness: Hamilton's (1972) life-for-life definition as employed by Trivers and Hare (1976) and Hamilton's (1970; 1972) regression definition as used in this chapter.

Sex of donor	Relationship of recipient to donor	Life-for-life	Regression
Female	Mother	0.5	0.5
	Father	0.5	1.0
	Sister	0.75	0.75
	Brother	0.25	0.5
	Daughter	0.5	0.5
	Son	0.5	1.0
Male	Mother	1.0	0.5
	'Father' (mother's mate)	0.0	0.0
	Sister	0.5	0.25
	Brother	0.5	0.5
	Daughter	1.0	0.5
	'Son' (mate's son)	0.0	0.0

Only outbred relationships are considered. Quick methods of calculating these values are (i) life-for-life: the fraction of the donor's genes that are identical by descent with any of the recipient's; (ii) regression: the fraction of the recipient's genes that are identical by descent with any of the donor's. For example, all of a male's genes are identical by descent with genes in his mother, so the life-for-life coefficient is 1. On the other hand, only one half of his mother's genes are identical by descent with genes in him, so the regression coefficient is one half. These methods do not work under inbreeding. Note that within-sex values are the same, and between-sex values are converted by multiplying by the ratio of ploidies of donor to recipient. Neither the life-for-life nor regression coefficient is *symmetric*. The regression relatedness of a son to his mother is one half, and of a mother to her son is one; these values are reversed for life-for-life coefficients.

does it have to do with common ancestry? An argument used by Charnov (1977) tells us. Consider the simple case of diploidy and random mating, with an infinite population at Hardy–Weinberg equilibrium not undergoing selection. Suppose we are concentrating on an allele *M*, and we agree to class all other alleles at the locus under the name *N*. A homozygote has genotype *MM*. What is the genotype of a full sib? The parents must have genotypes *M?* and *M?*, where '?' indicates that the allele is unspecified by knowledge of the focal individual's genotype. Because we have assumed random mating, Hardy–Weinberg equilibrium and no selection, the chance that '?' will be any particular allele is proportional to its frequency in the population. So '?' equals *M* with probability *p*, say, and *N* with

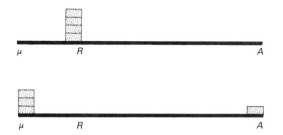

Fig. 1.5 The recipient is shown on the upper line as having four offspring. These are redistributed on the lower line, one to A and three to μ. The two sets of offspring will have the same expected effect on the gene frequency because they contain on average the same number of genes, and the same gene frequency. Any rearrangement that preserves the number of offspring and their 'centre of gravity' would have this property. The special point about this rearrangement is that offspring that contain the population average gene frequency do not contribute to a change in that frequency in the next generation. From the point of view of the direction of the change in gene frequency, therefore, the upper and lower arrangements have identical effects. Thus four children for R have the same effect as one child for A. This is Hamilton's rule. If R is a fraction λ of the distance from μ to A, then one offspring at R has the same centre of gravity as the combination of 1−λ offspring at μ, and λ offspring at A. Hence, in general, as well as in the special case of one quarter illustrated here, the factor by which A must discount offspring given to R is the fraction of the distance from μ to A at which R lies. Thus that fraction is the desired measure of relatedness.

probability $1-p$. A sib has a one quarter chance of being MM, half a chance of being M? (one quarter each from one parent having M and the other having ?), and a quarter chance of being ??. The frequency of M on average is therefore

$$\frac{1}{4}\times1+\frac{1}{2}\times\frac{1}{2}(1+p)+\frac{1}{4}\times p = \frac{1}{2}(1+p)$$

The average gene frequency in a sib is therefore $R = \frac{1}{2}(1+p)$, which is half-way between the gene frequency of the focus individual ($A = 1$) and the population average ($\mu = p$). The same result would apply if we had started with a heterozygote ($R = \frac{1}{2}(\frac{1}{2}+p)$; $A = \frac{1}{2}$; $\mu = p$) or the opposite homozygote ($R = \frac{1}{2}(0+p)$; $A = 0$; $\mu = p$). The same agreement between common ancestry and the regression relatedness works for all other relationships under the assumptions we have made.

This example makes clear an initially puzzling feature of the definition of relatedness. Under diploidy, every individual's gene frequency is either 0, $\frac{1}{2}$ or 1. Yet R is often assumed to take other values. For example, when μ is $\frac{1}{3}$ and A is 1, a full sib will be taken as having a gene frequency of $\frac{2}{3}$, which it is impossible for any one individual to have. This is because when A makes his decision, he is assumed to have only certain information available to him—in this case, of sibship—and this means that A has to assume an average position for R. The average position can be anywhere between 0 and 1, in just the same way as the population average can be.

This Charnovian calculation relies on selection being weak at the locus. To see why, consider an individual of genotype MM, whose parents are inferred to be $M?$, $M?$. The interpretation of the $?$ as M with probability p, N with probability $1-p$ depends on an assumption of Hardy–Weinberg equilibrium. For example, if MN is lethal, then we should infer that MM's parents are both MM. Even granted the $M?$, $M?$ parents, selection on the offspring of those parents may have altered the $MM : MN : NN$ genotypes among surviving sibs from the $1 : 2 : 1$ ratio needed for our computation.

One important point is that common ancestry turns out to cause the same relatedness for all alleles, at all autosomal loci, under our assumptions of random mating and weak selection in an infinite population. But under inbreeding, or strong selection, this definition can give different relatednesses, even for different alleles at the same locus! Causes of genetic similarity other than common ancestry are likely to bring about different relatednesses at different loci, even under random mating. Hamilton's rule still applies to each allele when relatednesses vary, but to each with its own appropriate relatedness. The behaviour of individuals may therefore be selected in different directions by the selection on different alleles. Different relatednesses are therefore likely to cause intragenomic conflict of the kind first considered by Hamilton (1967) in the context of sex ratios.

This possibility has an implication for Hamilton's model of social interactions. When relatednesses are the same for all alleles and loci, then the inclusive fitness of an individual is uniquely defined, and the organismal conclusion that individuals are under selection to maximize their inclusive fitness holds. Once complications arise that cause relatednesses to vary across alleles and loci, this simple picture breaks down.

The usual form of Hamilton's rule incorporates the equivalence illustrated in Fig. 1.5 into a rule about when a gene is selected for, that causes a potential actor to incur a cost, c, to itself, while conferring a benefit, b, to a potential recipient with relatedness r. The direct effect on own offspring is minus c, and the equivalent effect from helping R is rb. Selection favours the allele if $rb-c > 0$, which is Hamilton's rule.

Relatedness can help us to glimpse something of the more complicated world in which relatednesses vary between alleles and loci. For any given social action, for example helping a nest-mate, suppose we know the benefit to a recipient, b, and the cost to the actor, c. Then the critical relatedness above which the action will be favoured, that is, $rb-c > 0$, is c/b. Now if a large majority of alleles have a relatedness greater then c/b, we can expect the action to be favoured by selection; while if the large majority of relatednesses are less than c/b, then it will be selected out. So in an important class of social actions, the variability in relatedness will not matter. In intermediate cases, the results will be a complicated mess. The conclusion to be stressed is that the organismal level conclusions justified by inclusive fitness theory are threatened only for these intermediate cases. Most social actions will be unequivocally favoured or disfavoured: only a few will bring about genomic conflict.

The main purpose of this section was to provide a demonstration of Hamilton's rule. It was necessary at the same time to give an account of relatedness as a measure of genetic similarity that has peculiar significance for social behaviour, and to discuss the possibility that relatedness might vary between alleles and loci.

1.3.3 How not to measure inclusive fitness

Hamilton's (1964) verbal definition of inclusive fitness is:

> 'the animal's production of adult offspring . . . stripped of all components . . . due to the individual's social environment, leaving the fitness he would express if not exposed to any of the harms or benefits of that environment, . . . and augmented by certain fractions of the harm and benefit the individual himself causes to the fitnesses of his neighbours. The fractions in question are simply the coefficients of relationship . . .'

The picture of social interactions shown in Fig. 1.2 shows immediately that measuring inclusive fitness is a subtle business. It is not enough to measure how many offspring an individual produces, it is also necessary to be able to partition them. How many are the result of actions by others, and how many were the results of actions taken towards others?

There have been in the past various flawed attempts to measure inclusive fitness in nature that have ignored the partitioning of an individual's number of offspring. This leads to an exaggeration of the 'own offspring' component, by inclusion of help from others (the individual's own circles in Fig. 1.2), and it can also lead to a gross distortion of the 'others' offspring' component of inclusive fitness if *all* of the offspring of a relative are included instead of only the additional help the focal individual supplied (i.e. only the other individual's circles in Fig. 1.2 should be included, and only those caused by the focal individual).

Some readers may be now tempted to despair—it is hard enough to measure the number of offspring of an individual in the field, without managing the almost metaphysical task of deciding who was really responsible for them (the 'causal parent'). Part of this feeling is justified. Measuring the inclusive fitness of an individual is indeed a tall order, if it is even possible. Remember that the scheme of Fig. 1.2 has a mainly conceptual purpose. The right way to apply inclusive fitness theory to data is to apply Hamilton's rule, and this has been achieved to good effect with real data. For examples and references see Grafen (1984, section 3.3.3).

1.3.4 *How to use Hamilton's rule*

Hamilton's rule was derived in section 1.3.2. The purpose of applying it is to ask: would a given social action be favoured by selection? Hamilton's rule implies that only three quantities need be known to answer the question: the benefit to the recipient, the cost to the actor, and the relatedness. Under the assumption that the only cause of genetic similarity is recent common ancestry, the regression coefficients of relatedness from Table 1.1 are the right ones to use to make Hamilton's rule work.

There have been various quibbles about the validity of Hamilton's rule based on misunderstandings of the meaning of cost and benefit. The conceptual scheme of Fig. 1.2 can help here. Cost is the area of the triangle, the number of offspring the actor loses through performing the action. One way to estimate this from data would be to consider otherwise similar individuals, one of which performs the act and the other of which does not. Then the difference between their total number of offspring will represent the effect of the action. This method will be used below. More generally it is important to realize that the theory says the cost is a difference in numbers of offspring and is not, for example, a ratio. Inclusive fitness theory is a strong theory, in the sense that it dictates how its terms should be measured. It is not just a casually thrown together collection of symbols.

The same considerations can be applied to the benefit to the recipient, which could therefore be estimated by considering two otherwise identical individuals, one of which had the action performed to it while the other had not. Of course, in any application of this method, serious attention will need to be paid to the phrase 'otherwise identical'.

Having established what all three terms in Hamilton's rule mean, we move on to the first step in applying it in practice. This is to choose the decision we are interested in, being as explicit as possible about the alternative courses of action. Let us call performing the action Y, and not performing it X. To calculate b and c, we will need to estimate the difference it makes to lifetime number of offspring to do Y rather than X. If the actions are only vaguely defined, then we cannot estimate those differences.

A difference of c in the animal's lifetime number of offspring results from doing Y rather than X. Any consequences that would follow from doing Y and not X should be taken into account—decreased longevity, retribution and so on. It may seem at first sight that a simple way to estimate c from data is to take the difference in lifetime number of offspring between animals that do X and animals that do Y. However, this seemingly reasonable procedure may give the wrong answer. The reason is that animals that do X will have relatives who tend to do X, and animals that do Y will have relatives that tend to do Y. The simple difference in number of offspring between X-doers and Y-doers will therefore include the extra bs that a Y-doer can expect to receive from his Y-performing relatives, and therefore does not give a proper estimate of c. Of course, the same caution applies to measuring b.

The value of r has been assessed in various ways. Bertram (1976) modelled the structure of lion prides to arrive at relatednesses; Brown (1975) used simple ancestry; and Metcalf and Whitt (1977) used electrophoresis. These are all approximations to what r must be in principle, which is discussed in section 1.3.2.

Finally, Hamilton's rule in the form $rb-c > 0$ has definite advantages over the more popular form $b/c > 1/r$. For one thing, the second form is wrong if c or r is negative (though correct if both are). For another, in the common case that r is known on *a priori* grounds, then the sampling variance of $rb-c$ is calculable simply from the sampling variances of b and c, allowing confidence intervals to be constructed. In contrast, the ratio b/c has a sampling variance that depends in a more complicated way on the sampling distributions of b and c. Examples of applying Hamilton's rule are given in Chapters 10 and 11.

1.3.5 The validity of Hamilton's rule

The scope of inclusive fitness theory depends on the validity of the model of Fig. 1.2. We shall first look at an example of Charlesworth (1978a) and then draw some general conclusions.

In Charlesworth's example, a dominant allele causes its bearer to kill itself and feed itself to its sibs. An application of Hamilton's rule implies that this will be advantageous provided the sib gains more than twice as many offspring from the extra food as the bearer would itself have had. Yet a moment's reflection shows that if all bearers of the dominant allele kill themselves, then the allele will be extinguished immediately, no matter what advantage may be gained by the sibs of suicidal nest-mates. No bearers of the allele will survive to reproduce.

Hamilton's model of social interactions assumes that because the recipient is a sib, it has a relatedness of one half to the actor, and in effect computes the gene frequency of the created offspring on that basis. But because of the nature of its action, the suicidal allele cannot be present in a receiving sib. Hamilton's model fails to represent the social interactions correctly, and so the inclusive fitness principle does not apply.

This example lies at the intersection of three distinct general classes of exception, and can be looked at in three illuminating ways. We can look at Charlesworth's example as a case of multiplicative interactions of fitness, in place of the additive interactions the model requires. The action multiplies the actor's fitness by zero, and the recipients fitness by, say, three. Then multiplying a recipient's fitness by three when he is himself about to multiply it by zero is not a very effective way to help him.

The second general class of exception is where the benefits are genotype specific. In this case, the benefit to the non-suicidal genotypes is much greater than the benefit to suicidal genotypes. Hamilton's rule, on the other hand, assumes that the benefit is not systematically different for different recipient genotypes.

We move on to a third way in which Charlesworth's versatile example can be understood, as the result of strong selection at the controlling loci. The suicidal genotype helps its sibs, but owing to strong selection on the sibs (namely some of them have killed or are about to kill themselves), the relatedness to the recipients is less than common ancestry alone would predict.

What should we conclude from these three kinds of exception? Take, first, non-additivity of fitness interactions. During rapid change in a character there may well be favourable alleles with large effects. But during periods of equilibrium, when the character is close to its optimum, we can expect that evolution is slow, perhaps mainly stabilizing. In these circumstances, the relevant alleles have small effects because alleles of large effect are strongly disadvantageous. For alleles of small effect, fitness interactions will be additive to a sufficient approximation.

Let us see how this might work. The effects of different costs and benefits may not add up. For example, the first donated food item may save an animal's life; the second may enable it to have five offspring; and the third, even though of the same size, may raise the number of offspring only to six. This is a *prima facie* contradiction of the model. However, things look different when we consider an allele of small effect that tinkers with this system. Suppose the first and second donations are always made, and that the variability under selection is the third. Then the model will reflect correctly the existing variation, and so make the correct predictions about selection. Hamilton's rule works even for alleles of large effect if the decision to give the first, second or third donation is distinguished and separately subject to natural selection.

Thus we must admit that circumstances can be imagined in which the inclusive fitness principle breaks down because of non-additive fitness interactions. But if we accept that for most characters the relevant allelic effects are small most of the time, then the inclusive fitness principle will be upheld in the circumstances that matter most to us.

True genotype specificity of the magnitudes of benefits is hard to imagine if the action simply supplies food, or saves from a predator. If the benefit were a blood transfusion between vertebrates, on the other hand, then the benefit would be very different depending on the compatibility between donor and recipient. Accepting this as an exception in principle, it is probably fair to assume that genotype specificity will rarely be a problem in practice.

The last type of exception arose from strong selection, and indeed we noted in section 1.3.2 that the calculations linking relatedness to common ancestry break down then. The assumption of weak selection is therefore necessary to make Hamilton's rule work with ancestral relatednesses. But, as with additivity, even if sometimes strong selection does occur, it is reasonable to suppose that a character is perfected by natural selection under conditions of weak selection. The assumption of weak selection is therefore acceptable.

These three classes of exception are important in principle, particularly if you want to use Hamilton's rule to predict gene frequency changes in models with strong fitness effects, but do not threaten the practical value of Hamilton's rule. That value is to allow us to find organismal accounts of adaptations in social behaviour without doing genetics. If in any case we did need to do genetics, the study would almost certainly be too costly in time and effort to be worthwhile.

A good example of the value of inclusive fitness in modelling is provided by the disagreement between O'Connor (1978) and Stinson (1979) about the evolution of siblicide in birds. O'Connor used a naïve inclusive fitness argument and came to one set of conclusions, while Stinson used a population genetics model and came to another. Godfray and Harper (1990) present a more sophisticated population genetics model than Stinson's, incorporating the assumption of small genetic effects, and obtain O'Connor's 'naïve' results. This gives confidence that inclusive fitness arguments should be treated seriously, even when the actions in question involve large effects, and even when biologically naïve population genetics models cast doubt on them.

1.3.6 Conclusions on Hamilton's model of social interactions

Hamilton's model of social interactions has been outstandingly successful. Hamilton's rule has provided a practical tool with which social behaviour can be studied. The exchange rate concept of valuing others' offspring against one's own according to relatedness is invaluable in thinking about social traits. It is unrivalled in its scope as a model of social behaviour because it captures the biological essentials. At one time, it was a favourite pastime of theoretical biologists to prove Hamilton's rule incorrect, by setting up a model that broke one of the assumptions discussed above (reviewed by Grafen, 1985). Now, however, the value of Hamilton's model of social interactions is widely recognized.

1.4 A model of biological signals

One of the great virtues of evolutionarily stable strategy (ESS) modelling (see Maynard Smith, 1982) is that it helps to clarify muddy verbal arguments. Of the many topics discussed in previous editions of this book (Krebs & Davies, 1978; 1984), one in particular has since benefited from ESS modelling. This section describes the application of the phenotypic gambit of section 1.2.1 to biological signalling. Previous approaches will be discussed once the new ESS model has been described. (Handicap models of signalling are also discussed in section 7.3 and in sections 12.3 and 12.4.)

We will consider a simple imaginary species of beetle. Males vary in how much food they happen to find as larvae. For energetic and nutritional reasons, this affects various aspects of their ability to function as adults. In particular, better-fed males have more viable sperm. Males block the female reproductive tract with a plug after mating, and so females mate only once. Females therefore have a great interest in mating with a male with more viable sperm.

If females could perceive directly the viability of a male's sperm, there would be no signalling problem. Females would 'see' the sperm viability and choose a mate accordingly. Instead, let us make the reasonable assumption that females cannot perceive sperm viability.

Does a male know his own sperm viability? Translated out of metaphor, can there exist developmental rules that are flexibly expressed in male bodies, such that some other character can be made to co-vary with the viability of a male's sperm? We shall assume the answer is yes. This is reasonable because we have assumed that differences in sperm viability are caused by differences in larval nutrition. It is easy to imagine that some structures will be more fully developed in adults that were well fed as larvae than in those that were ill fed.

We have a set of females all wanting to know the sperm viabilities of the males they encounter, and a set of males each of whom knows his own sperm viability. The evolutionary signalling problem is this: is there an evolutionarily stable way in which males can convey the information females would like to have?

This might seem straightforward: the males have the information; the females want it. The problem is that each mate benefits if he is judged to have more viable sperm, and so, considered loosely, all want to signal that they are the best male. If all males do make the same signal, then females cannot distinguish between males at all, and so will not attend to it. At first sight, therefore, stable signalling cannot evolve. But our argument has been rather rough and ready. This is the kind of verbal impasse at which we should turn to ESS modelling. Here we will deliberately exclude mathematical symbols—but they were necessary for rigour in the papers on which the discussion is based (Grafen, 1990a,b).

We begin constructing the ESS model by assuming, for concreteness, that males have horns and that any communication about sperm viability takes place through the size of the male horns (although of course any other trait would have done if females could perceive it). We can now follow the phenotypic gambit of section 1.2.1. What is the strategy set for males? Let us assume that developmental rules exist that can establish any relationship between sperm viability and horn size. This means that, in terms of Fig. 1.6, a male strategy is any rule that specifies for each sperm viability a particular horn size.

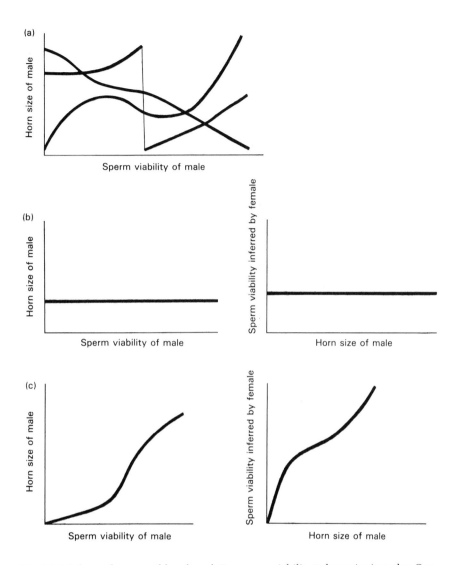

Fig. 1.6 (a) shows three possible rules relating sperm viability to horn size in males. One rule slopes downwards throughout, while another goes upwards overall but with a central downwards section. The third rule has a discontinuity half-way along the horizontal axis. Any rule is allowed in the model. (b) illustrates the non-signalling equilibrium. Males produce the minimum cost horn size, whatever their sperm viability. Females assess males, whatever their horn size, as having average sperm viability. No deviator can gain in this situation. (c) illustrates a signalling equilibrium. Males with greater sperm viabilities produce larger horns. Females assess males with larger horns as having greater sperm viabilities. Each rule is the reverse of the other. Start with any sperm viability, and look up the horn size from the male rule; then look up from the female rule what sperm viability is inferred from that horn size. The inferred sperm viability is the same as the orginal sperm viability, so that females make the correct assessment of sperm viability. Zahavi's handicap theory is developed from the supposition that an equilibrium like this exists. Grafen (1990a,b) shows that in a wide class of signalling models, an equilibrium like this does indeed exist.

Now what about the female strategies? A female has to infer from a male's horn size what his sperm viability is. Thus we can represent a female strategy as a rule specifying for each horn size an inferred sperm viability.

With the strategy sets in place, the next step is to consider the pay-off functions for males and females. For present purposes, we do not need to be precise. Assume that the size of a male's horns affects his fitness in some way, by using energy and so reducing survival from larval to adult stages, or by increasing the risk of predation as an adult. How the cost relates to horn size and sperm viability will turn out to be of the highest importance, but for the moment we leave it unspecified. So far as the signalling consequences of their horns are concerned, it will be enough to specify that a male is fitter, the higher his sperm viability is inferred to be. A female is assumed to be fitter the more accurately she infers males' sperm viabilities.

With these rather vague assumptions, we cannot prove that an equilibrium exists, which is anyway a rather technical affair. But we can ask: suppose there was an equilibrium, what would it be like?

The simplest equilibrium is the non-signalling one, illustrated in Fig. 1.6b. A male produces the same, cheapest, horn size irrespective of sperm viability. Females, with no information to go on, assess males as having the average sperm viability. Crucially, in the figure, the female rule assesses any rare mutant males as having the same average sperm viability.

It is an equilibrium because no rare mutant male or female can spread. A male that produces any other horn size will still be assessed as having average sperm viability, and so will gain no mating advantage. But other males are producing the cheapest horn size, and so the mutant must lose out in that respect, and so lose out overall. (Notice that a situation in which all males produce a horn size that is not the cheapest could not be an equilibrium.)

A female whose rule differs from the common rule in its assessment of the cheapest horn size will suffer, because she will think all males are better than they are, and so be too ready to mate; or think they are worse than they really are, and be too reluctant. We can imagine the female taking risks to mate, or mating at a suboptimal time because she assesses an encountered male as above average. Female mutants cannot gain either.

This discussion of the non-signalling equilibrium has shown the kinds of conclusion we can draw from the supposition of equilibrium. What can be inferred about a signalling equilibrium in which horn size increases with sperm viability?

We may immediately conclude what the female rule must be. As each horn size is produced only by males of a particular sperm viability, the best female rule is to draw the correct inferences and assess a given horn size as the sperm viability of the males that actually produce it. The equilibrium female rule must therefore be just the reverse of the equilibrium male rule, as illustrated in Fig. 1.6c. This is the first conclusion we reach on the supposition of equilibrium.

Now we turn to the males. Why does a male not have a uniformly high horn size, at the level played by the best males, no matter what his sperm viability? This would certainly be more advantageous from the point of view of obtaining matings. If the rule illustrated in Fig. 1.6c really is an equilibrium, then the other costs of having a higher horn size must out-weigh the mating advantage. It must therefore be more costly in fitness to have larger horns. Perhaps the horns make the male clumsy, rendering him at a disadvantage in fighting, or less able to avoid predators. The theory cannot say what the disadvantage is, but it does say that there must be one. Otherwise there can be no signalling equilibrium.

There is an additional conclusion that we can draw about the cost of horns. The differential cost of having larger horns must be greater for worse males. This follows because the assumption of equilibrium tells us that each quality of male produces its own optimal horn size. We have assumed that horn size alone signals quality. A male of any quality will therefore gain the same mating advantage from the same horn size. If good males and poor males have different optimal horn sizes, then it must be because the other relevant component of male fitness, the cost of the horns, is different for good and poor males. And if good males have a larger optimal horn size than poor males, it must be because the cost of having larger horns is less for good males.

These arguments are all based on the assumption that a signalling equilibrium exists, and are therefore worthless if such equilibria are im-possible. In fact, I have shown (Grafen, 1990a,b) that signalling equilibria do indeed exist in a wide range of models. Unfortunately, the arguments are too technical to give here, and we shall simply assume that signalling equilibria exist. From the supposition of a signalling equilibrium, we have therefore concluded that: (i) females make the correct deductions about male sperm viability from horn size; (ii) larger horns are costlier to males than smaller horns; and (iii) the differential cost of larger horns is greater for worse males than for good males.

These conclusions are the same as those of Zahavi (1975; 1977a). Zahavi claimed that signals were honest, our conclusion (i), that signals must be costly (ii), and that they were more costly for worse males (iii). We have just followed through, therefore, a vindication and clarification of Zahavi's handicap theory.

The discussion was made in terms of horns and sperm viability, to make it easier to follow. But the arguments hold more generally and are not restricted to mate choice. For example, consider a fight in red deer between a harem-holder and a challenger. Harem-holders eat very little and so lose strength. We can view the remaining strength of the harem-master as qual-ity, known to the harem-master and unknown but of great interest to the challenger. In the early stages of a fight, we can expect signals to convey information from the harem-master to the challenger about his strength. We could reach the analogous conclusions about these signals, assuming them to be in evolutionary equilibrium, as we did about the hypothetical insects' horn size. The signal will be correctly interpreted by the challenger; the signal must be costly; and the signal must be more costly to harem-masters with lower remaining strength. A signal requiring strength or stamina might well satisfy the second two conditions. There are complications in assessing the cost of a signal in cases of this sort (Grafen, 1990a), but the principle is just the same.

1.4.1 A comparison with previous approaches

How does this ESS model advance our understanding of signalling compared with previous approaches? The first comparison to be made must be made with Zahavi's work (Zahavi, 1975; 1977a; 1978). Here the advance is one of clarification. The ESS model takes Zahavi's ideas and, by placing them in a more formal context, makes explicit the assumptions and arguments. The conclusions remain the same.

The second comparison is with the work of Dawkins and Krebs (1978) and Krebs and Dawkins (1984). Krebs and Dawkins provide a general view of signalling that is very congenial, and in effect forms the background for the model given above. Their view about how signals involved in assessment can be stable is that the signals must be 'reliable indicators of RHP' (resource holding potential), which must be 'too costly to fake'. Zahavi's handicap principle is one way of making signals reliable, by making them too costly to be worth faking. The ESS models of Grafen (1990a,b) are fairly general models of signalling, and lend support to the view that the handicap principle may be the only way of making signals reliable. (It should be noted that this strong conclusion relies on what is meant by the term 'signal', which is also discussed in those papers.)

Krebs and Dawkins correctly identify selective forces. Signallers do wish to mislead receivers. Receivers do not wish to be tricked by signallers. Each party is concerned only about its own fitness. This inherent conflict of interest leads Krebs and Dawkins to expect continuing change and spiralling strategies. Pre-ESS thinking leads to similar expectations in other games (well discussed for the hawk–dove game, for example, by Dawkins (1989)). An ESS approach leads to meaningful and interesting conditions that emerge at equilibrium. The model of the previous section complies with all the requirements of Krebs and Dawkins' discussion, but when we insist on asking about what must be true in an equilibrium, we find the emergence of Zahavi's handicap principle. I believe that had Zahavi's principle been properly understood at the time, it would have taken a central part in Krebs and Dawkins' discussion.

Let us take apparently conflicting conclusions of the two approaches. At equilibrium, there is honesty. Why do males not lie as recommended by Krebs and Dawkins? One answer is that if they did lie, we could not be at equilibrium: females would have learned not to believe the signal. This just pushes the question one stage further back: how can it not be advantageous for males to lie? Because the signal they would have to make in order to lie would cost more than the mating benefit they would receive. If signals cannot be costly, then there can be no equilibrium. But the extravagance of signals is one of the facts that all the theories under consideration set out to explain, and the ESS approach brings out the vital conclusion about honesty at equilibrium.

There are other differences of emphasis between the approaches. The ESS model allows females not to use the information from signalling if it is unreliable, and takes this reluctance into account in studying the evolution of signals. Krebs and Dawkins assume that the females' only defence is to require even stronger signals. Simply ignoring signals is an option that would tend to defuse arms races. The signals must continue to convey important information if females are to be selected to continue to attend to them. What drives the ESS model is always the variability between males that matters to females, and signals are selected only according to relevant correlations. The pure arms race, sales resistance phenomenon that Dawkins and Krebs particularly stressed relies on the next difference we note between the approaches.

This lies in how signals acquire meaning. In the ESS model, a signal acquires a meaning only when receivers learn empirically (and possibly only over evolutionary time) that a certain signal is given by signallers of a certain quality. The signals have no inherent meaning. Female response to a 2 cm horn is determined only by previous experience of males with 2 cm horns. Krebs and Dawkins sometimes seem to assume that signals already possess a meaning before they are used. For example, if a man asserts that he is entirely free of parasites, this has meaning to a female even if she (or her ancestors) have never encountered such an assertion before, and even if there has never been a parasite-free man. This difference is in a way quite fundamental. Humans can continue to lie in some situations because their statements are given meaning by the context of language. It seems likely that peacocks are less likely to have meanings pre-assigned to different tail lengths or styles. Where signals are inherently meaningless, that is, they acquire meaning only empirically through the experiences of receivers, this too is likely to limit the scope for spiralling arms races. This difference may arise because Krebs and Dawkins focus on signals about intentions, while the model above concerns signals of quality.

The handicap principle does not assert that any possible signalling system is honest, costly, and has greater costs for poorer quality signallers. It merely asserts that any stable signalling system must have these properties. What happens to a signalling system without them? The system may cycle endlessly, with booms and busts, in a way that resembles an arms race much of the time, leading to a world like that envisaged by Krebs and Dawkins in which 'in actor–reactor coevolution both sides may gain the upper hand'. On the other hand it is also possible, and in view of earlier paragraphs I think more likely, that such a system would collapse. Females would be selected not to attend to the males' signals. The outcome could be settled for any hypothesized signalling system only with a model.

The model of the previous section is not intended as a template for every example of signalling, but to illustrate that the handicap principle can work. Even in our hypothetical example, we can see one omitted complication. We assumed that females gain information about male quality only through horn size. But if quality is dictated by success at larval feeding, then good males are likely to be simply bigger than worse males. Good and bad males will differ in many ways, and not just in horn size. Females will have many possible traits to detect that may correlate with quality. This would complicate our model of the handicap principle, as it would any model of female choice. The idea that quality shows through in many ways is the basis of the argument in the second paragraph of Zahavi (1975), that when real mate choice is going on, Fisher's runaway process is unlikely to occur. He argued that females probably use multiple cues to detect important variation in male quality, so if any one trait gets out of step it is likely to be ignored (see Chapters 7 and 12).

This discussion is based on models of Grafen (1990a,b) as these claim to provide a full and explicit justification of Zahavi's handicap principle. Previous relevant work on signals includes the first model of biological signals, by Enquist (1985), who refuted the still popular notion that animals cannot be selected to signal their intentions; a graphical exposition of the handicap principle by Nur and Hasson (1984) very close in spirit to Grafen's model; and Andersson's (1986) model of conditional handicaps.

1.4.2 Conclusion

The ESS signalling model clarifies Zahavi's original arguments and makes explicit a number of assumptions about how signals operate. The requirement of equilibrium made in the ESS model turns the deceit recommended by Dawkins and Krebs into the honesty of Zahavi. We can expect similar transmutations, counter-intuitive at first hearing but in reality quite reasonable, to arise from the application of ESS models in other areas, particularly other topics in signalling.

The model has various artificial restrictions, made to simplify the argument. The chief restriction is that the variation in male quality is assumed to be purely environmental. Greenough and Grafen (in preparation) present a model in which male quality is genetic, and study it by computer simulation. They show that advertising and preference work in the same way as when male quality is environmental. Other simplifications are discussed by Grafen (1990a,b).

The ESS signalling model described here is fairly complicated. A strategy is a function, loosely any curve drawn from left to right on the plane, allowing vertical jumps. The strategy set therefore has an infinite number of dimensions, a rather daunting fact that requires careful technique. Further, there are two sets of players, males and females, with different pay-off structures. These complications may help account for the delay between Zahavi's verbal proposal and an analytical justification.

I do not see these complexities as marking the boundary of the usefulness of ESS theory. Quite the reverse. The long arguments over the handicap principle show that verbal reasoning is even more adversely affected than analytical reasoning in complicated cases. If we are to understand communication, and other sophisticated games played by organisms, then we will need to become familiar with this kind of ESS model, grapple with its complexities and welcome its illuminations.

In broader terms, the modelling exercise shows that it is necessary to be wary of failures to model an idea. The handicap principle has been confidently refuted many times by modellers who did not understand it. The model described here shows that, as Zahavi maintains, the handicap principle is about the strategy of communication. It applies to human communication between governments and to interspecific communication in just the same way as it applies to sexual selection. It has therefore nothing to do with genetics. Zahavi (1987) goes so far as to claim that the distinction Darwin drew between natural and sexual selection is properly understood as a distinction between the selection of ordinary traits on the one hand and the selection of signalling traits on the other. Ordinary traits are selected for efficiency, while an essential part of the selection of signalling traits is that they are wasteful, the waste being the self-inflicted costs of the signallers. The model of the handicap principle offered above implies that Zahavi's far-reaching claim deserves serious attention.

2: The evolution of life histories

Catherine M. Lessells

2.1 Introduction

Imagine a female bruchid beetle belonging to the species *Callosobruchus maculatus* who has just found a black-eyed bean on which to lay her eggs. How many eggs should she lay on this bean? The choice she makes has important consequences because larvae are unable to move between beans and adults of this species do not feed, so inadequate larval nutrition cannot be compensated by later feeding. Should she lay a single egg, so that the larva has sole access to the resources, or should she lay more eggs, with the possible penalty that her offspring will compete among themselves within the bean? Moreover, should she be influenced in making her decision by the fact that by laying more eggs she may deplete her own resources and hence shorten her own lifespan?

Now imagine a female collared flycatcher about to lay a clutch of eggs. How many eggs should she lay, and what factors should influence her decision? At first sight, these seem very different problems: *C. maculatus* does not feed as an adult, whereas collared flycatchers must catch and consume insects regularly to survive. *C. maculatus* lays its eggs on the larval food supply and then abandons them to their fate, whereas collared flycatchers lay their eggs in a place that offers no sustenance and they must assiduously provision their nestlings for the first few weeks after they hatch. *C. maculatus* can lay on each bean it finds so that clutches may be irregularly spaced in time, whereas collared flycatchers are offered a breeding opportunity once a year by the progression of the seasons. Nevertheless, *C. maculatus*, collared flycatchers and all other organisms face the same two basic questions in coming to a decision over how many offspring to produce at one time; the first is what proportion of resources to devote to reproduction as opposed to maintenance or growth, that is what *reproductive effort* to make. The second question is how to divide resources that are devoted to reproduction between offspring; whether to produce a few lavishly provided for offspring or many poorly resourced young (see also Chapter 8). The answers to these two questions together determine the number of eggs or offspring produced in one bout of reproduction; neither of the two answers is sufficient in isolation.

The reason why these decisions have to be made at all is because of trade-offs between life history traits (Rose, 1983; Bell & Koufopanou, 1986; Stearns, 1989); one trait can be increased only at the expense of another. For instance, female *C. maculatus* beetles can increase the number of eggs they lay only by using up resources to the detriment of their own lifespan. Trade-offs result when two traits are limited by the same resources; time, energy, or any other resource can be spent only once (the 'principle of allocation'; Levins, 1968). Although trade-offs potentially occur between any pair of life history traits (see, for example, Stearns, 1989 for a list of the 45 pair-wise trade-offs between 10 life history traits), these other trade-offs can be subsumed into two major trade-offs which correspond to the two decisions already mentioned concerning reproductive effort and the division of resources between offspring. The first trade-off, known as the *cost of reproduction* (Williams, 1966) is between current and future reproduction. Current reproduction may not only use up energetic and other reserves, hence reducing future fecundity, but may also endanger the survival of the parent and hence its chance of reproducing in the future (Bell & Koufopanou, 1986). The second major trade-off involves the division of current reproductive effort between offspring, and will be referred to as the *trade-off between the number and fitness of offspring*. As the number of offspring increases, the amount of reserves than can be given to each egg, or the amount of provisioning or other parental care given to each offspring, must decrease. As a result, there will be a trade-off between the number of offspring produced and their survival (Lack, 1947). A knowledge of these two trade-offs is needed before answers can be given to the questions of how much reproductive effort should be made and how it should be divided, and hence in turn to the overall question of what clutch size should be laid.

This chapter concentrates on those areas of life history theory which have been confronted with comparative and experimental data. It starts by considering the two main approaches ('phenotypic' and 'genetic', see below) to life history evolution and describes the application of optimality models, which form the basic research tool used throughout the chapter, to life histories (see also Chapter 4 for a discussion of optimality models). It next considers the methods by which trade-offs can be measured; because of the central role of trade-offs in life history evolution, progress in testing optimality models can be made only when these are correctly measured. These methods are then applied to the two major life history trade-offs; the cost of reproduction and the trade-off between the number and fitness of offspring. Measurements of these trade-offs are then brought to bear on the evolution of clutch size, delayed maturity and semelparity (big bang reproduction). This is followed by a consideration of the use of comparative analyses in the study of life histories. The chapter finishes with a discussion of the phenotypic plasticity of life histories, that is the production of different phenotypes by the same genotype, depending on the environment.

2.1.1 Approaches to life history evolution

There are essentially two approaches to the current evolution of life history traits. The first is a phenotypic approach, which asks questions about the adaptive value of a trait in terms of the selection pressures acting on it and usually involves some kind of optimality modelling (see Chapter 4). The chief use of optimality models is in providing a method of gauging how complete any understanding of selection pressures is; a lack of fit between observed and predicted values suggests that there are additional selection pressures that are important to the maintenance of the trait in the population which have not been taken into account. Optimality models are phenotypic, in the sense that they ignore the possibility that the evolution of the optimal phenotype has been prevented by genetic constraints such as the lack of genetic variation (Maynard Smith, 1978; Lewontin, 1979; Grafen, 1984). This approach was first applied to life histories by Lack (1947), Cole (1954) and Medawar (1946; 1952). For the want of a better term, practitioners of this approach will be referred to as 'pheneticists'.

The genetic approach is more recent (Lande, 1982a,b; Via & Lande, 1985; de Jong, 1990) and also considers the effects of selection, but this time the way in which selection will affect gene frequencies. Because of this, these models are able to make predictions about the direction and speed of genetic change, and about equilibrium levels of genetic variation, something that phenotypic models are clearly incapable of doing. Geneticists have recently criticized pheneticists for their lack of knowledge of the genetic basis of the traits they study (e.g. Dingle & Hegmann, 1982; Rose *et al.*, 1987). However, rather than regarding one of the two approaches as superior to the other, it would be better to recognize them as different: phenotypic models cannot address questions about gene frequencies. But in attempting to understand why different life histories evolve in different environments, genetic models may add little except intractability to phenotypic models (Charnov, 1989a).

This chapter takes a primarily phenotypic approach, but genetics are not entirely neglected; first, genetical techniques may provide a means of measuring relationships for use in optimality models and, second, genetics may turn out to pose some of the most interesting (but not necessarily tractable) questions about life history evolution.

2.1.2 Optimality models of life histories

Optimality models are described in detail by Krebs and Kacelnik in Chapter 4, where they distinguish three main components of such models. This section describes how each of these components applies to optimality models of life histories.

DECISION VARIABLES: LIFE HISTORY TRAITS

Each optimality model deals with the evolution of one or more biological characters. These characters are referred to as the decision variables of that model. In life history models these are life history traits which, strictly speaking, consist of *age-specific values of fecundity and mortality* (Cole, 1954; Partridge & Harvey, 1988). Traits such as longevity (lifespan), age at first reproduction, and parity (the number of bouts of reproduction during a lifetime) summarize these age-specific values, and therefore also fall within the strict definition of a life history trait. Other traits that have a fairly direct effect on either survival or fecundity, such as growth rate, offspring size, timing of breeding within a season, size at maturity or metamorphosis, and dispersal are also often regarded as life history traits (e.g. Gadgil & Bossert, 1970; Horn, 1978; Stearns & Koella, 1986). Age-specific fecundity and survival differ from these other traits because they are components of fitness in the sense that an increase in any one of these values (while the others are held constant) increases fitness, whereas intermediate values of other traits, such as growth, may be associated with maximal fitness.

CURRENCY: INTRINSIC RATE OF INCREASE OR REPRODUCTIVE VALUE

Natural selection maximizes the rate of increase of an allele, but optimality models do not generally use this as a criterion for judging between alternative values of the decision variable. Instead, another currency is chosen that can be more easily measured. In life history models, the currency is relatively closely related to the rate of increase of an allele, because this will depend directly on fecundity and mortality. The two frequently used, and equivalent, currencies are r, the intrinsic rate of increase of a genotype (Charlesworth, 1980; Caswell, 1989) and RV, the reproductive value (Williams, 1966; Taylor *et al.*, 1974; Charlesworth, 1980; Yodzis, 1981; Goodman, 1982). The RV of a female is related to the number of descendants that a female of that age is expected to have in future generations, and RV at age x is given by

$$RV_x = \sum_{y=x}^{x} (l_y/l_x) \cdot m_y \cdot e^{-r(y-x+1)} \tag{2.1}$$

where x = age, l_x = probability of survival to age x, and m_x = fecundity (the number of daughters produced) at age x. Although the equation may look formidable, each term in the summation is simply the product of the probability of a female surviving from age x to age y (l_y/l_x), her fecundity at that age (m_y), and an expression ($e^{-r(y-x+1)}$) weighting offspring by how far in the future they are produced; in an expanding population offspring produced now are worth more than those produced at some time in the future, because that can 'accrue interest' (Fisher, 1930) by breeding sooner themselves. In populations that are neither growing nor shrinking, RV_0 equals R_0, the average number of offspring produced over an individual's lifespan. So in constant populations, R_0 may also be used as a currency (Charlesworth, 1980; Lande, 1982b).

The advantage of RV over r in experimental studies is that it can be readily broken down into components corresponding to reproduction at different ages. For instance, the right hand side of equation (2.1) can be split into two terms representing current and future reproduction:

$$RV_x = m_x \cdot e^{-r} + \sum_{y=x+1}^{\infty} (l_y/l_x) \cdot m_y \cdot e^{-r(y-x+1)} \qquad (2.2)$$

These two components of RV, the second of which is referred to as *residual RV* (Williams, 1966), represent the two traits traded off in the cost of reproduction.

CONSTRAINTS: FITNESS AND TRADE-OFF CURVES

Constraints relate the decision variable to the currency (Stephens & Krebs, 1986). Optimality models of life histories generally incorporate two kinds of constraint. The first is the direct relationship between fitness and the value of the trait. This relationship will be referred to as a *fitness curve*. For age-specific fecundity and survival the relationship follows directly from the equation for RV, but the relationship must be measured for other traits such as timing of breeding or egg size. The second kind of constraint is the relationship between different traits of the same individual which results from varying the allocation of resources between traits; in other words, a trade-off. The relationship between two traits that results from varying allo-cation of a constant amount of resource between the two traits will be re-ferred to as a *trade-off curve*. Trade-off curves cannot generally be deduced from energetic or other considerations, so must be measured. Both fitness curves and trade-off curves are measured, by definition, for individuals who do not differ in terms of their resources, their environment, or their other traits. All these other things must be equal in order for the fitness and trade-off curves to represent truly the consequences of the choice of a par-ticular value of the decision variable, and hence be the appropriate re-lationships for inclusion in optimality models. This condition of other things being equal sometimes creates problems in the measurement of these relationships, as explained in the following section.

2.2 Measuring constraints

2.2.1 Problems in measuring constraints

In order to predict the optimal clutch sizes for *C. maculatus* and collared flycatchers, we need first to know the constraints, in particular the trade-off curve for the cost of reproduction and the trade-off between the number and fitness of offspring. At first sight, this seems an easy task: measure each of the two traits for a range of individuals and plot the values against each

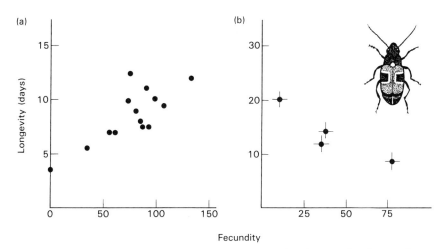

Fig. 2.1 Phenotypic correlations do not measure trade-off curves. (a) The phenotypic correlation between adult longevity and fecundity in female bruchid beetles *Callosobruchus maculatus* (K. Wilson, personal communication). (b) The trade-off between adult longevity and fecundity in female *C. maculatus* (Wilson, 1989). Points are means (± standard errors) for four experimental treatments manipulating access to mates and/or oviposition sites (beans).

other. If there is a trade-off the plotted relationship should be negative. Such relationships between measured phenotypic values are known as *phenotypic correlations*, and although the naïve expectation is a negative relationship, the opposite is often found (see, for example, Bell & Koufopanou, 1986; Partridge, 1989). For instance, the phenotypic correlation between adult longevity and fecundity in *C. maculatus* is positive (Fig. 2.1a). There are two reasons why phenotypic correlations cannot be used to measure trade-offs: lack of non-adaptive variation in allocation and adaptive variation in allocation (Högstedt, 1980; 1981; Reznick, 1985; Partridge & Harvey, 1988; Pease & Bull, 1988; Stearns, 1989).

LACK OF NON-ADAPTIVE VARIATION IN ALLOCATION

The measurement of trade-off curves requires variation in trait values. However, there will often be a single optimum way in which resources can be allocated, and if all individuals allocate resources in accord with this, there would be no variation with which to measure the trade-off.

ADAPTIVE VARIATION IN ALLOCATION

Individuals often differ in resources, and the optimal allocation of resources may then also vary. In these circumstances natural selection should equip organisms not with a single genetically determined allocation, but with a genetically determined *allocation rule* specifying allocation in relation to the resources. van Noordwijk and de Jong (1986) have provided a neat analogy: when money is a limiting resource, it can be spent on a car or a house, but not both, so that the trade-off has a negative slope. However, a survey of the values of cars and houses across households will generally reveal a positive phenotypic correlation. This is because households differ in the amount of resource (money) that they have, and those with more money generally choose to increase the amount spent on both, rather than on only one, of the two commodities.

2.2.2 How can constraints be measured?

Two general methods are currently considered as methods of measuring trade-offs: experimental manipulations and genetic correlations (Reznick, 1985; Partridge & Harvey, 1988).

EXPERIMENTAL MANIPULATIONS

Phenotypic correlations are not trade-off curves because phenotypes are non-randomly distributed with respect to environments or resources. Random allocation can often be achieved using experimental manipulations. For instance, female *C. maculatus* beetles can be discouraged from laying eggs by depriving them of mates or oviposition sites. These experimentally manipulated females show the expected trade-off between longevity and fecundity (Fig. 2.1b).

A distinction is often made (see, for example, Reznick, 1985) between direct manipulation of the trait itself, for instance clutch or brood size in birds, and indirect manipulation, in which manipulation of the environment provokes a change in the trait by the animal, for example the manipulation of clutch size by supplementary feeding (Kent, 1981). Indirect manipulations suffer from the same problems of disentangling causation as phenotypic correlations; the environmental change that provoked the trait change may also itself be responsible for the accompanying changes in fitness or other traits (Reznick, 1985; Bell & Koufopanou, 1986). Only direct manipulations appear to offer a means of measuring trade-offs. However, the above example for *C. maculatus* already suggests that the distinction between direct and indirect manipulations may not be clear-cut, and sections 2.3.1 and 2.4.1 offer further examples.

GENETIC CORRELATIONS

When there is a trade-off between two traits, the locus or loci determining the allocation of resources will affect both of the traits; a genetic increase in allocation to one of the traits will be accompanied by a genetic decrease in allocation to the other. In other words, there will tend to be a *negative genetic correlation* between the traits. (This situation is also referred to as *antagonistic pleiotropy*.) The genetic correlation will be measurable only if there is genetic variation at the locus. If both of the traits contribute positively to fitness, alleles that affect one of the two traits favourably will affect the other adversely, so that different alleles will differ little in overall fitness. Selection should fix these loci more slowly than those that affect both traits favourably or both traits adversely. Trade-offs are therefore expected to lead to negative genetic correlations (Lande, 1982a; Falconer, 1989).

Genetic correlations can be measured by using breeding experiments which create groups of genetically identical or similar individuals (clones or relatives) or by using selection experiments. For instance, breeding experiments have been used to estimate the genetic correlation coefficient between fecundity and longevity in *C. maculatus* (Møller *et al.*, 1989). The estimated genetic correlation is positive, which suggests that there is not a trade-off between these traits. The result is therefore at odds with the results of experimental manipulations (Fig. 2.1b; El-Sawaf, 1956; Wilson, 1989). Other genetic correlations between life history traits measured in the laboratory (Table 2.1) also do not fulfill the expectation of being predominantly negative, although, as noted by Bell and Koufopanou (1986), selection experiments tend to yield a higher proportion of negative values than breeding experiments. Although some positive coefficients can be

Table 2.1 Genetic correlations between life history traits. Correlations have been divided into those concerning the cost of reproduction (between a longevity component and a fecundity component, or between fecundity components) and other correlations (Appendices 2.1 and 2.2).

	Cost of reproduction				Other life history traits			
	No. of studies	Mean % of genetic correlation coefficients			No. of studies	Mean % of genetic correlation coefficients		
		−	0	+		−	0	+
Selection experiments on *Drosophila* species	10	56	40	4	4	58	33	8
Selection experiments on non-*Drosophila*, non-domestic species	3	33	50	17	3	0	0	100
Breeding experiments on *Drosophila* species	10	10	56	35	15	10	48	41
Breeding experiments on non-*Drosophila*, non-domestic species	14	9	70	21	10	25	60	14

explained away as the consequence of inbreeding (Rose, 1984b) or novel environments (Service & Rose, 1985), these results do not stimulate optimism over the usefulness of genetic correlation coefficients as a means of measuring trade-offs, especially because large sample sizes are needed to estimate genetic correlation coefficients with any degree of accuracy. Nevertheless, at present, they offer the only means of investigating trade-offs involving traits, such as egg size, which cannot easily be manipulated (see, for example, Lessells *et al.,* 1989).

2.3 The cost of reproduction

All models of the evolution of reproductive effort assume that current reproduction can be increased only at the expense of future survival or fecundity; in other words that there is a trade-off between current and residual RV. So by definition, the 'cost of reproduction' is paid for in reduced survival or fecundity, but it may be physiological or ecological in origin (Calow, 1979). A possible example of a physiological origin is the increased energy expenditure of wild birds as they increase the rate at which they provision the brood (Bryant, 1988). Ecological origins of costs are the result of diversion of resources away from the avoidance of environmental hazards, such as predation, and the increased use of environmentally hazardous environments, for instance to collect food to provision a brood. Snakes, for instance, are less mobile when gravid; viviparous checkered garter snakes, *Thamnophis marcianus*, tested on an experimental racetrack had longer lap times, moved a shorter distance overall and stopped after a shorter time during middle or late pregnancy than before or after pregnancy (Seigel *et al.,* 1987).

2.3.1 *Measurements of the cost of reproduction*

Genetic correlations between components of current and future reproductive value have been measured in captive animals and are far from yielding a consistent pattern (Table 2.1). In contrast, manipulation experiments have been carried out in both the laboratory and field and do give a consistent pattern. Laboratory studies were carried out mostly on species without parental care and in general show that suppression of oviposition increases longevity (Table 2.2). For example, female *C. maculatus* beetles live longer when deprived of oviposition sites or mates (Fig. 2.1b), and male *C. maculatus* beetles live longer when kept alone or with males than when kept with females (R.F. Holt, personal communication). Field studies have been carried out exclusively on birds and show that an experimental increase in brood size may delay or reduce the success rate of second broods, reduce the overwinter survival rate of the parents, or delay reproduction or reduce fecundity in the following year (Table 2.2). For instance,

Table 2.2 Experimental tests of the cost of reproduction.

	No. of studies	Effect on listed trait of higher levels of reproductive effort (% of studies)		
		−	0	+
*Laboratory studies of invertebrates**				
Adult longevity or survival				
Males	10	65	20	15
Females	27	89	7	4
Total or future fecundity				
Females	4	87	12	0
Field studies of birds†				
Probability of second brood	7	71	29	0
Interval to second brood	9	0	22	78
Clutch size of second brood	7	14	86	0
No. of fledglings in second brood	5	40	60	0
Weight of parents				
Males	8	25	75	0
Females	10	50	50	0
Both sexes	7	14	86	0
Timing of moult				
Males	2	0	100	0
Females	1	0	0	100
Survival to following year				
Males	7	14	86	0
Females	7	14	86	0
Both sexes	8	25	75	0
Laying date in following year	4	0	50	50
Clutch size in following year				
Males	1	0	100	0
Both sexes	9	28	72	0
No. of fledglings in following year				
Males	1	0	100	0
Both sexes	3	0	100	0

Sources of data: *Bell & Koufopanou, 1986 (review) plus El-Sawaf, 1956; Tallamy & Denno, 1982; Fitzpatrick & McNeil, 1989; Fowler & Partridge, 1989; Service, 1989; Wilson, 1989; R.F. Holt, personal communication; †Partridge, 1989; Dijkstra *et al.*, 1990 (reviews) plus Lessells, 1986; Lindén, 1988; Orell & Koivula, 1988; Pettifor, 1989; Orell, 1990; Torok & Toth, 1990; G. Högstedt, personal communication; H. Riley, personal communication; M. Thompson, personal communication.

the clutch size of female collared flycatchers is reduced in the year following an experimental increase in brood size, but the survival rate of the adults is unaffected (Fig. 2.2; Gustafsson & Sutherland, 1988).

These studies also demonstrate that the distinction between direct and indirect manipulation made in section 2.2.2 may not be as clear as it seems.

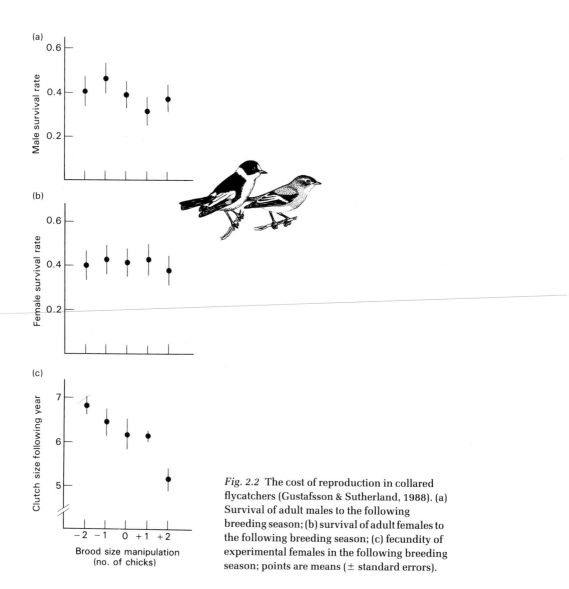

Fig. 2.2 The cost of reproduction in collared flycatchers (Gustafsson & Sutherland, 1988). (a) Survival of adult males to the following breeding season; (b) survival of adult females to the following breeding season; (c) fecundity of experimental females in the following breeding season; points are means (± standard errors).

Even manipulation of clutch size does not directly manipulate the trait of interest, reproductive effort. In fact, some species such as blue tits and glaucous-winged gulls increase their provisioning rates to increased broods (Nur, 1984a; Reid, 1987; Pettifor, 1989), but others such as Tengmalm's owls and great tits do not (Korpimäki, 1988; Smith *et al.*, 1988). Similarly, some indirect manipulations, such as supplementary feeding, are likely to have independent effects on fitness and other traits, while others, such as the withholding of mates or oviposition sites, are not. The result is that there is actually a spectrum from direct to indirect manipulation, underlining the need for careful confirmation that manipulation has had the expected effect and no others.

Responses to manipulation may be strategic decisions by animals, rather than something forced on them by the manipulation. The failure of some birds to initiate a second brood following experimental augmentation of their first brood may not be because birds are unable to produce such a clutch, but because a delayed start to the second clutch, or reduced parental condition, reduces the value of the brood to a point where it is not worth the costs involved (Smith *et al.*, 1987). A subtly different possibility is that individuals recalibrate their assessment of environmental suitability as a result of the manipulation. Some birds appear to modify their clutch size in response to the number of young they can raise (see section 2.8.3). This implies that they use some rule specifying clutch size to be laid as a function of an estimate of future environmental suitability made at or before the time of egg laying. Such a rule may well initially be genetically determined, but birds might be expected to recalibrate this rule if it appears to produce a clutch mismatched to the conditions for feeding young. Experimental manipulation may produce such a mismatch; in order to satisfy the young the parents must work harder than 'expected'. As a result females may recalibrate their egg-laying rule and produce a smaller clutch under given conditions during laying. This is one interpretation of the finding that collared flycatchers lay smaller clutches the year after experimental brood increases (Gustafsson & Sutherland, 1988) and that willow tits aged 1 year, but not older, appear to pay a cost of reproduction in terms of subsequent fecundity (Orell, 1990). These problems apply only to interpreting reduced fecundity as a cost of reproduction; it is difficult to see how an increased mortality rate could be a strategic decision or result from recalibration of a decision rule.

Field studies also encounter problems in keeping track of the whole population and of spatial and temporal variation. Högstedt (1981) has pointed out that researchers often measure only the 'local survival' of individuals, and do not distinguish between death and dispersal. Also, temporal variation may lead to the cost of reproduction being paid only in some years, necessitating longer-term studies to evaluate properly the costs of reproduction (Nur, 1988).

Lastly, both laboratory and field measurements of costs are often incomplete. Experimental suppression of egg laying does not stop the female from developing a reproductive tract or other structures necessary for reproduction, so that the true cost of reproduction may be underestimated. For instance, in birds, the energetic expense of incubation increases with the size of clutch incubated (Biebach, 1981; Haftorn & Reinertsen, 1985). Ideally, experiments would manipulate the number of eggs laid and incubated, as well as those fed; recent studies have begun to investigate the fitness costs of incubating experimentally manipulated clutches (Moreno *et al.*, 1991).

2.4 Trade-off between the number and fitness of offspring

All models of the evolution of family size assume that there is a trade-off between the number of progeny produced and the fitness of each of them. This trade-off is expected on the basis that reproductive effort is usually divided between the offspring, so that the number of young produced can be increased only at the expense of the fitness of each of them. In the simplest case, when there is no parental care and the young do not develop in a habitat patch discovered or created by their parents, the mother must divide the resources she has between the eggs that she produces. Under these circumstances, a trade-off between the number of eggs and the extent to which each is provisioned seems likely (see Chapter 8). Equally, when the young receive some form of parental care, the benefit that each derives will usually decrease as the family size increases. However, there are some forms of parental care, such as vigilance for predators, where young may accrue the same benefit independent of the number of other young in the family (Lazarus & Inglis, 1986). Many species are without parental care, but eggs are laid into a patch of resource that has been discoverd by the parents or actually created by them, as in the case of dung masses or balls produced by dung beetles (Klemperer, 1983). Although there may be no parental care in the usual sense, the time invested in searching for or producing a patch of suitable habitat is a form of reproductive effort, and the fitness of each of the offspring will often depend on the number of young between whom the effort is divided.

The number of young in a family may have other effects, both beneficial and deleterious, besides those resulting directly from the division of a limited amount of parental effort (see Godfray (1987) for a review for invertebrates). For instance, nestlings in larger broods may have lower energetic demands because of the beneficial thermal effects of the other nestlings (Royama, 1966).

Both the division of parental effort and the direct effects of family size on the fitness of individual offspring contribute to the relationship between offspring fitness and family size. It is this overall relationship that must be known in order to predict optimal reproductive tactics.

2.4.1 Measurements of the trade-off between the number and fitness of offspring

While a trade-off between egg and clutch size seems an inevitable consequence of limited resources for egg laying, the measurement of this trade-off is not easy. Neither clutch nor egg size laid can be directly manipulated, so genetic correlations appear to offer the only approach. However, the single study of a wild population yielded equivocal results (Lessells *et al.*, 1989), and laboratory studies have been restricted to domestic species. In domes-

tic poultry, which generally have a history of artificial selection for increased egg and clutch size, genetic correlations between these two traits
are generally negative (Wyatt, 1954; Jerome et al., 1956; Hogsett &
Nordskog, 1958; Jaffe, 1966; Emsley et al., 1977), but in domesticated rainbow trout there is no significant genetic correlation (Gall, 1975).

In contrast, manipulation studies have been carried out in both the field
and laboratory and do show a consistent pattern, with young in larger
families suffering from increased mortality rates, extended development,
and reduced size and fecundity (Table 2.3). These studies have been carried
out primarily on insects in the laboratory and on birds in the field. For
example, larvae of *C. maculatus* growing in crowded beans survive less
well to adult emergence, emerge at a lower weight, live for a shorter time as
adults and, if female, lay fewer eggs (Fig. 2.3: Bellows, 1982; Wilson, 1989.

One problem with many of the bird studies is that the total parental investment varies with brood size. This may occur even in birds which do not
feed their young; bar-headed geese, *Anser indicus*, appear to modify their
time budgets in relation to their brood sizes (Schindler & Lamprecht, 1987).
As a result, the measured relationships will tend to underestimate the true
costs on offspring fitness of increased family size, even in experimental

Table 2.3 Experimental tests of the trade-off between the number and fitness of offspring.

	No. of studies	Effect on listed trait of larger family sizes (% of studies)		
		−	0	+
*Laboratory studies of invertebrates**				
Larval survival	15	97	0	3
Adult longevity				
Females	2	100	0	0
Both sexes	1	100	0	0
Development period	4	0	50	50
Adult size	11	91	9	0
Female fecundity	7	100	0	0
Field studies of birds†				
Fledging weight	37	59	41	0
Survival to fledging	46	67	33	0
Fledgling survival	10	40	60	0
Fecundity of daughters	1	100	0	0

Sources of data: *Salt, 1940; Narayanan & Subba Rao, 1955; Wilkes, 1963; Chacko, 1964;
Wylie, 1965; Klomp & Teerink, 1967; Escalante & Rabinovich, 1979; Bellows, 1982; Danthanarayana et al., 1982; Quiring & McNeil, 1984a,b; Waage & Ng Sook Ming, 1984; Ikawa
& Okabe, 1985; Pallewatta in Waage & Godfray, 1985; Tagaki, 1985; Taylor, 1988; Wilson,
1989; †Dijkstra et al., 1990 (review) plus Safriel, 1975; Rohwer, 1985; Lessells, 1986.

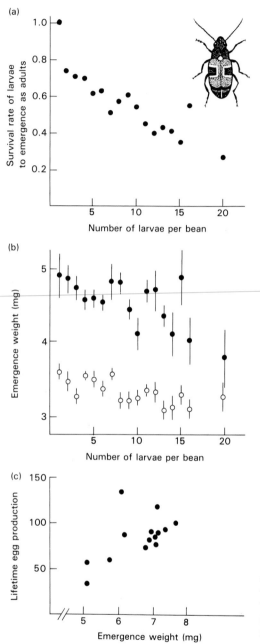

(a)

Survival rate of larvae to emergence as adults

Number of larvae per bean

(b)

Emergence weight (mg)

Number of larvae per bean

(c)

Lifetime egg production

Emergence weight (mg)

Fig. 2.3 The trade-off between the number and fitness of offspring in the bruchid beetle *Callosobruchus maculatus* (Wilson, 1989). (a) Survival to adult emergence; (b) emergence weight; ●, females; ○, males; vertical bars are standard errors; (c) lifetime egg production in relation to emergence weight.

studies. However, the measurements can be used in conjunction with those obtained from the same manipulations on the cost of reproduction to predict optimal clutch size. This involves making the tacit assumption that phenotypic plasticity in parental effort is adaptive (section 2.8.1).

2.5 Optimal clutch size in insects and birds

The clutch size of an individual depends on both its reproductive effort and on how it divides the resources between offspring. The first of these two decisions depends on how increased reproductive effort affects the parents' future prospects, in other words on the cost of reproduction. The second of the two decisions depends on the trade-off between the number and fitness of offspring. The previous sections showed that measurements of these trade-offs can be made experimentally. This section considers how the two trade-offs interact to determine optimal clutch size.

2.5.1 Lack's hypothesis

Lack (1947) was the first to propose a functional hypothesis for the evolution of clutch size in birds. He suggested that as clutch (and hence brood) size increased, each of the nestlings would receive less food and hence survive less well (Fig. 2.4a). As a result, an intermediate brood size might produce the greatest number of survivors (Fig. 2.4b). Lack proposed that the population would evolve this *most productive brood size* (Charnov & Krebs, 1974) which is also referred to as 'Lack's solution' (Godfray, 1987). In other words, the optimal clutch size is determined by the trade-off between the number and fitness of offspring.

In essence, Lack's hypothesis determines the optimal way of dividing resources by finding the optimal clutch size; the optimal investment per offspring follows from this. An alternative approach (Smith & Fretwell, 1974; see Chapter 8 for further details) is to find the optimal investment per offspring, from which follows the optimal clutch size. Certain restrictive assumptions are made in this second approach, in particular that the fitness of offspring does not depend on family size over and above the consequences of the division of resources between offspring.

Lack's hypothesis has been tested in both insects and birds. In insects the relationship between survival and brood size has been measured and used to predict the most productive brood size. When only the survival of offspring is taken into account, mean clutch size is consistently less than the most productive brood size. For instance, female *C. maculatus* beetles typically lay between two and six eggs on each black-eyed bean, while the most productive brood size is about 16 eggs (Wilson, 1989). Clutch size is also less than the most productive brood size in all seven species of parasitoids studied (Charnov & Skinner, 1984; 1985; Waage & Godfray, 1985). In birds, Lack's hypothesis has been tested by manipulating clutch or initial brood size both up and down. If Lack's hypothesis is correct, any experimental manipulation should lead to a reduction in the production of young. In collared flycatchers, an experimental increase in brood size led to an increase in the number of young fledging (Gustafsson & Sutherland, 1988), and similar increases in productivity, measured in terms of fledged young, have occurred in 35 out of 50 studies (Safriel, 1975; Rohwer, 1985; Lessells, 1986; Dijkstra *et al.*, 1990); so in birds, as in insects, the majority of species lay clutches smaller than the most productive brood size. Four kinds of explanation have been put forward to explain the general discrepancy between the most productive brood size and observed brood sizes.

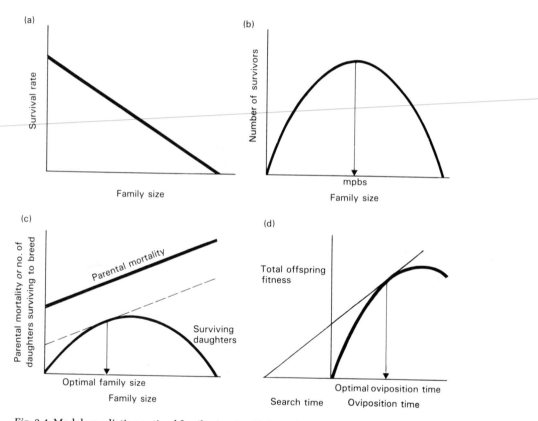

Fig. 2.4 Models predicting optimal family size. *Lack's hypothesis*: (a) survival of individual offspring declines with family size so that (b) an intermediate family size (the most productive brood size, mpbs) produces the greatest number of survivors. The *cost of reproduction* may reduce the optimal family size: (c) if parental mortality rate increases with family size, the optimal family size maximizes the difference between gains through reproduction and losses through adult mortality. The optimal family size is found where a curve parallel to the parental mortality curve is a tangent to the offspring production curve (Charnov & Krebs, 1974). (d) If the number of clutches produced is limited by the time available to search for oviposition sites and to oviposit, the optimal time to spend ovipositing per oviposition site (and hence the optimal clutch size) is found by constructing the tangent to the total offspring fitness curve from a point representing the search time (Charnov & Skinner, 1984; 1985; Skinner, 1985).

MEASUREMENT OF OFFSPRING FITNESS IS INCOMPLETE

The above studies examine brood productivity in terms of the number of survivors emerging from a patch or fledging from a brood, and neglect the effects of family size on subsequent survival or fecundity. Lack (1954) himself suggested that brood productivity should be assessed three months after fledging in order to include differential survival of offspring in the post-fledging period. In insects, inclusion of the effects of brood size on daughters' fecundity reduces the most productive brood size to that observed in *Trichogramma evanescens* (Waage & Godfray, 1985), but not in *C. maculatus* (Table 2.4; Wilson, 1989), *Trichogramma embryophagum* or *Nasonia vitripennis* (Charnov & Skinner, 1984). The discrepancy for the last three species might disappear if the fitness of daughters is more strongly dependent on size in the field than suggested by laboratory measurements, or if larval competition affects males (for whom the fitness consequences of size have not been measured) more strongly than females. In birds, inclusion of the effects of brood size on post-fledging survival reduces the most productive brood size to the actual clutch size in two out of eight species: clutch size equals the most productive brood size in collared flycatchers (Gustafsson & Sutherland, 1988) and great tits, (Pettifor *et al.*, 1988; Smith *et al.*, 1989; Tinbergen & Daan, 1990); clutch size is less than the most productive brood size in Canada geese (Lessells, 1986), kestrels (Dijkstra *et al.*, 1990), wood pigeons (Murton *et al.*, 1974), tree swallows (de Steven, 1980), pied flycatchers (von Haartman, 1954) and blue tits (Nur, 1984b).

MAXIMIZING FITNESS PER CLUTCH DOES NOT MAXIMIZE
LIFETIME FITNESS

If the number of clutches produced during an individual's lifetime is fixed, then maximizing fitness per clutch will also maximize lifetime fitness. But if there is a cost to reproduction, so that the number of clutches produced decreases as the current clutch size increases, a clutch size smaller than the most productive brood size could maximize lifetime fitness (Fig. 2.4c,d; Williams, 1966; Charnov & Krebs, 1974; Charnov & Skinner, 1984; 1985; Parker & Courtney, 1984; Skinner, 1985).

In insects, the cost of reproduction is seldom known in sufficient detail to make a quantitative prediction of the clutch size maximizing RV, but optimal clutch size can be predicted under varying assumptions concerning the limiting resource. For any particular limiting resource, the optimal clutch size is that which converts the resource most efficiently into total offspring fitness. For instance, if time or energy reserves are limiting, the optimal clutch size may be found using the marginal value theorem (Fig. 2.4d; Charnov & Skinner, 1984; 1985; Skinner, 1985), while if eggs are limiting, the optimal clutch size will equal one because this is the clutch size that

Table 2.4 Tests of constraint assumptions for optimal clutch size in the bruchid beetle *Callosobruchus maculatus* (Wilson, 1989). Predictions were made on the basis of experimental measurements of the cost of reproduction and trade-off between the number and fitness of offspring (including effects on daughters' fecundity). Predictions were tested by manipulating search time and observing the clutch size laid.

Constraint assumptions	Predictions	Tests and conclusions
Oviposition sites (beans) limiting	Clutch size of 15 (without multiple oviposition)	Clutch size of 2.3–5.5
	No increase in clutch size with search time (with or without multiple oviposition)	Increase in clutch size with search time
		Constraint rejected
Eggs limiting (with or without multiple oviposition)	Clutch size of one	Clutch size of 2.3–5.5
	No increase in clutch size with search time	Increase in clutch size with search time
		Constraint rejected
Reserves *or* eggs and time *or* eggs and oviposition sites limiting (with or without multiple oviposition)	Clutch size of one at short search times	Mean clutch size of 2.3 at short search times
		Contraints rejected
Time limiting (without multiple oviposition)	Clutch size of 1–3 at short search times	Clutch size of 2.3 at short search times
	Clutch size of 14–15 at long search times	Clutch size of 5.5 at long search times
		Constraint rejected
Time limiting (with multiple oviposition)	Clutch size of 1–2 at short search times	Clutch size of 2.3 at short search times
	Clutch size of 3–7 at long search times	Clutch size at 5.5 at long search times
		Constraint provisionally accepted

yields the highest fitness per egg. (See also Chapter 8 for a discussion of this problem.)

In birds, the cost of reproduction can be measured quantitatively, so that the clutch size maximizing RV can be predicted. The cost of reproduction accounts for the discrepancy between actual brood size and most productive brood size in two out of seven species: clutch size equals predicted optimum in kestrels (Daan *et al.*, 1990) and pied flycatchers (Askenmo, 1977; 1979); clutch size is less than the predicted optimum in Canada geese (Lessells, 1986), Tengmalm's owls (Korpimäki, 1988), tree swallows (de Steven, 1980), house wrens (Finke *et al.*, 1987) and blue tits (Nur, 1984b).

In one insect study and 17 bird studies, the actual clutch size equals the most productive brood size; in these studies the trade-off between the number and fitness of offspring 'explains' clutch size and therefore the cost of reproduction is apparently superfluous (see, for example, Boyce & Perrins, 1987). However, even in these situations, a cost of reproduction must be invoked to explain the overall level of reproductive effort at each reproductive episode in iteroparous species. Without it, there is no explanation why parents do not increase their most productive brood size by making a larger effort.

MULTIPLE OVIPOSITION

The above models assume that only one female lays in each resource patch or nest. When multiple oviposition occurs, the optimal clutch size is often reduced (Andersson & Eriksson, 1982; Parker & Courtney, 1984; Skinner, 1985; Smith & Lessells, 1985; Wilson, 1989), but may be increased (Ives, 1989).

In the absence of detailed measurements for insects of the costs of reproduction and frequency of multiple oviposition, a single prediction of the clutch size maximizing RV cannot be made. An alternative approach is to predict the optimal clutch size under a range of limiting resources and to use the fit between predictions and observations as a guide to the most important constraints on clutch size evolution. The most comprehensive use of this approach is by Wilson (1989) who investigated clutch size decisions in *C. maculatus* beetles ovipositing on black-eyed beans (Table 2.4). He considered females who were limited by oviposition sites, time, eggs or reserves, either singly or in combination and with or without multiple oviposition (Wilsons, 1989) and made quantitative predictions of optimal clutch size using experimental measurements of the cost of reproduction and the trade-off between the number and fitness of offspring. Females behaved as though they were limited by time and by other females ovipositing on the same seeds (Table 2.4).

STOCHASTIC VARIATION IN CLUTCH SIZE OR FITNESS

The above models predict optimal clutch size when females have perfect control of the family size they produce and there is no variation between clutches in optimal clutch size. However, stochastic variation may alter the optimum clutch size (see Godfray & Ives (1988) for a review of possible effects). If the peak in the relationship between RV and clutch size is asymmetrical, with a steeper drop (a 'cliff edge') for larger than for smaller clutch sizes, and females do not have perfect control over the family size they produce, then the optimal clutch size is smaller than the most productive

clutch size (Mountford, 1968). This is because clutches larger than the optimal clutch size will have much reduced fitness compared with those
smaller than the optimal clutch size. As a result, it pays females to lay a conservative mean clutch size which ensures that their larger clutches do not
fall off the cliff edge.

The optimal clutch size may also be smaller than the most productive
brood size if large clutches vary in productivity. Boyce and Perrins (1987)
have suggested that under such situations the geometric mean is a more appropriate long-term measure of fitness than the arithmetic mean, and in
great tits the actual clutch size is equal to the predicted optimum when the
geometric mean is used. However, the disparate conclusions reached by
this analysis and by Pettifor et al. (1988) from essentially the same data-set
have not yet been resolved.

2.6 Extremes of reproductive effort: semelparity and deferred maturity

All of the examples used in the previous section involve intermediate levels
of reproductive effort; sufficient resources are devoted to maintenance for
parents to survive reproduction, so that each individual potentially produces several clutches during its lifetime, a condition referred to as
iteroparity. If all resources are devoted to reproduction, the parents will
then die and so reproduce only once during their lifetime (semelparity or
'big bang' reproduction). At the other extremes, animals who make no reproductive effort will not reproduce. This usually occurs in younger age
classes, and so leads to deferred maturity, but in some animals, such as
elephants, there may be a post-reproductive period. What are the selection
pressures which favour these extremes of reproductive effort?

2.6.1 Semelparity and iteroparity

Interest in the evolution of semelparity and iteroparity (as did the terms
themselves!) began with Cole's (1954) mathematical models which showed
that a semelparous individual would leave as many descendants as an
iteroparous individual provided that its family size was one offspring
larger. This is known as Cole's paradox because, as species need to add
only one offspring to their family for semelparity to be favoured, it is
difficult to explain why there are so many iteroparous species. Cole's result
is easy to understand intuitively; the extra offspring produced replaces its
parent as a breeding individual in the next reproductive season (ignoring
the decrease in relatedness due to sexual reproduction) (Horn, 1978). This
also reveals the solution to the paradox: Cole neglected mortality in his
models (except for the death of semelparous individuals after reproduction) (Gadgil & Bossert, 1970; Charnov & Schaffer, 1973); in fact, a sexual
parent must produce $2P/Y$ extra young for semelparity to be favoured

(where P is parental survival and Y is the juvenile survival to the next breeding season) (Waller & Green, 1981; Horn & Rubenstein, 1984). As predicted, species with low P/Y values tend to be iteroparous (Stearns, 1976), but there are problems with this analysis (Partridge & Harvey, 1988). Stochastic variation is another factor favouring a relatively conservative reproductive policy, as it also did in the clutch size models (section 2.5.1). If juvenile mortality varies greatly between different bouts of reproduction, parents are selected to 'bet-hedge' and spread their reproduction over several reproductive bouts (Murphy, 1968).

Further analysis of the conditions favouring semelparity incorporating the cost of reproduction more explicitly was made by Pianka and Parker (1975) who pointed out that the shape of the curve relating residual RV to current reproduction determines whether iteroparity, semelparity or deferred reproduction are selected (Fig. 2.5). The shape of the curve may change with the age (or size) of the individual, so that deferred maturity may be followed by semelparity or iteroparity later in life. The shape of the curve is particularly likely to favour semelparity (i.e. be concave up) when the initial costs of reproduction are high, as for instance in the case of fish that undertake a long migration to spawning grounds, but where extra young can be produced at a relatively small cost. There are no reliable measurements of the shape of the whole curve for any species (Bell & Koufopanou, 1986) although experimental tests of optimal clutch size (section 2.5) imply that the curve is convex around the optimum. Measurement of the whole curve is virtually impossible because of the difficulties of experimentally manipulating animals to make extremes of reproductive effort. Hormonal

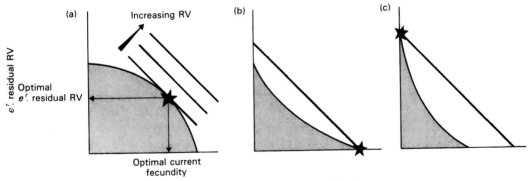

Fig. 2.5 The evolution of iteroparity, semelparity and deferred reproduction. The optimal reproductive effort depends on the shape of the curve relating $e^r \cdot$ residual RV to m_x, the current fecundity. When the curve is convex up (a), iteroparity is favoured; when the curve is concave up, semelparity (b) or deferred maturity (c) is favoured. Reproductive value at age x, $RV_x = m_x \cdot e^{-r} +$ residual RV (section 2.1.2), so that when $e^r \cdot$ residual RV is plotted against m_x, lines of slope -1 join points of equal RV. The optimal reproductive effort can therefore be found where the highest line of slope -1 intersects such a curve (Williams, 1966; Pianka & Parker, 1975; Bell, 1980).

manipulations have been suggested (Partridge, 1989), but these are likely to suffer from the same problems of interpretation as other indirect manipulations.

2.6.2 *Age at maturity*

Deferred maturity has been investigated in detail by considering the evolution of the age at maturity. Age at maturity will govern a trade-off between juvenile survival, which is highest for individuals who mature early because they suffer juvenile mortality for a shorter period, and RV at the age of maturity, which is highest in late maturing individuals because they will be largest and hence most fecund. Models incorporating this trade-off have

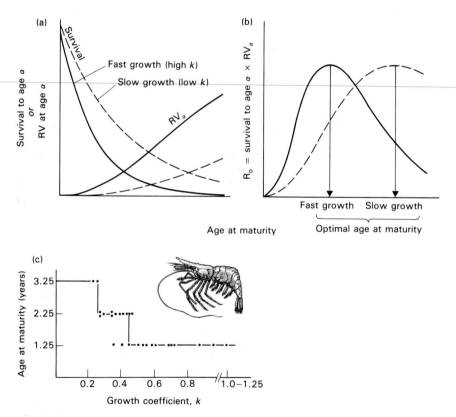

Fig. 2.6 Optimal age at maturity in pandalid shrimp (Charnov, 1989b). (a) Survival to reproduction decreases and RV at age of maturity, RV_α, increases with age at maturity, α. Survival decreases and RV_α increases as the growth coefficient, k, increases for a given age at maturity. (b) R_0, the average number of offspring produced over an individual's lifespan, is the product of survival and RV_α. The peak in the R_0 curve predicts the optimal age at maturity. (c) Observed ages at maturity in shrimp are close to those predicted. Points are measured values for different populations, and the stepped line is the predicted optimal age at maturity when this is constrained by seasonality to take a value which is an integer increment of 1.25 years.

been successful at predicting age at maturity in shrimps (Fig. 2.6; Charnov, 1989b), fish and other species (Roff, 1986; Stearns & Koella, 1986). The success of these models is more remarkable given that the trade-off curves are based on generalized relationships obtained from non-experimental data. These models are unlikely to be universally applicable; for instance many seabirds have deferred maturity despite having more or less completed their growth in their first year of life. In these species, deferred maturity is likely to result from RV at age of maturity being dependent on experience rather than size.

2.7 Comparative studies of life history traits

The tests of life history models described so far have generally involved precise quantitative predictions. Where such predictions can be made, the models can be tested on a single species, and tests over a range of populations or species (see, for example, Fig. 2.6) serve to strengthen the conclusions rather than offering the only means of testing the model. There are two situations in which studies of single populations are inadequate, and comparative studies therefore offer the only means of progress. The first is when only qualitative predictions can be made; this is particularly likely to be the case for traits such as dispersal, migration, diapause and semelparity for which constraints cannot easily be measured experimentally. The second situation is in exploring general patterns of the occurrence of life histories. In essence, many of the quantitative tests described above show that the life history of the animal is (or is not) better than other slightly different life histories, but they do not tell us much about why different species have radically different life histories. Comparative studies offer the main means of addressing such questions.

2.7.1 Habitat classifications

Habitats impose the selection pressures which cause particular life history traits to evolve, but a complete description of the features of a habitat which are relevant to the evolution of life histories would involve many different variables and become unwieldy in comparing different species. Habitat classifications are an attempt to summarize in just a few variables the kinds of selection pressures imposed by different habitats.

One of the earliest, and still probably best known, of the habitat classifications is r/K selection (MacArthur & Wilson, 1967). Species living at close to their equilibrium population densities are selected to maximize K, their carrying capacity, while species whose densities are typically low are selected to maximize r, their rate of increase without density limitation, where r and K are parameters of the logistic population growth equation. This classification was rapidly extended and generalized so that the ends of the continuum became characterized as stable, favourable environments selecting for competitive ability and hence for large body size, long generation time and low fecundity, while unstable harsh environments selected for rapid population growth and hence for small body size, short generation time and high fecundity (Pianka, 1970; Southwood, 1977; Horn, 1978; Horn & Rubenstein, 1984). Boyce (1984) has claimed that this is an unjustified generalization and argued for a return to a strict interpretation of r and K selection as density-dependent selection.

Since the r/K habitat classification, a number of other habitat classifications have been proposed (Grime, 1977; Southwood, 1977; 1988; Greenslade, 1983; Begon, 1985; Sibly & Calow, 1985; Hildrew & Townsend, 1987; Holm, 1988). Species can be fitted into these classifications with a greater or lesser degree of success, but a major problem in testing habitat classifications is measuring habitat variables independently of the life histories of the organisms inhabiting the habitats (Loehle, 1988); for instance, the stability of a habitat depends on the generation time of the species concerned (Southwood, 1977). An alternative approach to asking how many variables are needed to describe habitats is to ask how many variables are needed to describe life histories.

2.7.2 Relationships between life history traits

Just as habitat classifications attempt to summarize selection pressures in a few variables, a similar attempt can be made to summarize life history traits. In general, life history traits are more strongly related to each other than they are to the environment. For example, in a study of 14 life history traits and four ecological variables of primates, 86% of the 91 pair-wise relationships between life history traits were significant, but only 2% of the 56 relationships between a life history trait and an ecological variable were significant (Harvey & Clutton-Brock, 1985). To be fair, the latter relationships were controlled for body size, but even after similarly controlling for body size, 43% of 65 relationships between life history traits in eutherian mammals were significant (Read & Harvey, 1989). Stearns (1983b) found that about 70% of variation in life history traits of mammals could be summarized by one variable contrasting small, early maturing, short-lived species with large, late maturing, long-lived species. A second variable representing the range from altricial to precocial species explained about another 15% of the variation. The number of variables needed to summarize suites of life history traits sets a lower limit on the number needed to classify a habitat, so these comparative analyses suggest that habitats can be classified using relatively few variables. However, more variables may be needed to describe the life histories of taxonomically more diverse species, and the number of variables needed to describe life histories sets only a lower limit on the number needed to describe habitats. One challenge for the future is in assessing to what extent the small number of variables needed to describe life histories is a consequence of constraint and to what extent it is a consequence of adaptation (Read & Harvey, 1989).

Relationships between life history variables across species may stimulate research by demonstrating patterns that demand explanation (Gustafsson & Sutherland, 1988). For instance, age at maturity is directly proportional to adult lifespan, the proportionality constant ranging from 2 in fish to 1.5 in temperate snakes and 0.3–0.6 in mammals and birds (Charnov & Berrigan, 1990). This raises the question not only of why the relationship should be of direct proportionality, but also why the proportionality constant should vary between taxonomic groups.

Lastly, relationships between life history traits across species cannot be used to measure trade-offs. Just as phenotypic correlations cannot be used because individuals may vary in their resources, cross-species relationships cannot be used because species may similarly vary in their resources. Cross-species comparisons have been used, for instance, to investigate the trade-off between egg weight and clutch size in birds (Lack, 1968; Rohwer, 1988; Ekman & Johansson-Allende, 1990), but the lack of a relationship between these traits across species (Rohwer, 1988) does not demonstrate the lack of a trade-off any more than would the absence of any phenotypic correlation within a species. Cross-species comparisons claiming to investigate trade-offs are not uncommon (Zammuto, 1986; Schnebel & Grossfield, 1988; Read & Harvey, 1989). An additional problem exists when the trade-off investigated is between fecundity or survival traits, because the relationship between these traits is more or less fixed in populations that do not increase or decrease without limit (Sutherland *et al.*, 1986; Bennett & Harvey, 1988).

2.8 Phenotypic plasticity

Life history traits are frequently phenotypically plastic; that is, a single genotype produces a range of phenotypes depending on the environment. Phenotypic variation may be continuous, in which case the relationship between phenotype and the environment for each genotype is called a *reaction norm* (Woltereck, 1909), or gradual environmental change may be accompanied by sudden switches between discrete phenotypes. This phenomenon is known as polyphenism. Phenotypic plasticity may be irreversible, for instance when it involves developmental changes, or reversible, as in the case of clutch size variation in iteroparous species.

A large range of abiotic and biotic environmental features may invoke phenotypic plasticity. Growth rates are often phenotypically plastic (Stearns & Koella, 1986). Tadpoles of Couch's spadefoot toad, *Scaphopus couchii*, metamorphose earlier and at a smaller size in ponds that last for a shorter time before drying out (Newman, 1989). Colonies of the bryozoan *Membranipora membranacea* begin reproducing when their growth is limited by surrounding colonies, independently of their age or size at that stage (Harvell & Grosberg, 1988).

Phenotypic plasticity also occurs in response to predators and parasites. Many aquatic invertebrates have predator-induced morphological defences (Dodson, 1989). *Daphnia pulex* grown in the presence of predatory *Chaoborus americanus* larvae develop 'neck teeth', and the bryozoan *Membranipora membranacea* produces spines in reaction to grazing by the nudibranch *Doridella sternbergae* (Harvell, 1984). Morphology is not the only trait showing such phenotypic plasticity; the snail *Biomphalaria glabrata* increases its egg production when exposed to cercariae of its parasite *Schistosoma mansoni* (Minchella & Loverde, 1981). In some cases, such as the growth of spines in *M. membranacea* the response is directly to the predation, but in other cases chemical cues are used. The rotifer *Keratella testudo* develops spines in response to filtrates of at least 10 different species of predatory zooplankton (Stemberger & Gilbert, 1987).

2.8.1 Is phenotypic plasticity adaptive?

The most important question regarding such phenotypic plasticity is whether it is adaptive. The alternative interpretation is that plasticity is an inevitable consequence of the impact of the environment on the animal's physiology, and that phenotypic plasticity represents damage or stunting of the organism (Smith-Gill, 1983). The term 'adaptive' applied to phenotypic plasticity does not necessarily imply that individuals do equally well in all environments, but that resources are allocated in each environment to give the optimal life history for that environment. Some authors have implied that this is a restricted sense of the term 'adaptive' (Berven et al., 1979), but it is difficult to see any alternative meaning. Also, the existence of phenotypic plasticity, as opposed to genetic differentiation, does not imply that processes other than natural selection are responsible for adaptation; the allocation rule determining phenotypic plasticity is itself genetically determined and subject to natural selection (Wright, 1931; Dobzhansky, 1951; Mayr, 1963).

Any genotype that is phenotypically plastic and produces an adapted phenotype in a range of environments will out-compete genotypes which produce a fixed phenotype adapted to only one of those environments (Hamilton in Stearns, 1982). However, theoretical models suggest that evolution towards the optimal reaction norm may be slow, particularly when some of the environments are rare (Via & Lande, 1985). The question of the extent to which phenotypic plasticity is adaptive is therefore open to empirical investigation.

2.8.2 Qualitative evidence for adaptive phenotypic plasticity

The first line of evidence that phenotypic plasticity is adaptive comes, paradoxically, from the existence of homeostasis (Bradshaw, 1965). If phenotypic plasticity is the consequence of environmental interference in the physicochemical processes of the organism, phenotypes are expected to vary with the environment. The lack of such variation, homeostasis, implies that reaction norms have evolved to buffer the animal from environmental variation. Homeostasis is expected to evolve in those traits with the most important effects on fitness, while less critical traits are expected to vary with the environment (Bradshaw, 1965). For instance, larvae of the fly *Ephydra cinerea* are benthic feeders in saline lakes, a habitat free from predators. The larval period of *E. cinerea* increases more under poor feeding conditions than the larval periods of *Drosophila* or house flies, whose larvae are exposed to predation (Collins, 1980).

Further support for the idea that phenotypic plasticity is adaptive comes from situations in which the direction of phenotypic change can be predicted. This is the case for the examples of phenotypic plasticity given above: for example, Couch's spadefoot tadpoles benefit from early metamorphosis in rapidly drying ponds by avoiding desiccation. Ponds that last longer allow later metamorphosing tadpoles to survive, and these larger tadpoles then may benefit from increased terrestrial performance, higher juvenile survival and an earlier or larger size at maturity (Newman, 1989).

2.8.3 *Quantitative evidence for adaptive phenotypic plasticity*

Quantitative evidence for adaptive phenotypic plasticity requires a demonstration, first, that the organism displays the optimal life history in each of a range of environments and, second, that the differences in life histories between environments are the result of phenotypic plasticity rather than genetic differentiation. Given the difficulty of measuring fitness and trade-off curves in a single environment in order to predict a single optimum value, it is not surprising that such measurements are rare. However, several field studies of birds provide varying degrees of support for adaptive phenotypic plasticity.

If reaction norms are adaptive, then individuals should have traits that are optimal for their own individual circumstances. The existence of individual optimization of clutch size has been demonstrated in great tits, magpies, and collared flycatchers (Perrins & Moss, 1975; Högstedt, 1980; Gustafsson & Sutherland, 1988; Pettifor *et al.*, 1988; Tinbergen & Daan, 1990); in these species individuals laying the largest clutches have the highest reproductive success, but experimental increases or decreases of any size clutch result in a decrease in productivity for that individual. However, clutch size differences may be of genetic rather than environmental provenance. In fact, clutch size is heritable in many bird species, including great tits (Boag & van Noordwijk, 1987).

Slagsvold and Lifjeld's (1988) experiment on pied flycatchers showed that qualitatively adaptive variation in clutch size may be due to phenotypic plasticity. Females who had already laid their first clutch were handicapped by the removal of some flight feathers and were encouraged to re-nest by clutch removal. These females decreased their clutch size more than control females (who also had their clutches removed but who were not handicapped) and also fed their replacement brood at a lower rate.

Another avian field study both demonstrates individual optimization and appears to identify the relevant environmental variable. Daan *et al.* (1990) have predicted optimal combinations of clutch size and laying date for kestrels with territories of different quality. Territory quality is measured as the hunting yield (the number of prey caught per flight hour of hunting). Different territories show parallel seasonal increases in hunting yield. Information on hunting yield can be combined with that on the number of prey needed by a growing kestrel chick to predict the parental effort, in flight hours per day, that must be made for any given clutch size and laying date combination. The optimal combination can be found from the relationships between the RV of an egg and laying date, and the residual RV of the parent and hunting effort. The relationship between the RV of the egg and laying date has been based on observational data, but a causal relationship is currently being confirmed using releases of captively reared birds. The relationship between residual RV of parents and hunting effort is based on data from experimentally manipulated broods. Optimal clutch size and laying date combinations can be predicted for a range of territory qualities (for example, Fig. 2.7a); observed combinations of clutch size and laying date are close to these predictions, being, if anything, slightly larger

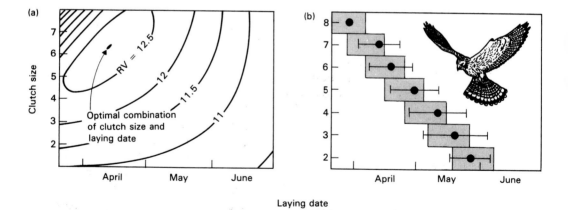

Fig. 2.7 Optimal clutch size and laying date in kestrels (Daan *et al.*, 1990). (a) Predicted reproductive value for a territory with high hunting yield. Contour lines join clutch size and laying date combinations with equal RV. The optimal combination is found at the peak in this RV landscape. Calculations are repeated for territories of different hunting yield in order to predict the relationship between clutch size and laying date. (b) Predicted (rectangles) and observed (points±standard deviations) combinations of clutch size in relation to laying date.

(or later) than predicted (Fig. 2.7b). Predictions made for individual territories on the basis of measured hunting yield are also close to the observed values. Moreover, the lack of a correlation between the laying dates of parents and offspring suggests that variation is due to phenotypic plasticity and is not of genetic provenance. Challenges for the future include confirming that hunting yield is the important environmental variable by experimental manipulation of individual territories.

2.8.4 Conditions favouring the evolution of phenotypic plasticity

Phenotypic plasticity is particularly advantageous when the environment changes on a small scale geographically or at the same or shorter interval than a generation time (Bradshaw, 1965). Conversely, genetic differentiation is more likely when associations between populations and environments persist over many generations. For example, populations of flightless thrips *Apterothrips secticornis* grow faster on their native clone of their host plant *Erigeron glaucus* than on other clones (Karban, 1989).

The occurrence of adaptive phenotypic plasticity also depends on the existence of environmental cues that accurately predict future conditions (Lively, 1986). For instance, a mismatch between the timing of hatching in great tits and peak caterpillar abundance probably occurs because cold weather, delaying the development of caterpillars, occurs after the onset of incubation when it is too late for the parent great tits to adjust the timing of hatching of their nestlings (van Noordwijk, 1990). Under these kinds of circumstances, alternative mechanisms adjusting the phenotype to the environment may evolve. Lack (1954; 1966) suggested that when the conditions for provisioning nestlings could not be predicted at the timing of laying allowing clutch size adjustment, birds should begin incubation before the clutch was complete so that individual eggs within a clutch hatched asynchronously. Under good conditions, the whole brood would still be raised, while under poor conditions the size hierarchy resulting from hatching asynchrony would result in the rapid deaths of the younger chicks without damaging the survival prospects of the older siblings. Support for this idea has recently been provided by Magrath (1989) who showed that control asynchronous broods of blackbird *Turdus merula* nestlings produced more surviving young than experimentally synchronized broods under poor, but not good, feeding conditions. However, there are at least 12 competing hypotheses (see Slagsvold & Lifjeld, 1989; Magrath, 1990; Slagsvold, 1990) and experiments which exclude all of these are difficult to design. For instance, the results for blackbirds are also consistent with the idea that staggering the hatching of the nestlings will reduce the peak demand of the brood (Hussell, 1972).

Whether phenotypic plasticity evolves will depend not only on the

reliability of environmental cues, but also on whether the phenotypic modification is reversible or not (Bradshaw, 1965). For instance, phenotypic
plasticity in *M. membranacea* is likely to have been favoured by the reversi-
bility of the decision; reproducing *M. membranacea* are able to restart
growth if space is made available by the removal of surrounding colonies
(Harvell & Grosberg, 1988).

2.8.5 *Individual optimization, directional selection and heritability*

While individual optimization (section 2.8.3) shows that individuals are
performing as well as they can for their own individual circumstances, it
does not explain why all individuals do not evolve the phenotype of the
best individuals. For instance, magpies who lay five eggs fledge about 0.7
young, as against those who lay eight eggs and fledge about 4.5 young
(Högstedt, 1980). One explanation is that individual differences are due to
environmental variation, so there is no genetic variation with which to respond to selection. But in some cases genetic variation is known to be present: the largest clutches laid by lesser snow geese consistently recruit the
most young to the breeding population (Rockwell *et al.*, 1987), and clutch
size is significantly heritable (Lessells *et al.*, 1989). The question is why, in
the presence of heritable variation, is there no response to selection?

Price *et al.* (1988) have suggested that the paradox is resolved if the trait
is partially heritable and partially determined by an environmental variable which independently affects fitness. Their hypothetical example is of
the timing of breeding in birds, which, in the absence of variation in the
nutritional state of individuals, is under stabilizing selection due to seasonal variation in weather and food availability. In addition, they suggest
that non-heritable high nutritional status both advances laying date and increases fecundity. As a result, although high fecundities are associated with
early breeding, giving a selection pressure curve suggesting selection for
early breeding, an individual cannot increase its fitness by breeding earlier.
In effect the selection is not for early breeding *per se*, but for those lucky
individuals which are well nourished and lay larger clutches; the genetic
variation that is seen is non-adaptive variation around the optimal laying
date.

The model of Price *et al.* does not involve individual optimization and
does not explain the response of laying date to nutritional status; earlier
breeding is proposed as a non-selected consequence of better nutrition.
Earlier breeding is required by the model to create a correlation between
earlier breeding and higher fecundity. Their explanation would also work
if the earlier breeding were an adaptive response, as is apparently the case
in kestrels (section 2.6.3). Non-adaptive genetic variation about the optimal
reaction norm for clutch size in relation to laying date would then lead to a
measurable heritability, but no response to selection; some genotypes

would lay clutches that were a little larger, and others a little smaller, than was optimal for the laying date.

An alternative explanation for the apparent lack of response to selection on heritable traits is that there is a genetic response, but the phenotypically expressed value of the trait depends on the genotypes of the rest of the population (Cooke *et al.*, 1990). For instance, ability to defend a territory may be improved genetically over a number of generations, but because the whole population is similarly improved there may be no corresponding increase in territory size.

2.8.6 Genetic variation in reaction norms and developmental and physiological control mechanisms as constraints on life histories

Optimality models generally predict a single reaction norm, but the commonness of genotype–environment interactions (Pani & Lasley, 1972; Bell, 1987) and crossing reaction norms (for example, Fig. 2.8; Parsons, 1977; Gupta & Lewontin, 1982; Groeters & Dingle, 1987; Bierbaum *et al.*, 1989) shows that reaction norms are often genetically variable. The existence of genetic variation in reaction norms may indicate that developmental or physiological control mechanisms (intermediate processes in the terminology of Stearns, 1989) are constraints on the evolution of life histories. If

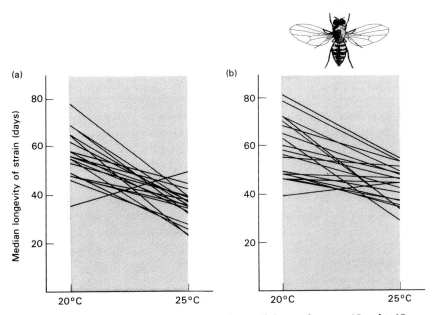

Fig. 2.8 Crossing reaction norms: longevity in *Drosophila simulans* at 20°C and 25°C (Parsons, 1977); (a) males; (b) females. Lines join the median longevity of strains at 20°C and 25°C.

different genotypes perform best in different environments, a genotype whose reaction norm consisted of the 'best' parts of other reaction norms would out-compete the other genotypes. However, such a genotype would not evolve if the metabolic costs of the mechanism producing the perfect reaction norm more than offset the benefit to be gained. As an example, Mead and Morton (1985) have suggested that hatching asynchrony is a consequence of the same hormone being used to terminate ovulation (a day or so before the last egg is laid) and to initiate incubation. Perhaps the extra metabolic costs of independently controlling the two processes more than offsets the benefit gained from the avoidance of deleterious effects of hatching asynchrony on the last chick. This argument is analagous to that used in foraging studies that the costs of extra sensory equipment, computational ability and memory needed to implement perfect decision rules may more than offset the benefit of using these rules rather than simpler 'rules-of-thumb' (Wilson, 1988). In principle, there is no difficulty in subjecting developmental and physiological processes to the same kind of cost–benefit analysis as the life history traits they control (see, for example, Baldwin & Krebs, 1981). In practice, measuring or estimating the costs and benefits of alternative control mechanisms for phenotypic plasticity in sufficient detail to predict optima is a truly formidable task.

2.9 Conclusions and prospects

Optimality models of any trait rely on measurements of constraints, including trade-off curves. For many years a confusion between these relationships and phenotypic correlations has obfuscated the measurement of constraints. The difference between these relationships is now clearly recognized, and there is no longer any excuse for mistakenly equating the relationships. This has left the problem of how constraints should be measured. Two candidates have presented themselves: experimental manipulation and measurement of genetic correlations. The debate between the proponents of these two methods has been clouded by a failure to recognize different research goals. The direct measurement of genetic constraints is clearly superior to experimental determination of trade-off curves if the aim is to understand genetic variation and predict evolutionary trajectories. If, on the other hand, the aim is to measure trade-offs for use in optimality models, genetic correlations represent a relatively indirect approach. For pheneticists the proof of the pudding may be in the eating: experimental manipulations have yielded intuitively reasonable and consistent patterns (see Tables 2.3, 2.4 and 2.5), whereas genetic correlations have painted a picture which is, at best, a little hazy (see Table 2.1). Experimental manipulations may, of course, be presenting a view that is consistently wrong; the disparate answers yielded by experimental manipulations and genetic correlations when the same trade-off has been investigated in both ways have yet to be resolved.

Experimental manipulations support the notion that trade-offs exist between life history traits, both between current and residual RV (the 'cost of reproduction') and between the number and fitness of offspring. The results from experimental manipulations in both the field and laboratory show that increased reproductive effort decreases the survival or future fecundity of parents, and that increased family size reduces the fitness of individual offspring. Challenges for the future include quantification of these relationships including all potential costs, measurement of other trade-offs, notably those involving growth, and determination of how fecundity and survival costs are incurred, particularly in the field.

Optimality models now give clear insights into the ways in which the trade-off between the number and fitness of offspring and the cost of reproduction can interact to determine optimal clutch size, and why the optimal clutch size may be less than the most productive brood size. In contrast to these theoretical insights, there are relatively few studies, except perhaps for birds, in which the relative importance of the costs of reproduction, multiple oviposition and stochastic variation as constraints on optimal clutch size have been investigated. Constraints must be measured for the environment in which the species has evolved; the fitness consequences of differential dispersal ability or vulnerability to predation may be almost impossible to quantify properly in the laboratory. As a result, the most successful studies have been of species, such as birds, that are easy to manipulate experimentally in the field, and other species, such as bruchid beetles or *Drosophila* where conditions provided in the laboratory can be regarded as close to their natural environment. An optimality approach has potential for investigating as yet relatively unexplored areas: phenotypic plasticity and physiological and developmental control.

This chapter has taken an explicitly phenotypic approach to the evolution of life histories. The alternative genetic approach is not intrinsically inferior or superior, merely different. In the long term, a complete understanding of the evolution of life histories will include both adaptive and genetic explanations. In the meantime it is as well to remember that trade-offs exist in science as in nature; different approaches may be most effective in addressing different kinds of questions.

Appendix 2.1 Genetic correlations between life history traits: *Drosophila* species. Correlations have been divided into those concerning the cost of reproduction (between a longevity component and a fecundity component, or between fecundity components) and other correlations.

	Genetic correlation coefficients						
	Cost of reproduction			Other life history traits			
	−	0	+	−	0	+	Reference
SELECTION EXPERIMENTS							
D. melanogaster		1					Lints & Hoste (1974)
		1					Lints & Hoste (1977)
		1					Lints et al. (1979)
	3		2		1		Rose (1984a); Rose & Charlesworth (1980; 1981a)
				1			Mueller & Ayala (1981)
	2						Luckinbull et al. (1984)
	1	1					Luckinbull & Clare (1985)
	1						Luckinbull et al. (1987)
	1	1		2		1	L. Partridge (personal communication)
D. pseudobscura	1						Wattiaux (1968a)
				4	2		Taylor & Condra (1980)
D. subobscura	1						Wattiaux (1986b)
NON-SELECTION EXPERIMENTS							
D. melanogaster			1				Hiraizumi (1961)
		1					Tantawy & Rakha (1964)
		1					Tantawy & El-Helw (1966)
					6	3	Temin (1966)
					1		Mukai & Yamazaki (1971)
		1	2			3	Giesel (1979)
		2	8		12	7	Giesel & Zettler (1980)
			1		1		Simmons et al. (1980)
					1	1	Mueller & Ayala (1981)
	2		5			4	Rose & Charlesworth (1981b)
		31	5	1	46	7	Giesel et al. (1982)
		7	29			8	Rose (1984b)
					1		Kosuda (1985)
					2	4	Giesel (1986)
	2		1	4	5		Tucić et al. (1988)
		12			6	2	Scheiner et al. (1989)
D. mercatorum				1		1	Gebhardt & Stearns (1988)
D. simulans		1					Tantawy & Rakha (1964)

Appendix 2.2 Genetic correlations between life history traits: non-*Drosophila*, non-domestic species. Correlations have been divided into those concerning the cost of reproduction (between a longevity component and a fecundity component, or between fecundity components) and other correlations.

Species	Cost of reproduction			Other life history traits			Reference
	−	0	+	−	0	+	
SELECTION EXPERIMENTS							
Escherichia coli						1	Luckinbull (1978)
Paramecium primaurelia						1	Luckinbull (1979)
Tribolium castaneum (flour beetle)	1						Sokal (1970)
						2	Bell & Burris (1973)
		2					Mertz (1975)
		1	1				Soliman (1982)
NON-SELECTION EXPERIMENTS							
Lymnaea peregra (pulmonate snail)	1			1	2		Lam & Calow (1989)
Aelosoma cf. *tenebrarum* (oligochaete)		2					Bell (1984a)
Pristina cf. *aequiseta* (naiad oligochaete)		2					
Phiolodina sp. (bdelloid rotifer)		2					
Cypridosis vidua (ostracod)		2					
Daphnia pulex (cladoceran)		2					
Daphnia pulex (cladoceran)			1	1	4	2	Lynch (1984)
Platyias patulus (rotifer)		21	1				Bell (1984b)
Oncopeltus fasciatus (milkweed bug)	1	2		8	24	12	Hegmann & Dingle (1982)
						1	Palmer & Dingle (1986)
				1	3		Groeters & Dingle (1987)

Continued

Appendix 2.2 Continued

Species	Genetic correlation coefficients						Reference
	Cost of reproduction			Other life history traits			
	−	0	+	−	0	+	
Liriomyza sativae (agromyzid leaf-miner)				1	1		Via (1984)
Tribolium castaneum (flour beetle)					7	13	Bell & Burris (1973)
Callosobruchus maculatus (bruchid beetle)			1	2			Møller *et al.* (1989)
Acanthoscelides obtectus (bruchid beetle)	1	13	6	4	25	7	Tucić *et al.* (1991)
Hyla crucifer (tree frog)	1		2				Travis *et al.* (1987)
Rana sylvatica (wood frog)	1	1					Berven (1987)
Gambusia affinis (mosquitofish)		1			19	1	Stearns (1983a)

3: Human behavioural ecology

Monique Borgerhoff Mulder

3.1 Introduction

This chapter has three aims: first, to review the principal areas where ideas from behavioural ecology have had some impact on explaining aspects of human variability in productive and reproductive enterprises; second, to introduce biologists to concepts developed by anthropologists that promise to be useful tools in future analyses of the demography and behaviour of traditional and historical populations; and third, to discuss current controversial topics arising from the study of humans that may challenge some conventional ideas within behavioural ecology.

HISTORY AND AIMS

Since the mid-1970s theoretical advances in behavioural ecology have increasingly influenced anthropological and archaeological research. Early investigators focused primarily on kinship and sex differences in reproductive behaviour, drawing on the concepts of inclusive fitness (see Chapter 1) and sexual selection (see Chapter 7) (Alexander, 1979; Chagnon & Irons, 1979; Symons, 1979; Daly & Wilson, 1983). Soon aspects of foraging and spatial organization were incorporated (Winterhalder & Smith, 1981).

Attention to ecological and environmental influences on behaviour was not new to anthropology. The North American 'cultural ecologists' emphasized technology and resource acquisition in explanations of variability between societies, and there were strong European schools examining individual behaviour and rationality. Nevertheless, it was entirely novel to link these two developments to the proposition that *reproductive competition* among individuals or kin groups might critically influence behavioural outcomes (reviewed by Smith, 1984).

The aim of modern human behavioural ecology is to determine how ecological and social factors affect behavioural variability within and between populations. In one sense its hypotheses are viewed as an alternative to the more traditional anthropological belief in an unspecified force of 'cultural' determination. In another sense, behavioural ecological anthropology can be seen as adding the study of function to investigations of causation, development and historical constraints that were already well established in the social sciences.

69

MODELS AND ASSUMPTIONS

A variety of *cultural* evolutionary models were considered from the start
(for example, Campbell, 1975; Cavalli-Sforza & Feldman, 1981; see review
in Boyd & Richerson, 1985, pp. 157–166), to explain the Lamarkian aspects
of cultural transmission so specific to humans; cultural traits (or 'memes';
Dawkins, 1976) were equated with genes, trait frequencies with gene
frequencies, and inventions and learning errors with mutation (section
3.4.3). Perhaps because of the complexity with which learning, genetic
transmission and cultural transmission were proposed to interact in the
determination of human behaviour, anthropologists preferred to use the
relatively simpler models based on genetic evolutionary processes that lie
at the heart of modern behavioural ecology. They founded their investigations on the loose assumption that human decisions are guided by complex
processes of observation, evaluation, recalled experience, experimentation
and strategizing which, as partially heritable psychological traits (see
section 3.4.1), have themselves been shaped by past selection pressures
(Alexander, 1979; Irons, 1979a). In short, most studies reviewed in this
chapter assume that natural selection moulds the rules for behavioural
change and stability, in so far as human motivations, satisfactions and
aspirations are assumed to serve some ultimate evolutionary end. Specific
problems arising in the study of humans, in particular whether behaviour
is adapted to current conditions and how to deal with the complexity of
social institutions, are discussed in sections 3.4.2 and 3.4.4.

3.2 Studies of foraging

3.2.1 Resource choice

Much of this discussion will be based on the Ache (Paraguay), the only
foraging population for whom data are available to test the effects of competing hypotheses for general and sex-specific patterns of resource utilization. The Ache now cultivate small gardens at a mission station and are no
longer full time foragers, but men, women and children make trips into the
forest (lasting from several days to several months) during which time they
depend solely on hunting and gathering. Although there is, for example, a
tendency for older men, and men with fewer dependents, to spend more
time in the forest than others (Hawkes *et al.*, 1987), behaviour on these trips
resembles the precontact life style in many ways (see Hill *et al.*, 1987;
Kaplan *et al.*, 1990). During the 1980–82 study period Ache regularly
exploited only 26 of the many hundreds of resources available for consumption in the tropical forest of Paraguay. What factors influence this
choice? Here I consider the maximization of caloric returns, specific nutritional demands and the sensitivity of foraging strategies to risk; for a fuller
review of the theory as applied to humans see Smith (1983), and for a recent

discussion of the extent to which critical assumptions are met see Kaplan and Hill (1991).

OPTIMAL DIET BREADTH

The first studies of Ache foraging (Hawkes *et al.*, 1982; Hill & Hawkes, 1983; Hill *et al.*, 1987) examined whether selective resource exploitation served to maximize caloric returns per time spent foraging, using optimal diet breadth or prey models (for example, Pyke *et al.*, 1977; Smith, 1983; see Chapter 4). These models (section 4.1.1) predict that on encounter an organism should pursue (or utilize) only those resources for which exploitation, on average, increases the organism's mean rate of energy acquisition; this is based on the assumptions that resources are randomly or uniformly distributed and are encountered at a predictable rate, and that time spent in the pursuit/handling of one resource is exclusive of time devoted to any other resource (see below). Results show that all but one of the resources commonly exploited yield returns upon encounter that are higher than the mean overall foraging caloric returns for men and women.

There is good evidence from other groups that foraging choices are determined by cost–benefit considerations as expressed in optimal foraging theory: first, in the optimal breadth for Alyawara women (Central Australia), several species of seeds are ranked low on account of the time required for collection and processing, and are included in or excluded from the diet depending on encounter rates with higher ranked resources, rates that are affected by season, rainfall and the availability of government hand-outs of flour (O'Connell & Hawkes, 1984). Second, using data on men's hunting behaviour in three Amazonian populations (and estimating preference rank only on the basis of prey size), Hames and Vickers (1982) showed that diet breadth increases as a function of the length of settlement of the village, and decreases as a function of distance from village, reflecting the chronological and radial depletion of highly ranked prey items in the vicinity of settlements. Finally Winterhalder (1981) predicted and found that an increase in search efficiency that results from the use of snowmobiles by boreal forest Cree (Canada) has led to an increased encounter rate with higher ranked prey, which thus now constitute a larger proportion of the diet. Guns, by contrast, increase pursuit efficiency, and their introduction can expand diet breadth; Hames' (1979) comparison between gun-bearing Ye'kwana and non-gun-bearing Yąnomamö in two adjacent villages in the Venezuelan Amazon partially supports this prediction with respect to the taking of birds (the Ye'kwana take many more than the Yąnomamö), but fails to support the prediction with respect to non-aerial prey; this may be because the ratio of pursuit time to search time is so low for aerial prey that technological innovations affecting pursuit time are minimal (cf. MacArthur & Pianka, 1966).

In general, these models have proved successful in explaining some aspects of resource utilization. Further considerations pertaining to tool availability (Hawkes *et al.*, 1982), the use of information gained from other foragers, seasonal variations, overlap in the pursuit or processing of multiple resources, and the possibility that processing (an activity that can be done at night) does not necessarily compete with pursuit time (Hill *et al.*, 1987) should improve predictive power (detailed discussion in Kaplan and Hill, 1991). Clearly problems arise because resource distributions are not measured: if resources are clumped search time is not shared equally among all prey, violating a key assumption of the prey choice model, necessitating the use of Stephens and Krebs' (1986) combined prey and patch model. Finally, apparent violations of optimal diet breadth models appear when male and female decisions over resource exploitation are considered separately; these have led to the consideration of at least two further factors in determining resource choice.

NUTRITIONAL DEMANDS

Men pass by several resources, such as palm fibre, despite the fact that exploitation would increase overall foraging returns (gross calories obtained per hour); similar observations are made for the savanna Hiwi foragers of Venezuela and the tropical forest Yora of Peru. Since each of the overlooked resources is vegetable matter, and in each of these cases men could markedly increase their foraging returns by taking more vegetable resources, there may be something about hunting, or its product meat, that is attractive to men (Hill, 1988). In addition hunting is dangerous and energetically expensive. Why do men do it?

Hill (1988) investigates the possibility that foragers are concerned not only with caloric returns, but with the macronutrients, particularly the protein and lipids in meat, through the use of indifference curves (Stephens & Krebs, 1986). Models built on data from three different South American foraging groups (including the Ache) and evidence on how meat and garden produce are exchanged in Zaïre suggest that the nutrient content of meat may offset the reduced caloric returns associated with hunting, as well as for the increased risk associated with hunting (see below). Nevertheless because *independently* derived indifference curves, based on the effects of macronutrients on physiology, growth and fitness are not yet available, there are still problems entailed in testing the importance of macronutrient constraints on diet choice. For an approach to this problem using linear programming, see Belovsky (1987).

RISK

During foraging trips Ache men provide 87% of all food consumed, 67% of the caloric content of which comes from game, yet hunting is risky: men acquire no game at all on 43.5% of all full days they forage, whereas foraging on vegetable matter is much less unpredictable. While risk *reduction* is generally assumed to be in a forager's interest, two empirical findings suggest that Ache men may favour risk-prone strategies (definitions of risk follow those of Stephens and Krebs, 1986; see also section 4.5). First, high-return hunters (averaged over several seasons' data) actually spend more time hunting than do low-return hunters (Hill & Hawkes, 1983), indicating that the provision of bonanzas may be advantageous. Second, high-return hunters appear to have higher fitness, measured as the probability of survival for their legitimate children and the frequency with which such men are named as extramarital lovers or fathers of 'illegitimate' children (Kaplan & Hill, 1985a). The reason for this second finding is not known, but it may be that in the context of band-wide sharing of meat in which there is no consumption advantage to the acquirer or his family (section 3.2.3), band members trade special favours (e.g. sex, particular attention to the welfare of his children) with a productive hunter to induce him to stay in the band; membership is very flexible.

3.2.2 *Foraging group size*

The emergence of cooperative groups is a key issue in human evolutionary history. As yet, the most rigorous and systematic analyses concern the foraging aggregations of hunter-gatherers.

To investigate whether Arctic Inuit (Canada) hunt and fish in optimal group sizes, Smith (1985) has calculated the relationship between group size and simple per capita returns (calories per hour of hunting) for 10 different types of hunting/fishing expedition. A comparison of optimal group size with the frequency distribution of observed groups shows that the modal group size is most efficient in four hunt types, but is suboptimal in four other hunt types, with two cases yielding indeterminate results, i.e. a significant correlation between the two measures but a lack of coincidence between the optimal and modal group size (for similarly equivocal results see Hill & Hawkes, 1983).

Considering the Inuit data further, Smith noted that where the optimal and modal group sizes converged (in ocean netting, lake jigging, the hunting of ptarmigan and geese), the optimum was always one. This suggests that the formation of groups of optimal size may be constrained by *social* factors. To investigate further the potential instability of any optimal group size when conflicts of interest are taken into account (cf. Pulliam & Caraco, 1984), Smith (see also Hill & Hawkes, 1983) examined the conflict between joiners and members. A joiner operates on the rule that joining is advantageous if the per capita returns in the group joined are greater than those of a single forager. A member, by contrast, will permit entry of a new member only if the per capita return of the group increases as a result. It follows that a conflict of interests will arise whenever $R_{n-1} > R_n > R_1$, where R is the per capita return rate for a group of size n, because a single forager has lower returns (R_1) than a forager in a group of n individuals (R_n), but by joining this group he reduces the returns of the original (unjoined) group (R_{n-1}). In the two hunt types where the optimum group size of joiners and members can be clearly discriminated, observed pursuit sizes approximate the interests of joiners rather than members, suggesting that joiners are able in some sense to outmanouevre members or that members are unable to exclude joiners.

Modification of these rules to incorporate the effects of relatedness on optimal group size provides no better fit with the data, suggesting that kin selection plays little or no role in determining the size of Inuit pursuit groups (for evidence of no effects of kin composition on hunting success of Efe in Zaire see Bailey & Aunger, 1989). This is not surprising; kin can benefit in other ways. For example, they can consume meat as a result of distribution within the domestic unit of close family members, without being invited to join (and perhaps spoil) the hunt. As stressed by Smith, very different factors will be involved in determining optimal group size depending on the permanence of the association. This calls into question much conventional anthropological dogma that variation in band and settlement size necessarily reflects foraging demands.

Smith's study indicates the potential in applying optimality models to temporary and permanent associations in foraging populations. New research is needed to investigate more systematically the social processes whereby groups are formed, as well as the conflicting constraints on search, pursuit and sharing decisions (Hill & Hawkes, 1983; Kaplan et al., 1990). The wide variety of hunting and fishing techniques and practices employed by foraging populations, even those groups that have adopted technological innovations such as guns, outboard motors and snowmobiles (section 3.2.1), provides ample material for comparative and experimental tests of the influences responsible for variable search and pursuit group sizes.

3.2.3 Food sharing

Sharing of food among family members, camps or bands is common in foraging populations, particularly where storage is unfeasible for reasons of climate or mobility. Often it is shared preferentially among kin (Betzig & Turke, 1986), although such evidence cannot support an inclusive fitness explanation for food sharing, in so far as the reasons for sharing (see below) may be different from those for selecting kin as partners with whom to share. Furthermore, the conditions for getting a tit-for-tat reciprocal food sharing strategy started (Axelrod & Hamilton, 1981) are often met in foraging populations (a collection of bands sharing a geographical zone), because overall population size is usually small and the expected number of future interactions is high.

A simple explanation for surrendering food is that differences in both fighting abilities and temporal asymmetries in the value of a food item among contestants (cf. Maynard Smith & Parker, 1976) will affect the cost that any individual is willing to incur in defence of food ('tolerated theft'; Blurton Jones, 1984). If we include among the threats to a possessor not simply violence but punishment, reprobation and perhaps ostracism from the group, this model may well be relevant to explanations for sharing in some human groups *if* membership of a group is advantageous for other reasons (e.g. village defence or cooperative labour). For example, Hames' (1990) discussion of the sharing of some food items among the Venezuelan Yąnomamö, who clearly live in groups for reasons of defence, shows that 'tolerated theft' may well be implicated. However, the tolerated theft model does not predict that individuals who give away more food should receive return favours from band members, a phenomenon observed in many studies of redistribution in foraging and agricultural societies; this suggests that food sharing should also be investigated from other perspectives. Here I focus on reciprocity and risk management.

Food sharing can be effective in reducing risks of starvation when there is high interindividual variability in foraging returns, variability that is not synchronized across individuals (Winterhalder, 1986). Support for this argument comes from the Ache, where the mean daily standard deviation across families in the amount of each resource acquired predicts the frequency with which that resource is shared, thus reducing variance in daily intake and improving the estimated caloric benefit to most individuals (Kaplin & Hill, 1985b; see also Hames, 1990). Average packet size is an even better predictor of whether a resource is shared, perhaps because this offers a simpler 'rule of thumb' for foragers than does the average difference between families (Kaplan *et al.*, 1990). If sharing stems from the use of unpredictable (risky) resources, we come back to the question of why any population should depend heavily on a risky resource if more reliable foods are available (section 3.2.1). Facing this question, Hawkes (1990) reverses the causality and proposes that sharing may be a *cause* of the pursuit of high-risk strategies by men rather than a consequence of the risks of starvation, on account of the trading of resources for sex. Without further data on the associations between resource distributions (over time and space), food sharing and reproductive differentials in other foraging contexts it is difficult to see quite how the causality might work. It is nevertheless clear from the Ache data that the relationships between resource availability and the sharing of resources may be critically mediated by reproductive as well as productive considerations.

3.2.4 Traditional conservationists or a Tragedy of the Commons?

Much speculation surrounds the question of whether traditional foragers deplete their common resources, or whether they regulate their impact on game populations in ways that incur short-term costs for long-term benefits. Focusing on the behavioural responses of hunters to game scarcity (hence lowered hunting efficiency), Hames (1988a) reviewed data from Amazonia to see whether hunters reduce their rate of hunting as game density declines (a 'conservationist' strategy) or whether they devote more time to hunting as densities decline (a short-term efficiency strategy, derived from optimality theory on the assumption that the value of animal protein is high, relatively inelastic and non-substitutable). Both within- and between-population comparisons indicate that time allocated to hunting increases as game densities decline, suggesting that Amazonian hunters do not conserve game; instead game seems to have sufficiently high value to induce greater efforts towards its acquisition compared with alternative uses of time.

Sparse as these data are, they suggest that Amazonian hunters adopt short-term maximizing strategies and are not willing to accept lower return rates to maximize long-term efficiency. Long-term conservationist strategies would be most likely to occur where local populations are territorial and where there are mechanisms for dealing with cheaters. As argued by Hames (1988a), neither of these conditions currently holds in much of the Amazon.

3.3 Studies of reproduction

3.3.1 Sex differences

SEXUAL DIMORPHISM IN SIZE

If size has stronger effects on the breeding success of males than females (often the case in mammals), we might expect polygynous mating systems to be associated with increased sexual dimorphism in body size; this reflects divergent selection pressures on the two sexes in polygynous species (Clutton-Brock, 1983; see Chapter 7).

Extrapolating this logic to a within-species comparison, Alexander et al. (1979) found that levels of dimorphism in stature are higher in polygynous than monogamous societies, using coded data from the Human Relations Area Files. Although the significant result was based on a highly tenuous distinction (see Gray & Wolfe, 1980) between societies that are monogamous for 'ecological' reasons and those in which monogamy has been 'socially imposed', this finding is commonly cited as evidence that sexual selection occurs in human societies in processes analogous to those in many other mammals. Supportive empirical evidence for intense and often violent competition among men over women can be found among the Yąnomamö horticulturalists of Venezuela (Chagnon, 1979a; 1988a). However, there is no evidence that men's height is greater in polygynous than monogamous breeding systems when regional biases are controlled (P.J.

Gray, personal communication), nor that stature has stronger effects on the breeding success of men than women.

In new analyses on a larger and more accurate data-set compiled by Gray and Wolfe (see below), Gaulin and Boster (1985) question Alexander *et al.*'s result, showing that height dimorphism varies primarily as a result of a statistical artefact, sample size (the effects of sample size were not examined in Alexander *et al.*'s sample). Gaulin and Boster note that levels of dimorphism (mean 1.073) are surprisingly invariable between populations. They suggest that cultural practices have failed to alter the phenotypic expression of size differences across human populations, proposing a number of possible explanations: among these, first, that marriage practices bear little relationship to actual mating behaviour and competition; second, that the current marriage system may have characterized the population for too short a time for natural selection to have altered gene frequencies.

Emphasis on the rate of genetic compared to cultural change does not rule out the importance of sexual selection from this discussion. Gray and Wolfe's (1980) analyses show that sexual dimorphism varies positively with protein availability, food security and abundance, suggesting that nutritional differences contribute to dimorphism differentials between populations. If food surpluses, or even favoured food items, are preferentially given to boys rather than girls in polygynous societies (cf. Hrdy, 1987), a *behavioural* response on the part of parents to food plenty could generate an association between sexual dimorphism and mating system, independent of genetic changes directly affecting size. Clearly, detailed analyses are needed to sort out the effects of mating system, nutrition and sample size. Finally, different kinds of polygyny will favour different competitive male traits (Borgerhoff Mulder, 1988a), with body size perhaps coming under weaker selection pressure than, for example, status-seeking or ambition (Low, 1989) in the resource-defence polygynous systems so characteristic of humans (section 3.3.2).

SEX DIFFERENCES IN BEHAVIOUR

A large proportion of human studies explore Darwin's and Trivers' ideas about sex differences with respect to the behaviour, psychology and development of males and females. Studies of the expression of violence (Daly & Wilson, 1988), intersexual behaviour (Symons, 1979) and mate preferences (Buss, 1988) demonstrate repeatedly that males are more competitive and less discriminating in their mating behaviour, whereas females are choosy and cautious. These studies (reviewed in Betzig, 1988) have stimulated great interest in the relevance of evolutionary concepts to human behaviour, but they may be oversimplistic (cf. Hrdy, 1986) and have made little progress in exploring the sources of inter- and intracultural variability in sex differences (see Borgerhoff Mulder, 1991). An examination of the ecological influences on the nature of parental care provides a useful perspective from which these differences can be investigated.

PROVISION OF PARENTAL CARE

The long dependency of human offspring, often entailing a *simultaneous* reliance of two or more dependents for food, training and defence on their parents (a pattern observed in few non-human primates), suggests to many (e.g. Lancaster & Lancaster, 1987) that fathers have been providers since early in human evolution. New ethnographic studies nevertheless reveal that the father's role is extremely variable between societies (Hewlett, 1991) and that it need not entail the direct provisioning of his family at all (Hawkes, 1990). Comparative study of sex differences in the provisioning of parental care can be based on an examination of the options available to males and females with respect to desertion (cf. Maynard Smith, 1977; see also Chapter 8), bringing ecological factors back into focus.

First we must consider the risks to an offspring of being raised by a single parent. There are at least three important socioecological dimensions:

(a) How safe is the environment? If the environment is dangerous, such that leaving a child unattended or arranging for substitute caretaking (see below) is risky, the stress on a single parent will be greater than it would be in safer environments. While difficult to measure, the rate and nature of accidents to infants can be determined in some circumstances (Borgerhoff Mulder & Milton, 1985). Moreover danger can be linked to patterns of parental behaviour: serious environmental hazards (snakes, parasite-infected biting insects, rain) in the temporary structureless forest-floor camps of Ache foragers are believed to contribute to the very intensive levels of maternal supervision of and contact with infants (Hurtado *et al.,* 1985) and the absence of sibling caretakers, as well as for the extremely slow development of gross motor abilities among 3- and 4-year-old Ache in comparison with other populations (Kaplan & Dove, 1987).

(b) What is the level of nutritional stress? Because children in poor nutritional status are more prone to morbidity and mortality than better-fed children, biparental food provisioning in poor environments can make the difference between life and death. This means that biparental care will be favoured in the kinds of environment where offspring survival is particularly sensitive to parental care (cf. Pennington & Harpending, 1988).

(c) Are alternative caretakers either available or suitable? Caretaking of dependents by sibs and other relatives is common in many human populations, and can relieve a single parent of the full stress of childraising. In some circumstances sibling caretakers demonstrate competence, as in the Kenyan agropastoral Kipsigis (Borgerhoff Mulder & Milton, 1985), whereas in others (Ache, see above) they are thought to be unsuitable. Relatives may also assist in the provisioning of offspring: among the horticultural Ye'kwana (Venezuela) nursing women can work less hard than non-nurs-

ing women because of help they receive from female kin with garden labour (Hames, 1988b). The availability of sibling caretakers, particularly a pair of first-born daughters, may have strong effects on a mother's subsequent reproductive performance, as shown among the Ifalukese (Micronesia) horticulturalists (Fig. 3.1).

Data confirming that uniparental care can entail reproductive failure come from several demographic sources. Among the Ache, a child's chances of mortality are significantly increased if its father dies before it reaches 15 years (43% versus 19%, $n = 171$); this is due at least in part to infanticide by other members of the band, who are not willing to provision a child whose father has died. A child whose mother dies also faces exacerbated risks of mortality, particularly among under 2-year-olds (Hill & Kaplan, 1988). The data indicate that both parents are important to a child's survivorship, and that a father's contribution becomes increasingly important after the age of 2 years. An investigation of the fate of half-orphans in eighteenth century Ostfriesland (Germany) shows a similarly age-dependent pattern, although loss of a father constitutes less of an overall mortality risk than does loss of a mother (Voland, 1988). Among Kipsigis (Kenya), agropastoralists by contrast, paternally orphaned children enjoy equal survival chances to children whose fathers survive until they reach 20 years (Borgerhoff Mulder, 1988b). The relative impact of some of the social and ecological factors discussed above on these interpopulational differences awaits comparative study of a larger number of traditional and historical populations.

Possible sex differences in the costs and benefits of deserting versus continuing with biparental care were suggested in the demographic

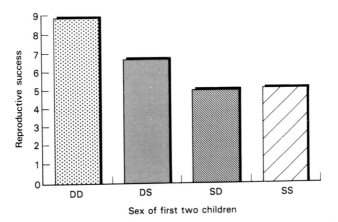

Fig. 3.1 Completed reproductive success of Ifalukese women according to the sex of their first two surviving children. DD (daughters first and second), $n = 7$; DS (daughter first, then son), $n = 6$; SD, $n = 4$; SS, $n = 11$. Mothers giving birth to daughters before sons have a higher RS because their daughters help in rearing subsequent offspring. From Turke (1988).

examples (above). These have been well examined in the theoretical literature (Maynard Smith, 1977), but their relevance to cross-cultural studies
of sex differences is still not well understood. Here I will pursue several
new lines of ecological analysis.

Where males contribute significantly to subsistence (in terms of calories
or labour time), their direct contributions to parental investment (holding,
carrying, supervising, etc.) are lower than where they make small subsistence contributions (reviewed in Hewlett, 1988). This is perhaps because
of the high opportunity costs incurred by males in providing direct care in
these types of productive systems. Unsurprisingly, variability in patterns of
paternal care among foraging populations cannot be explained by this
factor alone. For example, the extremely high level of direct paternal investment in the population of Aka pygmies (Central African Republic) studied
by Hewlett (1988) exceeds that of the Efe foragers of Zaïre and the !Kung
San of the Kalahari, where male contributions to the family are lower (considerably less than 50% of all calories consumed) than those of the Aka (approximately 50%). To examine this further, we can look at the compatibility of child-care with male subsistence activities. Because Aka are
unusual with respect to the amount of time husbands and wives spend net-
hunting together, direct male care is much more feasible in this population
than in those where male and female day ranges barely overlap.

The generally greater compatibility between child-care and *female*
subsistence activities has long been believed by anthropologists to account
for why women usually provide so much direct parental care (Brown,
1970), but again only recently has variability been explored. Hurtado *et al.*
(1985) showed that for Ache women nursing limits the amount of time they
can commit to foraging, and similar effects are found among the foraging
Hadza of Tanzania (Hawkes *et al.*, 1989). In a novel approach to this issue
of compatibility, Blurton Jones (1986) determined an optimal birth interval
for the !Kung, an interval that maximizes the number of surviving offspring
a mother can produce under the constraints of having to forage for heavy
loads of food with dependent toddlers to carry (Fig. 3.2). Reproductive
records confirm that the optimal interval curve centring on 48 months
matches the observed frequency distribution. Other predictions relating to
birth order and subsistence type (comparing foraging to settled !Kung, and
women of different parities) were confirmed (Blurton Jones, 1987).

Clearly the requirements of safety, travel, energetic exertion and the
availability of young or aged caretakers at the work area will all affect the
degree of compatibility between child-care and women's subsistence
activities. But until these relationships are more clearly delineated our
understanding of the socioecological sources of population differences in
parental roles remains sketchy.

Three provisional conclusions emerge. First, in examining variable
patterns of parental care we must consider sex-specific differences in the
compatibility between subsistence activities and direct parental care.

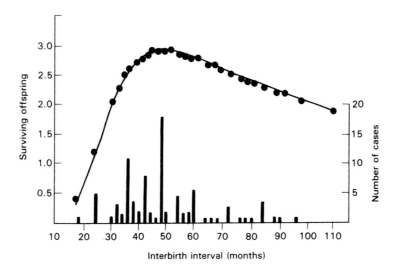

Fig. 3.2 The number of surviving offspring for mothers spacing their births at any specific interval can be calculated as (reproductive lifespan/interbirth interval)×probability of offspring surviving if born at that interval, and is denoted by the solid curve. The optimal interbirth interval, calculated in this way, is 50 months. The observed frequency distribution of interbirth intervals, denoted by vertical bars, has a mean value of 55 months, and a modal and median value of 48 months (based on 96 intervals for 65 women). After Blurton Jones (1986).

Second, we must examine the availability and suitability of alternate caretakers. Third, we must determine whether it is in a man's fitness interests to provision his own children rather than to procure goods that can be directed as mating effort to the wives of other men. Finally, discussion centred primarily on foraging populations, because of the problems in identifying and quantifying the effects of more indirect patterns of parental care (such as political activities that increase family status) (cf. Hames, 1988b) that characterize agricultural and pastoral populations.

3.3.2 *Mating systems*

RESOURCE-DEFENCE POLYGYNY

Marriage patterns are highly variable, and some clear analogues of non-human breeding systems can be identified (Fig. 3.3). First, men can compete for power or resources in order to attract (or force the monopolization of) women, effectively 'resource-defence polygyny' (cf. Emlen & Oring, 1977). This pattern characterizes many societies where differential control over resources commonly correlates with polygyny (e.g. Fig. 3.4; see also Chapter 9).

What ecological factors underlie resource-defence polygyny in humans? Obviously resources must be distributed in such a way as to be monopoliz-

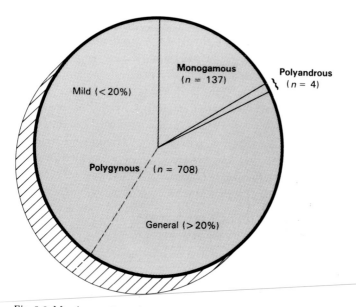

Fig. 3.3 Marriage patterns in 849 societies. Data from Murdock (1967); from Flinn & Low (1986).

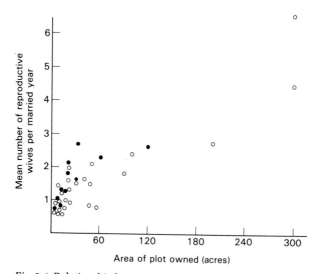

Fig. 3.4 Relationship between mean number of reproductive wives per married year and plot size (in acres) for 36 Kipsigis men ($r = 0.70$, $n = 11$, $p = 0.016$ for deceased individuals (solid circles); $r = 0.91$, $n = 25$, $p < 0.011$ for living individuals (open circles)). Cohort of men all aged approximately 15 years between 1930 and 1939. From Borgerhoff Mulder (1988b); see Betzig (1988) for a review of such studies.

able by a subset of men, but such distributions can be difficult to define in purely ecological terms. For example, Chagnon (1979b) attributes polygyny among the Yąnomamö to the abundance, predictability and ease of acquisition of subsistence resources in the Amazonian basin (horticultural produce, game and fish); men with large kin groups can finesse the obligations of younger dependent kin into helping provide for the requirements of a large polygynous household, such that the control of labour may be as important as the distribution of resources in time and space.

A more ecological approach to the study of polygyny was proposed by Dickemann (1979a,b), who noted that classic historical incidences of polygyny flourished in regimes faced with extreme environmental instability, such as drought, famine, war and plague. She argued that with high levels of mortality that were partially random and partially status-linked, polygynous marriage emphasizing preferential inheritance by the first sons of first wives would be favoured as a strategy for ensuring the existence of both a high-status lineage branch in the future and a large pool of potential heirs (should the heir-elect die). Betzig (1986) proposed an alternative environmental model: following Vehrencamp (1983), she suggested that men would tolerate extreme inequalities in mate number only in circumstances where the costs of leaving a socially defined unit exceeded the benefits of staying; this might result from geographical or political circumscription of a population. Both Dickemann's and Betzig's ideas remain untested.

In an attempt to determine the more general ecological correlates of polygyny, Low (1988, 1990) found a positive relation between polygyny and both the seasonality and unpredictability of rainfall, as well as between polygyny and the incidence of parasitic infection in the area. Low argues that these relationships are expected because fluctuations in major environmental variables (rainfall and parasitic infection) should (i) be associated with increased variability in male competitive ability, and (ii) favour the production of genetically variable offspring (cf. Hamilton & Zuk, 1982; see also Chapter 7). As yet it is unclear how to interpret these complicated relationships, and to what extent they reflect other confounding factors, such as subsistence patterns or geographical location, or simply plain non-heritable differences in luck.

Finally, if female choice occurs in the context of resource-defence polygyny, the polygyny threshold model may be relevant to some human populations (see below). Application of this model is both supported and complicated by the positive correlation across societies between the proportion of subsistence labour contributed by women and the level of polygyny (White & Burton, 1988; Borgerhoff Mulder, 1989a). This may indicate that polygynous marriage can be favoured by women, at least in contexts where male contributions to subsistence are of little significance (cf. Irons, 1983).

MONOGAMY

Despite the 16% of societies in the *Ethnographic Atlas* classified as monogamous, it is questionable whether any societies have ever been strictly monogamous, except those where the marriage rule is enforced by law and serial marriage is prohibited. Predominant monogamy is most commonly reported among hunter-gatherers, who at least nowadays tend to live in marginal or semi-marginal habitats; yet almost invariably men who are successful hunters or who have special skills as *de facto* leaders obtain access to multiple mates (Lee, 1979; Hill & Kaplan, 1988).

The emergence of monogamy in the stratified and agriculturally highly productive nations of western Europe is difficult to explain, inevitably reflecting a complex combination of historical, ecclesiastical and sociopolitical considerations beyond the scope of this chapter, and probably of behavioural ecological analysis. One important factor might be the development of a greater stability of rank among wealthy families, contingent on the establishment of a widely used currency into which productive surpluses could be invested and saved. Under such circumstances it might become reproductively advantageous to limit heirs (cf. Wrigley, 1978), at least with respect to long-term fitness interests (section 3.3.4).

POLYANDRY

Polyandry is a rare marriage system, and most polyandrous arrangements are fraternal. A long history of anthropological investigation demonstrates that polyandry usually occurs among landowners, and is generally the strategy of brothers who, on account of their poverty, combine their land to maintain a single family, thus ensuring that the family estate will not be fragmented among multiple heirs.

In a study of Tibetan peasants of the montane deserts of Ladakh, Crook and Crook (1988) identified the important ecological influences on polyandry: cultivation of highly restricted alluvial fans at the valley's edge; small estates that would become inviable if further subdivided; high labour demands required to make the estate productive; and an absence of alternative (i.e. off-farm) sources of income. Crook and Crook attributed Tibetan polyandry to a saturated environment, yet given the lack of polyandry in highly overpopulated areas such as Africa and New Guinea where the environment could also be classified as saturated, the very high labour demands required to provide subsistence in the arid Himalayan conditions may be more critical (see Chapter 9); the impact of heavy taxation by a foreign overlord may also be important, as Crook and Crook recognize. Finally, both polygyny and polyandry are reported for the Tibetans, indicating that a range of marital arrangements can occur within any one society, as in all the mating systems described above.

FEMALE CHOICE AND THE POLYGYNY THRESHOLD MODEL

Despite arguments over the relevance of female choice models to the analysis of human mating systems (discussed in Borgerhoff Mulder, 1991) there is now some evidence from at least two populations that females, or kin on their behalf, exercise marital preferences that can be shown (from demographic correlations) to enhance their reproductive success (see Voland & Engel, 1990).

First, a proviso: identifying female preferences from marital decisions is methodologically complex, particularly in studies of humans. Who actually makes the decision, and with what degree of autonomy? Is there a consensus among females over preferred mates? If so, does this lead to intrasexual competition that might *obscure* the preference with respect to mating or marital behaviour? For example, assortative mating may reflect either a true preference for mates similar to oneself or a single-type preference, such that the most preferred phenotypes *choose* to mate among themselves, the lesser preferred phenotypes being forced by default to mate among themselves (Burley, 1983).

In my own study of Kipsigis (Kenya) marriage, I showed that, among a group of pioneers who settled in new area in the 1930s and 1940s, Kipsigis women preferred men offering favourable breeding opportunities (Borgerhoff Mulder, 1990). In this polygynous society, men alone own the land used for cultivation and livestock raising, share it equally among each of their wives, and these shares are positively associated with women's reproductive success. Breeding opportunity can therefore be measured as the amount of land available to a prospective bride, namely the amount of land owned by a man divided by the number of his current wives (plus the prospective bride). Regression analysis shows independent and significant effects of two variables, size of breeding opportunity and current number of wives, on the probability of a man being chosen in any one year. If women's preferences can be deduced from their marital decisions (see above), they 'preferred' men offering larger breeding opportunities and men currently married to fewer wives. The preference for large breeding opportunities suggests that Kipsigis women are following an ideal free distribution with respect to resources, as predicted by the polygyny threshold model (Verner, 1964; Verner & Willson, 1966; Orians, 1969). However, the independent negative effect of number of co-wives on female preferences indicates that the costs of sharing a husband are not entirely compensated for by the larger breeding opportunity, and indeed such a cost is found in the reproductive data (Borgerhoff Mulder, 1990).

3.3.3 Kin selection

Given the centrality of kin groups at practically all levels of human social complexity, it is unsurprising that anthropologists have given so much attention to kin selection and inclusive fitness. Very simple tests show that relatedness affects, for example, patterns of adoption (Silk, 1990) and food sharing (Betzig & Turke, 1986), and have done much to dispel the bizarre view of some anthropologists that kinship systems are independent of biological relationship. Inclusive fitness consequences, however, will depend not only on the degree of relatedness between actors, but also on the costs and benefits to each actor from any interaction and on differences in their reproductive value. Tests that fail to take the latter factors into account are essentially inadequate (Hames, 1987).

While there have been no studies of humans that have investigated all the components of inclusive fitness that should be measured (cf. Grafen, 1984; see also Chapter 1), interesting conceptual and methodological developments have occurred.

RELATEDNESS TO SETS OF DEPENDENT OFFSPRING

Rather than considering individuals' relationships with one another through common descent, as is conventional in biology and anthropology, Hughes (1988) examined relatedness to sets of *dependent offspring*, generally full sib groups, the rationale being that these are individuals with particularly high reproductive value. To test the hypothesis that relatedness to juveniles is important for some aspects of social behaviour, Hughes examines well-documented ethnographic cases of household fissioning and the formation of kin group factions (Fig. 3.5). As predicted, individuals tend to assort themselves almost perfectly into the groups containing the set of dependent offspring to whom they are most closely related. This model is an alternative to that proposed by Chagnon and Bugos (1979), who argued that in a Yąnomamö intravillage fight sparked by an axe duel between two protagonists, individuals sided according to their relatedness to each of the two main fighters, not their dependents.

Hughes' ideas would seem to be of widespread relevance in so far as kin and residential groups so commonly contain affines (people related through marriage not descent). It also may be linked to the fact that in many societies marriage is not fully recognized until the birth of one or more children. It nevertheless raises the intriguing problem of why most people in traditional and modern societies still *think* of kinship in terms of descent.

MANIPULATION OF KINSHIP TERMINOLOGY

Chagnon (1988b) examined a uniquely human aspect to kinship behaviour—the use of language. He showed that Yąnomamö men manipulate their relationships with others in order to increase the number of 'marriageable' women available to them. Yąnomamö utilize a kin classification system (Iroquois-Dravidian) in which only cross-cousins are legitimate marriage partners (i.e. you must marry the child of a father's sister or a mother's brother, or a second or third cousin of this type). In reality, because of differences in the generation lengths of men and women and of repeated cross-cousin marriages over the generations, reckoning relationship becomes extremely complex. Furthermore, with high rates of

Household B   Household A

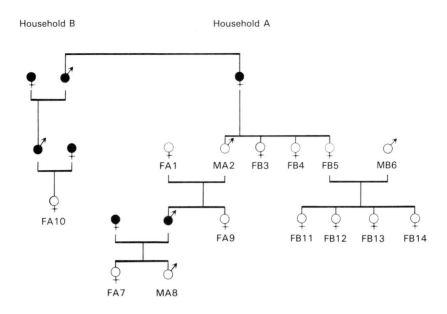

Fig. 3.5 Genealogy of a Rapan Polynesian household that fissioned with some acrimony to form two daughter households: A, open symbols, and B, solid symbols). Data from Hanson (1970). Individuals cannot be assigned to daughter households on the basis of average relatedness. FA10 is more closely related to household B ($r = 0.3138$) which she did not join than to daughter household A ($r = 0.0250$) which she joined. Membership of the two daughter households can be perfectly discriminated on the basis of individual relationships to the two sets of terminal descendants (FA7, MA8 and FB11–14). From Hughes (1988).

mortality and divorce, only 18% of Yąnomamö men reaching the average age of first marriage (20 years) can rely on co-resident biological parents to help them in securing a wife.

    Against this background, and in the context of high levels of competition among men because of polygyny, Chagnon predicted that men should have greater knowledge and articulative use of genealogy than women, and that the terminological manipulations that occur (e.g. between categories of cousins and categories of nieces and nephews) should increase the numbers of individuals falling in the marriageable categories. In a series of innovative experiments, in which 100 respondents were asked to categorize (terminologically) their kin (as shown in ID photographs), men were found to be faster in classifying kin than were women. Furthermore adult males were found consistently to reclassify non-marriageable women as marriageable women, and not vice versa. While many controls should be tested in this type of experiment before sound conclusions can be drawn, such specific biases to the manipulative use of terminology had not previously been suspected by anthropologists, and might be predicted to vary according to kinship systems and the extent of inter- and intrasexual competition.

This discussion has focused primarily on new developments within the study of kin selection in human populations. More generally, with respect to the role of relatedness in shaping human social organization, little is well understood. The association between low paternity certainty and matrilineal inheritance (investment in one's sisters' rather than one's own children) (Alexander, 1979) is now well established (Table 3.1) and can perhaps be interpreted as a male strategy avoiding investment in 'illegitimate' offspring. Nevertheless there are problems with this interpretation. First, how reliable are estimates of paternity certainty from the cross-cultural ethnographic sources (Gray, 1985)? Second, are paternity probability levels sufficiently low in any human societies to favour matrilineal inheritance (0.268) (Hartung, 1985)? Third, what are the socioecological bases of high rates of divorce and marital infidelity that contribute to low relatedness between fathers and co-resident children (Flinn, 1981)? Fourth, it is not yet clear why, in some societies, residential arrangements (that probably affect paternity certainty levels) are primarily patrilocal (core related through men) and in others they are matrilocal (core related through women). Fruitful future research lies in determining whether this variability reflects sex differences in the costs of transfer (cf. Greenwood, 1980) or sex-specific benefits in cooperation with kin (Irons, 1979b).

Table 3.1 Inheritance practice in relation to probability of paternity in 70 societies. Where paternity certainty is low, inheritance follows the matriline. Where it is high, inheritance is patrilineal. From Hartung (1985).

Inheritance practice	Probability of paternity	
	Moderate to low	High
Matrilineal	17	3
Patrilineal	5	45

3.3.4 Parental investment

TARGETS OF INVESTMENT

Complete or near-complete curtailing of parental investment in some offspring has been investigated intensively in studies of infanticide and child abuse in western and traditional populations (for example, Daly & Wilson, 1984). Evolutionarily based predictions are generally well supported, in that parents tend to neglect, abuse or kill offspring of low reproductive value, typically those who are deformed, orphaned by one parent (see section 3.3.1) or born into poor economic circumstances (reviewed in Betzig, 1988; Borgerhoff Mulder, 1991).

Recently there has been a spate of empirical studies that demonstrate within-society differences in the attention, care and resources allocated to sons and daughters (reviewed in Hrdy, 1987; Sieff, 1990). Commonly it is shown that parents provide better care for males than females (inferred from mortality differentials) in contexts where sons can be expected to outreproduce daughters, as predicted by Trivers and Willard (1973); favouritism may occur as a result of rational choice rather than as a direct consequence of physiological condition. There have been no systematic attempts to test *alternative* explanations for the data, although Sieff (1990) argues that where parental rank or socioeconomic changes affect the potential for cooperation and competition within the family, models based on local resource competition (Clark, 1978) or enhancement (Gowaty & Lennartz, 1985) provide more satisfactory alternatives. Thus the rapid fall in childhood mortality rates for males but not for females in rapidly modernizing rural areas of seventeenth to nineteenth century Europe (Johansson, 1984) may reflect a lowering of the overall costs of raising sons once young men have wage-earning opportunities.

The effects of competition and cooperation among sibs and their parents are likely to be particularly strong and variable in human societies, on account of marked sex differences in the labour tasks, in entitlement to inherited property and in postmarital residence arrangements, that often vary according to wealth and social status.

MEASUREMENT OF PARENTAL INVESTMENT

As in studies of non-humans it is difficult to find a currency of parental investment that is useful in comparisons between populations or even between the sexes (see Chapter 8). Parental investment can take multiple forms, such as breast milk, food provisioning, education, inherited resources and status, that can only to a limited degree be reduced to measures of caloric, labour or time inputs. Furthermore, investment extends, in at least some populations, over the lifetime, such that focusing on any one period of offspring dependence can be problematic: wills or premortem bequests, for example, constitute a major parental contribution to offspring fitness (and might therefore be expected to vary in an adaptive manner), but may also compensate for previous inequities and chance occurrences that are usually unknown to the investigator; this difficulty arises in all studies that cover only a segment of parental dependence.

Faced with this problem in a study of social change and sex-biased investment in the Kipsigis, I resorted to examining the correlations between parental resources and the age-specific fitness of married offspring, assuming that the associations were in part due to investment (Borgerhoff Mulder, 1989b). Restricting comments here to the effects of land inheritance only,

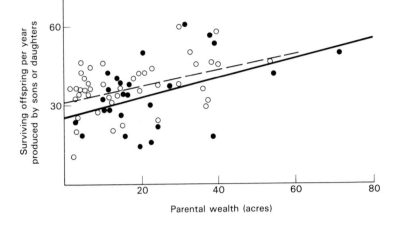

Fig. 3.6 Effects of parental wealth (in acres) on the number of surviving offspring produced per year by sons (solid circles and solid regression line) and daughters (open circles and dashed regression line) in the Kipsigis. Both regressions are significant and there is no significant difference between the steepness of the two slopes. After Borgerhoff Mulder (1989b).

results showed that even with a strict patrilineal rule (Kipsigis sons have exclusive rights to their parents' land, whereas daughters emigrate to their husband's home) parental resources and daughters' fitness were positively correlated (Fig. 3.6). Thus, despite the inheritance rule, daughters seem to benefit from parental land wealth. Further analyses revealed that this was primarily due to the earlier menarche of girls brought up on larger farms, perhaps reflecting nutritional conditions during development (Frisch, 1984), which leads to higher lifetime reproductive success (Borgerhoff Mulder, 1989c). Indeed physiological evidence (for example, Apter & Vihko, 1983) indicates that the ovarian function of late maturers may never reach the same level of ovulatory frequency achieved by early maturers (reviewed in Ellison, 1991).

These results show that in the Kipsigis daughters benefit from parental wealth when they are young (premenarcheal) whereas sons benefit when they are older, when they can attract secondary wives (section 3.3.2). They demonstrate the need for longitudinal studies of the full extent of parental investment in human populations.

PARENTAL INVESTMENT AND DEMOGRAPHIC TRANSITION

The modern demographic transition refers to the radical decline in marital fertility levels that characterized many western nations between 1870 and 1940, and parts of the developing world over the last 40 years. Very broadly, the economic context in which these transitions occurred was one of increased competition over new possibilities for social advancement in professional, industrial and service sectors for which education and training was a prerequisite. Many historical demographers have pointed out that these conditions would encourage the production of fewer more competitive offspring, giving rise to the 'costs of children' concept in the social sciences. Furthermore, they note that this occurred at a time when extended families were being broken down by new demands on individual mobility, thus reducing the availability of potential caretakers (reviewed in Borgerhoff Mulder, 1991).

There are several sociobiological variants on these arguments. Some focus on the decline in availability of alternative (non-parental) caretakers and providers, on whom parents can depend for assistance with parental care (Turke, 1988); a reduction in available caretakers might lead to a *perception* of resource shortages amongst parents (Lancaster & Lancaster, 1987), despite the fact that resources are probably more plentiful in most post-demographic transition populations than in pre-demographic transition populations.

Others, guided by arguments about individual variability and optimal clutch size (e.g. Högstedt, 1980), suggest that by limiting the number of first generation offspring in a highly competitive environment parents might, through focused investment, be able to enhance the reproductive value of their offspring (cf. Hartung, 1985). Rogers (1990) models this process, in an attempt to determine the reproductive decisions that maximize the ultimate rate of descendants' increase. Using a model that assumes a haploid genetic system and no density-dependent population regulation, the effects of different reproductive decisions on longterm fitness are investigated. Results show that at the lower wealth ranges of a population, using wealth to increase fertility maximizes longterm fitness, but that beyond this range fertility is only weakly correlated with longterm fitness. Consequently it may pay, in terms of overall fitness to limit fertility, at least among the rich.

As Rogers recognizes, these results are affected by distribution of wealth in the population, the threshold below which reproduction is impossible, the relative importance of inherited to earned wealth, and initial assumptions about the effects of wealth on fertility. Furthermore, for such a model to represent reality accurately, it is critical to ascertain the predictability of wealth ownership over time. Nevertheless the model should stimulate the collection of longterm demographic and genealogical records. In addition it should lead to the demise of the entirely circular assumption that selection maximizes the number of surviving offspring, which tells us nothing about process or variability. Finally, it may have some relevance to evolutionary explanations for the demographic transition.

3.3.5 *Reproductive strategy and social change*

There is now considerable evidence that reproductive interests may be an important source of social change. One instance will briefly be described.

Until the 1900s, the Mukogodo (Kenya) were cave-living foragers, but in the third and fourth decades of this century began herding cattle and small stock. Cronk (1989) proposes that this was a consequence of mate competition. Traditionally Mukogodo men acquired wives through the payment (bridewealth) of beehives to the bride's father. Using both entnographic and archival evidence, Cronk documents the sources of bridewealth inflation during this period. Most important was the colonial government's attempts to remove pastoral Maasai and Samburu from the highland areas around Mt Kenya through forced relocation to southern Kenya. In face of this threat, Maasai and Samburu saw intermarriage with the local Mukogodo foragers, who were not harrassed by the colonial power, as a way of validating their claims to continued residence in the area. This lead to a documented rise in the number of Mukogodo girls married to Samburu men, an infiltration of livestock (the Maasai bridewealth currency) into the economy of the Mukogodo, inflation in bridewealth payments, and a shortage of brides for Mukogodo men who were uncompetitive with respect to the size of their herds. Mukogodo fathers were essentially forced to give their daughters to pastoralists, in order to obtain the stock they now required for their sons' marriages. Cronk can to some extent exclude alternative explanations for this subsistence change, for example that foraging was no longer feasible after the 1930s, as there is no evidence of game or vegetable depletion in the area at that period; nor was there any direct colonial pressure on the Mukogodo to adopt pastoralism. In sum, reproductive strategy appears to have driven a major subsistence shift.

3.4 Overview

3.4.1 Critiques and limitations

Since its inception the sociobiological enterprise in the study of humans has been bombarded with the criticism that it entails genetic determinism. Very little is actually known about heritable differences underlying any of the behavioural differences discussed in this chapter. The common and conservative position is that most behavioural variability probably reflects flexible responses of similar genotypes to varying conditions (Irons, 1979a); this contrasts sharply with the views of Eibl-Eibesfeldt (1989) who seeks universal, developmentally fixed, behaviour patterns.

Politically motivated criticisms have also appeared: namely that an evolutionary biological approach to human affairs can lead to justifications of racial, sexual and class inequities (Kitcher, 1985; Lewontin et al., 1985; Maynard Smith, 1985). The conventional defence against this critique is that science is a descriptive, not a prescriptive enterprise, but slogans cannot allay such deeply held fears. There are indeed legitimate grounds for fear and (debatably) censorship of any science that might be inflammatory and misused for political ends, but on these grounds the whole discipline cannot be discounted. The research reported in this chapter, in so far as it aims at understanding the social, ecological and historical sources of human variability, is no more 'dangerous' than any other social science. Unfortunately, a recent unsubstantiated claim that racial differences in physique, morality and character reflect different reproductive strategies on the r–K continuum (Rushton, 1988) have rekindled the dying embers of political fires that raged in the wake of Wilson's (1975) tome on sociobiology.

Substantive theoretical and methodological weaknesses in human evolutionary biology can nevertheless be identified: shoddy data, reliance on plausibility and casual assertions of advocacy, inconsistent methodology, and little formal hypothesis testing (cf. Gray, 1985; Hinde, 1987). However, before demolishing the whole edifice for being 'rotten at every rung' (cf. Kitcher, 1985), it is worth pointing out that, as in any field, there is wide variability in quality of research; this review focuses on studies that seem least flawed. It is also worth recognizing that human behavioural ecology faces serious methodological challenges: consider the investigation of mating systems in a species that exhibits the full range from polyandry to polygyny, monogamy to promiscuity (cf. Davies & Houston, 1986), and every combination thereof. The number of ecological, demographic and social variables that should be measured is enormous. While the first empirical investigations reported here can be seen as inadequate on many counts, they do at least indicate what kinds of variables should be measured in the future. Furthermore, they indicate areas where mainstream behavioural ecological thinking may be challenged in the future.

3.4.2 Cultural complexity

Hinde (1987), although generally supportive of human behavioural ecology, expresses doubts as to whether the study of individually adaptive outcomes will ultimately shed much light on the structure and evolution of human society. This position can be seen as an alternative to the view that human society is simply the cumulative product of individually adaptive decisions over the entire course of history (Alexander, 1979). Let us look in more detail at Hinde's position, at how it questions previous studies, and see what can be learnt about the similarities and differences between human and non-human behavioural ecology.

First, while the study of adaptiveness concerns the investigation of outcomes, evolutionary change is a consequence of numerous effects that may not reflect the optimal outcome for all, or even any individual (Hinde, 1987). Resolution of this issue is no simple matter, but it is a challenge facing those who study non-humans and humans alike. Indeed the use of game theory modelling in the non-human literature has shown that evolutionary stability and change can result in situations in which every individual's pay-off is compromised by the strategies of others. Use of game theoretical models in human studies (for example, Smith & Boyd, 1990) should help to tackle this problem.

Hinde's second warning stems from an appreciation of the complexity of human societies, specifically the functional interrelationships between institutions regulating kinship, labour regulation, leadership, laws, religious beliefs etc. As Dunbar (1988a) cautions, in applying an adaptationist perspective to, for example, a society's marriage system we risk searching for optimal solutions to a particular problem while ignoring the fact that any optimal solution is necessarily constrained by other features of the system, such as its economic or political institutions; this too is not a problem unique to the study of humans (Dunbar, 1988b). Recognition of an increasing number of constraints on foraging decisions in the non-human literature (see Chapter 4) reflects a growing awareness of the complex selective pressures on behaviour and the difficulty in identifying appropriate currencies for optimality models, but does not suggest that optimality theory should be dismissed altogether. Dunbar's point is nevertheless important in alerting us to a careful consideration of which behavioural patterns are likely to be the result of adaptive decisions, and which are more epiphenomenal. A similar problem is faced in the study of life history variation: which aspects of variability should we focus on, single traits or adapted complexes (Read & Harvey, 1989a)?

Third, a social system (just as an organism) bears the print of previous adaptation and organization, in the form of its history (cf. Foley & Lee, 1989). Because of this, any study benefits from knowledge of selective pressures in the past. Separating studies of function from those of evolution, as so commonly occurs in the study of animal behaviour (Tinbergen, 1963), to some extent blinds the investigator to a key source of interpopulational variability—history (cf. Gould & Lewontin, 1979). The importance of history in the study of human population variability is thrown into relief on account of the fact that morphological or physiological phylogenetic constraints are less relevant. Nevertheless, as evolutionary biologists begin to integrate studies of function and evolution, the conceptual advances human scientists make in studying the interactions between historical incidents, historical constraints and day-to-day selective processes may prove useful.

In sum, problems raised by the cultural complexity of humans do not separate human behavioural ecology from the rest of the discipline. Rather, they parallel challenges that biologists already face: the further development of game theory, how to deal with complex optimality models where multiple currencies are at stake, and the incorporation of historical processes and constraints into explanatory models.

3.4.3 Cultural evolution

The most fully developed model of cultural evolution (Boyd & Richerson, 1985) investigates the relative importance of three processes in the generation of human behavioural variability: differential replication of genes, individual learning (an acquired phenotypic adaptation to environmental contingencies that perishes with the individual learner; cf. Staddon, 1983), and cultural transmission (information acquired from conspecifics through imitation; cf. Zentall & Galef, 1988).

The transmission properties of culture can be studied with the same basic Darwinian methods used in the study of genetic evolution, drift and mutation, with a focus on the natural selection of the decision-making rules that govern the acquisition of cultural traits. There are also novel processes that may be involved in the spread of traits, which Boyd and Richerson (1985) explore mathematically. These include: (i) *direct bias,* which refers to the ability of an individual to choose from among many behavioural variants a few traits to imitate; this process can greatly enhance the speed at which novel traits spread through a population, in comparison with those spread through individual learning; (ii) *frequency-dependent bias,* which refers to the tendency of individuals to imitate common traits; this simple rule increases an individual's chances of acquiring a locally favoured trait; and (iii) *indirect bias,* referring to the copying of a whole suite of traits exhibited by an individual who is chosen for imitation because of, for example, his or her success, without the imitator knowing precisely *which* trait contributes to success. These latter two biases have complex effects that may or may not result in straightforward adaptive behaviour (see below).

An evolutionarily stable strategy model is used to assess the relative importance of social and individual learning in a range of different environments. Results show that the optimal form of information acquisition depends on two factors: the difficulties entailed in accurate individual learning and the degree of average environmental similarity between generations. If there is a reasonable degree of resemblance between parental and offspring environments, and if accurate individual learning is expensive, social learning can be favoured (Boyd & Richerson, 1988; 1989).

Here I can only outline some of the important issues that the cultural transmission models may help us to address (see Richerson & Boyd, 1991). First, choice of a 'cultural parent' (for example, a very productive herder) may lead to the spread of other variants that happen to be correlated with productive pastoralism but are not *functionally related* to productive pastoralism (indirect bias), such as extravagant dress or extreme obesity. This process has dynamics very similar to the potential role of mate choice in driving runaway sexual selection (Bradbury & Andersson, 1987), suggesting the intriguing possibility of runaway *cultural* selection (Boyd & Richerson, 1985). Second, with a frequency-dependent bias that encourages individuals to adopt the commonest variant, group-selected traits may occur in human populations; this is because the rule 'when in Rome do as the Romans do' discriminates against rare variants. While interpopulation mobility in most human populations is extensive, the *effects* of such mobility on the pool of cultural variants will be modified by discrimination against rare traits. This, Richerson and Boyd (1991) suggest, may help to account for the cooperation found in very large human populations that cannot easily be explained in terms of kin selection or reciprocal altruism.

These developments in the models of cultural evolution may become important in the study of non-humans, particularly as the evolutionary ecology of social learning in non-humans is better understood.

3.4.4 The study of humans in changing environments

Have the changes in ecological and social environments since the end of the Pleistocene been so rapid as to render the study of the functional con-sequences of behaviour irrelevant? Investigators differ in their views on this. Generally it is only anthropologists working with traditional and his-torical populations who have retained an interest in measuring the foraging and/or fitness returns from different behaviour patterns, studies that have been the focus of this chapter.

Other investigators, particularly those studying humans living in modern and highly atypical circumstances, have shifted their attention away from the direct study of function towards the more general psychological mechanisms that underlie decision-making (Symons, 1987), broadening their research strategies as a result. Thus Cosmides and Tooby (1989; Tooby & Cosmides, 1989) make an ambitious attempt to unravel the nature of these evolved psychological mechanisms, through pioneering evolutionary interpretations of cognitive tests that aim to identify the prob-lem-solving rules (algorithms) that govern our activities and reasoning in various domains.

Some psychologists investigating evolutionary issues in modern societies have strongly criticized behavioural ecological anthropology, arguing that *studies of function* in contemporary (any post-Pleistocene population) fail to test any hypothesis about evolutionary processes or the human nature that these processes have produced (Symons, 1987). Thus, in criticizing the claim that polyandry is adaptive in the Tibetan montane desert (3.3.2), Symons (1989) argues that because agriculture, land shortages and taxation did not exist during the Pleistocene, there could not have been selection for the mechanisms that are implied by Crook and Crook's hypothesis, namely a suite of flexible behavioural adaptations to resource and labour shortages. This is equivalent to saying that stepping out of the way of a large moving vehicle is fitness enhancing, but because none of the neural mechanisms involved were designed specifically for this purpose (there were no lorries in the Pleistocene, only large and dangerous megaherbivores) the behaviour is not an adaptation. On the basis of this reasoning many now argue that psychological mechanisms should be the exclusive focus of evolutionarily biologically oriented social scientists.

Some aspects of these recent critiques remain contentious. Here the issues can be addressed only summarily.

1   Natural selection operates on outcomes. Mechanisms are therefore selected only in so far as they reliably generate behaviour that influences fitness. Hence, until psychological mechanisms become behaviourally visible, they cannot be affected by natural selection. To exclude categorically any consideration of behaviour and its fitness consequences from the study of adaptation therefore makes little sense.

2   The psychological mechanisms (proclivities, algorithms, propensities, etc.) that psychologists refer to are as yet poorly defined and understood, necessitating the use of hazy language and idiosyncratic terminology. They would be extremely hard to identify in field conditions where psychological testing is usually inappropriate. Hence, anthropologists will continue to look at behaviour and to speculate about mechanism (for example, Betzig, 1989). Nevertheless, more careful attempts to tease out the mechanistic processes involved in, for example, studies of the correlation between cultural and reproductive success, should be central to any study of function.

3   Ruling out the plausibility of adaptive explanations on the basis of the *specificity* of the mechanisms underlying, for example, polyandry in the Himalayas seems unwarranted without more understanding of the mechanisms themselves; after all, resource and labour shortages must also have challenged some of our Pleistocene ancestors. Again, studies of function and mechanism should be seen as complementary and not as opposed to one another.

4   In general, the fitness consequences of behaviour in modern contexts are not well understood. To assume that the previous fitness-enhancing effects of behaviour will not be observed in contemporary populations not only begs the question, but impedes investigation of how selective optima might have changed (Caro & Borgerhoff Mulder, 1987). Furthermore, how do we know when to make the cut-off—the Neolithic Revolution, the rise of early modern commercial societies, the Industrial Revolution?

The assumption that all human behaviour is a consequence of selective pressures in the Pleistocene raises more specific problems.

5   What do we mean by the Pleistocene; was it a sufficiently homogeneous period to support such generalizations; and how can we derive our knowledge about its selective conditions? First, we know from climatic and mammalian evidence that Pleistocene selective conditions were highly variable (Foley, 1987). Second, important differences between hominids, anatomically modern humans and contemporary hunter-gatherers with respect to their physiology, ecology, foraging and social behaviour (Foley, 1988) render any generalizations about Pleistocene conditions on the basis

of modern hunter-gatherers highly suspect. Constructing a hypothetical 'environment of evolutionary adaptedness' (cf. Symons, 1979) in the face of such complexity invites the wildest of unsupportable speculations (Kitcher, 1985), as revealed in a recent critique of sociobiological arguments for the function of human fat deposition (Caro & Sellen, 1990).

6   Finally, behaviour patterns have changed radically since the Neolithic Revolution. While it is unlikely that there have been significant corresponding changes in gene frequencies, behaviour governed by facultative decision-making processes with respect to productive and reproductive ends still characterizes much of the traditional and developing world, as reviewed in this chapter. While the much vaunted flexibility of human adaptive responses (cf. Alexander, 1979) is as much a statement of advocacy as is the 'Pleistocene fixation' exhibited by some evolutionary psychologists, the former position offers greater research potential in exploring the limits of flexibility in human behavioural adaptation.

Note that disagreement on these issues arises primarily from studying a species that has experienced rapid environmental and social change for several millenia. Most investigators would probably agree with Williams (1985, p. 1) that the goal of the adaptationist's programme is 'to recognize certain of [the organism's] features as components of some special problem-solving machinery'. These features are necessarily adaptations to past environments and may or may not promote current fitness interests, depending on the similarities between past and present environments. Furthermore, in so far as all ecological communities are now being disturbed by human encroachment, pollution, temperature change, etc. to some degree, the issue of how to cope both conceptually and methodologically with the study of intense change can no longer be ignored in behavioural ecology.

PART 2

EXPLOITATION OF RESOURCES

Introduction

Acquisition of resources, especially food, is a central topic of study for behavioural ecologists, partly as a result of the simple fact that many animals spend much of their active time throughout the year engaged in foraging. Therefore data on resource acquisition are readily obtainable. Furthermore, in comparison with, for example, mating it is relatively easy to measure the benefits of food intake—calories or other nutrients—as well as the costs in terms of time and energy spent harvesting resources.

In the 1970s and early 1980s there was an exponential (Stephens & Krebs, 1986) expansion of the literature on 'optimal foraging'. The great excitement of this research was that for the first time quantitative, *a priori* predictions were used successfully to account for behaviour in the laboratory and the field. Imagine the thrill of calculating on the back of the proverbial envelope how long a starling should spend foraging in a patch, and then going into the field and finding that the animals behave as predicted to within a few per cent of the expected answer! As Krebs and Kacelnik show in their chapter, the apparent success of predictions in earlier studies may in some, or perhaps most, instances have been only the beginning of unravelling the process of decision-making by foragers. In the example of starlings exploiting patches, consideration of both psychological mechanisms such as learning and memory, and the fitness consequences of foraging (currencies) reveal a much richer story than simply concluding that 'starlings obey the marginal value theorem'.

In many ways it is surprising that simple rate-maximization models such as the marginal value theorem have any success in predicting an animal's behaviour. After all, animals must balance conflicting demands of foraging, vigilance, mating and so on and, furthermore, the benefits of maximizing rate of food intake are likely to vary with hunger. Krebs and Kacelnik discuss several methods for bringing greater realism (and, inevitably, complexity) into models of decision-making. In this context an important technique that has emerged in the past few years is *stochastic dynamic modelling*.

In stochastic dynamic models, not only are the fitness consequences of different kinds of behaviour (e.g. foraging and vigilance) combined, but also the effects of hunger and other components of internal state can be included. To be weighed against these advantages is the fact that stochastic dynamic models are often complex in their structure and do not have the transparency of, say, the marginal value theorem that appeals to the empiricist. One of the exciting challenges in the next few years for behavioural ecologists will be to try to bring stochastic dynamic models into play in generating predictions for field or laboratory tests.

There are now sufficient studies of optimal decision-making, not only in the context of foraging but also in relation to mating, life history and other aspects of behaviour, for many behavioural ecologists to accept that optimality modelling is a powerful tool in the analysis of behaviour. How should this tool be used in the future? One development, already referred to above, is to use optimality models, essentially concerned with outcomes of behavioural decisions, as a way into investigating mechanisms at the psychological or physiological level underlying decisions. A second use of optimality models is to investigate how behavioural decisions by individuals affect population or community phenomena.

Milinski and Parker expand on this second approach in their discussion of the ideal free distribution (IFD). The IFD describes how a population of rate-maximizing competitors should be distributed between habitat patches. As with the marginal value model, the IFD has had surprising success, in view of its great simplicity, in accounting for the behaviour of animals in laboratory experiments. So much so, in fact, that it is now often used as a starting point for more elaborate experiments, such as those involving trade-offs between foraging and danger of predation (Abrahams & Dill, 1989). Milinski and Parker lead us through a range of refinements of the classical model including models with unequal competitive abilities, varying degrees of average competitive interference, costs of migration and kleptoparasitism. These models lead to many as yet untested, but easily testable, predictions. For example, if one class of individuals (say juveniles) is more severely affected by competition than another, such as adults, then the two classes should segregate in a predictable fashion into different habitats.

Milinski and Parker also emphasize that many seemingly disparate phenomena can be treated as examples of the IFD, especially when one considers that the IFD can apply in the temporal as well as spatial domain. Thus, they suggest, it is possible to look for equilibrium distributions of seasonal emergence of insects in relation to the emergence of mates or the growth of host plants, for equilibrium distributions of sex change in sequential hermaphrodites, or of metamorphosis in insects, and equilibrium distributions of habitat selection during ontogeny, as Werner et al. (1983) have done for fish.

Finally, Milinski and Parker suggest how investigation of interspecific competition and community structure might be approached from the IFD, a theme more fully developed by Rosenzweig (1981; 1985). If behavioural ecologists can succeed in bringing studies of individual decisions to bear on community and population processes, one of the original objectives of optimal foraging theory (MacArthur & Pianka, 1966) will have been fulfilled.

In his chapter, Endler extends the discussion of foraging to new regions by considering the evolutionary responses of prey organisms. An important theme in his discussion is that in order to understand the evolution of prey defences such as crypsis and warning coloration, behavioural ecologists need to know about the perceptual properties of the predators against which the defences are directed. This echoes the conclusion of Krebs and Kacelnik that a full understanding of foraging decisions needs a greater background of animal psychology than behavioural ecologists traditionally accept. A similar argument could also be made in relation to studies of sexual selection and mate choice.

In the final section of Chapter 6, Endler re-examines the concept of 'coevolution' of predators and prey, adaptation in one lineage leading to specific counteradaptations in the other. He finds little evidence for tightly linked coevolution, but suggests that there may be looser coevolutionary interactions between predator and prey lineages resulting in overall increases in investment in antipredator defences (e.g. shell thickness in molluscs). Although there is some empirical evidence for gradual accretion of defences during evolution, the theoretical possibility of evolutionary arms-races is still a matter of debate (Abrams, 1986).

4: Decision-making

John R. Krebs and Alejandro Kacelnik

4.1 Introduction

If you spend a few minutes watching a small bird such as a European robin
in the spring, you will see it performing a variety of activities. Perhaps,
when you first see it, it is searching for food on the ground. Soon it flies into
a tree and sings, then after a short period of singing it stops and preens or
simply sits 'resting'. If you extend your observations to cover a series of
whole days, daily patterns will emerge. For example, there will be more
singing in the early morning and more resting in the middle of the day. If,
on the other hand, you focus your observations on just one kind of activity,
for example foraging, you will discover that it too can be broken down into
a series of components. The robin may sometimes forage on lawns, sometimes on flower borders, sometimes it captures worms and sometimes it
catches flying insects. In this chapter we will present a framework for describing, understanding and predicting these sorts of patterns. We shall use
the metaphor of the animal as a 'decision-maker'. Without implying any
conscious choice, the robin can be thought of as 'deciding' whether to sing
or feed, whether to feed on worms or on insects, whether to search for food
on the grass or on the flower bed. We shall see how these decisions can be
analysed in terms of the costs and benefits of alternative courses of action.
Costs and benefits are ultimately measured in terms of Darwinian fitness
(survival and reproduction), and may, in many instances, be measured in
terms of some other more immediate metric such as energy expenditure,
food intake, or amount of body reserves. As will become apparent later on,
analysing decisions in terms of their costs and benefits cannot be done
without also taking into consideration physiological and psychological
features that might act as constraints on an animal's performance. The
fitness consequences of decisions, and the various constraints that limit
an animal's options, can be brought together in a single framework using
optimality modelling.

4.1.1 Optimality modelling

The main tool for helping us to analyse decisions in terms of their costs and
benefits is optimality modelling. The logic of this approach in biology (as
opposed to economics or decision theory) is a Darwinian one. Selection, it
is argued, is an iterative and competitive process, so that eventually it will
tend to produce outcomes (phenotypes) that represent the best achievable

balance of costs and benefits. These adaptations are often so exquisite (an obvious example being the match of many cryptic insects to their background) that in pre-Darwinian days they were taken as evidence for a divine creator (Paley, 1828). Nowadays, they are seen as the outcome of natural selection and as suitable material for some kind of optimality analysis. This sketchy and simplistic view of natural selection and adaptations is not one to which all authorities would subscribe (Gould & Lewontin, 1979; Gray, 1987; Lewontin, 1987). We will return to this issue at the end of the chapter, but for the moment we will accept the general argument as a basis for using optimality models.

It is, of course, possible to think about costs and benefits without building formal models. Indeed, many of the chapters in this book use verbal arguments to analyse adaptations in terms of their costs and benefits. However, the advantage of a formal mathematical analysis of the problem is that it helps to clarify exactly what has been assumed, and exactly what predictions arise from the assumptions. This point can be illustrated by referring to an example. Goss-Custard (1977) recorded the foraging behaviour of redshank *Tringa totanus*, a shorebird. One of the redshank's main prey was the polychaete worm *Nereis diversicolor*. These worms occur in different sizes, and redshank are sometimes seen to feed exclusively on large worms and sometimes to take both large and small ones. This is not simply because of variation in availability of small worms: they are always plentiful in the mud. What determines the redshank's decision to eat or reject small worms? An intuitive argument is that large worms are more 'profitable', that is to say they yield a higher amount of energy per unit of pursuit and handling time. When large worms are sufficiently common, it never pays to miss opportunities of catching them while pursuing or handling less profitable small worms. However, when large worms are rare, the small, unprofitable, worms are worth eating when they are encountered because the alternative is to spend a long time searching for the occasional large item. In other words, the 'lost opportunities' while performing a behaviour determine whether the decision to indulge in it is profitable or not. This intuitive argument, called the principle of lost opportunity by Stephens and Krebs (1986), may explain the change in redshank behaviour with respect to small worms in terms of changes in the encounter rate with large worms.

But can we be sure that selective foraging by redshanks really reflects a strategy of maximizing rate of energy procurement? One way to increase our certainty that we have correctly interpreted the cost–benefit relationships underlying the redshank's foraging decision is to try to make a more precise prediction. If, for example, we could predict *precisely* the encounter rate with large worms at which the redshank should change from selective to non-selective foraging in order to maximize intake rate, we could compare this value with the observed switch in selectivity, and if agreement was good we would be more confident that our account of redshank foraging decisions was in fact correct. Where there are two classes of worm, large and small, it is possible to derive a precise quantitative prediction about when redshank should and should not take small worms. The crucial variables, namely the energy values of the large and small prey types and their handling times, are all measurable, so in principle it is possible to test the quantitative prediction. Goss-Custard did this and found that he was able to account for some aspects of changes in the selectivity of foraging of redshank upon *Nereis* (for a summary of many other similar studies see Stephens & Krebs, 1986).

This example illustrates two advantages of a formal mathematical model. First, the assumptions underlying the hypothesis are made unambiguous and, second, the predictions, and their relationship to the assumptions, are clear and precise (Table 4.1). The advantage of precise predictions is not only that they increase our confidence in the model if substantiated, but they also allow us to see ways in which data do not fit the predictions. Only by making a formal model with precise predictions was it possible to recognize that there were differences between predicted and observed behaviour of redshank. For example, the redshank did not show the step-wise switch to completely selective foraging that might be expected. Instead they showed 'partial preferences', a phenomenon also seen in all other studies of optimal prey choice. Several hypotheses to account for partial preferences are discussed by Stephens and Krebs (1986), Krebs and McCleery (1984) and McNamara and Houston (1987a). Rather than treating this discrepancy as a fatal flaw in optimality modelling, as implied by some

Table 4.1 Assumptions and predictions of the classical optimal diet model.

Assumptions

Prey value is measurable net energy or some other comparable single dimension

Handling time is a fixed constraint

Handling and searching cannot be done at the same time. If this were not true there would be no need for the animal to forage selectively

Prey are recognized instantaneously and with no errors

Prey are encountered sequentially and randomly: the expected time to find the next prey of type i is always $1/\lambda_i$

Energetic costs per second of handling are similar for different prey

Predators are designed to maximize rate of energy intake (or other measure of value)

Predictions

The highest ranking prey (in terms of profitability = energy/handling time) should never be ignored

Low-ranking prey should be ignored according to the rule specified in the model

The exclusion of low-ranking prey should be all-or-nothing

The exclusion of a low-ranking prey does not depend on the animal's encounter rate with that type. This point can be grasped intuitively by realizing that when type 1 are sufficiently abundant, 'time out' from search to handle type 2 is not profitable. In other words E_2/h_2 is less than what could be obtained by searching for and consuming the next type 1, so the abundance of type 2 is irrelevant

---

authors (for example, Gray, 1987), we prefer to see it as an advantage of the approach that the discrepancy is recognized and needs further explanation or experimental analysis (for example, Rechten *et al.*, 1983).

The redshank–worm example also highlights the three main components of an optimality model.

1   *The currency.* In this example the currency is maximizing rate of energy intake. Other alternatives might include minimizing risk of starvation (section 4.5) or achieving an optimal balance between danger of predation and the risk of starvation (section 4.3).

2   *The constraints.* In the redshank example, handling time and search time were treated as fixed constraints, reflecting a combination of properties of the environment and properties of the animal. It is not always straightforward to decide what to include as a constraint and what to treat as a decision variable. For example, in other studies both handling time (Krebs, 1980; Formanowicz, 1984; Newman *et al.*, 1988; see Chapter 6) and search speed (Gendron & Staddon, 1983) are treated as variables rather than constraints.

3   *The decision variable.* In the redshank example, the decision variable is whether or not to eat small worms when encountered, and the variable has two states: eat or reject. In other models the variable might be how long to spend foraging in a particular patch, how much time to spend scanning for predators, how much to invest in mate guarding, and so on. The choice of decision variable is crucial because it defines the problem embodied in the model. It is usually the biological intuition of the researcher which determines the aspects of an animal's behaviour that may be profitably considered as decision variables and those that are taken as fixed constraints.

Finally, the example illustrates one of the difficulties with optimality modelling. The components of optimality models can be seen as hypotheses. If all components were completely characterized, it would be possible to predict the animal's behaviour in terms of costs and benefits perfectly, and optimality modelling would be superfluous. If, on the other hand, none was fully known (as in the redshank case) we might be left in the difficult situation of not knowing where to start (Lewontin, 1987). If, for example, the model fails to account for observed behaviour, as in the occurrence of partial preferences, and we need to revise it, should we alter the currency assumptions, the constraints, the decision variable or all of them? Rechten *et al.* (1983) explained partial selectivity shown by great tits, *Parus major*, in a laboratory situation formally analogous to the redshank's problem in terms of a constraint: the birds made imperfect discriminations between size classes of prey. For the same problem, Houston and McNamara (1985) offered an explanation in terms of the currency: if the animal minimizes the risk of starvation it should sometimes take only large items and sometimes take both large and small, depending on its energy reserves (section 4.4.3): if the energy reserves change during the course of an observation period, the animal will appear to show partial selectivity. There is no simple recipe that can be offered as a solution to the problem of which hypothesis should be changed first (Cheverton *et al.*, 1985; Kacelnik & Cuthill, 1987). One important consideration nevertheless is that models are valuable mostly in terms of what one does with them. If one modification leads to 'interesting' new questions and experiments, and it is based on knowledge of the biology of the system under study, while another leads to a dead end, one would naturally prefer the former.

4.2 Constraints and currencies: an example

The interplay between constraints and currencies in foraging behaviour can be illustrated by referring to a well-studied example: starlings foraging in a patchy environment (Tinbergen, 1981; Kacelnik, 1984; 1988; Cuthill, 1985; Kacelnik & Cuthill, 1987; 1990; Cuthill & Kacelnik, 1990; Cuthill *et al.*, 1990; Kacelnik *et al.*, 1990). The study started from an application of what is probably the best known model of 'optimal foraging', the so-called marginal value theorem (Fig. 4.1a,b), to the foraging behaviour of parent starlings collecting food for their young. In the natural foraging cycle, the starling flies from its nest to a foraging site (often a short-grass meadow) and collects a beak-load of prey before delivering it to its young. Within each foraging trip, diminishing returns (Fig. 4.1a,b) arise because, as it loads up with prey, the starling's efficiency at collecting further prey is progressively diminished. This is sometimes referred to as the 'loading curve' (Orians & Pearson, 1979). Kacelnik (1984) generated a loading curve by training individual parents to collect mealworms from a food dispenser which delivered prey according to a 'progressive-interval' schedule, in which the interval between successive worms became progressively longer. When the parent flew back to the nest, delivered its food and returned to the feeder, the progressive-interval schedule returned to its starting value. Thus within each trip the starling experienced diminishing returns and the shape of the loading curve was completely standardized. Figure 4.1c shows that when the feeder was presented at different distances from the nest, the bird's response was to vary load size in a way which both in qualitative pattern (increasing loads with increasing distance) and in quantitative detail was as predicted by a version of the marginal value theorem. When several different versions of the model with different currencies were compared, it was found that the best quantitative fit was when the currency was taken to be net energy gain to the family.

The results in Fig. 4.1c appear to show that the marginal value theorem provides a satisfactory account of foraging by parent starlings. As we shall see in a moment, this conclusion is only the first stepping stone; but for the moment let us accept it at face value. What conclusions can we draw about the biology of starlings? First, the results do *not* show that starlings are clever, any more than a demonstration that starlings can fly shows that they are able to solve aerodynamic equations. The marginal value theorem and

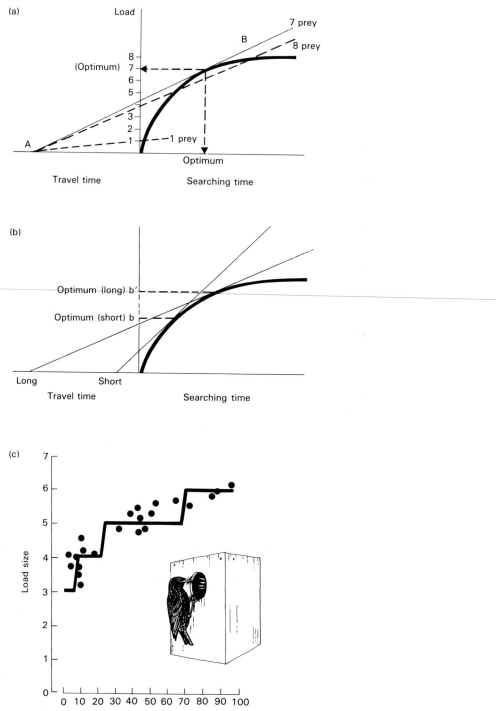

(a)

Load

7 prey

B

8 prey

8
(Optimum) 7
6
5
3
2
1   ──1 prey

A

Optimum

Travel time   Searching time

(b)

Optimum (long) b'

Optimum (short) b

Long   Short

Travel time   Searching time

(c)

Load size

Round trip time (s)

the data used to test it only show the outcome of behaviour and not the mechanisms by which the outcome was achieved. Second, the results do *not* show that starlings are optimal. If we mean by optimal 'maximizing lifetime reproductive success', then we have not shown this because there may be other ways of foraging that would yield a higher fitness gain. Finally, the results *do* show that the assumptions about currency and con- straints incorporated into the marginal value theorem are helpful in explaining the behaviour of starlings in the seminatural experimental set up. As we shall see in the following sections, a closer scrutiny of the birds' behaviour reveals many complications.

4.2.1 *Currencies*

Although Kacelnik (1984) found that the fit to the model was good when net energy gain to the whole family was used as a currency, this currency can be only approximately correct. A full analysis of the fitness consequences of foraging would separate the effects of load size on parental and offspring fitness (Houston, 1987). Both of these are likely to depend on the time spent in the patch and therefore on load size. If, for example, offspring survival is a function of rate of delivery of food to the nest, then offspring fitness will be maximal when patch time corresponds to the gross rate-maximizing load (as in the simplest version of the marginal value theorem). However, parental fitness in the starling also depends on offspring other than those from the current batch (future production or residual reproductive value; see Chapter 2), and this component may be affected by foraging behaviour. It might, for example, increase with average time in the patch, because foraging in the patch is less costly in terms of energy expenditure and hence parental survival. In this case, the optimal patch time for the parent may be longer than that predicted by rate-maximizing for the family. The close fit to net energy gain observed by Kacelnik might indicate that these addi- tional refinements do not produce measurable differences to the predic-

Fig. 4.1 (Opposite) (a) The essential features of the marginal value theorem are as follows. The predator spends time travelling between or foraging within patches; within the patch, the predator experiences diminishing returns (shown here as the slope of the 'gain curve' getting shallower). In the version of the model shown here, the gain curve is smooth and monotonically decreasing, but it does not have to be so: the decision variable is the length of time spent foraging in a patch; the currency to be maximized is long-term average rate of food intake. Average rate of intake is represented by the slopes of lines traced between A and the intersection with the gain curve. The maximum can be found graphically by the tangent solution shown. When the predator takes seven prey per visit it achieves the highest possible intake rate (slope). The slope resulting from two other values of the decision variable (take one prey and take eight prey) are shown for comparison. (b) When travel time is increased, the rate-maximizing patch residence time also increases. (c) Kacelnik's (1984) results showing how load size varies with travel time in parent starlings collecting food for their young. (—) Predicted, (●) observed.

tions, but they also illustrate that good fit to one model does not exclude the possibility that other models could lead to similar predictions.

4.2.2 Constraints: information

The marginal value theorem assumes complete knowledge. The predator 'knows' the shape of the gain curve, the travel time between patches, and therefore the average rate of intake for the environment. This could be seen as a fatal weakness of the model, because the assumption of omniscience is unrealistic, but McNamara (1985) and McNamara and Houston (1985) have shown that in principle a predator starting off with an inaccurate estimate of γ^*, the maximum average rate for the environment, and following a strict 'marginal rate' rule could eventually converge on the correct value (and hence on the correct patch residence time) if it uses an algorithm in which the time spent in a patch is adjusted according to experienced average gains. In their model, the animal starts off with an arbitrary estimate of γ^*—call it γ_0; it leaves the patch when local rate of gain equals this estimate and in consequence (on average) it experiences a certain γ, different from the expected γ_0. As a consequence of this discrepancy, it corrects its estimate, which will now be γ_1. This new estimate is then used as the basis for the next patch visit and for a comparison with the obtained rate that follows. If an appropriate updating rule is used, only when $\gamma = \gamma^*$ will the obtained and expected rates of gain coincide and the estimate stabilize. Thus, at least in theory, an animal starting off in ignorance of properties of the environment could end up with the optimal solution predicted by the marginal value theorem.

Evidence on how starlings actually adjust their patch residence time in relation to past experience is presented by Cuthill et al. (1990) who studied starlings in a laboratory set-up designed to mimic exploitation of depleting patches. Within each patch visit, the bird experienced diminishing returns as a result of hopping on a perch. Food was delivered according to a schedule of reinforcement in which each successive food item required more hops than the preceding one (a 'progressive ratio schedule'). In order to gain access to the feeding patch, the bird had to travel by flying a number of times between two 'travel' perches. The number of flights required varied at random between two values, 'long' and 'short', resulting in a bimodal distribution of travel times. The number of prey taken per visit (equivalent to patch residence time) was taken as the dependent variable and multiple regression analysis was used to analyse to what extent residence time was influenced by the preceding travel requirement, the penultimate travel, and so on. The results showed that within such an environment patch residence time is strongly influenced by the most recent travel experienced, but hardly at all by earlier travels. The starlings behaved within this experiment as if they remembered only the last travel time.

Cuthill et al. (1990) considered the implications of this result for maximization of intake rate. They pointed out that although Stephens and Krebs (1986) had argued that adjusting patch residence time in relation to previous travel time ('tracking') is not the rate-maximizing strategy, this conclusion is not correct for all environments. Depending on how the environment changes, for example whether changes occur on average every n minutes or every n patch visits, tracking (responding to the last travel) may or may not be optimal for rate maximizing. By simulating the intake rate that would result from tracking and the opposite extreme, 'averaging' (responding only to the mean travel time), for a range of different environments, Cuthill *et al.* concluded that tracking is a robust strategy in that it gave a higher pay-off in a range of conditions, although the particular experimental environment studied was one in which it gave a lower pay-off than averaging would have done. Thus, for the experimental environment, the starlings' response to the last travel time only was a constraint on rate maximization, but, viewed in a wider perspective, it might be an adaptive use of past information for a rate maximizer.

4.2.3 *Alternative explanations*

Although the fit between observed and predicted behaviour in the example shown in Fig. 4.1 is very close, it is worth considering whether other hypotheses might predict a similar result. Kacelnik and Cuthill (1987) and Cuthill and Kacelnik (1990) carried out experiments in which they presented parent starlings in the field with prey at different distances from the nest. Instead of using a progressive-interval schedule to mimic depletion within a patch visit, they used a 'fixed-interval' schedule in which the rate of arrival of prey remained constant up to a maximum value within any one patch visit. As shown in Fig. 4.2a, the marginal value theorem now predicts that there should be no relationship between load size and distance. Instead, the forager should always take the maximum load. However, the starlings in Cuthill and Kacelnik's experiment still carried home larger loads from greater distances (Fig. 4.2b)! Thus a qualitatively similar result to that explained by gradual depletion was also obtained under circumstances where there is no gradual depletion. Cuthill and Kacelnik (in press) suggest that the explanation of a load–distance effect in the experiment without gradual depletion may lie in the energy or time cost of transporting differently sized loads (see also Cheverton *et al.*, 1985; Schmid-Hempel *et al.*, 1985; Kacelnik *et al.*, 1986). How could time or energy costs of carrying a load result in the prediction of a load size–distance effect? There are at least three ways in which these costs could change the prediction of Fig. 4.2a into one like that of Fig. 4.1b. In each case the effect of the cost is to transform the linear gain function of Fig. 4.2a into one that gradually levels off like that in Fig. 4.1b First, consider energetic costs. Suppose energetic expenditure in the patch increases in proportion to amount of food collected: the *net* rate of gain will drop with time in the patch even

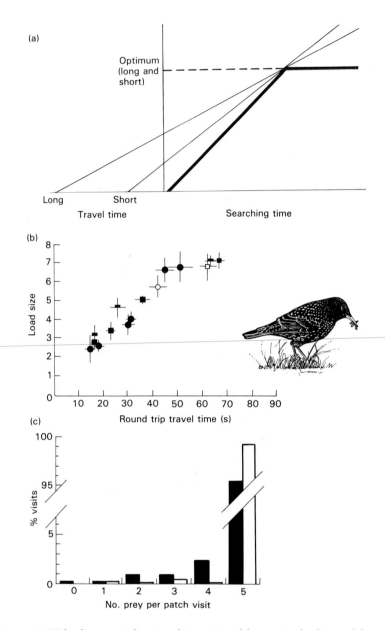

Fig. 4.2 (a) With a linear gain function this version of the marginal value model predicts no effect of travel time on load size. The optimal solution is always to take the maximum load. (b) Starlings show a load size–distance effect with a linear gain function (Cuthill, 1985; Kacelnik & Cuthill, 1987). Different symbols refer to different individual birds. The bars are standard deviations. This result may be explicable in terms of maximizing net energy gains, if the energetic cost per trip is an increasing function of load size. (c) This interpretation is supported by the observation that when starlings do not have to carry a load back to the nest (they forage for themselves), they nearly always take the maximum number of prey, as predicted by the version of marginal value model in (a). The results shown are from a laboratory experiment in which starlings foraging in patches with a linear gain function following long (white) or short (black) travel times. The maximum number of prey obtainable per patch visit (the asymptote of (a)) is five. Irrespective of travel time, the birds take five prey on more than 95% of patch visits (Cuthill *et al.*, 1990).

though the *gross* rate of gain may be constant. Metabolic expenditure during the return trip to the nest may also increase in proportion to the load carried home, and this will also have the effect of transforming a linear gross intake curve into a curve of decelerating net gain. Now consider time rather than energy: the velocity of the starling during its return journey to the nest may decrease in proportion to the size of the transported load, thus introducing a time cost (Kacelnik *et al.*, 1986; Wetterer, 1989; Cuthill & Kacelnik, 1990). This will also have the effect of translating a linear gross intake curve into one with decelerating net gains.

If any of or all these effects occur, the marginal value theorem still explains the load size–distance relationship, but in a modified form. The diminishing returns within a patch arise not from changing rate of encounter with prey, but from changing net gains. This result does not invalidate Kacelnik's original conclusion since, although the qualitative pattern could be explained by either hypothesis, the quantititive details in the original study were better explained by the hypothesis of changing rate of encounter with prey. There are two general points here. The first is that qualitatively similar results may be the product of different mechanisms and only careful experimental analysis can distinguish such possibilities. The second is that often optimality models are useful to guide research into animal behaviour even when they fail to predict results. In this case the correlation between load size and distance with a linear gain function was interesting precisely because it was against the predictions of the available model. This discrepancy caused further developments of the theory and suggested specific experiments to study, for example the relation between load and flight velocity. The original model was found to be at fault because of its oversimplification in ignoring the costs associated with bigger loads. Support for Cuthill and Kacelnik's (1990) interpretation in terms of the cost of carrying a load comes from further laboratory experiments reported in the same paper in which the birds did not carry a load but foraged for themselves in an environment with linear gain functions. Here the birds took the maximum load irrespective of distance travelled, as predicted by rate maximization (Fig. 4.2c).

4.2.4 *What about non-patchy environments?*

The world modelled by the marginal value theorem is one in which the animal is either travelling between patches or foraging within patches. But, for many animals, food may fluctuate in abundance with space, but not occur in discrete patches. Returning to our starting example of the robin, it is likely that some parts of the lawn contain more worms than other parts, but there are no discrete patches of worms with empty areas between them. Arditi and d'Acorogna (1988) analyse the rate-maximizing foraging strategy of an animal in an environment in which there are spatial variations in prey density but no discrete patches. They show that the animal should, in such an environment, travel through areas of low prey density and exploit only the high density peaks. Thus, even when the world is not made up of discrete patches, a rate-maximizing forager should divide its behaviour into travel and exploitation as though the world were patchy. In other words, the simplified abstraction of the world represented by the marginal value theorem may be of wider applicability than appears at first sight.

4.2.5 *Summary*

We have dealt with the case history of the starling in detail in order to illustrate some general points. To summarize, these are as follows: (i) simple foraging models based on rate-maximizing are a valuable research device. Many studies have used the marginal value theorem and corresponding models of prey choice as a tool for guiding studies of foraging behaviour both in the laboratory and in the field. In general, the qualitative predictions of rate-maximizing models are borne out (for fuller discussion see Stephens & Krebs, 1986; Gray, 1987; Schoener, 1987); (ii) rate-maximizing, whilst clearly incomplete as a model of behaviour (see next three sections), nevertheless does a good job of predicting foraging decisions and may therefore be a useful basis for describing foraging behaviour; (iii) qualitative fit to rate-maximizing models must be interpreted cautiously since they may well not distinguish between alternative hypotheses; and (iv) to understand behavioural decisions in greater detail it is necessary to analyse mechanistic questions such as memory and time perception. There is a close interplay between optimality analyses and analyses of behavioural mechanisms, with the ultimate goal of incorporating both into the same framework.

4.3 The trade-off between foraging and danger of predation

In spite of the fact that 'classical' rate-maximizing models such as the marginal value theorem are often successful in predicting behavioural decisions, they are inadequate as representations of decision-making in three major respects: (i) they do not consider how an animal might trade-off alternative activities, such as foraging, vigilance and guarding a mate; (ii) they do not analyse how the animal's behaviour might change with internal state. If there is a compromise between foraging and other activities, eating food immediately and/or foraging efficiently may take higher priority when body reserves are low and the animal is near starvation than if body reserves are high; and (iii) they assume that the environment is deterministic, whilst in reality ecological processes such as encountering food and being attacked by predators occur in a stochastic fashion. In this section we focus on trade-offs, and in subsequent sections we deal with state and with stochasticity.

4.3.1 Evidence for trade-offs

Numerous studies have shown that animals sacrifice food intake in face of potential danger from predators. They can be summarized under four headings.

CHOICE OF PATCHES OR HABITATS DEPENDS ON DANGER AS WELL AS FOOD

In a field experiment, Newman and Caraco (1987) presented grey squirrels *Sciurus carolinensis* with two feeding 'patches' (trays containing sunflower seeds), one large and one small. When the patches were 5 m from the cover of nearby trees, more individuals fed in the bigger patch and the distribution of individuals between patches was such that average intake rate was equal in the two sites (the 'ideal free distribution'; see Chapter 5). However, when the bigger patch was moved to 15 m from cover, fewer squirrels foraged there than before and the rate of food intake for those who did was now higher than that for the small patch. One likely interpretation of this result is that, for the squirrels, the perceived danger of feeding in a patch is positively correlated with distance from cover and that the animals were trading off danger and rate of food intake in their choice of where to feed. This interpretation has been tested more directly in other studies in which the distribution of foraging effort has been manipulated by placing a predator near one of two patches and showing that animals will shift towards feeding in less profitable (in terms of food intake rate) patches to avoid the predator (Gilliam & Fraser, 1987; Abrahams & Dill, 1989; reviews in Dill, 1987; Sih, 1987).

BEHAVIOUR WITHIN A PATCH CHANGES IN RELATION TO DANGER

The classical rate-maximizing models treat handling time, travel time between patches, and encounter rate with prey items as fixed constraints. However, these aspects of behaviour may be decision variables reflecting a trade-off between foraging and danger (section 4.1.1). The expression of this trade-off may vary between situations. For example, Krebs (1980) reported that great tits increase their handling time on a given-sized prey following brief exposure to a stuffed hawk and that this increase was due to an increase in 'looking up' during handling, presumably scanning for predators. Newman *et al.* (1988) showed that when grey squirrels forage at a greater distance from cover they travel between patches more rapidly and have a steeper gain function within a patch, reflecting more rapid search and handling. The likely interpretation is that squirrels alter their travel and search to reduce the time spent far from cover.

   Although these kinds of study show that danger affects foraging behaviour, they do not reveal the nature of the trade-offs involved. Why, for example, do squirrels not travel between patches at a higher velocity in the absence of danger when they are clearly capable of doing so? It could be that there is an optimum travel velocity from the energetic point of view and that in the face of danger the animals exceed this optimum.

QUANTITATIVE PREDICTIONS OF TRADE-OFFS

Two approaches have been taken in making quantitative predictions about trade-offs. Abrahams and Dill (1989) adopted a descriptive or inverse optimality (McFarland, 1977) approach. They first measured the sacrifice in terms of food intake made by guppies in order to avoid feeding near a predator. This was done by placing a group of ten guppies in a tank divided into two halves by a mesh through which the guppies could swim. A feeder at each end delivered prey at the same rate. In this experiment, the guppies distributed themselves in two groups of five with equal feeding rates in the two halves of the tank (see Chapter 5). Then a cichlid predator in a small enclosure was introduced into one side of the tank. Now more guppies foraged on the 'safe' than on the 'dangerous' side, as a consequence of which rate of food intake per guppy was lower on the safe side. The difference in rate of food intake on the two sides was taken as an indicator of the 'energetic equivalent' of danger—how much the guppies that avoid the dangerous side would pay in terms of food intake to avoid the predator. This is the same logic as is used in studies of consumer choice to estimate the value placed by consumers on different commodities (reviewed in Stephens & Krebs, 1986). Abrahams and Dill (1989) went on to use this estimate of the energetic equivalent of danger to predict by how much they would have to increase the food supply in the dangerous half of the tank to persuade the guppies to return to an equal distribution (5:5) on the two sides of the mesh partition. This prediction was successful for female but not male guppies.

This experiment is descriptive in the sense that Abrahams and Dill did not start from an *a priori* theory of how the fish should trade off danger and food from the point of view of maximizing lifetime fitness. Instead they showed that the behaviour (at least of females) in one situation can be predicted from that in another. In the psychological literature this has sometimes been referred to as a test of 'transituationality'. In contrast, Gilliam and Fraser (1987) start from a model of the fitness consequences of foraging and danger. They assume that the animal has to eat a certain amount of food per time period (e.g. day) in order to survive and that, beyond this amount, extra food is of no fitness benefit (note that this would not be an appropriate assumption if extra food could be used for growth or reproduction). Different patches or habitats can be characterized by their rate of food intake and their mortality rate due to predators. The forager should choose the habitat (or combination of habitats) which allows it to meet its daily requirements with the lowest rate of mortality (Fig. 4.3). This is equivalent to saying that the forager should choose the patch with the lowest ratio of mortality rate (m) to feeding rate (h) subject to the constraint that it is above the level re-

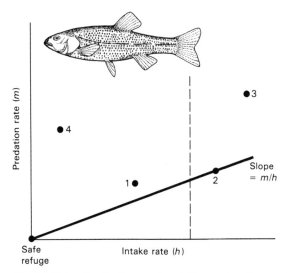

Fig. 4.3 Balancing the danger of predation against the rate of food intake. Each habitat or patch is characterized by a combination of mortality rate due to predation (m) and rate of food intake (h). The predator has a minimum food requirement per time period (vertical dotted line). In Gilliam and Fraser's (1987) model the predator chooses the patch or combination of patches with the lowest m/h ratio, subject to the constraint of meeting the minimum requirement. In this case, patch 2 is chosen. It is assumed that there is also a completely safe refuge with no food (dot at the origin).

quired to survive. Gilliam and Fraser showed that this simple model is sufficient to account for the changes in behaviour of juvenile creek chub *Semotilus atromaculatus* in the presence of predators (adults of the same species). They measured food intake as a function of prey density and mortality rate as a function of predator density and then offered juvenile chub the choice of foraging at two ends of a stream tank with different combinations of prey and predators resulting in known (m/h) ratios. In all six treatments, the chub chose, on average, to forage in the patch with the lower ratio of m/h, as predicted.

THE RESPONSE OF FORAGERS TO A TRADE-OFF DEPENDS ON STATE

In the next section we will deal with state-dependent decisions in more detail. Here we simply note that a number of studies have shown that when there is a trade-off between feeding and danger, the choice of feeding site or behaviour within a site depend on the animal's state. In a classic study, Milinski and Heller (1978) (see also Milinski, 1984) compared the behaviour of hungry and well-fed sticklebacks. The former were more likely to feed in a patch where rate of food intake was potentially higher, but danger of predation was also higher. Parallel state-dependent trade-offs have been reported by Krebs (1980); Dill and Fraser (1984) and Werner and Gilliam (1984).

4.4 The importance of state: stochastic dynamic modelling

A technique which has grown in popularity in the last few years is *stochastic dynamic programming* (Houston *et al.*, 1988; Mangel & Clark, 1988). In this section we will first outline the essence of the stochastic dynamic programming method and then give some examples of how it has been used to explain and predict decision-making. It is perhaps worth emphasizing that stochastic dynamic programming is a technique and that the biological question being addressed is how to analyse optimal decision-making in a way that incorporates state (for example, how hungry the animal is), stochasticity (the unpredictable nature of events) and trade-offs. The 'dynamic' aspect of the models refers to the fact that the state of an animal changes dynamically as a consequence of its decisions, and this in turn influences the optimal decisions in a feedback loop. For an alternative method see McFarland (1977).

4.4.1 Stochastic dynamic programming

A stochastic dynamic programming (SDP) model attempts to evaluate how short-term behavioural decisions contribute to lifetime fitness. Usually, rather than trying to model the whole life of an individual, the fitness consequences of behaviour are considered over a fixed time period. A central concept in SDP is the *terminal reward function*, a hypothetical mathematical relationship between lifetime fitness and state at the end of the period under study. If, for example, the fixed time period over which behaviour was modelled was a single day, the terminal reward would define the animal's future fitness as a function of its state at the end of the day. It transpires that the exact details of the terminal reward function are only critical in affecting the predictions of a SDP model near the end of the time period (McNamara *et al.*, 1990). Therefore, although there is an element of guesswork in choosing the terminal reward function, it often does not matter much if the guess is slightly inaccurate given that the general shape of the function is correct.

In order to identify optimal decision rules, a SDP model works by dividing the period under consideration into a number of discrete time intervals (these can be thought of as representing days, hours or minutes according to the problem under study). In each time period, the animal is able to choose between alternative actions and each action has fitness consequences for the animal. These consequences could be in terms of immediate probability of survival during the time interval, for example resting and foraging may each incur a characteristic danger of death due to predation in each time interval. Alternatively, the fitness consequences could be a result of changes in the animal's state, which could have both immediate effects, the risk of starvation during the current time interval, and longer-term effects by influencing the state and therefore the terminal reward at the end of the period that is being modelled. If, for example, the animal chooses not to eat in the present time interval it may condemn itself to starve at some time later on if it does not encounter sufficient food. It is intuitively easy to see that there is a problem with calculating this kind of long-term fitness consequence of an action, namely, that these consequences depend on what may happen in the future. To take an everyday example, whether or not you should eat a large meal at midday depends not only on how hungry you are, but also on whether there will be another opportunity for eating later on in the day. For this reason, it is computationally most efficient to calculate the consequences of alternative actions starting at the end of the time period and working backwards. Going back to the example of the mid-day meal, if you knew that you were going to eat a large meal in the evening, you may decide that it is not worth taking time out to eat a large meal in the middle of the day.

   Table 4.2 shows an example of doing this kind of calculation for a simple foraging decision in which there are two patches: a 'good patch' with a high probability of finding food but also with a risk of being killed by a predator, and a 'poor patch' with a low probability of food but no predators. The model represents and analyses the state-dependent trade-off between foraging and predation discussed briefly in section 4.3.1. Assume that searching for food costs the animal a unit of energy, whichever patch it chooses, and that finding an item yields two units of energy. Given these values, it is possible to compute the terminal reward resulting from different courses of action at time $(T-1)$ in different states. Two examples are given. When reserves at time $(T-1)$ are high $(x = 5)$ it pays to choose the safe, poor patch whilst if reserves are low $(x = 3)$ it pays to choose the dangerous patch with a higher food probability, as reported by Milinski and Heller (1978). At $(T-2, x = 4)$ it pays to choose the safe patch, followed by either the safe or the dangerous patch at $T-1$ (for a full analysis of the problem see McNamara & Houston, 1990a).

   A complete SDP model would work backwards from the final time (T) through successive time periods computing at each stage the course of action that will maximize survival (the index of fitness in this model) given the future course of action in the remaining time periods of the day. The end result of completing these calculations is a matrix of choices that will maximize survival probability for each state and each time interval. This is referred to as the *optimal policy*. In Table 4.1 we showed the optimal policy for the long-term average rate-maximizer choosing between two prey types. In order to translate this into predicted behaviour it would be necessary to measure the parameters of the equation for selectivity and find out if the animal is in the 'take large only' or 'take both' region of parameter space. In a similar way it is necessary to translate the optimal policy of a SDP model into an optimal course of action by finding out which state the animal will be in at each time period. For the example in Table 4.2, we would have to know the most frequently observed state at $(t-1)$ to predict

Table 4.2 State-dependent choice of patches.

In patch 1 there is a probability of 0.5 of finding food in a time interval, a probability of 0.5 of finding no food, and a probability of 0.0 of being killed by a predator. The corresponding parameters for patch 2 are 0.75, 0.10 and 0.15. Thus patch 1 has a lower average rate of intake but is also safe. The table shows the optimal decision at each of the last two time intervals of the period under consideration for two values of state. State in this case means energy reserves: if state falls to zero, the animal dies. A choice of patch is made at each time interval and the choice costs 1 unit of reserves. A food item yields 2 units of reserves. The animal should maximize its terminal reward, which in this example is simply state at the end (time T). Calculations are shown for sample states at two time intervals, $t = T-1$, $t = T-2$. The calculations show the expected terminal reward. In any particular instance the animal will experience only one of the possible outcomes.

1 *At T−1* $(3+2-1)0.5$
   (a) $x = 3$
       Choose patch 1: Terminal reward = 0.5×4.0   +   0.5×2.0 = 3.0
                                     (find food)   (no food) $(1 - 0.15)$
       Choose patch 2: Terminal reward $=(0.75×4.0$ + $0.1×2.0) + 0.15×0$ = 3.2 2.72
                                     (find food)   (no food)   (death by
                                                               predation)

       Terminal reward ($x = 3$) is greater if patch 2 is chosen at $T−1$

   (b) $x = 5$
       Choose patch 1: Terminal reward = 0.5×6.0+0.5×4.0 = 5.0
       Choose patch 2: Terminal reward = 0.75×6.0+0.1×4.0+0.15×0 = 4.9
       Terminal reward ($x = 5$) is greater if patch 1 is chosen at $T−1$

   (c) Summary of similar calculations for other states

$x(T−1)$	Optimal choice	$x(T)$ = terminal reward
5	1	5.00
4	2	4.05
3	2	3.2
2	2	2.35
1	2	1.50

2 *At T−2*
   It is assumed that the animal will make the optimal choice at $T−1$, so pay-offs are calculated substituting $x(T)$ for $x(T−1)$, e.g. if $x = 4$ at $T−2$
       Choose patch 1: Terminal reward = 0.5×5.0+0.5×3.2 = 4.10
       Choose patch 2: Terminal reward = 0.75×5.0+0.1×3.2 = 4.07
       If $x = 4$ at $T−2$ it pays to choose patch 1. The optimal choice at $T−1$ depends on the outcome of $T−2$. If food is found, $x(T−1) = 5$ and the optimal choice is patch 1. If no food is found, $x(T−1) = 3$ and patch 2 should be chosen. On average it pays to choose patch 2 at $T−1$.

---

if animals should, on the whole, choose the good or the poor patch at this time. This is done by running the model as a simulation, starting off with the animal in an arbitrary state and allowing it to follow the optimal policy at each successive time step. The process is reiterated until the animal's average state as a function of time remains the same as one adds successive 'runs' of the model (e.g. on each 'day', if the model represents a daily cycle).

4.4.2 *Applications of stochastic dynamic programming*

Here the example of daily routines is taken to show how SDP models might be applied to decision-making; for further examples see Mangel and Clark (1988) and Houston *et al.* (1988).

McNamara *et al.* (1987) modelled the dawn chorus of songbirds and produced a novel explanation of this widespread phenomenon. The observation to be explained is that many songbirds show a peak of singing activity in the early morning. At first sight this seems paradoxical, because it might be thought that birds would wake up with a deficit in body reserves and need to forage intensively rather than sing just after dawn. Previous functional explanations for the dawn chorus in terms of ecological factors have suggested that daily variations in sound transmission (Henwood & Fabrik, 1979), food availability (Kacelnik & Krebs, 1983) or intruder pressure (in the case of territorial song) (Kacelnik & Krebs, 1983) are important. McNamara *et al.* (1987), in their SDP model, show that the dawn chorus could arise without *any* circadian variation in ecological factors other than the day–night cycle, simply as a consequence of the way in which birds regulate their body reserves.

The essence of their argument is that small birds, rather than maximizing the body reserves they carry, optimize their reserves to reflect a trade-off between the advantage of high reserves to counter starvation and the cost of carrying high reserves (King, 1972; Lima, 1986; McNamara & Houston, 1990b). The cost of carrying body reserves may arise as follows. If metabolism is mass-dependent, extra reserves result in additional energy expenditure and therefore additional time spent foraging to replace lost energy and, if foraging is more dangerous in terms of predation risk than, for example, resting, maintaining extra reserves will result in a higher death rate from predation. A second way in which reserves might be disadvantageous is if they actually slow the animal down so that it is more vulnerable to predator attack because its escape speed is reduced (this was not included in McNamara *et al.*'s model). Evidence that birds do not maximize their body reserves comes from observations of both seasonal and daily patterns of change in body mass. On a seasonal basis, many birds reach their highest mass on days when food availability might be expected to be lowest (e.g. extremely cold days in winter). This means that on other days, when food is more plentiful, the birds cannot be maximizing their body reserves. With regard to daily patterns, most small birds lose about 10–15%

of their body mass overnight and do not fully regain these reserves until late in the afternoon, even if supplied with *ad libitum* food, implying that for much of the day they maintain reserves below the maximum level.

How does regulation of body reserves explain the dawn chorus? McNamara *et al.* suggest the following explanation. If overnight expenditure of energy is entirely predictable, birds should put on just enough reserves each afternoon to survive the night. If, however, overnight expenditure is unpredictable, the bird will have to put on more reserves than the average needed to survive the night, in order to ensure survival on nights when expenditure is abnormally high. This leads to the paradoxical situation in which, on many mornings, the birds start the day with an excess of body reserves. In their model, singing was a major alternative activity to foraging, so that when the bird had more than the required level of reserves, it would stop foraging and sing instead. Thus the consequence of optimal regulation of body reserves and unpredictable overnight expenditure is a peak of singing activity in the early morning. Consistent with this hypothesis, some studies (Garson & Hunter, 1979; Reid, 1987) have shown that the amount of singing in the dawn chorus is inversely related to overnight temperature. Whilst McNamara *et al.* would not claim that their model is a complete explanation of the dawn chorus, it makes the important general point that daily routines in behaviour, such as the dawn chorus, could be generated without any variation through the course of the day in ecological factors that influence the costs and benefits of different activities.

Regulation of body reserves may also be important in understanding another daily routine, that of food storing in tits *Parus spp.* Some species of tits (such as the marsh tit *Parus palustris*, willow tit *Parus montanus*, crested tit *Parus cristatus* and black-capped chickadee *Parus atricapillus*) store large numbers of food items, especially during the autumn and winter. Pravosudov (1985) estimates that willow tits may store as many as half a million items per season. Unlike certain species of corvids, such as jays and nutcrackers, tits appear to store and retrieve food on a short-term basis. The interval between storage and retrieval is often in the order of hours to days rather than weeks to months. Field and laboratory evidence (Stevens & Krebs, 1986; McNamara *et al.*, 1990) indicate that there is a daily routine of storing and retrieval. Tits tend to store early in the day and retrieve late in the day. Using a SDP model, McNamara *et al.* (1990) (see also Lucas, 1990) account for this daily routine in terms of daily variation in reserves. They consider food storing to be an alternative to storing reserves on the body. Therefore the food-storing bird builds up its reserves by storing food in the environment during the early part of the day and then transfers these reserves to its body for overnight survival by retrieving its hoards near the end of the day. Thus, as in the case of the dawn chorus, the daily routine of storage and retrieval is not dependent on fluctuations in the availability of food, but arises from physiological costs of carrying fat reserves. Consistent with this, under laboratory conditions of constant average food availability, marsh tits still show a daily routine of storage and retrieval (McNamara *et al.*, 1990). The model also allows McNamara *et al.* to predict the conditions under which storage of food is advantageous. Among other things, when the cost of metabolism increases sharply with body reserves, storing in the environment is more advantageous than storing on the body. One, as yet untested, prediction made by the McNamara *et al.* model is that, all other things being equal, storing and non-storing species of tits should show different daily patterns of fluctuation in body mass.

4.4.3 *Comparing stochastic dynamic programming and rate-maximizing models*

Mangel and Clark's (1988) analysis of oviposition decisions in parasitoids shows how a SDP model can make predictions beyond those made by a simple rate-maximizing analysis, in that case the marginal value theorem. Houston and McNamara (1985) compare the predictions of a rate-maximizing model of prey choice with that of a SDP model. With two prey types (large and small), the rate-maximizing model predicts that the predator should either take large prey only or take both, according to the rate of encounter with large prey. In a SDP model, in which the currency to be minimized is risk of starvation, optimal prey choice depends not only on the encounter rate with large prey, but also on the time of day and level of energy reserves. In conditions where rate maximizing would predict selective foraging, the SDP model may predict non-selective foraging to minimize the probabality of starvation during the day. This is because, although selective foraging results in a higher mean, it also results in a greater probability of not eating anything during a particular time period. If the animal is close to a lethal boundary below which it will die from lack of reserves, it is worth foraging non-selectively and sacrificing gains in terms of rate in order to reduce the probability of not eating anything during a particular time period and thus reduce the probability of crossing the lethal boundary. Time of day also has an effect. Near the end of the day, increasing reserves to survive the night is of paramount importance; i.e. the animal needs to build up energy reserves and therefore it pays to choose a higher rate, even if this results in a slightly higher probability of starvation whilst foraging. Thus, in general, the SDP model predicts an increase in selectivity towards dusk in conditions where the rate-maximizing model would predict selectivity throughout. Although this latter prediction has not been tested, several studies have shown that the level of deprivation (body reserves) does affect prey choice. Hungrier animals, with lower reserves, are less selective (Rechten *et al.*, 1983; Snyderman, 1983) as predicted by the SDP model.

   Thus, in both the example of prey choice and in the analysis of oviposition in insects, we have seen that SDP models can incorporate more of the complexity of behaviour and its consequences than rate-maximizing models. In other words, they have greater realism, primarily because they incorporate state, and allow decisions to be state-dependent. The sacrifice that is made in exchange for the greater realism of SDP models is their greater complexity, which may lead to greater difficulty in understanding exactly why results emerge as they do. The greater complexity of SDP models also means that they require more detailed physiological knowledge (e.g. in the case of hoarding and the dawn chorus, knowledge of mass-dependent metabolic costs, which are not available in the literature). This may be seen as an advantage of SDP models, since they highlight the kinds of information that would be necessary to carry out a full optimality analysis of decision-making.

It seems likely that there will continue to be a role for both approaches. Rate-maximizing models may give a simple intuitive feel of what is going on and often generate successful predictions, whilst SDP models with their greater realism and detail may be able to explain more of the complexity of decision-making. So far, there have not been many empirical studies using SDP models as a basis for predictions and here there is clearly a need for more work to explore the potential of this approach.

4.5 Risk sensitivity: economic, psychological and biological models of choice

In the previous section, we introduced the idea that foraging decisions may be influenced not just by long-term mean rate, but also by variability in rate of intake. In Houston and McNamara's (1985) model of prey choice, minimizing the probability of starvation during the day was achieved by choosing the less variable of two options (unselective foraging). The first study which explicitly examined whether or not foragers discriminate between alternatives offering the same *mean* amount of food per unit time but differing in *variability* of amount was that of Caraco *et al.* (1980). They offered juncos choices between a fixed and a variable number of seeds with the same mean (e.g. 2 seeds versus 0 or 4 with a probability of 0.5, or 3 versus 0 or 6 with a probability of 0.5); each bird was offered a series of such choices during a single session. In one experimental condition, the birds were deprived for an hour before testing and for each seed eaten during a particular trial, 30 s delay was added to the interval before the next trial (thus if the bird ate three seeds in a trial it had to wait for 90 seconds before its next choice), whilst if it ate no seeds, the next choice followed immediately. The purpose of this device was to ensure that the overall mean rate was independent of the number of seeds obtained in a trial. In the other treatment, the birds were tested after 4 hours' deprivation and the delay between trials was incremented by 60 seconds for each seed eaten. In the first treatment, the birds preferred the fixed over the variable amount, whilst in the second treatment they preferred the variable reward.

Caraco *et al.* (1980) explained their results with reference to utility theory (Keeney & Raiffa, 1976). In economic models of decision-making (for example, Thaler, 1980), utility is used to refer to the value of an outcome. In economics, utility is often defined in terms of the preferences of consumers (i.e. it describes the outcome), but in a Darwinian context it could be thought of as reflecting the fitness consequences of an outcome. As Fig. 4.4 shows, depending on the shape of the utility function, i.e. the function relating utility to amount of reward obtained, the decision-maker aiming to maximize utility would prefer a variable option ('risk prone') behaviour or a certain option ('risk averse') behaviour. In Caraco *et al.*'s experiments, deprivation appeared to change the shape of the utility function. High deprivation combined with a low rate of gain (60 seconds delay per seed) led to a 'concave-up' utility function, whilst the low deprivation condition led to a 'concave-down' utility function (Fig. 4.4). The biological interpretation of this is in terms of expected daily energy budgets. In the high deprivation condition, the birds could not expect to meet their daily energy requirements (the reserves needed to survive the night) by choosing the fixed reward, whilst in the low deprivation they could. In the former case, the animal's chance of survival would have been lower if it chose the fixed option whilst it stood a higher chance of survival if it chose the variable option and obtained the high pay-off. In contrast, in the latter treatment, by choosing the fixed option, the animal's chance of survival was higher than the average of the variable option (see Fig. 4.4). The general rule, therefore, is 'be risk averse if expected energy budget positive, be risk prone if expected energy budget negative' (Stephens, 1981). Note that this model of risk sensitivity is state-dependent but, unlike the SDP models discussed in section 4.4, it does not incorporate dynamic feedback between state and decisions.

This is a simplified form of the theory and it is deficient in several ways:

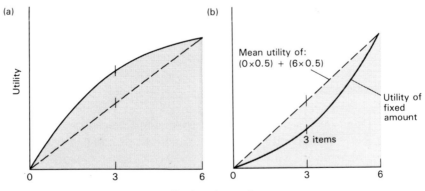

Fig. 4.4 With a bowed up utility function (a), the utility of a fixed reward of three items is higher than the mean utility of a variable reward of zero and six, each with a probability of 0.5. When the utility function is bowed down, this is reversed.

1   The argument is based on a single choice rather than a sequence. If the animal makes a sequence of choices as in Caraco *et al*'s experiments, the pattern of risk sensitivity may change with time. If, for example, the animal starts off risk prone, but as a result of obtaining a high reward in early choices it moves into positive expected energy budget, it may switch to risk-averse behaviour later in the day (Houston & McNamara, 1982). A further question arising from the situation in which there is a sequence of choices is that the time interval over which mean and variance are calculated by the animal is undefined (Possingham *et al*., 1990). Caraco *et al*. (1980) adjusted the interval between trials so that the birds obtained the same mean reward over a whole session irrespective of the choices they made. If, however, the animal calculated mean reward rate over a single trial or over a small number of trials, it may not have perceived the two options as offering the same mean rate.
2   As Houston and McNamara (1985) have shown, daytime starvation and night-time starvation can have different consequences for risk sensitivity. In their model of prey choice, daytime starvation is minimized by choosing the less variable option whilst night-time starvation is minimized by becoming risk prone near dusk (see above).
3   Finally, the theoretical interpretation of Caraco *et al*.'s (1980) result outlined above considers only variability in *amounts* and not variability in *delay*. However, in Caraco *et al*.'s experiments the variable option varied in both amount and delay; longer delays followed larger rewards. As discussed below (section 4.5.2), a different theoretical framework may be needed to explain response to variability in delay to reward.

4.5.1 Evidence for risk sensitivity in relation to reward size

Since Caraco *et al*.'s (1980) original study, many experiments have investigated the phenomenon of risk sensitivity and a few have attempted to explore the conditions under which animals switch from risk-averse to risk-prone behaviour (reviews in Real & Caraco, 1986; Stephens, in press; see also Moore & Simm, 1986; Stephens & Paton, 1986; Gillespie & Caraco, 1987; Wunderle *et al*., 1987; Caraco *et al*., 1990; Cartar & Dill, 1990; Ha *et al*., 1990). The two most convincing studies are those of Caraco *et al*. (1990) and Cartar and Dill (1990).

Caraco *et al*. (1990) tested juncos for risk sensitivity at two temperatures, 1°C and 19°C. As in their earlier experiment, Caraco *et al*. offered the birds a sequence of choices between a fixed reward (three or four seeds) and a variable reward with the same mean. At 1°C, birds preferred the variable reward, whilst at 19°C they preferred the fixed alternative. The effect of increasing the temperature was to shift the birds from negative to positive expected energy budgets. In this experiment, the interval between trials was held constant so that there was no confounding effect of variability in delay and amount.

Cartar and Dill (1990) studied risk sensitivity in bumble bees. In earlier work, Real (1981) and Real *et al.* (1982) had shown that bumble bees were always risk averse. Further, Harder and Real (1987) had interpreted this result in terms of the bee's maximizing mean rate of reward per flower visited. They showed that the net rate of energy intake per 'visit cycle' (travel to a flower and extraction of nectar) is non-linearly related to nectar volume, so that as nectar volume increases the marginal increase in rate gets smaller. Thus, in a choice between a fixed volume of 2 μl or a variable of 1 and 3 μl, the mean rate per flower is higher on the fixed option. In Cartar and Dill's experiment, the mean rate from the variable and fixed options was adjusted to be the same by using a fixed value of 2 μl and a variable option in which half the flowers had 5 μl and half had 1 μl. The energy budget manipulated was not that of the individual bumble bee worker, but of the nest. They either depleted the sugar store of the nest or augmented it, and in the former condition found that the bees were risk prone, whilst in the latter condition they were risk averse. Thus the bees adjusted their risk sensitivity in relation to the needs of the colony as a whole.

Whilst both of these experiments provide convincing evidence of shifts in risk sensitivity in relation to expected energy budgets, they both leave open the question of the time interval over which the individual forager calculates the mean and variance of alternative options.

4.5.2 *Variability in delay: future discounting*

A well-established finding in the psychological literature is that animals prefer variable delays to reinforcement over a fixed delay if the mean time to reinforcement for the two options is the same (Herrnstein, 1964; Fantino, 1969). This has been investigated in detail by Mazur (1984) and Mazur *et al.* (1985) using a titration procedure to calculate the point of indifference between a fixed time to reinforcement and a mixture of two times, long and short. In the titration procedure, when the animal prefers the variable delay to reinforcement, the duration of the fixed delay is decreased progressively until indifference is achieved. The indifference points can be used to assess how the animal values rewards that occur after different delays. Mazur found that his results could be best described by assuming that the animal devalues rewards in a hyperbolic fashion in relation to delay. In other words, a reward of a given size is worth more if it occurs immediately than if it occurs, say, 10 seconds into the future. An interpretation of 'future discounting' in terms of fitness is that rewards occurring after a delay might never be obtained because the animal is interrupted before they occur (Kagel *et al.*, 1986a,b; McNamara & Houston, 1987c). Thus, sensitivity to variation in delay may have a different functional explanation from sensitivity to variation in reward amount. Future discounting could also be involved in other results, both in the psychological literature such as 'failure to show self-control' (Logue, 1988) and in the behavioural ecology literature, such a bias in favour of prey with low energy value and short handling time which may not be predicted by rate maximization (Barkan & Withiam, 1989). Gibbon *et al.* (1988) provide a mechanistic account of preference for variable delays in terms of memory for time intervals.

4.6 Psychological studies of choice and learning

The parallels between studies of choice and learning by psychologists studying rats and pigeons responding to levers or keys in a Skinner box and studies of foraging decisions by behavioural ecologists is well established (Shettleworth, 1988; Fantino & Abarca, 1985; Kamil & Roitblat, 1985; Commons *et al.*, 1987). Psychologists working on operant conditioning are concerned with the ways in which patterns of reinforcement influence behaviour, which is similar to the same question as that addressed by foraging theorists, if one substitutes 'prey availability' for 'patterns of reinforcement' and 'foraging' for 'behaviour' or 'choice'. An essential difference between the two approaches is that behavioural ecologists attempt to derive their models of choice from considerations of fitness whereas psychologists are mainly interested in mechanisms of choice. Interactions between the two fields can be considered under three headings.

1 *Techniques.* Behavioural ecologists have over the past 10 years increasingly used the powerful techniques of operant psychology (controlled schedules of reinforcement; automated recording and control) to generate laboratory simulations of foraging decisions.

2 *Application of foraging theory to psychological studies.* Theories of risk sensitivity and future discounting (see above) are clear examples of how phenomena revealed in studies of operant conditioning can be explained within a functional framework derived from behavioural ecology. Other examples include response to progressive ratio schedules ('patch depletion') (section 4.2) and 'sampling' behaviour (section 4.6.1).

3 *Application of psychological theory to behavioural ecology.* Rules of choice and response to reinforcement derived by psychologists may help to explain aspects of behaviour which could not be accounted for in terms of conventional fitness-maximizing models. Examples include constraints on time perception (Kacelnik *et al.*, 1990) and effects of immediate versus global rate of reinforcement (see below). It is important to remember that the kinds of explanation offered by behavioural ecologists (in terms of costs and benefits) and by operant psychologists (in terms of proximate causes of behaviour) are complementary rather than alternatives.

4.6.1 *Matching, melioration, maximizing and learning*

One of the central issues in studies of operant conditioning is the analysis of rules governing choice between simultaneously available sources of reinforcement. In a typical experiment, a pigeon or a rat is confronted with two keys or levers offering reinforcement at different average rates. Rate may be determined on a per response basis (so called ratio schedules) or on a per time basis (interval schedules). This choice situation can be thought of as being analogous to choice between patches by a foraging animal.

In many studies of choice, overall allocation of behaviour among two alternatives is described by the 'matching law' (Herrnstein, 1961). In its generalized form, summarizing results across experimental treatments (Baum, 1974), the matching law can be summarized as follows:

$$\log (B_1/B_2) = \log k + b \log (r_1/r_2)$$

where B_1 and B_2 are the times allocated (over a whole time period) to alternatives 1 and 2; r_1 and r_2 are the attained rates of reinforcement from the two alternatives; and k and b are scaling parameters fitted to each set of data. In other words, the matching law states that the animal allocates its behaviour to two alternatives in proportion to the rewards it has obtained from them. Three points about the matching law are worth noting; (i) it is an empirical description of the overall distribution of responses ('molar pattern') not a rule for choice; (ii) it is a very flexible description and not a unique solution. Any point along a line described by the above equation would be counted as matching and, given that the slope and the intercept of the line can be varied by the constants k and b, many outcomes can be described by the matching relationship; and (iii) the question of whether or not matching corresponds to maximizing overall rate of reinforcement through the time period has been hotly debated (Heyman & Luce, 1979; Rachlin *et al.*, 1981; Prelec, 1982; Hinson & Staddon,, 1983; Houston, 1983). The general conclusion of this debate is that in many circumstances following the matching rule is almost equivalent to maximizing total reward, but the details of the reinforcement schedule can have an important influence on whether or not this holds.

Herrnstein and Vaughan (1980) and Vaughan and Herrnstein (1987) proposed a moment-to-moment rule that might produce matching as an overall outcome: the rule is called 'melioration'. Consider a series of short time intervals. An animal following the melioration rule allocates more behavioural responses in the next time interval to the alternative that has offered the higher rate of reward in the preceding interval. In other words, melioration is a rule whereby the animal adjusts its allocation of behaviour until it reaches an equilibrium in which the rate of reinforcement attained in both alternatives is equal, formally analogous to the 'ideal free distribution' of distribution of competitors (see Chapter 5). Perhaps the most striking demonstration of this local equilibrium-seeking behaviour comes from studies of choice between schedules of reinforcement in which melioration leads the animal to move far away from the global optimum of maximizing overall reinforcement rate. For example, if the animal has to respond to option A (which itself sometimes gives rewards) to 'set up' rewards at option B it ought, in order to maximize overall rate of reinforcement, to choose A most of the time and visit B simply to collect the rewards as they are set up. In actual fact, animals choose B most of the time (as predicted by melioration) and thereby lose most of the potential rewards they could obtain (Vaughan & Herrnstein, 1987). Melioration is a 'molecular' (moment-to-moment) rule of choice, but it still does not specify exactly how experience determines choice. It does not specify the exact time interval over which local rates are measured, nor does it specify how past experience is stored in memory to influence future choices.

Various 'dynamic' models of choice have been proposed which attempt to take the analysis of choice to this finer level of detail (for example, Bush & Mosteller, 1951; Myerson & Miezin, 1980; Hinson & Staddon, 1983; Gibbon et al., 1988; Staddon & Horner, 1989). Although the details vary, two features of many such models are that the animal has an estimate of rate of reinforcement from each option stored in memory (for example, as an exponentially weighted average of past experience, or as a remembered distribution of intervals between reinforcements) and at each choice point it responds to the option with the higher expected rate of reinforcement or shorter expected delay to reward.

4.6.2 *Application of operant theories to foraging decisions*

Probably the most useful level of analysis for application to foraging behaviour is the third of the levels listed above, because this actually specifies rules governing choice and how they are related to past experience. An example is the study of Shettleworth et al. (1988) (see also Tamm, 1987; Inman, 1990). These authors studied 'sampling' as a problem in foraging. The animal (in Shettleworth et al.'s experiment, a pigeon in a large Skinner box) had a choice between two patches. One of these offered a constant, low rate of reward per response, whilst the other fluctuated at random between a good and a bad state. In the good state it was much better than the constant alternative, whilst in the bad state it was much worse. This might be considered as a laboratory analogue of the general problem of how a predator should sample parts of its home range to track local changes in food availability. 'Sampling' in this context means attending to the fluctuating place when on the last visit it was in its bad state. Stephens (1987) analysed sampling in terms of its costs and benefits; the benefit is reflected in extra rewards obtainable by detecting the change to the good state of the fluctuating patch, whilst the cost is reflected in lost opportunities to forage in the stable alternative whilst sampling. Shettleworth et al. manipulated these costs and benefits by altering the relative values of the good state of the fluctuating patch and the stable patch. They compared the response of pigeons both with the Stephens' model, based purely on costs and benefits of foraging decisions, and with Gibbon et al.'s (1988) model based purely on mechanisms of choice. Both models qualitatively explained many aspects of the data. However, Gibbon et al.'s, model gave a better quantitative fit and accounted for more of the qualitative trends. One possible conclusion to be drawn from this study is that the actual choice rule is similar to that proposed by Gibbon *et al.* and that one selective advantage of this choice rule is that it approximates the behaviour predicted by a cost–benefit analysis of sampling.

4.7 Conclusions

4.7.1 'Tests' of optimality models: what do they mean?

Testing optimality models is not an end in itself, but a means to gaining insight into the behaviour of individuals. In any particular study using optimality modelling a 'negative' result means that the constraint and/or currency assumptions must be modified, whilst a 'positive' result means that the constraint and currency assumptions have not so far failed to account for observed behaviour. However, a 'positive' result never means that *all* details of behaviour have been explained, so that the way forward is to try to account for the remaining variance that is not explained by the initial model. This was illustrated in the starling study, whereby chipping away with successive optimality models, more was learned about the details of starling behaviour even though in some respects the original model (the marginal value theorem) became obsolete as experiments proceeded. At a more general level, the accumulation of evidence from many studies using either simple rate-maximizing or more complex models has led to a growth in confidence in the optimality approach in general as well as in the value of the particular constraint and currency assumptions embodied in the majority of models used (Stephens & Krebs, 1986).

The understanding of behaviour that arises from optimality modelling could be of value in one or more of three ways: (i) optimality models might provide general organizing principles (general rules) for predicting and describing behaviour and a coherent framework for designing experiments; (ii) optimality models might be valuable as a device for investigating psychological or physiological mechanisms underlying behaviour and their interaction with currencies, as for example illustrated in the starling example described earlier; and (c) in the particular case of resource exploitation discussed in this chapter, optimality models of individual behaviour might provide a basis for ecological models of population and community level phenomena.

4.7.2 Ecological consequences of individual decisions

Ecological phenomena such as birth and death rates, species diversity, population or community stability and competitive exclusion are ultimately the consequence of individual behaviour. Therefore a full understanding of these phenomena depends on understanding how they are affected by individual decisions (Hassell & May, 1985; Schoener, 1986; Murdoch & Stewart Oaten, 1989; see also Chapter 5). Three examples serve to illustrate this point.

1   McNamara and Houston (1987b) show how the optimal trade-off between body reserves and danger of predation can lead to surprising consequences for mortality rates at the population level. A frequent debate amongst population ecologists is whether food or predators limit population size (Sinclair, 1989). McNamara and Houston show that even if food is the limiting resource, most individuals in a population may die from predation. This is because individuals following the optimal policy with respect to body reserves regulate their reserves at a level where they expose themselves to the danger of predation but avoid starvation. As food becomes

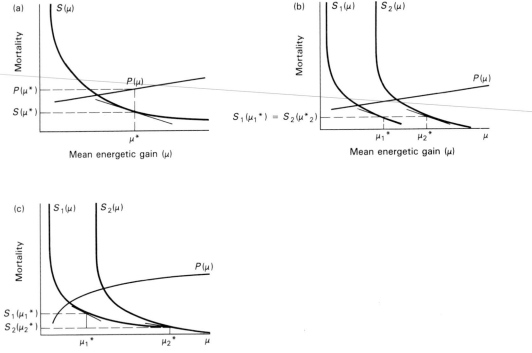

Fig. 4.5 Graphical representation of the McNamara and Houston (1987b) model of the optimal trade-off between starvation and predation. (a) The optimal level of food intake (μ^*) occurs where the rate of increase in predation mortality exactly balances the rate of decrease of starvation mortality, that is to say, where the slopes of the starvation and predation functions ($S(\mu)$ and $P(\mu)$, respectively) are equal and opposite (the slope of the $S(\mu)$ line is indicated by a tangent to the curve). This optimal point is indicated by the value of energetic gain μ^*. The corresponding starvation and predation rates are $S(\mu^*)$ and $P(\mu^*)$. (b) Increased environmental harshness can be depicted by moving the $S(\mu)$ curve to the right ($S_1(\mu)$ to $S_2(\mu)$). When the slope of the $P(\mu)$ line is constant, μ^* increases ($\mu_2^* > \mu_1^*$) but the value of $S(\mu^*)$ does not change. (c) When $P(\mu)$ is a decelerating function, increased environmental harshness results in a decrease in $S(\mu^*) \cdot (S_2(\mu_2^*) < S_1(\mu_1^*))$.

scarcer, starvation does not necessarily increase but predation may (Fig. 4.5).

2   Bernstein *et al.* (1988; in press) analyse how individual decisions based on learning can affect the spatial distribution of prey or host mortality. Spatial variation in mortality is thought to be an important factor affecting the stability of predator–prey and parasitoid–host interactions (Pacala *et al.*, 1990). Bernstein *et al.* show that learning by individual predators can lead to lags in predator redistribution that affect the spatial distribution of mortality. Similarly, the cost of migration between patches and the 'grain of the environment' affect individual decisions in a way that can produce a range of patterns of spatial distribution of mortality.

3   van Alphen and Visser (1990) explain 'superparasitism', an at first sight puzzling behaviour of many insect parasitoids, in optimality terms. 'Superparasitism' occurs when a parasitoid lays an egg on an already parasitized host; this reduces the chance of survival of the larva as a result of competition. It would at first sight, therefore, seem adaptive for a parasitoid to avoid superparasitism, especially as it is known that parasitoids can detect the presence of a prior egg or larva. However, if the parasitoid is 'foraging' with a limited time horizon (for example, it is unlikely to reach another patch in its lifetime), it can be shown that the optimal rule is to avoid superparasitism early in life, but to accept it later on (see also section 4.4.2). As van Alphen and Visser show, this can have significant consequences for the spatial distribution of parasitoid searching effort between patches and hence an impact on the stability of parasitoid–host interactions.

These three examples illustrate the beginning of what we expect to see as a major growth area in the next 5 years: the integration of individual decision-making into population and, eventually, community ecology.

4.7.3 Criticisms of the optimality approach

As mentioned earlier, the optimality approach, used not only in this chapter but throughout this book, has been strongly criticized by some authors (Gould & Lewontin, 1979; Lewontin, 1979); for example, Pierce and Ollason (1987) entitle their article 'Eight reasons why optimal foraging theory is a complete waste of time'. These criticisms and responses to them are discussed in detail by Stephens and Krebs (1986), Stearns and Schmid-Hempel (1987), Dupré (1987) and Krebs and Houston (1989). In brief, four points are worth noting:

1   Much of the criticism is aimed at 'naïve' optimality modelling in which, according to the critics, a 'Panglossian' approach is taken in which it is assumed that organisms are free to evolve any adaptation specified by an optimality model. Clearly, natural selection is not free to go in any direction; it can work only on the material at hand. Therefore phylogenetic, ontogenetic and other constraints need to be taken into account. As we have seen in

this chapter, 'sophisticated' optimality models do indeed take into account constraints and do not have a Panglossian flavour.

2   According to some authors (for example, Gray, 1987), the evidence from studies of foraging do not support the predictions of optimality models. Most of this argument boils down to the question of what constitutes 'support'. For example, Gray argued that the optimal diet model predicts an all-or-nothing switch from non-selective to selective foraging, a result that has not been observed in any study. However, this is only one version of a diet model, the simplest one. As discussed earlier in the chapter, there are many possible testable modifications of the original model which would predict partial preferences (Rechten *et al.*, 1983; Stephens, 1985; McNamara & Houston, 1987a; Recer & Caraco, 1989).

3   A third criticism is that modifications of the original models (for example, such as those which would account for partial preferences) are *ad hoc* and reduce the approach to a circular argument. However, since each modification in itself generates testable predictions, this criticism does not seem to have much substance.

4   Finally, some authors (for example, Ollason, 1980) argue that optimality models are unnecessary because a descriptive account of mechanism could equally well explain behaviour. Mechanistic and functional accounts of behaviour are not, however, alternatives, but complementary ways of understanding individual decisions. Further, as a research strategy, functional models, as used in all the chapters of this book, provide an efficient tool for generating organizing principles around which to explain existing data and generate predictions for new observations or experiments.

5: Competition for resources

Manfred Milinski and Geoffrey A. Parker

5.1 Introduction

One of the essential components of Darwin's theory of natural selection is that only a small fraction of each generation survives until maturity as a result of competition for scarce resources. The winners in the struggle for existence are therefore most probably individuals with superior competitive abilities. An individual's competitive success is determined not only by its physical strength but also by the behavioural strategies it uses when competing with others. The best strategy to adopt often depends on what its competitors do. Our aim in this chapter is to review how competition for resources affects an individual's choice of strategy. We first describe how competition arises and then consider the behavioural rules that natural selection may have favoured in these situations.

5.1.1 *How competition arises*

Imagine two birds searching for insect larvae in a meadow. If the foraging success of one bird is reduced when a competitor exists, then the two birds must have interfered in some way. Interference can be defined generally as the reduction of a given individual's intake rate as a result of competition. It may arise through a number of different mechanisms, such as simple exploitation (called 'pseudo-interference' by Free *et al.*, 1977), scramble or contest interactions.

Both birds may have a lowered success rate because they cannot avoid foraging sometimes in areas that have already been exploited by the competitor. If the prey items are randomly distributed over a large area, the two birds do not often suffer from exploitation competition (Fig. 5.1a). This is different when the prey density is higher in some areas than in others, and the birds can recognize the most rewarding high density patches (Fig. 5.1b). Now the probability that both of them will try to exploit the same patch is increased. Thus, if the resource items have a heterogeneous distribution in the habitat, the probability of competition is enhanced (for example, Monaghan & Metcalfe, 1985). Of course each bird's foraging success would drop further with increasing number of competitors in the same patch (Fig. 5.1c). As long as there is some spatial distribution of resource items within a patch, the competitors need not interfere directly with each other over each single item, especially when they have started to exploit a rich patch. But if several competitors decide simultaneously to eat the same item, only one can be successful, at the expense of the others. In such scramble competition as opposed to contest competition the competitors do not interact aggressively with each other—everybody tries to be the fastest. The unsuccessful competitors suffer not only from the exploitation effect when someone else has been faster but also from having wasted time and energy to catch the item. Many competitors deplete a patch faster than a few do, thus the probability that two or more individuals interfere over the same items is increased.

Renewable resources are also distributed in time, for example by hatching or migration. The highest potential for competition occurs when resource items are densely clumped in space, and are evenly distributed

Fig. 5.1 Distribution of resource items in space and time. (a) Resource items are widely dispersed in space. Any two foraging birds do not suffer much from competition because they are unlikely to exploit the same area. (b) Resource items are clumped in space and occur in discrete patches in the habitat. Now the probability that both birds will attempt to exploit the same patch is increased. (c) In this patchy habitat, each bird's foraging success would drop further with increasing number of competitors in the same patch. (d) Birds are fed at a given point in space; therefore resource items are highly clumped in space and are evenly distributed through time. Each bird interferes with every other bird over each single piece of bread the woman drops.

through time. An example would be when all resource items appear from one spot, one after the other at a constant rate. Since we are talking about scarce resources, this continuous steady input of resource items must not exceed demand. Imagine someone feeding tiny pieces of bread to a flock of sparrows by throwing one piece after the other. Here each bird interferes with each other bird over each single piece of bread (Fig. 5.1d); on every occasion only one is successful, the rest are not. This would be different if the food items were clumped in time. When all the food is thrown at once, interference is dramatically reduced. Now each bird has a probability of almost 1 of consuming the next item it tries to pick up—until the bread becomes significantly depleted. In fact, were it not for this depletion, there would be little if any competition. Thus, clumping of resource items in space increases competition whereas their clumping in time reduces competition. Experiments on the interaction of these effects would be valuable.

Until now we have assumed that resource items are all alike, which is rarely so under natural conditions. The pieces of bread thrown to the sparrows would certainly vary in size and thus in profitability. A meadow may contain not only differently sized larvae of the same species, but also other arthropods of varying profitabilities for bird predators. Optimal diet theory (for a recent review see Stephens & Krebs, 1986) predicts—depending on a number of variables—that the predator should sometimes concentrate on the most rewarding prey type only, and sometimes that it should eat several different prey types unselectively to maximize its net energy intake (see Chapter 4). Obviously competition is enhanced when all the competitors concentrate on the same small fraction of the resource spectrum (Schoener, 1974), for example the most profitable prey size.

Also the competitors are usually not equal. Some may be stronger, quicker or more experienced than others, perhaps simply because they are older, or have experienced some simple chance advantage. Being competitive in one situation need not correlate with competitive advantage in another: for instance, when competing for food a male with extensive secondary sexual ornaments such as a long tail may be handicapped, whereas he may be very successful when competing for choosy mates (Møller, 1989). There are many reasons why we should expect competitors to vary in their ability to compete for a specific resource. Not only the ability but also the motivation to compete may vary among different individuals. If one of the sparrows the woman is feeding (Fig. 5.1d) is hungrier than its competitors, it may try harder to grab the next piece of bread—probably at increased costs in other respects, such as risk of predation. Better or more motivated competitors may suffer less severely from competition than poorer or less motivated competitors.

5.1.2 Concepts about the outcome of competitive interactions

In the simplest form of competition, exploitation, the competitors do not need to meet or even see each other (Fig. 5.1a). Each competitor can use optimal foraging rules to decide whether and for how long to exploit a resource, irrespective of whether the resource has been exploited already or was poor anyway. In scramble competition, the competitors see each other exploiting the same resource. Each competitor can increase their effort to get as much as possible of the resource only by trying to exploit it. They may increase their effort dependent on what their competitors do. No aggression is involved. If nobody is prepared to fight for a resource all competitors are free to enter a patch and to engage in the scramble competition going on (Figs 5.1b–d). As a consequence, each competitor should go where the expected gain is highest. If two people throw bread at similar rates, then each bird arriving anew should join the one that is surrounded by the smaller number of birds. As we shall later see, no bird can do better by deviating from this rule. However, the outcome is difficult to predict when the competitors have different competitive abilities.

If one bird can chase any other bird away from the point at which a woman throws the pieces of bread, and if the costs of chasing (and sometimes fighting) are lower than the benefit of monopolizing all the woman's bread, then this woman is an 'economically defendable resource'. The bird would gain by becoming a despotic territory owner. This type of interference is called 'contest competition'. The concept of economic defendability of a territory was developed by Brown (1964). It has been shown to describe a number of competitive outcomes rather precisely (for example, Gill & Wolf, 1975; for a review see Davies & Houston, 1984).

The costs of fighting are not a fixed quantity: they will depend critically on the strategy played by one's opponent. The appropriate logic for analysing such situations (and all others in which fitness pay-offs depend on the strategies played by other competitors) is that of the evolutionarily stable strategy, a concept introduced by Maynard Smith and Price (1973), and which is derived from the mathematics of game theory. A strategy is an evolutionarily stable strategy or ESS if, when fixed in a population, there is no alternative strategy that can invade as a rare mutant. The formal rules of strategic decisions in pair-wise contests have been investigated in depth in recent years. Briefly, contests may be symmetric (opponents are exactly equal) or they may be asymmetric (opponents differ in one or more features, such as fighting ability, who was first in the territory, or in the value of the resource to each). When there are asymmetries present in a contest, these may be defined as either 'pay-off relevant', for example fighting ability or resource value, or 'pay-off irrelevant' or 'arbitrary', for example arrival order (Maynard Smith & Parker, 1976; Hammerstein, 1981). In general there is a higher likelihood for outcomes of contests to be 'common-sense'—for example, the individual with the higher fighting ability or with the higher resource value wins—than 'paradoxical' (for example, the individual with the lower fighting ability or with the lower resource value wins), although paradoxical solutions are possible under some circumstances. An extensive literature now exists on pair-wise contests, which are mathematically complex and beyond the scope of the present chapter. The reader is referred to Maynard Smith (1982).

Contests are likely to be restricted to situations when resources cannot be shared, i.e. when they are economically defendable units. In this chapter, however, we concentrate on the outcome of competitive interactions when resources can be shared between competitors, especially for cases where competitors have different competitive abilities.

5.2 The ideal free distribution

The term 'ideal free distribution' was first proposed by Fretwell and Lucas (1970) and Fretwell (1972). It is one of a series of terms they used to describe the possible distribution of animals in a patchy habitat. If all individuals were 'free' to move to alternative patches without any constraint or restriction, then 'ideally' each individual goes to the place where its gains will be highest. However, as more individuals move in, the value of a habitat patch to each individual would eventually decline. So if all of the competitors were to pour into the best resource patch, each of them may fare badly; it would pay some of them to move to the next best patch, and so on. The resulting distribution is one in which several (or even all) of the patches may be exploited, with more individuals in the better patches. The number of competitors in each becomes balanced so that each competitor does equally well, whatever the patch, and no individual can profit by moving elsewhere. This equilibrium is the ideal free distribution.

Other biologists independently achieved essentially the same insight at around the same time. Orians (1969) argued that females would distribute themselves amongst male territories (resource patches for the females) in a manner that is clearly ideal free in concept. Parker (1970; 1974) used the term 'equilibrium position' to describe the expected spatial and temporal distribution of male dungflies around cow droppings (patches where males search for females). His assumptions were exactly parallel to those made independently by Fretwell, and the observations conformed well with the predicted 'equilibrium position' (i.e. ideal free distribution). Brown (1969) looked for the optimal distribution of great tits between two habitats. Although his optimization criterion was the maximization of the total production of the population, some qualitative aspects of his predictions resemble those of ideal free. Since its inception two decades ago, ideal free philosophy became one of the central concepts of behavioural ecology, and one that is currently expanding rapidly, both theoretically and empirically.

5.2.1 Assumptions, predictions and scope

In testing any biological model, the assumptions must be matched as well

as the predictions; there is a tendency for empiricists to concentrate on predictions. The assumptions of the ideal free model are as follows: (i) a number
of individuals (the competitors) of identical competitive ability are found in
a habitat; (ii) the habitat contains a series of resource patches that vary in
fitness value to the competitors; (iii) the competitors can move freely without cost or constraint to any patch within the habitat. Each competitor goes
where its expected gain is highest; and (iv) the fitness value of a given patch
to any given competitor declines (at least eventually) with increase in
the number of individuals exploiting that patch. In most ideal free models,
it has been assumed that there is a continuous, monotonic decline in fitness with numbers in the patch, mainly in the interests of mathematical
tractability.

The predictions of the ideal free model are as follows: (i) all competitors
experience equal gains, whatever patch they are in; and (ii) average gain
rates are equal in all patches.

It is clear that most real systems will violate some of the above assumptions, and we shall later explore some of the consequences of altering these
assumptions. Nevertheless, we shall proceed with faith in the heuristic
value of the theory: competition in nature for limited resources appears to
approximate closely enough to the assumptions to make ideal free a concept worthy of further consideration. The scope indeed appears to be very
broad, with potential applications from water fleas (Jakobsen & Johnsen,
1987) to people (Krebs & Davies, 1987).

So far, the model has been couched in terms of space—how individuals
should be distributed in space at a given time. A strategy is an animal's
choice of where to forage at a given time. In addition, we can use versions
of ideal free theory to predict how individuals should be distributed
through time in a given place, for example a strategy may be how long to
stay in a given place. At its most general level we can use ideal free theory
to make predictions about any strategic choice in which the fitness consequences of adopting a particular strategy decline with the number of
other individuals exploiting the same strategy. The concept is generally a
rather robust one with wide potential application, as we shall see.

5.2.2 The simplest model for equal competitors

Consider N equal competitors in a habitat containing several resource
patches. Resource items arrive into each patch at different rates and are
consumed immediately as they arrive by the competitors waiting there;
there is a scramble competition for each item that arrives. Suppose that in
each unit of time, Q_1 resource items arrive into patch 1, Q_2 arrive to patch
2, Q_3 to patch 3, etc. We want to predict the number of competitors in each
patch: n_1 in patch 1, n_2 in patch 2, n_3 in patch 3 . . . such that
$N = n_1 + n_2 + n_3$ . . . Milinski (1979; 1984), Harper (1982) and Godin and
Keenleyside (1984) have studied this situation experimentally.

We now define the number of resource items gained by each competitor in patch i. This gain, G_i, is a function of (i.e. depends on) the number of competitors in the patch (it is lowered by increasing n_i), and also a function of the input rate to the patch (it is raised by increasing Q_i). If the ideal free distribution applies, the gains must be equal for each competitor, whatever patch it is in. If this is not so then it will pay competitors to move from patches where gains are lower to the patch where they are highest. So the ideal free distribution will satisfy:

$$G_1 = G_2 = G_3 \ldots = G_i = C \text{ (constant) for all patches} \tag{1}$$

This is the 'fitnesses must be equal, whatever patch' rule: gain is assumed to be proportional to fitness. In other words, the average gain rate, G, to each individual in a given place (call it patch i) will be equal to the input rate, Q, in that place divided by the number of competitors. That is:

$$G_i = Q_i/n_i$$

and since we know from equation (1) that G_i is constant for all patches,

$$Q_1/n_1 = Q_2/n_2 \ldots = Q_i/n_i = C$$

i.e. each competitor does equally well whatever the patch. Thus the number of competitors in a given patch i is:

$$n_i = Q_i/C \text{ for all i} \tag{2}$$

Stated verbally, we expect the number of competitors in each resource patch to be directly proportional to the input rate to that patch. It has been termed the 'input matching rule' (Parker, 1978) or 'habitat matching rule' (Pulliam & Caraco, 1984).

This is really the simplest ideal free model: it has been termed the 'continuous input' model (Sutherland & Parker, 1985; Parker & Sutherland, 1986) to describe the concept of continuously arriving resource items which are consumed instantly by the waiting competitors. There are several examples, both natural and experimental, in which the ideal free prediction (2) appears to be met (Fig. 5.2).

These are merely three of many (for reviews see Parker & Sutherland, 1986; Milinski, 1988). However, in each case there is a basic violation of the assumptions: differences in competitive ability or dominance cause individuals to experience consistent differences in pay-off (Fig. 5.3). Thus, although the overall prediction is met that *average* gain rates will be equal in all patches, the prediction that *all individuals* should experience equal gains is not—all competitors are not equal.

5.2.3 *The continuous input model with unequal competitors*

What happens if we change the model to take account of the fact that the competitors will differ? How will the presence of a mix of phenotypes of

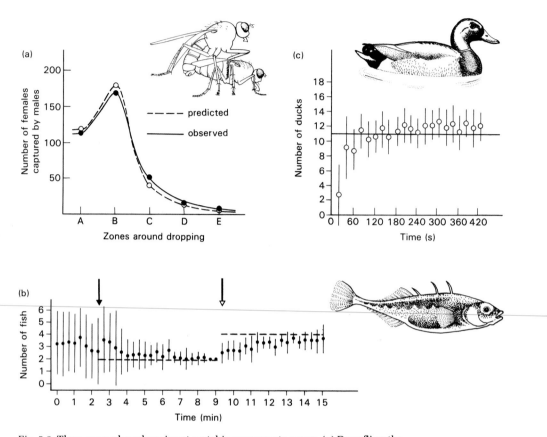

Fig. 5.2 Three examples where input matching appears to occur. (a) Dungflies: the observed number of females captured (●——●) by male dungflies in a series of zones A–E on or around a cow dropping fits well with the number from the input matching rule (○– – –○) (from Parker, 1978). (b) Sticklebacks: 6 fish in a tank consume Daphnia introduced at two ends with an initial profitability ratio of 2:1; after the start of feeding (first arrow) the observed number of fish in patch 2 quickly approaches the number predicted from the input matching rule (indicated by dotted lines). At the second arrow, the patch profitabilities are reversed, and the fish distribution soon approaches the new prediction (from Milinski, 1979). (c) Ducks: in a similar experiment in which 33 ducks were fed pieces of bread from two stations (with profitability ratio of 2:1) around a pond, the number of ducks found at the poorer quickly approaches that predicted by the input matching rule (from Harper, 1982).

differing competitive abilities affect the gain of a given individual in a given patch? It is clear that there is no simple answer to this problem. Consider an individual of phenotype A in patch i. Its gains will depend on the number of each of the possible phenotypes (A, B, C, D, . . . etc.) also in patch i. But there is no obvious explicit expression that could account for all the possible ways in which varieties of numerical mixtures of A, B, C, D, . . . etc. individuals could affect the pay-offs of our chosen individual of phenotype A.

One approach to this problem has been suggested by Parker & Sutherland (1986). They ascribe what they termed 'competitive weights' to each

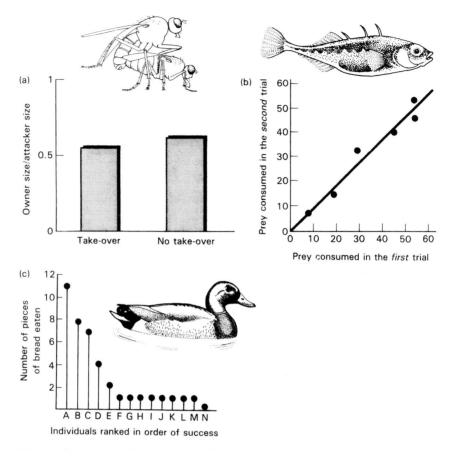

Fig. 5.3 The examples of apparent fit to the input matching rule shown in Fig. 5.2 all violate the rule because some individuals were consistently more successful than others. (a) Dungflies: bigger males tend to have an advantage in 'take-overs', i.e. a struggle in which an attacking male takes a female away from its original owner. The probability of take-over depends on the relative sizes of the two male flies. Struggles for a female usually occur only when the attacking male is larger than the owner. When there is a take-over, the ratio of the two male sizes (owner/attacker) is significantly lower than when there is not (drawn from data in Sigurjònsdòttir & Parker, 1981). (b) Sticklebacks: in experiments of the type in Fig. 5.2b, each of the six sticklebacks caught a similar number of prey items in one trial as in the next trial. Differences in pay-offs between individuals remained consistent across trials (from Milinski, 1984). (c) Ducks: in the experiment in Fig. 5.2c, in one patch some of the ducks were far more successful than others at snatching up pieces of bread; again, pay-offs were not equal (from Davies, 1982; Harper, 1982).

phenotype. Pay-offs are in some way related to the competitive weight of an individual relative to the average or total competitive weight of all individuals in the same patch. For example, suppose that there are only two phenotypes A and B, and that each individual A has a competitive weight of 2 and each individual B has a competitive weight of 1. Thus, each individual A is twice as successful as each individual B in a given patch. If there are two individuals of phenotype A and four of phenotype B in a patch, then the total competitive weight of all individuals in that patch will be 8. If we assume that each competitor's share of the input of resource items is equal to its share of the total competitive weight, then the gain achieved by an individual of phenotype A will be 2/8 and each B will get 1/8 of the input to that patch.

Parker and Sutherland analysed this model of unequal competition using the ideal free rule that any competitor must be free to move to any patch where the gains are higher. They also used the principle that if individuals of the same phenotype were to be found in more than one patch, they must achieve the same pay-off whatever the patch or they would move (intuitively: individuals of the same phenotype are *equal* competitors and must therefore obey the rule for ideal free with individuals equal). They obtained two quite distinct types of result, depending on whether competitive weights remain constant (e.g. the competitive weight of phenotype A is the same in patch 1, 2, . . . etc.) or whether they change between patches. More strictly, it depends on whether the *relative pay-offs* (= relative gains) remain constant (W.J. Sutherland & G.A. Parker, unpublished analysis); in the present model relative gains remain constant if competitive weights remain constant. We shall deal separately with the two cases.

RELATIVE PAY-OFFS STAY CONSTANT

What do we mean by this? If, say, phenotype A does twice as well as phenotype B in patch i, then it must do twice as well (whatever the type and number of competitors) in patch j, and also in patches k, l, m, . . . etc. The solution here is that there are many possible equilibrium distributions of competitors. Where competitive weights remain constant across patches, the feature common to each equilibrium is that the ratio of input/total competitive weight is equalized for all patches, i.e. competitors must be added to a patch until the total competitive weight exactly balances the input rate for each patch.

It is easiest to demonstrate the logic of this conclusion with a hypothetical example based on the stickleback experiments (Fig. 5.4). There are just two phenotypes: big and small. In all cases, the big fish have twice the competitive weight of the small fish. Figure 5.4 shows various alternative equilibrium distributions. Each obeys the conclusion above because the total competitive weight on the left is always twice that on the right, balancing the input rates. Each is stable in the sense that it cannot pay any fish to shift unilaterally to the opposite end of the tank, and big fish always gain a pay-off of 2 wherever they wait, whilst small fish always gain a pay-off of 1.

However, the *average pay-off per individual* in each site is not equal in each case, except in distribution (c). The maximum discrepancy between the average pay-offs occurs in (a), where on the left it will be 2 versus 1 on the right. Now we come to the problem. The ideal free prediction for equal competitors was that the average pay-off should be the same whatever the patch. Many real examples seem to fit this prediction, although they clearly also show that competitors are unequal. So is there some reason to suppose that distributions of type (c) are much more likely to be achieved than any of the other equilibria?

The characteristic of the type (c) distribution is that the frequencies of phenotypes in each patch are the same as the frequencies of phenotypes

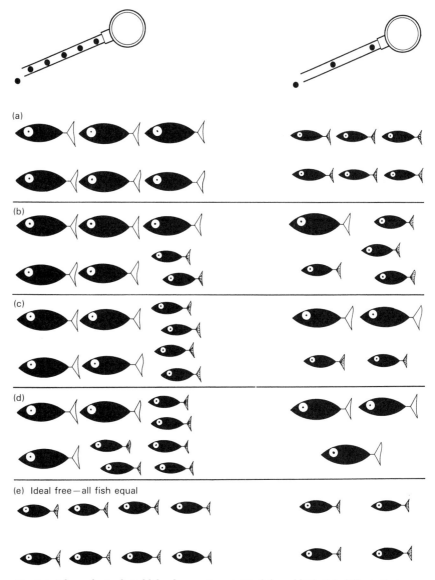

Fig. 5.4 A hypothetical stickleback experiment. Big fish and little fish differ in their competitive weights such that each of the six big fish always captures twice as much as each of the six little fish. For further explanation see text (modified from Sutherland & Parker, 1985; Milinski, 1988).

in the entire population, and such a distribution must always form part of the equilibrium set. Parker and Sutherland suggested that a possible mechanism for achieving these type (c) distributions is that the competitors distribute themselves randomly with respect to phenotype, and simply match numbers with input rate; plausibly, each competitor may be quick at assessing the input rate of a patch and the competitor density, but much poorer at assessing its relative competitive weight. Alternatively, competitors distribute ideal free fashion with respect to their own competitive rank, ignoring all other ranks (N.B. Davies, personal communication). Furthermore, if a food item is competed for only by the nearest two individuals in a patch, then a type (c) distribution is the unique ESS (Korona, 1989).

Houston and McNamara (1988) analysed Parker and Sutherland's proposition by calculating the exact probabilities of each possible distribution in models with varying total numbers of competitors. They sought the number of ways in which individual good and poor competitors could be assigned to the two patches for each of the solutions (a) to (d). For example, suppose that each fish in solution (c) has a name. How many different mixtures of names are possible under the solution (c) condition, that four good and four poor competitors have to be in the better patch? The answer is 225. There are 20 different ways for solution (d), 90 for (b) and one for (a). Thus solution (c) is the one most likely to occur by chance. Average pay-offs in better and poorer sides respectively would be for solution (a) 2:1, for (b) 1.714:1.2, for (c) 1.5:1.5, and for (d) 1.33:2. If we were to repeat the experiment a huge number of times, the weighted mean pay-off (probability times pay-off, summed over the four solutions) to better and poorer sides would be 1.54:1.45, i.e. very close to the expectation of 1.5:1.5 for equal competitors, and probably indistinguishable from it in practice. In a further stickleback experiment (Milinski, 1984a) solution (c) exactly was in fact predominant (Milinski, 1986).

Sometimes there will be only one distribution in which the sums of competitive weights match the input rates to the alternative patches. This is most likely when there are few competitors and each has a different competitive weight. Suppose, for example, that the competitive weights of six fish have values of 1, 1.3, 1.9, 2.5, 2.9 and 3.6 respectively. In this example input matching in the two patches can be achieved by one distribution only, i.e. in the poor patch, 1.9+2.5 (= 4.4); in the good patch, 1+1.3+2.9+3.6 (= 8.8). With few competitors, it will usually be impossible to balance input rate and total competitive weight exactly in a patch. Some imbalanced systems can clearly lead to stable distributions; others may not.

Is there any evidence that relative pay-offs remain constant across patches in real situations? If they do, we can probably explain why the data show a superficial fit with the predictions of ideal free for equal competitors, even though we know that competitors are generally not equal.

Although there is some evidence for constant relative pay-offs in stickle-backs (Milinski, 1986), experiments on goldfish indicated that relative pay-offs of individuals differed between patches (Sutherland *et al.*, 1988). Rigorous experimental tests of the constancy of relative pay-offs are badly needed. Whether these differ or remain constant across patches is perhaps more likely to reflect ecological properties of food distribution rather than species differences, but, as we shall now see, this will exert a major influence on the distribution of phenotypes across patches.

RELATIVE PAY-OFFS CHANGE ACROSS PATCHES

In contrast, it seems quite possible that relative pay-offs may not remain constant. For instance, phenotype A may do twice as well as phenotype B in one patch, but only 1.5 times as well in another patch. The effect of changing relative pay-offs leads to a quite different sort of solution. In the models they studied, Parker and Sutherland found that typically there was just one stable equilibrium distribution, not a set of equilibria. In what ways might we expect relative pay-offs to change across patches? There are clearly a vast series of biological possibilities, but we shall first consider just one, chosen now for its mathematical simplicity. It is again a continuous input model like the fish, fly and duck examples used above.

Imagine two sticklebacks waiting for food, which is drifting towards them. Both scan the water volume in front of them. The smaller fish (with lower competitive weight) scans a smaller domain, which is entirely within the domain of the larger fish. In the domain of overlap, the two fish scramble to obtain food items, but the larger fish gains extra items from the non-overlapping part of its domain. In another patch, the visual range of the larger fish is constrained by the turbidity of the water, so that the ratio of their pay-offs is more similar. Hence the ratio of the pay-offs of the two fish changes across patches. It seems quite possible that the ratio of pay-offs within each patch is unaffected by increasing the number of competitors, although this will of course affect the absolute pay-offs.

These are precisely the assumptions of the model we are now about to investigate, but for this model we further assume that there is a very large number of competitors. Each differs in some feature (e.g. size) that is related to competitive ability. The frequency distribution of size (and hence competitive weight) can be represented as a continuous probability distribution because of the large number of competitors. Figure 5.5a shows a purely hypothetical distribution of size for a population of competitors in a habitat. Suppose that there are three patches (1, 2, 3) in this habitat. (The model can be extended to include any number of patches.) Figure 5.5b shows how size is correlated with competitive weights in each of the three patches. Imagine first that each patch contains the full range of sizes. The smallest individual in each patch is arbitrarily assigned a competitive weight of 1, and the lines show how the competitive weight increases with size in a given patch. In patch type 1, the slope of the line is not very steep, i.e. the competitive weight of the largest competitors is not much different from that of the smallest competitors. Patch type 2 is intermediate, and in patch 3 competitive weights change steeply with size so that the biggest competitors do much better than the smallest. Finally, assume (as in the previous continuous input models) that an individual's pay-off is equal to

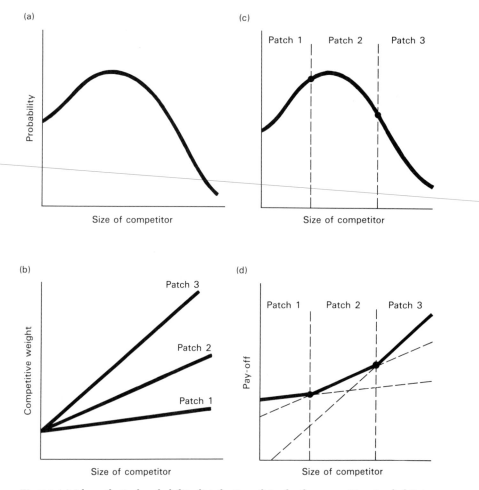

Fig. 5.5 (a) A hypothetical probability distribution of size for the competitors in a habitat. (b) The competitive weight always increases linearly with size in each of patches, 1, 2 and 3 in the habitat. However, the slope of each line differs across patches; it is steepest in patch 3 and least steep in patch 1. Bigger competitors have a relatively higher competitive weight in patch 3 than in patch 1. In the model described in the text, the relative competitive weights in a given patch are not affected by competitor density. (c) The solution to this model is that the size distribution is split up (truncated) among the three patches, with the smallest competitors in patch 1 and the biggest in patch 3. (d) The relationship between pay-off and size in each of the three patches at the ESS (solid line). Notice that the boundary phenotypes (solid dots) have the same pay-offs in each of the two adjacent patch types. The pay-offs of mutants (dotted lines) are always lower.

the input rate to the patch multiplied by self's competitive weight/the total competitive weight in the patch.

Let us summarize so far. In this model an individual's size is a fixed phenotypic character. Its relative competitive weight relates linearly to size. The *ratio* of pay-offs of any two given phenotypes in a patch is assumed to be independent of the number of competitors, but this ratio alters across patches. Of course, the *absolute* pay-off achieved by a competitor depends on the number of competitors in the patch, its phenotype (size) and the patch that it is in. The result of this model (Parker & Sutherland, 1986) is that the distribution of phenotypes should be truncated between patches (Fig. 5.5c), so that the smallest competitors occur in patch 1, where the effects of size are least important. Medium-sized competitors occur in patch 2. The largest competitors should be found in patch 3, the patch where the effects of size are most critical. This type of distribution is the unique ESS for this model. It has been termed 'the truncated phenotype distribution'.

How we arrive at such an ESS distribution of phenotypes among patches is outlined in Fig. 5.5d. Since the explanation of this figure has proven to be a rather complicated enterprise, we now shall take a short excursion (Fig. 5.6). Take the simpler case of just two patch types, 1 and 2. We shall be searching for the ESS point at which the range of sizes is split between the two patches. Call this boundary size S^*. Figure 5.6a shows what would happen if the size range was split at a boundary size, S^-, below S^*. Although larger individuals have relatively higher competitive weights in patch 2 compared with patch 1 (see Fig. 5.5b), their absolute pay-offs are lower in patch 2 in our example (Fig. 5.6a) because too many of them are found there. The total competitive weight in patch 2 is high so that pay-offs are relatively low. Because of this, any given individual in patch 2 shifting to patch 1, i.e. the broken line is moved to the right (open arrow), would profit markedly (closed arrow). Truncation of the size range at S^- cannot therefore be an ESS. Now consider the other extreme (Fig. 5.6c), at which the size range is truncated between patches at a size S^+, which is greater than S^* (see Fig. 5.6b). The opposite effect occurs: given individuals in patch 1 can profit by switching to patch 2, and this cannot be an ESS either. We achieve an ESS only when the boundary (or truncation) size is S^* (Fig. 5.6b). First notice here that the boundary phenotype, S^*, has the same pay-off whatever patch it occurs in; it neither profits nor loses by switching. Second, notice that any 'mutant' individual below or above size S^* will lose by switching to the other patch. Its pay-off would drop to the respective dotted line. At the ESS, it is clear that all dotted lines (which represent 'mutant' pay-offs) must fall below the solid line (which represents the ESS pay-offs for each size). Further, this ESS line is continuous because pay-offs must be equalized at the boundary phenotype.

Hopefully, Fig. 5.5d is now more readily understood as the ESS solution to the three patch game, although it could be expanded to cover *n* patches.

At the ESS, how will pay-offs vary in relation to the size and strategy (patch exploited) of a given individual? First, pay-offs will always increase with size (and hence competitive ability). They increase with size both within each patch, and across the entire set of patches, so that (in the present model) the slope of the 'pay-off line' increases as we go from patch 1 to patch 3. Second, we now have two boundary phenotypes (their pay-offs are represented by the large dots in Fig. 5.5d). The importance of the boundary phenotype rule is that it enables us to calculate exactly how the competitor distribution will be truncated for any given case for which we know the input rate to each patch and how competitive weight increases with size in

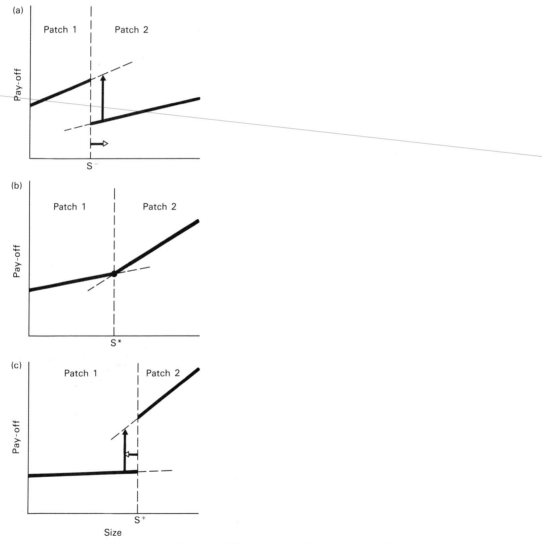

Fig. 5.6 Explanation of a truncated phenotype ESS (evolutionarily stable strategy) between two patches, for the type of model developed in Fig. 5.5. For further details see text.

each patch (Fig. 5.5b). Note that the patch qualities, in terms of input rates, tell us nothing about which patch to expect which competitors to be in. The model simply predicts that individuals with the highest competitive ability will be in the patch in which the effects of competitive ability are strongest, and the poorest competitors will be in the patch where the effects are weakest. In the next section, in a model where relative pay-offs in each patch are altered by competitor density, we find a plausible reason why the biggest competitors may occur in the best patches (highest input rate Q). But the present model does not formally include such an effect, because it assumes that the slope of each line in Fig. 5.5b remains constant whatever the competition in a patch.

Ideal free distributions with unequal competitors have neither the property that all competitors achieve equal pay-offs nor the property that the average pay-off in one patch is the same as in every other patch. A biological summary of the present model would be that we expect the phenotype distribution to be split up between patches, with the best competitors—and the highest average pay-offs— occurring in the patch where the competitive differences are greatest.

5.3 Interference

The term interference has been used in a variety of contexts to describe the effects that the density of competitors has on the rate at which resources are found and exploited. In a sense, interference can be seen as expressing the extent of competition: with zero interference there is zero competition. Interference can be defined generally as the reduction of a given individual's intake rate as a result of competition (see section 5.1). The continuous input model (see previous section) has the property that each resource item is taken up as soon as it is presented to the competitors in a patch. Interference here is such that for n equal competitors, the gains to each individual reduce as $1/n$. But this is only one possible magnitude of interference, as we shall see later. In nature, several situations, particularly in mate searching, may obey continuous input rules, but many food foraging competitions will not. In the present section we examine a more general way to describe and model competition for resources; it is one that allows the magnitude of interference to vary continuously.

5.3.1 Equal competitors

Interference may often be negligible. Imagine a shoal of fish in a river foraging for drift food items. The individual fish are well spaced out and do not interact, so that the intake rate of resource items by each fish is independent of the shoal size, and depends only on the density of prey items. It is clear that interference in such a situation is zero. When this is so, we should expect all the fish to forage in the patch where the prey density is highest. This offers the highest intake rate, and any fish moving to a less profitable patch (less dense in prey items) would do worse than the rest. Only when there is some interference which increases with competitor density might we expect some fish to move to less profitable patches. If all fish are equal competitors, the ideal free rule would now be that the reduction in interference must exactly offset the poorer prey density, so that all individuals do

---

Box 5.1 Sutherland's interference model

Assume that all competitors are equal. Let the gain rate G_i of an individual in patch i where there are n_i competitors be

$$G_i(n_i) = Q_i/n_i^m$$

Again, Q_i is a measure of the intrinsic quality (input rate) of patch i. In the fish example, it is proportional to the density of prey items in patch i. Now suppose that $m = 1$. Then $G_i(n_i) = Q_i/n_i$; in other words each individual's intake rate decreases as $1/n_i$, exactly as for the continuous input case. Suppose instead that $m = 0$. Then $G_i(n_i) = Q_i$, there is no effect of interference, and so whatever the number of competitors, the intake rate remains the same. Hence the above equation allows us to model the influence of interference continuously by varying m. Equally, keeping m constant (> 0), we can see that a given individual's intake will always decline with the number of competitors in the same patch. Individuals can therefore profit by switching to another patch if the competitor density in the present patch is high. Sutherland therefore expected that for ideal free foraging, the pay-offs must be constant for all competitors for all patches. Therefore

$$G_i(n_i) = Q_i/n_i^m = C \text{ (constant) for all patches,}$$

so that the number of competitors in a given patch i must be

$$n_i = (Q_i/C)^{1/m} \tag{4}$$

(Sutherland, 1983). Equation (4) defines the ESS distribution; it defines the equilibrium number of competitors for each patch, i, j, . . . etc. It can be compared with equation (2) for continuous input; as expected it is the same if $m = 1$. Thus, under Sutherland's model, continuous input becomes a special case in a continuum in which interference can range from negligible to severe by altering the interference constant m. Further, the value of this constant can be (and has been) estimated from field data by a linear version of the equation, $G_i(n_i) = Q_i/n_i^m$. The linear version is

$$\log G_i = \log Q_i - m(\log n_i) \tag{5}$$

It is often easy to measure intake rates and competitor numbers in the field and in experimental situations, although we cannot measure m (or, in the field, also possibly Q_i) directly. The great advantage of using equation (5) is that m and Q_i can be estimated by plotting the log of intake rate ($\log G_i$) against the log of competitor number ($\log n_i$). The slope of the resulting regression line is $-m$, and the intercept on the G_i axis is Q_i. In natural foraging situations, m has been estimated to vary between 0 and 1.13 (Sutherland, 1983). m can exceed 1.0 when some resource items can be lost as a result of interference, for example resource items (such as prey or mates) may flee.

equally well whatever patch they search in. In other words, as long as interference in the better patch reduces the individual uptake rate there to a
value below that achieved by individuals in the poorer patch, it would pay
competitors in the better patch to move to the poorer patch. The equilibrium distribution is achieved when the influence of food abundance and
interference balance out.

Therefore, without interference, all competitors should be in the best
patch. With some interference, all but a few competitors should choose the
best site, and with strong interference, such that all competitors scramble
for the same items, the competitors should be distributed in the ratio of
patch profitabilities, i.e. they should follow the input matching rule. Notice
that the ideal free distribution does not necessarily imply that competitors
are distributed among patches according to patch profitabilities; this is a
special case for strong interference. Sutherland (1983) was the first
explicitly to propose a model for variable interference in an ideal free context. Following Hassell and Varley (1969) he incorporated the 'interference
constant' m (see Box 5.1 for details). m varies usually between 0 (no interference) and 1 (high interference). In the continuous input model where
intake is inversely proportional to number of competitors, m is exactly 1.
The continuous input case can therefore be defined by $m = 1$.

5.3.2 *Unequal competitors*

Good competitors have sometimes been found to suffer less from an increase in competitor density than poorer competitors (for example, Coates,
1980; Rubenstein, 1981). Remember that this feature was not included in
the model for Fig. 5.5, where the *relative* pay-offs to A and B in a given
patch were independent of the number of competitors, although competitor numbers did affect *absolute* pay-offs in a patch.

How can Sutherland's ideal free interference model be used to study the
situation depicted by Fig. 5.7? A and B can be made to suffer differently
from interference with increasing competitor density by giving them different interference constants, m_A and m_B. Figure 5.7 shows how intake rates of
A and B may reduce in a patch where the ratio of A to B remains constant
whilst the numbers increase from n' to n''. In simulations for many
phenotypes, in which m_A, m_B . . . etc. were proportional to the relative
competitive weights of A, B . . . etc. in a patch, it was found that the stable
solution was always a truncated phenotype distribution, provided that the
patch qualities differed (Sutherland & Parker, 1985; Parker & Sutherland,
1986). The probable reason relates again to the notion of changing relative
pay-offs across patches. In the former model, changes in the ratio of pay-
offs of A and B were a property of the patch type, whereas in the present
model it is a property of competitor density (other things being equal). Note
that the ratio of pay-offs at n' differs from the ratio at n''.

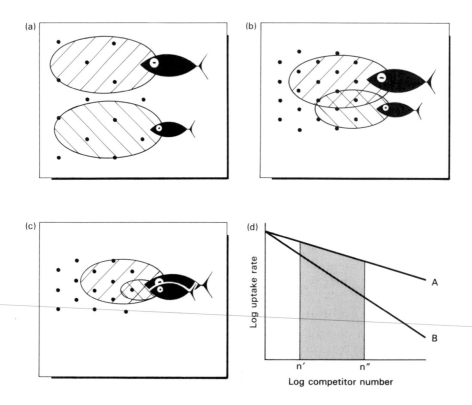

Fig. 5.7 Hypothetical example of how competitor density might change *relative* pay-offs across patches. Two unequal fish, A and B, in a hypothetical shoal, search the water drifting towards them for prey items. At a low density of competitors (a), there is no interference, and thus the poor competitor B gets the same gain as the good competitor A. (b) As competitor density increases, both get less gain than before, but B's gain is reduced more than that of A. (c) At even higher competitor densities, the relative difference between the gain rates becomes even greater. (d) This situation can be represented by a plot of the log of uptake rate (log G_i) against the log of competitor number (log n_i) for the two competitors (see equation (5)). A and B have different interference constants, m_A and m_B. The figure shows how uptake rates of A and B may reduce in a patch whilst the number of competitors increases, for example from n' to n''. It is assumed that the ratio of A:B phenotypes remains constant during this increase.

5.4 Instability

The models we have considered so far have all shown the characteristic that increasing the competitor density reduces each individual's gain rate in a given patch. But some types of interaction between phenotypes generate instability, rather than a stable distribution in a habitat. This typically occurs if sharing a resource is detrimental to one individual and advantageous to another. Imagine a habitat containing two birds and two prey patches (see also Pulliam & Caraco, 1984). One bird is dominant over the other. The patches contain equal densities of prey items. If the birds search entirely independently, they achieve equal intake rates (proportional to the prey density). However, the dominant bird can take some prey items from the subordinate by kleptoparasitism if they occur in the same patch. It obviously pays the subordinate to be on its own, but the dominant to occur with the subordinate. There is no stable distribution, instead the dominant chases the subordinate, which would alternate between the two patches. Stability can be achieved only if the patch qualities differ sufficiently: the subordinate must gain more in the better patch in the presence of the dominant than it would gain on its own in the poorer patch. This situation has been developed theoretically and tested experimentally on pairs of juncos by Caraco *et al.* (1989) (see also Caraco, 1987).

Parker and Sutherland (1986) examined a more extensive version of this model by computer simulation. The habitat contained 100 competitors, i.e. 10 individuals of each of 10 different phenotypes. The phenotypes obeyed a linear dominance hierarchy. Each individual received the pay-off set by the patch quality, plus its gains from subordinates, minus its losses from dominants. The sum of all the gains in a patch equalled the sum of all the losses. If patch qualities were highly disparate, all individuals exploited the best patch. But if the patch qualities were sufficiently similar, no stable distribution could be found. There was continuous change of phenotype numbers across patches, and fitnesses (of individuals of a given phenotype) never equilibrated. However, there was not an equal degree of switching in all phenotypes. The most dominant phenotype tended to be found in the best patch. The next most dominant phenotypes tended to be represented most often across the two best patch types, whilst the lower-ranking individuals switched continuously across all three patch types. Fitnesses declined with decreasing rank. Schwinning and Rosenzweig (1990) have simulated a rather similar example involving a three-population prey–predator game with two patches, one of which was a relative refuge from predation. Their models also typically generated oscillations, although these could be stabilized by increasing the protection provided by the refuge or by increasing intraspecific competition among the prey.

5.5 Suggestions for experiments

In the previous sections we tried to explain the theoretical framework of the behavioural ecology of competition. The theoreticians' intention for developing the different models was to cover a broad range of ecological conditions which may influence the outcome of competitive interactions over sharable resources. Therefore the various models differ in their assumptions about how competition arises, for example competitors are either equal or unequal, relative pay-offs of unequal competitors are either constant or vary across patches, the degree of interference is either constantly strong or varies across patches, etc. We know neither whether this range of assumptions covers most natural situations—perhaps a major constraint is missing—nor which model is the most realistic one for helping us to understand competitive interactions more than others. We shall have achieved our goal in writing this chapter if we can stimulate researchers into investigating the key variables determining which strategy a given competitor chooses to adopt. This is a virtually unexploited habitat for behavioural ecologists working empirically.

5.5.1 Resources and behaviour—the ecology of interference

One strategy for research would be to investigate under field conditions which model's assumptions are met and to compare its predictions with the observed behaviour of the competitors. If the predictions are not verified, additional constraints must be identified and the theoreticians must be persuaded to adjust the model and to generate new predictions. Another strategy would be to generate the conditions of one model experimentally in the lab and then to change them gradually until they fulfil those of another model. This should be done with a species for which both situations occur under natural conditions. For example, in an experiment which fulfils the assumptions of the continuous input model with unequal competitors whose relative pay-offs are constant between patches, we would expect input matching and preferentially an equilibrium distribution of the competitors resembling that shown in Fig. 5.4c. This was in fact what Milinski (1979; 1984; 1986) observed. This distribution should be changed into one where the better and poorer competitors are truncated between patches when, for example, the condition of constant relative pay-offs of good and poor competitors between patches is not fulfilled. This can be achieved by changing the visibility of the prey items (Milinski, 1987) in only one of the patches, which has not yet been done. Another way to generate a truncated phenotype distribution would be to reduce interference in one of the patches by pipetting the prey items on a rotating plate so that they are distributed over a larger area than in the other patch (most easily done with ducks—one site has bread thrown at one spot, the other site has bread thrown in at random). Now we would expect to find the better competitors where they gain most by their competitive superiority, i.e. in the patch with all prey items arriving at the same spot, and the poorer competitors where they suffer least from their competitive inferiority, i.e. in the patch with prey items distributed over a large area. The ESS distribution is predicted by Figs 5.5d and 5.6. The boundary phenotypes should be the fish which continue to switch between patches. (In a three-patch set-up one can predict between which two patches they switch.) Such an experiment has not yet been attempted. The reader is invited to think about an experiment testing the assumptions and predictions shown in Fig. 5.7.

We have proposed these experiments in order to show how the step can be made from *understanding* a model to *testing* its assumptions and predictions experimentally. A mathematical model is nothing more than a hypothesis formulated quantitatively, which without using algebra can be expressed only qualitatively in words. When a situation is as complex as the one unequal competitors find themselves in, we need a model to formulate hypotheses for research.

5.5.2 All other things are not equal—a challenge to understand real life

'All other things being equal' is the statement with which a theoretician is able to concentrate on one relationship only. A well-designed laboratory experiment can approach this condition. The field-worker, however, has to face the whole complexity of real life. Ideally, models and laboratory experiments should be expanded until they meet real life, the first steps being just a prerequisite for later achievements. The models discussed in the previous sections assume that every competitor has complete information about all parameter values such as input rate to each patch, etc. In the real world, which this time includes the laboratory experiment, each competitor has to sample and learn at least something (see Bernstein *et al.*, 1988; see Chapter 4). In experiments where we expect input matching, i.e. the competitors distribute themselves between the patches in the ratio of patch profitabilities, this should lead to a relative overuse of the poorer patch (Regelmann, 1984; Abrahams, 1986; Houston & McNamara, 1987; see Chapter 4). This phenomenon is indeed usually observed in this kind of experiment, and certainly needs further experimental investigation.

Recently, ideal free distributions have been used to investigate a possible trade-off between foraging and predator avoidance (see also Chapter 4). When the risk of predation is increased in one of the food patches, both minnows (Pitcher *et al.*, 1988) and guppies (Abrahams & Dill, 1989) balance the risk of being eaten against the benefit of getting food. When both patches had equal food but one contained a predator of guppies, there was a certain preference for the predator-free site. By adding food to the risky patch, risk could be titrated against energy until both patches had equal value from the guppies' point of view, so that they returned to a 1:1 distribution. By this method different currencies of fitness become quantitatively comparable. Because there are usually individual differences in boldness towards predators (for a review see Magurran, 1986), it would alter the predictions of ideal free foraging models if these differences do not correlate with those in competitive ability. It would be interesting to know who takes the risk of predation under reduced competition, and who flees from the predator but accepts strong competition. Each competitor's decision should be influenced also by its internal state, for example by its hunger level. The hungrier an animal is, the more effort it should be prepared to invest in competition for food—at the expense of fulfilling other needs such as avoiding predation. The resulting ESS would be a 'state-dependent ideal free distribution' (McNamara & Houston, 1990c; see Chapter 4), a straightforward theoretical concept which calls for experiments.

Another fact that adds to the complexity of the field situation is that resource patches are usually not homogeneous with regard to the distribution of resource items. This may specifically prevent distributions of different phenotypes which are strictly truncated *between* patches, because there may be a truncation *within* each patch as well (for example, Whitham, 1980). Although there are field observations of better competitors concentrating in better patches and poorer competitors in poorer patches (herring gulls: Monaghan, 1980; oystercatchers competing for food: Goss-Custard *et al.*, 1984; male black grouse lekking for mates: R. Alatalo personal communication), these distributions were not strictly truncated. Again it is easy to think of an experiment designed to investigate the effect of patch heterogeneity on the probability of truncated phenotype distributions.

5.6 Ecological consequences

The ecological consequences of ideal free philosophy are extremely wide ranging. The present section serves only to give a brief introduction to some of the broader issues (see also section 4.7.2).

5.6.1 Ideal free distributions in time

So far, we have discussed only spatial distributions. Over the periods considered, the quality, Q, of a patch remained constant. We also assumed zero mortality, and zero costs of moving from patch to patch. What happens if these assumptions are relaxed?

SHORT TIME PERIODS—NO SIGNIFICANT MORTALITY

Imagine a habitat with a series of patches—1, 2, 3 . . . etc. as before. Through time t the patches vary in input rates $Q_1(t)$, $Q_2(t)$, $Q_3(t)$ . . . etc. If equal competitors know all the conditions and there is no cost of moving between patches, then from equation (2) the number of competitors in patch i must be ideal free at time t:

$$n_i(t) = Q_i(t)/C \tag{6}$$

for all patch types i, where C is the average gain from the entire habitat.

Switching between patches cannot be entirely costless, and most patches will change in value with time. A specific case has been investigated and applied to mate searching in dungflies (Parker & Stuart, 1976). The patches (cow pats) decay in resource input rate (females) from time zero: females arrive fastest to new droppings; the arrival rate then shows a negative exponential decline as the cow pat surface becomes drier. There is a fixed travel time cost in moving from one patch to another, and new patches are assumed to be available continuously.

The results of this model are that: (i) competitors should leave one patch only to go to a new patch; (ii) initially, gain rates in a new patch will exceed constant C, so that all the competitors should exploit the patch until the time t_{crit}, when an individual's gain rate reduces to C; and (iii) after t_{crit}, competitors should leave gradually to go to new patches, and the number of competitors remaining after t_{crit} is given by the input matching rule in equation (6). The costs constrain free movement by making it advantageous for all competitors to stay in a patch until its value drops to a critical value, $Q_i(t_{crit})$. After this, individuals must depart at different times, but each achieves the same pay-off, whatever time it leaves. In dungflies, a good fit to these predictions was found, and all departure times appeared to yield equal pay-offs (Parker, 1970; 1984; Parker & Stuart, 1976), although this was not tested statistically (Curtsinger, 1986). Re-analysis showed no significant difference between prediction and observation (Parker & Maynard Smith, 1987).

This model of competitive resource use in a merger of ideal free theory and Charnov's marginal value theorem (Charnov, 1976; Parker & Stuart, 1976) for patch use without competition. If there were just one competitor, it should leave at t_{crit}, just as the first of n competitors should leave in the competitive model. t_{crit} can be solved by the tangent method as used in the marginal value theorem (Parker & Stuart, 1976).

Iwasa and Obara (1989) sought an ESS pattern of daily mate-searching activity in male butterflies. They used a different type of time constraint: each male can spend only a fixed time each day on mate searching or it would pay to be active all day to maximize matings. The females' daily emergence pattern typically peaks during the morning. If all males searched during this peak, some individuals could profit by switching to other periods. If females are captured soon after emergence, males could avoid searching in the early morning and late afternoon/evening. In between, the number searching should follow equation (6) above. The ESS becomes more complex if males cannot find females quickly.

LONG TIME PERIODS—MORTALITY IS SIGNIFICANT

Two major applications of ideal free theory for long-term changes in resource quality are seasonal incidence (Bulmer, 1983; Iwasa et al., 1983; Parker & Courtney, 1983) and seasonal migration (Baker, 1978). Mortality now becomes important in the models. How can ideal free theory be applied to study the distribution of competitors through a season, i.e. seasonal incidence? The following example relates specifically to male butterflies, but the approach is quite general. Imagine a butterfly in which females have a characteristic seasonal emergence pattern: in a given place f females emerge on date t. This pattern, $f(t)$, is determined by the availability of a larval food plant. Females mate on emergence.

We now seek an ESS emergence pattern, $m(t)$, for the male butterflies. If there is no mortality risk (or decline in competitive ability), all the males

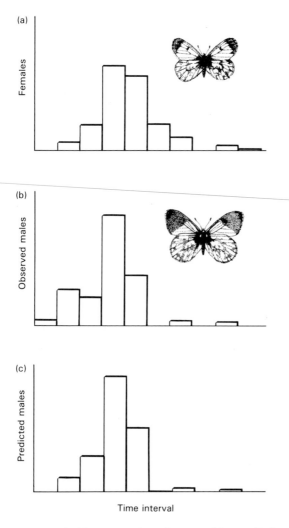

Fig. 5.8 Ideal free seasonal incidence model as applied to male emergence pattern in the orange tip butterfly. *Anthocaris cardamis* (L.). Data are from Durham, between 19 May and 28 June 1977; each time interval equals 4 days. (a) Observed female emergence, and (b) observed male emergence, both from Fisher–Ford estimates; (c) is the predicted male emergence pattern from the ideal free model, assuming that there is 'restricted polygamy' (where surviving females remate after two 4-day intervals) as appears to apply in this species. This gives a better fit between observation and prediction than when monogamy or full polygamy (females remate at each interval) is assumed (from Parker & Courtney, 1983).

should emerge at the start of the season ($t = 0$). Emerging early has no cost, so the best strategy is to emerge as soon as possible and to 'cover' the entire season. At the opposite extreme, the mortality rate is so high that a male emerging on one day is unlikely to survive the next. The ESS is again obvious: in order that all males achieve equal pay-offs, the number of males emerging must balance the number of females emerging, i.e. $m(t) = f(t)$. With an intermediate mortality rate, the ESS emergence distribution for $m(t)$ is skewed to the left of the $f(t)$ distribution: males tend to emerge before females ('protandry'; see Parker, 1985). Protandry is increased by decreasing the mortality rate.

Many male butterflies do show protandry. Fagerstrom and Wiklund (1982) constructed a male–male competition model for the optimal mean date of emergence, in which the variation around this optimum was not shaped by selection. However, for two species, the male emergence was found to fit well the ideal free model (Iwasa et al., 1983; Parker & Courtney, 1983), indicating that the variation itself may be adaptive. Figure 5.8 shows Parker and Courtney's results.

Further studies of seasonal incidence games—both theoretical and empirical—would be very useful. They might examine the following type of problem. Call the seasonal incidence of a food plant $p(t)$. This plant is exploited by an insect pest, which has the seasonal incidence $u(t)$, whilst that of an insect predator of pest is $v(t)$. Each distribution is positively influenced by that of the resource it exploits, but is negatively influenced by that of the predator for which it itself forms a resource. For instance, $u(t)$ is influenced positively by $p(t)$, but any peak in $u(t)$ would be exploited by the emergence timing of the predator, $v(t)$. Whether such games lead to stable solutions or to cyclical dynamics is not yet known; the latter seems quite likely.

MIGRATION

Many bird species migrate a long distance—sometimes to another continent—to spend the winter in a more favourable place. In several bird species, for example in some warblers, the European robin and the European blackbird, some birds stay at home whereas others migrate, a phenomenon called partial migration (for example, Berthold, 1988). Under harsh winter conditions food becomes an even scarcer resource. Migrating birds, however, have to pay the costs of moving in terms of energy and risk of mortality. If all birds which use the same waning resource stayed at home during the winter season, after some time they would certainly experience a lower pay-off per individual than a 'mutant' bird that had migrated. The pay-off of staying is thus frequency dependent and we expect an ideal free distribution of the birds between the strategies of staying and migrating (Baker, 1978; Lundberg, 1988). The number of birds that must stay at home is given by the input matching rule. Since there is evidence that partial migration is controlled by a genetic polymorphism in a number of bird species (for a review see Berthold, 1988), we can expect to find that migrants and stayers have equal fitness, at the polymorphic frequency averaged over a number of years.

5.6.2 Ideal free distributions in ontogeny

When we looked for possible equilibrium distributions of unequal competitors, we assumed that phenotypes differ in some feature, such as size, that is related to competitive ability (Figs 5.5 & 5.6). Because size is usually not a phenotypic trait which is constant over an individual's lifetime but changes during its ontogeny, we can read the truncated ESS distribution of phenotypes which differ in size equally well as a distribution of phenotypes which differ in age. The type of patch, or more generally the type of strategy, which provides the highest possible pay-off for an individual of a given age can be called its 'ontogenetic niche' (Werner & Gilliam, 1984).

Ontogenetic niches can be parts of the habitat which differ, for example in food content and/or predation risk. We expect to find a truncated distribution of age classes between the different parts of the habitat, with the youngest individuals in the part where the effect of size on fitness is lowest and the oldest individuals where size is most important. For example, 1-year-old bluegill sunfish preferred the littoral zone of an experimental pond where low foraging opportunities combined with low predation pressure offered a higher chance to reach reproductive age than the benthos with more food and high predation risk. The benthos was, however, preferred by the 2-year-old fish which were too big to be caught by the common predator and which therefore could achieve a higher fitness by using the better feeding opportunities there (Werner et al., 1983). The switch age would be predicted by the boundary phenotype rule. Ontogenetic niches can also be strategies such as 'male' and 'female' in sequential hermaphrodites and 'larva' and 'adult' in amphibians and holometabolous insects which have to decide when to start their metamorphosis. Ideal free theory can be applied to predict at which age individuals should switch from one strategy to the other.

SEQUENTIAL SEX CHANGES AND ALTERNATIVE MATING STRATEGIES

A sequential hermaphrodite begins reproduction as one sex and later switches to the other. Ghiselin (1969) suggested that this system may be favoured when an individual's reproductive success correlates with size or age, and where this relationship is different for each sex (see also Charnov, 1982).

Sequential sex change is in fact a form of a truncated phenotype distribution (see Parker, 1984). Consider again Fig. 5.6: a strategy is now a 'choice' of a sex rather than a choice of patch. Whereas previously gain rates declined with numbers in a patch, we now have, from Fisher's (1930) principle for sex ratio, that pay-offs decrease with numbers 'playing' the same sex. Following Ghiselin, suppose that pay-offs correlate with size (= age), but that this is steeper in males than in females. Individuals should first reproduce as females. The boundary phenotype rule defines the size (= age) for switching to male. Conversely, if female reproductive success is more strongly affected by size, the ESS will be to be male first. Sex reversal must follow a 'commonsense' rule: the sex in which size exerts a *weaker* positive effect on reproductive success must be exploited *first* (as in Fig. 5.6c). The alternative sequence is unstable. Similarly, if the effect of size is negative in one sex and positive in the other, the sex which must be first must be that which shows the negative effect. Case studies appear to fit the commonsense ESS. For example, in Pandalid shrimp, egg production increases with size, but size may not influence male success; sex reversal is from male to female and the switch occurs at the correct stage (Charnov, 1982).

The same argument can be applied to alternative mating strategies (Parker, 1982). Imagine two male mating strategies: 'sneaks', which gain matings opportunistically and where size is unimportant, and 'guarders', where success at guarding females is increased by size. Suppose that pay-offs decline as more males adopt a given strategy. There is again a formal analogy to the two patch case (Fig. 5.6). The ESS is for males first to play sneak, switching to guarding as their size increases. The switch age follows the boundary phenotype rule. There is some evidence for this sort of pattern in mammals (for example, Dunbar, 1982). In insects, the size of adult may be fixed on emergence from the pupa. In such cases, there can be 'switch size' but not a 'switch age' from one strategy to another. In the bee, *Centris pallida*, 'patrollers' search out and fight over virgin females emerging from their burrows and 'hoverers' wait at feeding sites. Size is important only for patrollers, and they are indeed larger on average than the hoverers (Alcock, 1979).

METAMORPHOSIS IN INSECTS AND AMPHIBIA

The metamorphosis from larva to adult state in holometabolous insects and amphibians is analogous to the sequential sex change in hermaphrodites. In each case there is a size-dependent switch from one pure strategy to another during ontogeny. We can use ideal free theory to predict the boundary phenotype, i.e. the size or age at which individuals should start metamorphosis. In both insects and amphibians there is a dramatic shift in habitat and resource use at metamorphosis. Suppose we regard 'larva' and 'adult' as two independent exploitative strategies and that, although reproduction must occur in the later part of life, either strategy could hypothetically be second. If growth rate and survival in the larval stage is less size dependent than survival and reproduction in the adult stage (for which evidence exists in amphibians; Werner, 1986), then an individual should indeed start as a larva, and by the boundary phenotype rule we can again define the size (= age) at which it should later switch to the adult stage (Fig. 5.6). The result is again a (size-) phenotype distribution which is truncated between the strategies 'larva' and 'adult'. Although the size (or age) at which meta-morphosis occurs is roughly species specific (Werner, 1986), it may well be that individuals can measure actual pay-offs shaped by competition and predation pressure in the larval niche and shift their switch age accordingly (Wilbur, 1980; Werner & Gilliam, 1984; Werner, 1986).

5.6.3 Despotic distributions and territoriality

Fretwell (1972) termed any distribution in which animals showed ag-gressive guarding of resources a 'despotic' distribution. With despotic guarding, individuals are not 'free' to move about the habitat, and ideal free predictions may not apply.

To summarize the implications of despotism, consider the following two extremes. In one there is no exclusion of individuals from resource patches, in the other some competitors are entirely excluded from a patch. The extreme no-exclusion case can generate an exact parallel of the ideal free distribution. Imagine a set of equal competitors, and that as their density in the habitat increases, so does their density in each of the patches. The territory size of each individual simply becomes compressed as more individuals pack in. Let the total size of patch i be T_i; then if there are n_i equal competitors in i, each competitor gains a territory of size T_i/n_i and a fitness of $W(T_i/n_i)$. From ideal free theory, we would expect the fitness of all individuals to equalize across all patches, i, j, . . . Hence $W(T_i n_i) = W(T_j/n_j)$ . . . etc., for all patch types. If fitness increases monotonically with territ-ory size, and patches vary only in size (and not in the value of each unit of area in them) it is obvious that

$$T_i/n_i = T_j/n_j \dots \text{etc.} = C \text{ (constant), for all patch types,} \qquad (7)$$

i.e. all territories are the same size and the number of competitors relates directly to patch size. This 'habitat matching rule' (7) has been derived by Fagen (1987), who defines T much more generally than a patch to be divided into territories: T is the patch's total resource, whatever it is. There is some danger in being quite so general, for example if T_i is the density of prey items hidden in patch i, then, as we have already seen, only inter-ference can prevent all competitors from exploiting the best patch.

Thus with perfect compression of territories so that all competitors fit into patches, rule (7) applies and despotic behaviour will not have altered the ideal free distribution. The situation is very different if some com-petitors are excluded from holding territories. The extreme case here has each competitor taking its optimal-sized territory, starting with arrival order or dominance. As the competitor density increases, the number of territories in a patch increases linearly until all patches are full; the remaining competitors become 'floaters' (see Davies, 1978). Provided that there are some floaters, then equation (7) applies trivially for the territory holders. If competition is low and all individuals gain territories, many distributions are possible.

These two extremes serve simply to outline the boundaries of the possibilities. In nature, territoriality is likely to lie between the two extremes, and we need to consider effects of competitive differences. Much work—both theoretical and empirical—concentrates on optimal territory size (see reviews in Davies, 1978; Davies & Houston, 1984). Benefits of increasing territory size are traded off against rising costs of defence; for example, Myers *et al.* (1981) examined cost–benefit models of sanderling territories, and Parker and Knowlton (1980) made ESS models of territory size for the above two extreme cases (including unequal competitive ability for the no-exclusion model). However, there has been rather little theoretical analysis of the effects of territoriality on animal distributions since Fretwell's (1972) 'ideal despotic distribution' (his despotic version of ideal free). More analyses for unequal competitors would be useful.

5.6.4 *Ideal free distributions and item variation in a resource*

Suppose the woman in Fig. 5.1d throws the whole contents of her bag to the sparrows at once. While the birds are depleting the bread on the ground, it should occur often that several birds try to grab the same piece. If, however, the woman has cut two different sorts of bread, perhaps brown and white, each bird could specialize in eating one colour type. The number of birds specializing in one colour type may match the frequency of that type, thus achieving an ideal free distribution between the two sorts of bread.

A similar scenario in everyday life is salmon farming, where many fish are kept in a tank into which a large number of food pellets of similar size and colour are released all at once (an even distribution of the pellets through time would produce strong interference). Pellets that have reached the bottom are not regarded as food. When the salmon were fed on a mixture of brown and yellow pellets, more pellets were eaten and the fish grew faster than those in control groups which were fed either on brown or yellow pellets (Jakobsen *et al.*, 1987). When pigeons that had a preference for the same food type (of a mixture of two) were presented in pairs with the mixed diet, they modified their initial individual preferences in the direction of resource partitioning (Inman *et al.*, 1987), i.e. an ideal free distribution between food types. However, when food items differ in profitability and competitors differ in their ability to compete for the more profitable items, we expect a truncated distribution of good and poor competitors between food types. An experiment with sticklebacks supported this prediction (Milinski, 1982).

5.6.5 *Niche separation and competitive release*

Ideal free theory can also be used to understand the outcome of competitive interactions between different species (Rosenzweig, 1981; 1985). We earlier discussed partial migration in birds where migrants and stayers are members of the same species. However, we can apply the same logic to explain why one bird species stays at home during the winter season whereas another species migrates. An obvious condition is that both species compete for the *same* resource, which becomes scarcer during winter. In a two-phenotype (or two-species) two-patch model, Parker and Sutherland (1986) found one possible ESS to consist of one phenotype in one patch and the other phenotype in the other patch. This could relate to the case where one bird species shows total migration and the other shows total residency. The other ESS consisted of one phenotype alone in one patch and a mix of the other two phenotypes in the other. The obvious parallel here would be where one bird species shows total residency and the other shows partial migration or, alternatively, one species shows total migration and the other shows partial migration.

The widespread existence of resource partitioning between species (Schoener, 1974) indicates the potential application of the models that we have discussed in this chapter. Species-specific foraging niches may be expanded when a competing species is removed, a phenomenon called 'competitive release' (for example, Fjeldså, 1983; Alatalo *et al.*, 1987). This provides suggestive evidence for truncated phenotype distributions. We would expect here that the competitive ability of an individual of one species relative to that of an individual of the other species varies between niches, and that the niche in which competitive abilities differs most is dominated by the superior species. The power of ideal free theory is that it can predict the exact form of the truncated distribution, for example whether the boundary phenotype is within one species or a hypothetical one in between the two species. Possibly the truncation between species leads to further niche differentiation, especially in the competitively inferior species.

No single theory can explain everything. Ideal free theory seems to be more general and more diverse in its applications than many.

6: Interactions between predators and prey

John A. Endler

6.1 Introduction

All animals are at risk from predation, and many are themselves predators. As diverse as predator and prey species are, successful predation events characteristically follow a sequence of six stages: encounter, detection, identification, approach, subjugation and consumption (Table 6.1). Prey wish to interrupt this sequence as soon as possible by means of defences, whereas predators wish to reach the conclusion quickly by means of counter-defences. Although there have been several reviews of anti-predator defences (Edmunds, 1974; Curio, 1976; Endler, 1986a; Vermeij, 1987; Greene, 1988), there are no systematic reviews of corresponding predator counter-defences. However, it is clear that defences and counter-defences can act at any stage of an encounter (Table 6.1). A study of many of these processes reveals a number of interesting and unsolved behavioural and evolutionary problems, and this chapter will briefly discuss some of them. How do predators detect prey? What is the evolutionary effect of prey abundance and predator behaviour? How does this affect prey density and prey variation? How do prey evolve in response to predation? When and how are colour patterns cryptic? Why and when should they be conspicuous or mimetic? What defences should be used by prey against predators, and how are they overcome by predators? What is the evidence for coevolution between predator and prey?

6.2 Detecting prey: apostatic selection, rarity and polymorphisms

6.2.1 Apostatic selection

Virtually all predators eat more than one prey type, be it different species or phenotypes within species. For brevity, let 'morph' apply to any prey type regardless of whether it is distinct at the species level. Many predators do not attack morphs in proportion to their availability; instead, they prey differentially upon more common phenotypes or species. This is called *apostatic selection*, and it has been demonstrated in both vertebrate and invertebrate predators (Holling, 1965; 1966; Murdoch, 1969; Curio, 1976; Allen, 1988; Endler, 1986b; 1988). An example is shown in Fig. 6.1. Apostatic predation will increase the protective effects of rarity (and apparent rarity; see Table 6.1) because attack probability declines faster than prey

Table 6.1 Stages of predation and the corresponding antipredator defences (after Endler 1986a).

1 ENCOUNTER, or get within a distance from which predator can detect prey
  A *Rarity*: reduces the random encounter rate between predator and prey, and reduces the risk still further if predators exhibit apostatic behaviour (Curio, 1976). Rarity makes predator specialization unlikely
  B *Apparent rarity*: similar effects to rarity, but without the mate-finding and other costs (Endler, 1986a)
    (i) Differences between predator and prey in activity times or seasons
    (ii) Hiding or inconspicuous resting places
    (iii) Polymorphism
    (iv) Seasonal changes in colour patterns or other signals
  C *One-upmanship*: greater detection distance of predators by prey than vice versa

2 DETECTION of prey as objects which are distinct from the background
  A *Immobility*: for any sensory mode which detects motion (Cott, 1940; Edmunds, 1974); may depend upon seasonal changes in colour patterns (Radabaugh, 1989)
  B *Crypsis*: reduces signal/noise ratio of prey in predator's sensory field (Endler, 1978; 1984; 1986a)
  C *Confusion*: makes detection of a single individual more difficult, or makes it difficult to 'fix' on a single individual for long enough to identify it as edible (Cott, 1940; Edmunds, 1974; Curio, 1976)
    (i) Random or unpredictable movement; may also shift predator's attention to other objects or other prey species (Edmunds, 1974; Curio, 1976; Greene, 1988)
    (ii) Movement between contrasting sensory backgrounds (Endler, 1978)
    (iii) Random or unpredictable sensory effects (Endler, 1978; 1986a), especially when enhanced by, and genetically correlated with, colour patterns (Brodie, 1989)
    (iv) Extreme abundance: predator saturation, schooling or other concerted behaviour (Edmunds, 1974; Curio, 1976; Turchin & Kareiva, 1989)
    (v) Polymorphism (Endler, 1986a)
  D *Sensory limits and perception* (Endler, 1978; 1986a)
    (i) Minimum distance for non-detection of any spot or pattern element
    (ii) Minimum distance for detection of colour
    (iii) Flicker fusion
    (iv) 'Private wavelengths' (see also Lythgoe, 1979; Endler, 1983)
    (v) Sealed shells to prevent leakage of chemical cues (Vermeij, 1987)

3 IDENTIFICATION as profitable or edible prey and decision to attack
  A *Masquerade* (special resemblance to inedible objects; Cott, 1940; Endler, 1981; 1986a)
  B *Confusion* (also 2C)
  C *Aposematism* (conspicuousness associated with distastefulness; Cott, 1940; Edmunds, 1974; Ford, 1975; Guilford, 1989)
  D *Mullerian mimicry* (less distasteful species strongly resembling more distasteful species; Wickler, 1968; Edmunds, 1974; Ford, 1975)
  E *Batesian mimicry* (palatable species strongly resembling distasteful species; Cott, 1940; Wickler, 1968; Ford, 1975; Edmunds, 1974)
  F *Honest signalling of unprofitability* (Fitzgibbon & Fanshawe, 1988)

4 APPROACH (attack)
  A *Mode of fleeing*
    (i) Speed
    (ii) Sprint to cover
    (iii) Different style than predator (flying, running, swimming; Endler, 1986a; Greene, 1988)

Table 6.1   Continued

    B *Unpredictable behaviour* (also 2C and 2B) (Edmunds, 1974; Curio, 1976)

    C *Rush for cover* or other predator-inaccessible microhabitat (Endler, 1986a; Greene, 1988)

    D *Startle, bluffing and threat behaviour* (Edmunds, 1974; Greene, 1988)

    E *Redirection* (Cott, 1940; Edmunds, 1974; Powell, 1982; Wourms & Wasserman, 1985; Greene, 1988)

    F *Encouraging premature attack* (stotting, etc.) (Edmunds, 1974)

    G *Aggregation and predator saturation* (Edmunds, 1974; Turchin & Kareiva, 1989)

5 SUBJUGATION (prevent escape)

    A *Strength to escape* (Endler, 1986a)

    B *Mechanical methods* (Edmunds, 1974; Endler, 1986a)

       (i) Physical toughness to withstand handling (e.g. tics), or hard shells (Vermeij, 1987)

       (ii) Mucus or slime

       (iii) Autotomy of body parts (e.g., salamander and lizard tails: Greene, 1988; nudibranch slugs: Edmunds, 1974; razor clam siphons and brittle stars: Vermeij, 1987)

       (iv) Spines or other structures

    C *Noxiousness* (Edmunds, 1974)

       (i) Spines and prickles

       (ii) Jaws and claws (bite and scratch at predator)

       (iii) Bad tastes, toxins, stings

    D *Lethality* (Endler, 1986a)

    E *Group defence, mobbing, etc.* (Edmunds, 1974)

    F   Resistance to venom (Poran *et al.*, 1987; Towers & Cross, 1990)

6 CONSUMPTION (Wickler, 1968; Edmunds, 1974; Ford, 1975; Endler, 1986a)

    A *Safe passage through the gut* (by bivalves: Vermeij, 1987; by snails: Norton 1988)

    B *Emetic*

    C *Poisonous*

    D *Lethal*

       (B, C, amd D may only benefit prey through 'kin' or 'green beard' selection (see text) if predator density is reduced by lethality or predators are selected to ignore prey through predator kin selection)

density. In addition, if a prey species is rare enough, an apostatic predator may switch to other more common prey species.

The reasons for apostatic behaviour are controversial because several possible mechanisms produce similar behaviour (Greenwood, 1984; Guilford & Dawkins, 1987; Endler, 1988; Kamil, 1989). One explanation is the formation of search images, or transitory enhancement of detection ability for particular cryptic (inconspicuous) prey types (Ford, 1975; Allen, 1988; Kamil, 1989) (Fig. 6.2a). The apostatic effect can arise from image formation as well as from interference between images of different prey (Endler, 1988). If the search image for a given morph is created faster and retained longer for higher encounter and detection rates, then that morph's fitness declines with increasing density. This is a matter of learning and experience. If the formation of the search image of one morph reduces the quality or retention time of the search image of another morph (for example, selective attention), then the fitness of a morph will depend upon both its density

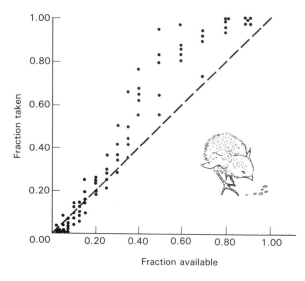

Fig. 6.1 An example of apostatic (frequency-dependent) predation, in this case the fraction of brown pellets presented to and taken by chicks from a mixed population of brown and green pellets. The dashed line is the line of equality; a point falling on this line would indicate that a chick took brown in the same frequency that it was present in the population. Brown is clearly favoured when common (above 25%) and is possibly discriminated against when rare (redrawn after Fullick & Greenwood, 1979).

and frequency. Although there is no direct evidence for interference be-tween search images, it is usually assumed to be true merely because of finite brain size (Kamil, 1989). Search images will cause apostatic preda-tion regardless of interference, but apostacy is more likely to arise from in-terference as the number of morphs increases. This needs to be investigated experimentally.

Foraging theory suggests that apostatic predation should occur even in the absence of changes in perception because foraging behaviour should change in response to local prey density (Holling, 1965; 1966; Murdoch, 1969; Stephens & Krebs, 1986; see Chapter 4). If a predator encounters a habitat patch with a locally higher prey density, then it should spend more time searching that patch for prey and less time in patches with a lower dis-covery rate; this is the optimal search rate hypothesis (Gendron & Staddon, 1983). There is some evidence for this (for example, Fig. 6.3 and Gendron, 1986). Provided that different prey are differentially abundant in different patches, apostatic predation will result as patches are depleted and the predators switch patches. For example, if patch A contains 90% prey type 1 and patch B contains 90% type 2, and initially patch A has many more prey than B, then predators will concentrate on patch A, and consequently reduce the frequency of prey type 1 and increase 2. As a consequence, prey in patch B may increase, until their density causes predators to switch over to that patch, reversing the process, until the densities equalize (Murdoch,

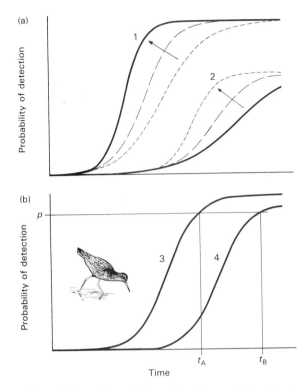

Fig. 6.2 Effects of search image formation and stare duration on the probability of
detection of cryptic prey. It is assumed that the visual system needs some time to recognize
the prey item, so the probability of detection increases logistically with time spent viewing
a prey against a background. More cryptic prey (curves 2 and 4) will take longer to detect
than less cryptic prey (1 and 3). (a) Search image hypothesis: the formation of a search
image for a particular species (1 or 2) results in a more rapid detection of that species
(arrows), but more cryptic species (2) will still take longer to detect than less cryptic
species (1). However, the formation of a search image for one species may inhibit the
formation of a search image for another species (matched line styles). (b) Stare duration
hypothesis: it takes a minimum amount of time (t_A or t_B) to detect a particular prey item (3
or 4) with a given probability (p). Viewing the background for a short time (t_A) may result
in only the more conspicuous prey (3) being detected. Viewing for longer (t_B) will allow
detection of more and more cryptic prey (3 and 4). The minimum viewing time is also
reduced for prey for which there is a search image.

1969; Stephens & Krebs, 1986). The effect will be apostatic predation on the
two prey types, simply as a result of changing search rates (see Gendron &
Staddon, 1983).

The apparent dichotomy between the search image and optimal search
rate hypothesis is complicated by mechanisms of perception and detection
of cryptic prey. For example, it is possible that more cryptic prey require
longer viewing times before detection than less cryptic prey (Endler, 1986a;
Guilford & Dawkins, 1987; Fig. 6.2). If a predator increases its mean viewing time within a habitat patch, it will increase the probability of detecting

more cryptic prey (Fig. 6.2b). This will also reduce its search speed because more time will be spent staring and less time will be spent moving between stares and among patches. This will result in the observed behaviour appearing very similar to that expected from the optimal search rate hypothesis. Indeed, the data in Fig. 6.3 (Gendron, 1986) can be interpreted either way. This similarity resulted in this phenomenon initially being lumped with the optimal search rate hypothesis (Guilford & Dawkins, 1987), but it should really be called the 'stare duration hypothesis' (T. Guilford, personal communication, 1989). In summary, predators can alter their return in two ways: (i) spend more time in patches, which increases the probability of randomly encountering prey in that patch (optimal search rate hypothesis); and (ii) spend more time viewing any particular part of a patch, which increases the probability of detecting a prey that is already in the field of view (stare duration hypothesis). Both foraging modes increase the time spent in more productive patches (Figs 6.2 & 6.3; see also Gendron, 1986). If patches differ in prey type frequencies, then both modes can result in apostatic selection. But stare duration may be a better way to detect more cryptic prey, so the stare duration mode may induce stronger apostatic selection than the optimal search rate mode.

Much of the experimental evidence cannot distinguish between the search image, search rate, and stare duration hypotheses because they can yield the same outcome (Guilford & Dawkins, 1987). For example, consider Fig. 6.3. In this experiment (Gendron, 1986), bobwhite quail searched for crumbs of varying degrees of conspicuousness although only one coloured

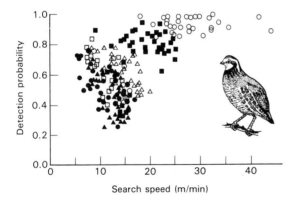

Fig. 6.3 Probability that bobwhite quail (*Colinus virginianus*) will detect baits as a function of search speed through an arena filled with prey at 9 m^{-2} (data from Gendron, 1986, experiment series 2). Symbols indicate baits of varying degrees of conspicuousness; in order of increasing conspicuousness they are: closed triangles, open squares, closed circles, open triangles, closed squares, and open circles. Note that a slower search speed (longer time per patch) is required to detect more cryptic prey, and that more cryptic prey are rarely as much at risk as conspicuous prey no matter what the search rate. If a patch contains cryptic prey, a greater return will result if a predator spends more time in it.

bait was available at any one trial. Over all prey types, more prey were discovered for higher search rates, as expected from the optimal search rate hypothesis. Discovery consists of two steps: (i) encounter, or entry of the prey and background into the visual field; and (ii) detection of the prey against the background; an encounter may not necessarily lead to detection. (Beware, the literature can be confusing: sometimes 'encounter' refers to the complete predation event, and often 'detection' is synonymous with discovery.) The search rate chosen by an individual reflects a balance ('trade-off') between an increase in encounter frequency and a reduction in detection probability as search rate increases. So, because the mean detection probability of more cryptic prey is always lower than that for less cryptic prey, and it takes longer to detect them (Fig. 6.2), it is not surprising that the quail slowed their search rate as crypsis increased (Fig. 6.3). Within each prey type (except for the most conspicuous one) *fewer* were detected as the birds moved among patches more rapidly, and this negative relationship was stronger for more cryptic prey (Fig. 6.3). This could occur because: (i) it takes longer to find cryptic prey in a patch (optimal search rate hypothesis); (ii) it takes longer for the retina and brain to detect more cryptic prey (stare duration hypothesis); or (iii) because search images take longer to form for more cryptic prey. On the other hand, there is no reason why all three hypotheses cannot be true simultaneously.

There are two ways in which it may be possible to distinguish among some of these hypotheses. First, unlike the search image hypothesis, the search rate and stare duration hypotheses will not necessarily result in apostatic predation. This will arise if all patches have the same proportions of different prey types; adjustment of search rate or stare duration will have the same effect on all patches, so relative frequencies will not change. Testing this requires detailed knowledge of both the predator behaviour and the distribution of its prey at all times. Second, both Gendron (1986) and Kamil (1989) pointed out that if search images are being formed, then predators become more accurate and rapid at detecting cryptic prey during the course of foraging; this is not expected under the optimal search rate or stare duration hypotheses. This does seem to be true; in a variety of experiments, quail (Gendron, 1986), jays, and pigeons (Kamil, 1989) do become more accurate at detecting cryptic prey. Unfortunately there is an ambiguity here in what is meant by speed (rate) and accuracy. If the rate of detection is measured from the time the prey comes into the visual field (encounter, or start of stare duration), then both speed and accuracy of detection will increase only if search images are being formed. However, rate of detection can be measured from the consumption of the last prey item, including the time between last predation and the entry of a new prey into the visual field (encounter). In this case, if the predator learns to spend more time in a patch (optimal search rate) and to stare for longer at each visual field, then both detection rate (prey/min) and accuracy (fraction discovered) will increase during the course of foraging as the predator learns to adjust its search rate and stare duration. Consequently, critical tests require knowing times between potential detection (encounter) and actual detection. Most of the existing studies are ambiguous, but, as in Gendron's (1986) experiment (Fig. 6.3), there is no reason why all three hypotheses cannot be true simultaneously.

Apostatic selection can also arise as a result of predator behaviour after detection. If attack and killing skills depend upon practice, then the more frequent morphs will be at an additional relative disadvantage because a morph gives a predator more practice as it becomes more common; the level of capture skills may be higher for common than for rare prey (Lawton et al., 1974; Bergelson, 1985). In addition, and as with search images, there may be interference between skills acquired to attack and subdue different prey types. Apostatic selection can also result from learning to recognize new food types, aversion to rare forms, and other processes (Murdoch, 1969; Greenwood, 1984; Bergelson, 1985; Endler, 1988).

There can be interesting interactions between pre- and post-detection defences which cause or affect apostatic selection. For example, the degree of crypsis can affect predator search costs, altering the frequency-dependent relationship. Gendron and Staddon (1983) showed that if we assume the probability of detecting a prey is inversely related to search rate as well as to the degree of crypsis, then the optimum search rate represents a balance between encounter rate and degree of crypsis, which together determine the discovery rate. The degree of crypsis affects the optimal search style in that, as the prey becomes more cryptic, the search rate drops (the search rate hypothesis). If there are two morphs with differing crypsis, then the optimal search rate increases with the frequency of the less cryptic morph, which in turn may result in a rapid drop in the probability of taking the more cryptic morph. This interaction between detection efficiency and foraging decisions can feed back on the evolution of the colour patterns in interesting but as yet unexplored ways.

6.2.2 Polymorphism

Polymorphism is the presence of two or more distinct forms (morphs) in the same population of a single species (examples in Ford, 1975). Polymorphisms can be advantageous in their own right. For species with apostatic predators, polymorphism reduces the overall risk per individual because there are two or more rare forms rather than a single more common form. In addition, the increased apparent rarity caused by polymorphism may cause an apostatic predator to switch over to equally abundant but monomorphic prey species, because the latter are apparently more abundant (Murdoch, 1969). Even in the absence of apostatic selection, polymorphism may be confusing to predators if the morphs are simultaneously within view. If a predator has too many choices with different appearances, it may hesitate to attack any of them, resulting in a lower attack rate compared with a monomorphic species (Curio, 1976). Such confusion may provide an additional source of apostatic selection if learning to recognize or attack one form interferes with the ability to deal with others. All of these factors suggest that an individual with polymorphic offspring may have more grandchildren than one with monomorphic offspring. Even without these ancillary advantages, polymorphisms can be actively maintained by apostatic predation, no matter what the cause of either the apostacy or polymorphism. Apostatic predation will favour polymorphisms because rarer forms are at an advantage compared with common forms (see Fig. 6.1). Rare forms will therefore increase until they are no longer rare. If a form becomes too common, it will be differentially preyed upon and will become less common. Eventually a balanced polymorphism will occur as the fitnesses of all morphs equalize (Ford, 1975; Greenwood, 1984; Allen, 1988; Endler, 1978; 1988).

6.3 When is a colour pattern cryptic?

A prey's colour pattern is cryptic if it resembles a random sample of the visual background as perceived by the predator at the time and place at which the prey is most vulnerable to predation. Colour patterns can be regarded as mosaics of patches which vary in size, shape, brightness (total reflectance) and colour (reflectance spectrum shape). Therefore an animal is cryptic if the distribution of these four attributes could have been drawn at random from the visual background against which the animal is seen by its most dangerous predators. (One can get an idea of these distributions by making a wire frame the same size and shape of the prey's body outline and throwing it repeatedly on the visual background, sampling the colour patterns within the frame each time.) This definition not only allows us to measure the degree of crypsis, but also to predict colour pattern parameters (Endler, 1978; 1983; 1984; 1986a; 1988). It also shows that our intuitive feel for colour patterns can be wrong if our vision and viewing conditions do not match those of the predator. By the same reasoning, if the vision and viewing conditions and visual background are different during intraspecific communication and predation, then the same colour pattern can be used both as an antipredator defence and for other purposes such as in social behaviour or active thermoregulation.

The definition of crypsis implies that there need be no best single cryptic colour pattern for any particular combination of predator vision, visual conditions and visual background. Since there are many different random samples of the same background under the same conditions, there are many different ways to be cryptic; it is only the colour pattern parameters which matter (Endler, 1978; 1984). For example, leaf litter is a complex mosaic of patches with predictable parameters; moths with different colour patterns can resemble pieces of broken leaves (*Mamestra renigera*), part of a leaf including its midrib (*Abbotana clemataria*), a leaf edge and shadow

(*Pero honestarius*), or even a broken twig fragment and shadow (*Morrisonia sectilis*), and different patterns can resemble different examples of these background elements (for discussion see Endler, 1984). Because all of these patterns are found in the background, different animal colour patterns can be equally cryptic, so long as they contain no rare background colour pattern elements. As a result, species can be polymorphic with all morphs equally cryptic, and these cryptic morphs can be selectively neutral with respect to each other. Of course, if predators exhibit apostatic behaviour, such polymorphisms can be actively maintained. Similarly, different species seen under the same conditions by the same predators can have different equally cryptic colour patterns. Given that different species will have different suites of predators, and be seen under different conditions, divergent colour patterns are even more likely. This allows much evolutionary leeway in the particular colour patterns each species uses in species recognition and sexual selection.

The definition also implies that prey behaviour and phenology can affect crypsis. Crypsis may be aided by special behaviour which aids the cryptic effect (as in leaf-mimicking fishes and insects; Edmunds, 1974), ensures the proper background (pipe-fish, some moths and grasshoppers; Edmunds, 1974; Endler, 1984), the proper alignment between the animal and background patterns (many moths and fish; see especially Cott, 1940; Sargent, 1969; Edmunds, 1974), or even the proper light levels (guppies; Endler, 1987). Slightly different behaviour, such as displaying the same colour pattern at different times, places and lighting conditions, may result in a conspicuous appearance, which can be used in intraspecific communication. This dual function would not necessarily reduce full crypsis at other times and places. In habitats where backgrounds change with season, prey phenology is critical to crypsis. For example, moth species living in an oak-hickory forest in New Jersey were more cryptic against backgrounds available during their normal flight dates than if they emerged earlier or later in the year (Endler, 1984). Crypsis is context dependent.

Crypsis does not necessarily imply 'dull' coloration (Endler, 1978; 1984; Endler & Lyles, 1989). For example, in their natural habitats, supposedly 'bright' birds such as parrots, orioles, tanagers, and titmice are remarkably difficult to see against their 'bright' backgrounds. Similarly, a supposedly 'dull' pattern may be conspicuous against a 'bright' background. This can be aided by properties of sensory systems. When forest-dwelling butterflies with highly contrasting colour patterns fly in and out of sun flecks, their appearance radically changes because the rapid changes in light intensity affect the perception of their different colours differently. This makes it very difficult for one to follow the butterfly's flight path (Papageorgis, 1975; Endler, 1978). The effect would be much smaller if their colours did not contrast so much. If the sun flecks are large, the transition between sun and shade may be made more difficult because the predator may become light adapted in the sun patch, making vision in the dark shade inefficient, and vice versa. Similar phenomena may occur in other sensory modes. For example, pulses of chemicals by aquatic animals in streams might affect sensory contrast for chemosensory predators. Certain modes of swimming in certain habitats may affect contrast for predators using lateral line systems. On land, water and air, certain forms of movement may affect auditory contrast for a sound-detecting predator, as in moths avoiding bats (Edmunds, 1974). These mechanisms work better if the signal or colour pattern is complex or bright rather than simple or 'dull', and also if the background is complex and bright.

Species differ widely in their colour vision (Endler, 1978; 1986a; Lythgoe, 1979). An apparently conspicuous colour pattern to us may be cryptic to a predator, and vice versa (Endler, 1990; 1991). Similarly, if the vision of predator and prey differ, the same colour pattern can be cryptic to the predators while conspicuous to conspecifics (Endler, 1978; 1983; 1986a). One example is the use of red by two Poeciliid fishes to communicate in a way that is difficult or impossible to detect by 'red-blind' invertebrate predators (Endler, 1983). There are many ways in which colour patterns can be used both as antipredator defences and for other purposes, depending upon the vision of the predators and conspecifics, varying ambient light conditions and backgrounds (Endler, 1978; 1986a; 1988; 1991). A good knowledge of physiological and biophysical conditions during signalling may lead to predictions about the direction of signal evolution (Endler & McLellan, 1988); for example, divergent water colour conditions favour the use of blue/yellow signals in clear tropical marine water and red/green signals in temperate lakes and streams (Lythgoe, 1979).

Colour patterns may be conspicuous at a short distance even though they are cryptic at a distance (reviewed in Endler, 1978). For example, if the grain size or scale of an animal's colour pattern is smaller than the predator's minimum resolvable angle at the normal detection distance, then spots will blend together, making the pattern less conspicuous. The same is true for resolution of colour; below a certain size, the predator will not see separate colours and these may blend to match the background (Endler, 1978; Schultz & Bernard, 1989). These effects are stronger at lower overall light intensities, so a combination of short-distance courtship or territorial display at low light levels (small spots) may simultaneously afford protection against predators hunting from greater distances, as in guppies (*Poecilia reticulata*; Endler, 1987).

Even though a colour pattern can be made to look relatively cryptic at some times and relatively conspicuous at others, if it is used both in crypsis and intraspecific communication it will necessarily evolve to be a compromise which depends upon the relative importance of the two functions. The colour patterns of guppies are a good example (Fig. 6.4; see also Endler, 1978; 1983; Houde, 1987; 1988). They consist of a mosaic of patches varying in colour, size and brightness. The colour patterns in any one population represent a balance between selection by predators for crypsis and

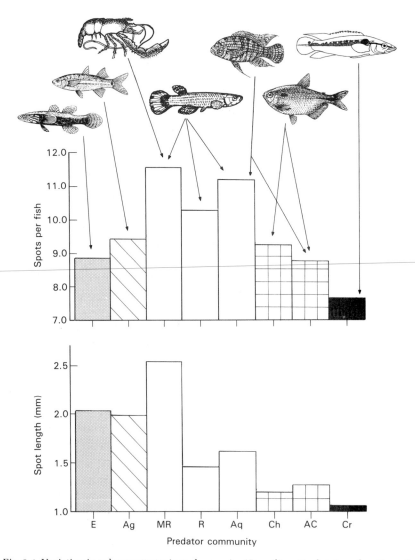

Fig. 6.4 Variation in colour patterns in male guppies (*Poecilia reticulata*) as a function of predation intensity and predator communities (data from Endler, 1983). Predator communities are arranged as a transect from the north coast of Trinidad up (E–MR) the Northern Range Mountains, and down their southern slope (R–Cr). There is more risk to predation at lower elevations; predation intensity is indicated by increasing hatching or shading, with lowest predation in MR, R and Ag, and highest in E and Cr. Predator communities are named after their chief guppy predator (illustrated): E, *Eleotris pisonis*; Ag, *Agonosotmus monticola*; MR, *Macrobrachium crenulatum* and *Rivulus hartii*; R, *R. hartii*; Aq, *Aequidens pulchur*; Ch, Characins (*Astyanax bimaculatus* (illustrated) and/or *Hemibrycon dentatum*); AC, *A. pulchur* and characins; Cr, *Crenicichla alta* and other species. Increased predation is associated with smaller and fewer spots (review in Endler, 1983). Guppies with larger and more spots are more conspicuous and are differentially preyed upon, but they are also at a mating advantage (Endler, 1983; Houde, 1987; 1988).

selection by females for conspicuousness. Field and laboratory experiments have shown that colour patterns with larger spots, more reflective spots and a greater diversity of colours are more conspicuous to both predators and females. Where predation is weak the effects of sexual selection predominate and the males are more conspicuous. Where predation is intense males are relatively inconspicuous; they have fewer, less reflective, smaller, and less colourful spots (Fig. 6.4; Endler, 1983). Crypsis does not have to be absolute; often animals are cryptic enough to minimize predation without seriously compromising other needs.

6.4 Identifying prey: aposematic coloration

Aposematic coloration combines conspicuous coloration with noxiousness (including unpalatability). It can also be associated with sounds (as in bees, rattlesnakes), odours (skunks, stink bugs) and other stimuli, but this has been little studied (Guilford et al., 1987). For example, Arctiid moths, which are distasteful to bats, emit ultrasonic pulses in response to bat sonar. This either serves a direct warning function or confuses the echo received by the bat. In experiments, bats learn to avoid moths which give off the clicks (Edmunds, 1974).

Aposematism results in predators associating noxiousness with the colour pattern (or other signal) and therefore avoiding the aposematic form in future encounters; protection is at the identification stage of predation (Table 6.1). Protection increases with the density of the aposematic signal, so, unlike cryptic coloration, selection is antiapostatic (Greenwood, 1984). Predator training is more efficient at higher densities of fewer and more similar colour patterns. It is also stronger with clumped food (Fig. 6.5) or higher local densities (Allen, 1988; Endler, 1988). These density effects on learning result in strong stabilizing selection; rarer forms will either disappear, or modifier genes which increase their similarity to common forms will spread, leading to monomorphism (review in Endler, 1988). The function of aposematism is experimentally well known, but the causes of its evolution are problematic (summary in Guilford, 1990). There are three basic problems: how does noxiousness evolve; why use conspicuousness for a warning; and how does conspicuousness evolve?

6.4.1 Evolution of unpalatability

Palatability is very common; most species in most genera and families are eaten by a large number of predators. Unpalatability is comparatively rare, but shows up in a variety of families and genera. This implies that unpalatability is a derived condition, and lack of noxiousness is primitive; unpalatability may have evolved from palatability. But since noxiousness often operates during the last two stages of predation there is a serious risk that the prey will not survive the attack. So how can noxiousness or un-

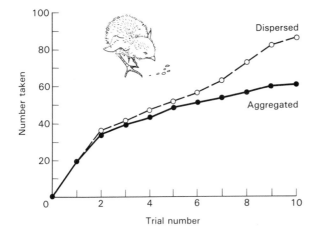

Fig. 6.5 The advantage of aggregation in an aposematic colour pattern. Cumulative number of unpalatable crumbs taken by chicks as a function of trial number (redrawn from Guilford, 1989). If the crumbs are aggregated the chicks learn to avoid them earlier and take fewer than if the crumbs are not aggregated.

palatability evolve? If the predators are inefficient killers, or the defence works well enough so that some prey escape to reproduce, then natural selection will favour noxiousness. But what about distastefulness, which often requires killing the prey to be effective? Wicklund and Jarvi (1982) suggest that because many aposematic species are tough and difficult to kill (Cott, 1940; Edmunds, 1974), toughness will reduce the risk of lethal attack and allow enough distasteful individuals to escape to favour distasteful-ness. But this begs the question of how toughness evolves. The problem is especially difficult if the defence effect is emetic or poisonous, and does not act until several hours or days after ingestion (Table 6.1). The fundamental question is how can natural selection favour a defence that operates during or after the subjugation and consumption phases of predation?

Because many distasteful species aggregate, kin selection is a possible explanation (Fisher, 1930). For example, if a bird attacks an aggregation of distasteful caterpillars and eats a few, it may learn to associate the appear-ance of the caterpillars with the bad taste. Even without kin selection, being in an aggregation of distasteful prey may be advantageous; there is evi-dence that learning to avoid noxious prey is faster when they are clumped (Fig. 6.5). Because aggregations of caterpillars usually consist of individu-als from the same egg mass, the death of one or two of the group benefits their siblings, and their differential survival will increase the frequency of the distasteful trait in the population. However, there are two objections. First, a phylogenetic analysis of many butterfly species shows that gregari-ousness usually evolves after unpalatability, and never before (Sillén-Tullberg, 1988). In addition, many species which are unpalatable do not

aggregate. Examples are marine species with planktonic larvae (Rosenberg, 1989) and lycid beetles. Marine larvae mix in and settle out of the
plankton at random, so it is difficult or impossible for kin to aggregate. So
kin selection is not a good explanation for distastefulness. The second ob-
jection applies to warning coloration rather than distastefulness; predator
training will occur whether or not the caterpillars are kin (Guilford, 1985;
1990). For example, mullerian mimics are protected by each other even
when they are in separate insect orders, and many distasteful species (such
as lycid beetles) do not aggregate. All that is required is that the predator
learns to associate a particular inherited colour pattern (or other signal)
with the noxiousness; if this happens, then all carriers of the pattern will
have increased fitness, regardless of relatedness. Guilford (1985) calls this
'green beard' selection after a process described by Richard Dawkins
(1982), but a clearer name is 'synergistic selection' (Maynard Smith, 1989).
There is still another possibility (Wicklund & Jarvi, 1982). If there are
already post-attack defences present (such as toughness), then neither kin
selection nor 'green beard' selection is necessary to promote the spread of
distastefulness, although they will certainly help. Common post-attack
defences can reinforce the spread of rare defences by individual selection
alone (Wicklund & Jarvi, 1982; Endler, 1988). However, these hypotheses
are not mutually exclusive and probably vary only in relative importance
among systems (Guilford, 1990). Each can encourage the evolution of
warning coloration without kin selection, but neither explains the evolution of distastefulness.

6.4.2 Brightness of aposematic colours

Why use conspicuousness for a warning? Guilford (1990) provides four not
necessarily mutually exclusive hypotheses. First, predators learn to associate distastefulness with a conspicuous colour pattern more rapidly
than with a cryptic colour pattern. The evidence for this is quite strong (Fig.
6.6; Gittleman & Harvey, 1980; Roper & Wistow, 1986; Roper & Redston,
1987). The second hypothesis is that certain specific colour patterns are
easier to associate with distastefulness than others, and these are conspicuous. Presumably this is a matter of ease of pattern recognition rather than
simply of detection and decision-making (as was the case in the first
hypothesis). Guilford does not mention the possibility that simpler colour
patterns may be easier to learn than more complex ones. Hypothesis 2 may
really be more a matter of complexity than conspicuousness because
aposematic colour patterns are simpler on average than cryptic ones (Cott,
1940). Although the effects of pattern complexity have been examined, no
experiments have simultaneously controlled for both complexity and conspicuousness, and probably both are important in aposematism. The third
hypothesis is that new patterns are easier to learn. Since there are fewer
conspicuous than cryptic colour patterns, conspicuous patterns will be

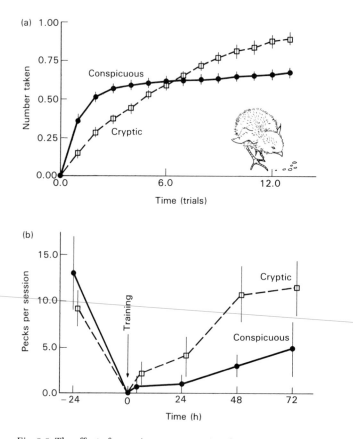

Fig. 6.6 The effect of conspicuous or cryptic colour patterns on the rate of learning (a) and forgetting (b) of chicks feeding on baits. (a) Cumulative number of distasteful baits taken as a function of number of trials (redrawn from Gittleman & Harvey. 1980). Note how the chicks learned to avoid the conspicuous distasteful baits more rapidly than the cryptic distasteful baits. (b) Pecks per session at palatable baits before and after a single session of training them on distasteful baits (redrawn from Roper & Redston, 1987). Note how they retained their learned avoidance much longer when the baits were conspicuous.

new, and will also be examined more closely—both leading to faster learning. There is no direct evidence for this interesting idea. The fourth hypothesis is that conspicuous coloration will allow fewer recognition errors than crypsis (Guilford, 1985). This could be related to the simplicity of patterns and pattern recognition implicit in hypothesis 2, but Guilford (1990) presents another idea, the detection distance model. Conspicuous colour patterns will be detected at a greater distance from the predator than cryptic patterns, hence the predator will have more time, on average, to change its mind about attacking the prey if the prey is conspicuous; some experimental evidence supports this (Guilford, 1990). All four hypotheses could operate to favour conspicuous over inconspicuous warning coloration; a good aposematic signal (visual or otherwise) should be easy to detect, easy to recall, and easy to associate with the defence (Endler, 1988; Guilford, 1990).

6.4.3 Evolution of bright colours

How does conspicuousness evolve? Because most species in a genus or family are cryptic and palatable rather than aposematic, it is reasonable to assume that crypsis is ancestral. The initial spread of conspicuous noxious morphs is probematical because at low frequencies there will be too few to train the predator, and their increased conspicuousness will cause a proportionally higher mortality rate by naïve predators. Once the frequency increases enough to make predator training efficient, then the antiapostatic dynamics will ensure rapid spread, even if there are several predators with different search modes and sensory abilities (Endler, 1988). There are at least six ways to get beyond the problem of initial low frequencies (see also Endler, 1988): (i) random processes may increase the frequency of the initially rare conspicuous morph (Mallet & Singer, 1987); (ii) aggregations of kin (Leimar et al., 1986) or similar genotypes carrying the rare allele (Guilford, 1985) may increase the effective local density of the rare morph; (iii) reluctance of predators to take novel and rare forms and other apostatic effects may increase the frequency of the rare morph. Apostatic selection (though not search images) can apply to aposematically coloured prey as well as to cryptic prey; the decision to attack after detection is all that is affected by the noxiousness; (iv) predators may recognize both the new and old phenotypes as the same defended species because they differ only slightly at the early stages of the evolution of aposematic coloration; (v) natural populations may vary enough in crypsis (as in moths; Endler, 1984) so that initially the less cryptic morph may already be at a high enough density to make predator training possible; and (vi) shifts by insects on to new host plants may be associated with very different visual backgrounds as well as new secondary compounds with which to create noxiousness. Such host shifts may fortuitously result in conspicuous coloration because a colour pattern which is cryptic on the old host plant may be conspicuous on the new host. These possibilities are virtually unexplored.

6.5 Mimicry

The two most common forms of mimicry are Batesian and Mullerian mimicry. In Batesian mimicry a palatable species resembles an unpalatable species, whereas in Mullerian mimicry a less strongly unpalatable species resembles a more strongly unpalatable species. The mimicry can be extensive in morphology and behaviour; some insect examples are shown in Fig. 6.7. The evidence for and the principles of mimicry have been reviewed in Wickler (1968), Edmunds (1974) and Turner (1984; Turner et al,, 1984).

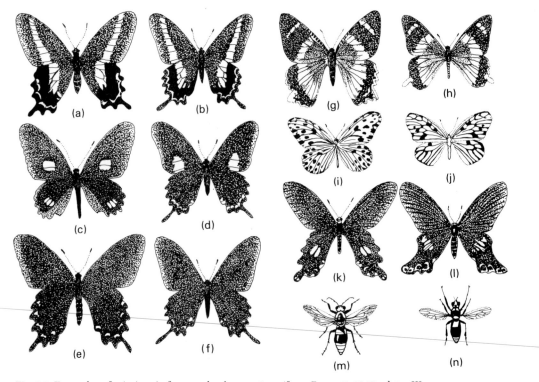

Fig. 6.7 Examples of mimicry in form and colour pattern (from Punnett, 1915, plates III and XIII; and Poulton, 1898, plate 41). In each case the model is to the left and the mimic is to the right. Note that (b), (d) and (f) are different forms of a single species (*Papilio lysithous*) each mimicking a different species of *Papilio*, and that there is one case of a butterfly mimicking a moth (g, h). (a) *Papilio mephalion*, mimicked by (b) *P. lysithous lysithous*; (c) *P. chamissonia*, mimicked by (d) *P. lysithous rurik*; (e) *P. perrhebus*, mimicked by (f) *P. lysithous pomponius*; (g) *Alcidis agathrysus* (a moth), mimicked by (h) *P. laglaizei*; (i) *Ideopsis daos*, mimicked by (j) *Cyclosia hestinoides* (a moth); (k) *Papilio bootes*, mimicked by (l) *Epicopeia polydora* (a moth); (m) *Abispa australis* (a wasp), mimicked by (n) *Dasypogon* sp. (a robber fly).

Mullerian mimicry is an extension of 'ordinary' aposematic coloration involving two or more species, and most of the principles of aposematic coloration apply. In aposematic coloration there is strong stabilizing selection, promoting monomorphism within species (see section 6.4). The same will be true for two or more species. If they initially are similar enough so that predators can generalize between their two colour patterns, then modifier genes will spread if they make the two species more similar in appearance. The result is convergence of colour patterns and several species being Mullerian mimics of each other; any colour pattern deviants are selected against (Fig. 6.8). The principles are not limited to visual signals; there is some evidence that several genera of Actiid moths, differing in distastefulness to bats, have converged on warning ultrasonic pulses (Edmunds, 1974). Both the predator and all mimic species benefit from the presence of the similar colour patterns so the system will be stable and resistant to change (Turner, 1984; Turner *et al.*, 1984; Endler, 1988). However, the system can change if a new more noxious form becomes common; this will follow the requirements for the formation of a new aposematic pattern (see

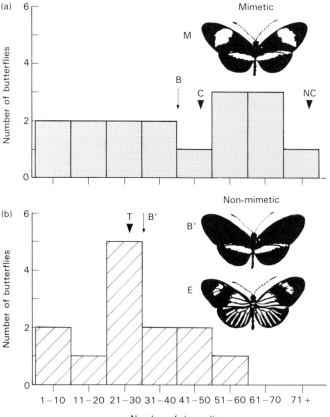

Fig. 6.8 Mean number of days alive in capture–recapture experiments with *Heliconius erato* from Costa Rica (Benson, 1972) and Peru (Mallet & Barton, 1989). Butterfly M is variety *favorinus* (used by Mallet and Barton) and is very similar to *petiverana* (used by Benson). Benson (1972) painted *petiverana* to make them non-mimetic (butterfly B′). Mallet and Barton transferred butterflies across a hybrid zone between *favorinius* (butterfly M) and *emma* (Butterfly E); the transferred butterflies were not mimetic in the area of the other geographical variety. The histograms show the distribution of data from individual butterflies in Benson's (1972) experiment in Costa Rica in which he altered the colour pattern of some butterflies (B′) but not others (M). Arrows (labelled B and B′) indicate the means for Benson's two experiments (Benson, 1972, table 1). Butterflies with their colour pattern altered (B′, non-mimetic) did not survive as long as controls (M, mimetic). Inverted triangles are the estimated life expectancies of Peruvian butterflies in Mallet and Barton's (1989, table 2) experiments in which they compared the survivorship of butterflies transferred (T) to the other side of the hybrid zone to butterflies transferred between localities on the same side of the zone (C). An additional control was butterflies that were marked and released at the same site (NC). The butterflies transferred across the hybrid zone (T) did not have the same colour pattern as the host population, and showed a much lower survivorship than both controls (C and NC).

previous section), at least within populations. Another possibility for change arises if the degree of noxiousness of one species is considerably less than that of the other. In this case the less noxious species will benefit more from the Mullerian mimicry than the more noxious species. As a result, stabilizing selection will be weaker on the more protected species relative to the less protected species, and any changes in the more protected species will be quickly followed by convergence by the less noxious form. This is a weak form of what happens in Batesian mimicry, but the effect would presumably not be as large, because both species benefit from each other's distastefulness.

Alternatively, if the colour patterns of aposematic species are so different that predators cannot generalize between them, then they will probably evolve independently. Any given place can have a mixture of similarly and differently coloured species; here the most similar ones will converge on each other, leaving a few classes of colour patterns which are sufficiently different so that they cannot converge. These groups of similar species are known as mimicry rings, and can involve species from many different families and orders (Turner, 1984). An example is shown in Fig. 6.9.

One unsolved problem is that mimicry rings frequently show parallel geographical variation; the colour patterns within a given ring change abruptly in certain places, and these places often correspond to changes in other mimicry rings. There are several possible explanations including ancestral independent differentiation in formerly isolated patches of habitat, geographical variation in the most distasteful species hence the direction of divergence within the same ring in different palaces, and the movement and aggregation of the clines between the zones of different colour patterns (Turner, 1984; Turner et al., 1984; Mallet, 1986). The dynamics of mimicry rings are essentially unexplored.

Batesian mimicry is rather different in mechanism and evolutionary dynamics from Mullerian mimicry and aposematic coloration (Turner et al., 1984). Here only the mimic benefits while both the model (increased attack rate) and predator (lost food) lose. As a result of the asymmetry, genes which cause the mimic to resemble the model more closely will spread, but so will genes which cause the model to look different from the mimic (so long as they do not look so different that they no longer are recognizable as the same model). As a result, Batesian mimicry can be unstable and potentially subject to rapid directional evolution. This is an additional possible explanation for geographical variation in mimicry rings, since they often include Batesian as well as Mullerian mimics. For detailed discussion see Charlesworth and Charlesworth (1975) and Turner (1984; Turner et al., 1984).

It is not clear how Batesian mimicry originates (Wickler, 1968; Turner, 1984; Turner et al,, 1984). When a new Batesian mimic first appears it probably resembles the model only slightly. The model may already have enough genetic variation to diverge from the new mimic because it presumably has had a history of divergence from other mimics. Consequently, the new mimic may not get a chance to 'catch up' and converge on the model by the spread of colour pattern modifier genes. On the other hand, a new mimic may be initially rare, and therefore cause only a slight drop in the model's fitness, and therefore only a very slow rate of divergence from the

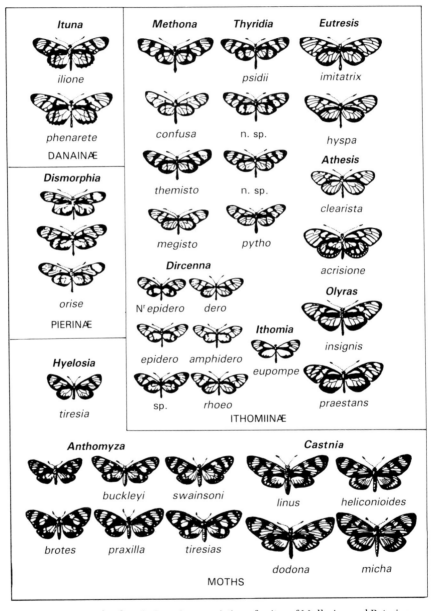

Fig. 6.9 An example of a mimicry ring consisting of suites of Mullerian and Batesian mimics belonging to many species in different families of butterflies and moths (from Poulton, 1898, plate 42). This is the clear-wing ring of South America.

mimic. A second barrier to the origin of new Batesian mimics is more difficult to overcome. The new mimic which partially resembles a model will experience a significant fitness loss because it shifted from a cryptic to a conspicuous colour pattern but is not noxious. In fact the new mimic can spread only if this fitness loss is more than offset by the new protection by the model. Since the degree of dissimilarity in appearance between mimic and model, which will still be protected by the model, is directly proportional to the degree of noxiousness, we would expect Batesian mimicry to evolve more frequently with very noxious prey. This would be difficult to test because noxiousness itself may evolve and, in addition, noxiousness is known to vary within and between food plants in insect models.

The origin of Mullerian mimicry does not require more stringent conditions than aposematic coloration, but Batesian mimicry does. New Batesian mimics face not only the problems faced by new aposematic forms (rarity, hence poor predator training), but also the problem of no protection by noxiousness. They face additional problems if their new model is rare, or effectively rare because it does not closely resemble the mimics. This makes Batesian mimicry seem very unlikely if we assume it always evolves from crypsis. On the other hand, it is relatively more likely that Batesian mimicry evolved from Mullerian mimicry; becoming palatable may allow resources formerly allocated to being noxious to be reallocated towards producing more young. Testing this would require a detailed phylogenetic analysis of both model and mimic.

6.6 Which defences should be used

Which defences are actually used depends upon their relative costs and benefits and the evolutionary history of the group, but this is an essentially unexplored problem (Endler, 1986a). It is to the advantage of the prey to interrupt the predation sequence as early as possible for four reasons: risk, current energy, future fitness and relative frequency of events.

Risk increases with predation stage for at least three reasons: (i) even if the probability of surviving between one stage and the next is high, the probability of getting through all stages is low. If s_i is the probability of passing through stage i, then the probability of surviving all stages is $S = s_1 s_2 s_3 s_4 s_5 s_6$. For example if each $s_i = 0.9$ then $S = 0.9^6 = 0.53$, whereas the probability of passing through the first two stages is 0.81; (ii) the predator is closer with later stages, so escape is less likely; and (iii) defences used later in predation require a greater and more rapid energy expenditure than defences used earlier (Table 6.1), so the probability of exhaustion, and hence 'unfair' capture, increases with predation stage. All three factors increase risk with stage.

The rate and cumulative energy used on all defences increase with stage. As a result, an animal that has escaped late in the sequence will have less energy available for defence in future encounters than if it had escaped at an earlier stage. Early escape also saves energy which could be allocated towards growth and reproduction, so it will be favoured by both viability and fertility selection.

A fourth reason for early termination arises from the frequency of occurrence of each stage. Predation and predator defence go back at least to the early Cambrian (Vermeij, 1987), so we can assume that prey have evolved many methods to terminate predation as early as possible. As a result, we would expect the frequency of predator encounters terminating early to outnumber the frequency of encounters terminating later. The cost C_j of a defence j is proportional to the product of its energetic requirement E_j and the frequency with which it is used f_j: $C_j = E_j f_j$. The total cost of surviving an encounter with a predator is $C = \Sigma C_j = \Sigma E_j f_j$ ($j = 1,2,\ldots,n$ for all n defences), so selection will favour prey that minimize C. If costs per defence were equal among defence mechanisms, then selection should differentially affect the more frequent defences, by making them more efficient (lower E_j) or less frequent (lower f_j). Given that later defences are less frequent and more costly, selection for greater efficiency of early defences, and hence earlier termination of predator encounters, is even greater.

These considerations alone may not be sufficient to explain the use of particular defences because other factors may interact with them. Sociality, which can evolve for a variety of reasons which have nothing to do with predation, results in more effective vigilance towards predators than is possible for solitary animals (Bertram, 1978), automatically increasing the effectiveness of defences at the earlier stages. Even in the absence of sociality, other factors may affect which defence is used. For example, the degree of development of the defence may depend strongly upon the local risk of predation, leading to geographical variation in defences such as colour patterns (Endler, 1978), escape behaviour (Magurran, 1989) and resistance to snake venom (Poran et al., 1987; Towers & Coss, 1990). Reproductive or physiological state may determine whether to 'freeze' or flee, given the same predator (Bauwens & Thoen, 1981), as may the degree of crypsis (Heatwole, 1968; Radabaugh, 1989). In order to feed (or mate), prey may have to move into areas of greater predator risk, and this may also depend upon the current condition or age of the prey (Abrahams & Dill, 1989). Finally, there may be developmental and biophysical limits to what defences a particular lineage may adopt. For example, armoured fishes are less likely to flee and do so at shorter distances than weakly armoured or unarmoured species (McLean & Godin, 1989). It is difficult to separate out cause and effect here: either (i) armour may reduce the need to flee predators; or (ii) fleeing may be more difficult if one is heavily armoured. Either reason (or both) can constrain the subsequent evolution of entire suites of defence mechanisms. Even if there are no constraints of this kind, the fact that different suites of defences may be equally effective (equal C), may allow considerable diversity of defences among species. This would repay further study in a variety of phyla and classes.

6.7 Defence mitigation by predators

Apart from descriptive studies, the methods that predators use to mitigate prey defences have received little attention. The only thorough treatment from the predator's viewpoint is that of optimal foraging (Stephens & Krebs, 1986) and related approaches (Griffiths, 1980; Leimar et al., 1986; Endler, 1988). But these approaches deal primarily with time budgets and the choice of prey on the basis of overall profitability. They rarely (Griffiths, 1980) address the question of what kinds of prey defences it would be most profitable to overcome, and what kinds of predatory behaviour would be best for a given suite of defences.

One pattern which does seem to emerge is that there is a tendency for generalized predation methods to be used early and specialized methods later in the predation sequence. Methods used early in the sequence apply to most prey, such as good and diverse sensory systems, speed, strength, learning and so on. Methods used late in the sequence are more specific, and include long teeth for slippery prey (slug, fish and skink-eating predators), disarticulating jaws for large prey (snakes), venom for large, swift or dangerous prey (snakes, spiders, scorpions, cone shells), and large flat teeth and other devices (fishes, lizards, crabs) for crushing prey with shells (see also Curio, 1976; Vermeij, 1987). Lures (fish, turtles, snakes) are one exception; they may attract any kind of prey and eliminate the detection and approach stages and probably the recognition stage too. Strength is another exception, as it is a benefit at all but the first two stages.

There are two possible reasons why generalized predatory methods should operate early and specialized ones later. First, because defences that operate early in the predation sequence are by their nature very generalized and not that variable among species; only generalized methods are required to overcome them. For example, detecting crypsis, deciding to attack, and approaching the prey are based upon very general sensory, cognitive and physiological properties rather than species-specific details, although, for example, sensory systems will be specific to specific habitats (Endler, 1986a). By the same argument, methods of attack, subjugation and killing will depend upon specific details of prey defences, morphology and habitat, and the relative size of predator and prey (for example, Griffiths, 1980), and so must be more prey-specific, as in an oystercatcher's bill or a planktivore's or piscivore's mouth.

A second possibility arises from a consideration of relative costs of different defence mitigating mechanisms, similar to that given in the previous section. Generalized methods may be less expensive energetically than specialized methods because they are used continuously at a low rate rather than intensely at a high rate. For example, vision is probably less costly than subduing strong or rapid prey. In addition, generalized methods can also be used for other purposes such as finding mates and holding territories (vision, learning, etc.), something relatively or absolutely impractical for specialized predatory mechanisms. This reduces the costs of generalized relative to specialized predatory mechanisms even further. Because defences acting early in the predation sequence are more generalized, more frequent and energetically less costly to overcome than late defences, generalized predator mechanisms are favoured early and specialized mechanisms late in the predation sequence.

6.8 Arms races and coevolution

The long-term evolution of species involved in a predator–prey interaction has frequently been regarded as an arms race or a case of coevolution. Coevolution requires a specific evolutionary response by both species: specific new defences by the prey must be continually counteracted by specific new defence-breakers in the predators and vice versa (Janzen, 1980). Such specific mutual adaptation is known in host–parasite systems (references in Endler, 1986b), as well as brood parasite–host systems (such as cuckoos, cowbirds, slave-making ants and ant inquilines; Davies et al., 1989). For example, in cuckoos (Cuculus canorus) and cowbirds (Molothrus ater) laying behaviour is specifically adapted to defeat host defences, the degree of perfection of egg mimicry varies with the degree of host discrimination, host populations outside the range of the parasite show less strong egg rejection behaviour, and species unfamiliar with the parasites show little egg rejection (Davies et al., 1989). But true coevolution (Janzen, 1980) seems remarkably rare in predator–prey systems (Peckarsky, 1985). Two very broad reviews (Edmunds, 1974; Vermeij, 1987) discuss many cases of specialized defences, and specialized predatory counter-defences, but between them there seems to be only one possible case of specific coevolution. Araneus cereolus of Ghana is unusual among orb web spiders in that it builds its web at dusk and takes it down at dawn, presumably a specific defence against diurnal spider-hunting wasps, which apparently hunt by sight. But Sceliphron spider wasps have specialized on A. cereolus by becoming crepuscular (Edmunds, 1974).

There is a fair number of cases of specific predator and prey adaptations (Edmunds, 1974; Curio, 1976; Vermeij, 1987), but it is not clear in these cases that both predator and prey specifically evolved under the influence of each other more than one step each, or if some of the adaptations or counter-adaptations result from generalized defence and counter-defence (Janzen, 1980). Examples are the specific escape behaviour by the gastropod genera Struthiolaria and Nassarius when attacked by starfish but not by other predators; false burrows in digger wasp nest holes which are known to reduce parasitism rates; the specialized ultrasonic sound detectors of Celerio moths and lacewings, which allows them to be particularly adept at avoiding sonar-hunting bats (Edmunds, 1974); and the genetically distinct 'rover' and 'sitter' larval behaviour in Drosophila which allows each genotype to escape from different species of parasitoid wasps (Carton

& David, 1985; Sokolowski, 1985; Y. Carton, personal communication). There are many cases of specialized predators, and specialized prey defences, but very few cases in which both are found together in the same system. Generally, what seems to be found is groups of prey types coevolving with groups of predators, a form of 'diffuse coevolution' (Janzen, 1980); this is best documented in the fossil record of molluscs and other shelled animals, as in Fig. 6.10 (Vermeij, 1987).

Even within this pattern it is more common to find more prey evolving more specialized defences and fewer predators evolving more specialized defence mitigators. A good example is in fish and hard-shelled prey. There are many fish which eat hard-shelled prey by crushing them, and their prey

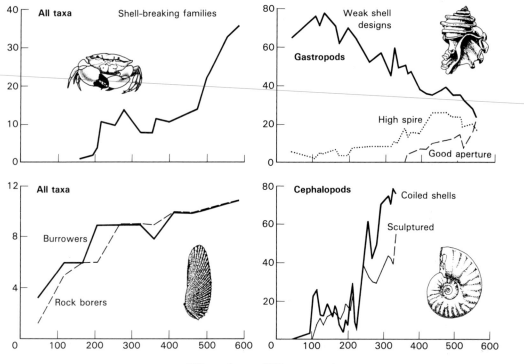

Fig. 6.10 Diffuse coevolution of predators and prey in the fossil record (data from Vermeij, 1987). Except for number of shell-breaking families, the vertical axes are the percentage of taxa with the noted traits. High-spired gastropod shells are more difficult to break. Gastropods with good apertures are those where the aperture is internally thickened or narrowed, both excellent defences against predators. Tightly coiled and sculptured cephalopod shells are also excellent defences against predation. Note how as the number of shell-breaking families increased, the fraction of taxa that burrow in soft or hard substrates increases, the fraction of taxa with predation-resistant shell designs increases, and the fraction of taxa (gastropods) with weak shell designs (umbilicate or loosely coiled) declines. The frequency of damaged and drilled shells has also increased in time (not shown).

probably evolved hard shells against all predators (Vermeij, 1987). But there are hardly any fish species which specialize on hard-shelled molluscs (Norton, 1988). One which does (the Cottid *Asemichthys taylori*) has specialized teeth to puncture rather than crush the snails, but this is unusual (Norton, 1988). Similarly, there are hardly any species (mostly snakes) which specialize on scincid lizards. Skinks have very smooth scales and their bodies are very flexible, so they are difficult to grasp, even for snakes with long sharp teeth. But the Colubrid snake *Psammodynastes pulverulentus* and a handful of other species specialize on skinks (Greene, 1989). This is aided by alternating groups of long and short teeth, which act like a ratchet and trap the 'slippery, writhing prey' in the mouth. They also have enlarged rear fangs which deliver venom from Duvernoy's glands, quickly immobilizing the prey. These species have evolved these mechanisms in parallel, but they are a small fraction of the world snake fauna (Greene, 1989).

Generally, it appears that there is an assymetry in the rates of evolution of prey defences and predator mechanisms. Therefore it would be much better to refer to joint evolution of predators and prey as in 'arms race' (Dawkins & Krebs, 1979), 'diffuse coevolution' (Janzen, 1980) or 'escalation' (Vermeij, 1987) rather than coevolution (Janzen, 1980). This does not imply specific detailed mutual evolutionary responses but only that each species responds in some way to the others.

Why is precise coevolution of the type found in some host–parasite systems rare or absent in predator–prey systems? Why are there so many more specialized defences by prey than specialized methods in predators? Here are some speculative reasons: (i) artificial selection experiments (Falconer, 1981) and studies of evolved pesticide resistance (Endler, 1986b) have shown that it is easier to select for one trait than many. Most predators prey on several prey species. This makes it difficult for predators to specialize on particular species. On the other hand, parasites are most often specific to their hosts, inducing a more consistent and simpler selection pressure, making specialization much more likely; (ii) apostatic effects make specialization by predators still more difficult. If partial specialization on one prey species results in the reduction of that prey's density, then the predator may switch to another species, minimizing selection for specialization on both (Dawkins & Krebs, 1979); (iii) specialized defences are normally found later in the predation sequence than generalized defences, so predators experience generalized defences more frequently than specialized defences. This may favour more frequent adaptation by predators to the more frequent generalized defences. This might only allow specialization later in the sequence if detection, recognition and approach are energetically trivial parts of the sequence (see also Griffiths, 1980); (iv) the life/dinner principle emphasizes the fact that a missed predation attempt saves the life of the prey, but loses only a meal for a predator (Dawkins & Krebs, 1979). This implies stronger selection for defences than for counter-defences in the same way that there is stronger selection on Batesian mimics to converge on their model than on the model to diverge from the mimic (section 6.2.4); (v) it is a truism that predators have lower densities than prey. This might be associated with lower effective population sizes of predators compared with prey. This results in a higher rate of genetic drift and inbreeding, a lower degree of genetic variation, and a slower response to directional natural selection in predators compared with prey. The same kind of argument may also explain why host–parasite systems are more tightly coevolved; parasite numbers may be equal to or greater than that of their hosts; allowing parasites to evolve at least as fast as their hosts; and (vi) in many systems there are several generations of prey to every generation of predators; this may allow faster evolution by prey than predators (Dawkins & Krebs, 1979). All of these ideas are in desperate need of study.

PART 3

SEXUAL SELECTION AND REPRODUCTIVE STRATEGIES

Introduction

Whereas during the early days of behavioural ecology, in the late 1960s and the 1970s, most studies were concerned with developing and testing models of resource exploitation, particularly foraging, recent years have seen a remarkable surge of interest in sexual selection and mating systems. Field-workers have been attracted to this field because of the opportunities for direct measurement of the influence of traits on an individual's fitness, for example numbers of mates gained and offspring produced. New molecular techniques for assigning maternity and paternity (e.g. DNA fingerprinting) have provided powerful tools for testing how behaviour patterns influence reproductive success.

At the same time there has been a marked increase in the number of theoretical studies devoted to the evolution of mate choice, bizarre male adornments and pattern of parental care. The three chapters in this section review these recent theoretical and empirical results.

Darwin's theory of sexual selection is concerned with the evolution of characters involved in competition for mates. In Chapter 7, Harvey and Bradbury review the various models of sexual selection. There has been little controversy over the process of *intrasexual selection*, where traits such as horns and antlers have clearly evolved in relation to direct combat between males for mates. It is the second process, namely *intersexual selection* or female choice, which has attracted the most attention. Some aspects of female choice are clearly simply concerned with choosing a mate of the right species, or choosing a conspecific who will provide good resources in the form of parental care or a good territory. However in other cases males provide females only with sperm, yet females still appear to exercise careful choice. It is this last process which has attracted controversy and is exemplified by the problem of how the tail of the male peacock could have evolved.

Chapter 7 summarizes recent models of Fisher's 'runaway process', in which female preference and the male trait coevolve to an equilibrium, as well as various models derived from Amotz Zahavi's idea that traits such as a peacock's tail are handicaps which act as honest advertisements of a male's viability, thus enabling females to choose for good genes. When Zahavi first proposed his idea, theoreticians were rather quick to dismiss it out of hand. However, recent work by Grafen (1990a,b; section 1.4, this volume) (see also Pomiankowski, 1987b and Kirkpatrick, 1987b) has shown that handicaps can indeed evolve under certain conditions. This is a nice illustration of the general point that predictions from models vary depending on their precise assumptions, as well as showing that it is worthwhile making explicit models of verbal arguments however plausible or implausible they may at first seem!

One of the problems of 'good gene' models of female choice is that of maintaining genetic variability; if females choose to mate only with the longest-tailed males, for example, then the genetic variance for tail length diminishes so that after several generations it no longer pays females to choose. Hamilton and Zuk (1982) proposed a novel solution to this problem. Their idea is that male ornaments and displays are 'revealing handicaps' which signal genetic resistance to parasite infection. Coevolution between parasite and host can result in constantly changing selective pressures so that the best host genotype varies with time. This results in maintenance of genetic variability and so maintains the advantage of female choice. Harvey and Bradbury assess the comparative and experimental tests of this idea. One of the problems to be considered is that it may pay a female to avoid mating with a diseased male simply to avoid getting the disease herself, even in the absence of heritable disease resistance.

In Chapter 8, Clutton-Brock and Godfray discuss parental investment, particularly the problem of how a parent should best allocate resources among offspring. They consider whether a parent should care or desert, if it cares how much care and whether different amounts of investment should be devoted to sons versus daughters. Maynard Smith's (1977) well-known model specifies the conditions under which no care, uniparental male or female care, or biparental care can be evolutionarily stable. The chapter shows how comparative data support the idea that parental care patterns are influenced by both the costs and benefits of desertion by either sex. However, Clutton-Brock and Godfray point out that tests of Maynard Smith's model in contemporary populations may not tell us much about the initial evolution of parental care because characteristics of populations today may have evolved away from initial conditions. Nevertheless, there is excellent scope for experimental work in species where either sex may care for the young (e.g. cichlid fish). The chapter goes on to consider how biparental care can be stable when each sex has a 'reaction' to changes in investment by its partner. In the final section the authors discuss adaptive variations in sex allocation. The most striking examples come from parasitic wasps and social hymenoptera, where a female can control the sex ratio of her progeny precisely by either fertilizing an egg or not (fertilized eggs develop into females, unfertilized eggs into males). However, even in vertebrates where sex is determined by sex chromosomes, there is good evidence for adaptive sex-ratio bias, for example in relation to maternal condition, though the mechanisms involved are unknown.

In the final chapter of the section, Davies shows how two themes can help to explain the great variability in animal mating systems. The first is the idea that because female reproductive success tends to be limited by resources while male reproductive success tends to be limited by access to females, resources influence female distribution and female distribution, in turn, influences male distribution. Comparative studies of mammals, lizards, amphibians and insects show how different mating systems may emerge depending on how females can best be monopolized by males in space and time. In birds, where males often provide parental care, female dispersion is influenced by both resources (e.g. food, nest sites) and males. The second main theme in that the mating system is often an outcome of patterns of desertion by either sex. Two points are emphasized. The first is that there are a variety of models which can explain mating systems, such as leks and resource-defence polygyny, and different models may apply not only to different species but to different populations within a species, or even different individuals within a population. The second is that it is important to focus on the alternative options facing each individual in a system and to recognize that, with conflicts of interest, not all individuals will achieve their preferred option.

7: Sexual selection

Paul H. Harvey and Jack W. Bradbury

7.1 Introduction

These are exciting and challenging days for students of sexual selection. From Darwin's time until the 1960s, with the notable exception of R.A. Fisher and Julian Huxley's work in the 1930s, the subject lay dormant. Since the 1960s there has been ever-increasing effort put into solving the problems in sexual selection that Darwin left us. That effort has paid off with a number of important developments which come from new theoretical insights, key observational and experimental studies, and cross-taxonomic comparisons. Our aim in this chapter is not to survey the whole field of sexual selection, but rather to focus on these recent developments.

Sexual selection was sketched in the *Origin of Species* (Darwin, 1859) and variously elaborated in its later editions, but it was only developed as a topic in its own right sometime later in *The Descent of Man* and *Selection in Relation to Sex* (Darwin, 1871; 1874). Darwin was puzzled by the selective forces responsible for the evolution of characters, often sexually dimorphic characters, that seemed to hinder an organism's chances of survival and which were necessary for sexual reproduction only in the presence of competition for mates. Such characters were considered to evolve by sexual selection which 'depends on the advantage which certain individuals have over other individuals of the same sex and species, in exclusive relation to reproduction' (Darwin, 1871, Pt. 1, p. 256). By defining the process this way, Darwin was able to exclude primary sexual characters such as genitalia, reproductive tracts and mammary glands. He was able to focus on what 'the illustrious Hunter' (Darwin, 1871, Pt. I, p. 273) had called secondary sexual characters (Hunter, 1837; 1861). The key issue in Darwin's process was competition for mates.

To make the point about competition for mates clear, Darwin considered the evolution of prehensile organs used by male moths and crabs during mating. He distinguished between (i) the forces of natural selection responsible for the evolution of prehensile organs which are necessary for a male to hold a female during copulation, and (ii) the forces of sexual selection that may have taken over to prevent either (a) the female choosing to leave and mate with other males or (b) the male being dislodged from the female by another suitor. Darwin despaired that 'in most cases it is scarcely possible to distinguish between the effects of natural and sexual selection' (i.e. between (i) and (ii) in the above example). This problem persists in many modern studies.

Fig. 7.1 Two drawings taken from Darwin (1871). (a) The plumage of male birds of paradise, such as this *Paradisea rubra*, was considered to have evolved by female choice. (b) The horns of the kudu were thought to have been selected for by male combat, although Darwin considered that their beauty may have been enhanced by female choice.

A second distinction made by Darwin was between sexual selection due to female choice and that due to male–male competition (e.g. (a) versus (b) above: see Fig. 7.1). This also has proved difficult to apply in practice. For example, consider pheasants. First, the males have long ornate tails, a classic case of a character considered to have evolved through female choice among males. Second, the males have tarsal spurs that are known to be used in male–male combat. Reference to Fig. 7.2, drawn from a photograph, shows how the tail is used by the golden pheasant (*Chrysolophus pictus*) during fights between males over mating access for females. On the other hand, the observational and experimental work by von Schantz *et al.* (1989) suggests that ring-necked pheasant (*C. colchicus*) hens do not choose mates on the basis of plumage characters but on the length of their spurs! We do not doubt that female choice is in part responsible for the evolution of larger tails in males or that male combat is in part responsible for the evolution of effective spurs. However, the relative extent to which each process has been involved in the evolution of tails and spurs is unknown.

This chapter reviews recent progress in our understanding of sexual selection and contrasts results of theoretical models, observational and experimental analyses, and the comparative method. We focus in turn on intrasexual selection (section 7.2), intersexual selection (section 7.3), sperm

Fig. 7.2 Two male golden pheasants (*Chrysolophus pictus*) using their elaborate tails for support during fighting (drawn from a photograph in McMahon & Bonner, 1983).

competition (section 7.4) and sexual selection in plants (section 7.5). These last two topics were not considered seriously by Darwin under the rubric of sexual selection. Yet, just as there is competition among males for mating access to females, so there is also competition among the sperm of different males for access to fertile eggs. Sexual selection occurs in plants just as it does in animals (Darwin relegated sexual selection in plants to a footnote), and the experimental manipulation of flowers can be easier than manipulating the sexually selected organs of animals.

7.2 Intrasexual selection

7.2.1 *Models of intrasexual selection*

Conflict between members of the same sex over mates is so common in nature that Darwin referred to it as the 'Law of Battle'. He argued that the consequent intrasexual selection was bound to favour special weapons and defensive organs, sexual differences in body size and shape, and the elaboration of various communicatory organs to threaten opponents. The fact that some of these structures might be costly to the males bearing them, either energetically or in terms of predator risk, was not a problem if the gains in reproductive success more than made up for the losses in viability. We might expect intrasexual contests to lead to arms races because mating success depends not on the absolute size or efficacy of such structures, but on their relative values in the current population. If the traits were at all heritable, selection favouring larger males or those with nastier horns would shift the mean in the next generation to higher values and, given associated costs, thereby lower the viability of the average male. We can imagine three outcomes to this process: (i) the arms race continues unabated until adult males are so rare that the species goes extinct; (ii) the population cycles with ever larger forms invading until adult males become so rare that the smaller forms can reinvade and thus reinitiate the process; or (iii) the population goes to some stable distribution of male sizes at which the costs and benefits of exaggerated traits are balanced.

Intrasexual arms races have been modelled by a number of authors. Some models are based on game theory with haploid reproduction, whereas others are more formal genetic models. The general result is the same: when stability exists, it consists of a polymorphic distribution of male trait values with the mean displaced from the male ecological optimum. Stable male distributions are likely to occur if the costs of armaments increase in an accelerating fashion with each additional increment in size and if there is some environmentally induced variation in the production of such armaments and size (Gadgil, 1972; Haigh & Rose, 1980; Parker, 1983; Charlesworth, 1984; Maynard Smith & Brown, 1986). In a discrete genetic model, Maynard Smith and Brown (1986) even obtained polymorphic stability without environmental variability if costs were sufficiently high. If the two conditions are not met, cycling or extinction are likely outcomes. Since both conditions are probably satisfied in many natural circumstances, these models support Darwin's proposition that intrasexual competition over mates can lead to relatively stable exaggerations of male traits even though they reduce mean male viability.

In addition, Parker (1983) provided a theoretical justification for several predictions made by Darwin, and added a few of his own. Specifically, his evolutionarily stable strategy models predicted that (i) the evolutionarily stable male investment in size or armaments should be higher when the ratio of defended females to defending males is high; (ii) the stable male investment should be lower when the environmentally induced phenotypic variance in male armaments is higher because, as Parker puts it, 'the less the variance in armament, the greater the value of a unit increase in armament' (Parker, 1983, p. 634); and (iii) the stable male investment should be higher when the phenotypic distribution of male armaments is skewed to lower values than when the opposite is true. In vertebrates, where many male traits such as body size are age dependent, older males are usually rarer than younger ones and thus the distribution of male trait sizes is skewed to small values; in insects, most adults are normally distributed in size or even skewed to larger sizes. Thus, Parker predicts relatively higher mean investment in male intrasexual traits among vertebrates than insects.

7.2.2 Observational and experimental studies of intrasexual selection

If the models of intrasexual selection are relevant, we might expect to find significant positive correlations between intrasexually exaggerated male traits and male reproductive success, and opposite but similarly sized negative correlations between these traits and male viability. In natural populations, it has so far proved difficult to measure these components with any accuracy in the same system. Even if field measurements were made, the results could be difficult to interpret. In particular, individual differences in the ability to acquire resources can have important consequences. For example, males reared in good territories may be able to allocate more resources to both survival and reproduction, leading to positive correlations between sexually selected male traits and male viability among individuals. This problem can be ameliorated by experimentally transferring young between parents.

Furthermore, it is often possible that the costs of producing large horns or growing a large body during maturation are reflected in unrecorded differential survival among immature stages. Laboratory systems have the advantage that it is possible to measure selection differentials, effective heritabilities, and responses to selection. Wilkinson (1987) examined the relationship between male wing length, mating success, and larval mortality in laboratory *Drosophila melanogaster*. In this experimental system, there appeared to be little sexual selection through female choice and differentials among males probably resulted from intrasexual competition. Wilkinson found that the response of wing length to sexual selection was positive whereas the response of larval viability to selection on adult male wing length was, as theory predicts, opposite in sign but of similar magnitude to the sexual selection effect. He suggested that longer winged flies required longer periods in the risky larval medium and this reduced their juvenile survival.

Many field studies have partitioned the variance in lifetime reproductive success into sexual and viability components, but few have linked these to specific male traits, and even fewer to male traits used primarily in intrasexual selection (see examples in Clutton-Brock, 1988). Price (1984a,b) found that beak depth in male Darwin's finches had similarly signed (instead of the predicted opposite) selection gradients for both reproductive success and for adult survival, even after correcting for correlations of beak depth with body weight, territory size, plumage patterns, and other beak measures. (Selection gradient refers to the slopes of the graphs relating, in this case, beak depth to reproductive success and adult survival.) The degree to which male reproductive success is due to intrasexual versus intersexual factors is unclear in these monogamous birds. It is possible, therefore, that the intrasexual model is inappropriate in this case. Conner (1988) demonstrated significant positive selection gradients between horn size in fungus beetles and mating success controlling for correlations between horn size and other body size measures. Horns were clearly involved in male combat and not in female choice (Conner, 1989). However, as with Price's finches, the selection gradients associated with viability were also large and positive, instead of negative as predicted by the models.

These examples demonstrate the difficulty in testing many of the quantitative predictions of intrasexual selection theory in the field. The cost of sexually selected armaments or larger body size may well be higher mortality during periods or stages other than those monitored to measure reproductive success. It is often difficult to monitor free-ranging organisms at all life stages, and there is always the problem of determining what the adult trait values of dead immatures would have been had they matured. In addition, to compute selection gradients accurately, one must collect concomitant data on many sexually dimorphic male traits in a sufficiently large sample that the necessary multiple regressions can sort out the various effects. While these conditions are not impossible to meet, they are sufficiently severe that progress using the observational approach has been very slow. Even if the conditions were met, positive rather than negative associations between viability and the development of sexually selected traits (as found with Price and Conner's studies) could still result from confounding effects of environmental differences during rearing discussed earlier.

7.2.3 Comparative studies

While the predicted dynamics of intrasexual selection remain difficult to test, many of the other more general predictions have been examined using comparative methods. As noted above, Parker's (1983) models predict different degrees of male exaggeration depending on the skew in the mating distribution and the shape of the male trait distribution. Given no concurrent selection on females, the degree of sexual dimorphism in a species can be used as one measure of the magnitude of the intrasexual exaggeration of male traits. This suggests that comparative analyses of sexual dimorphism versus mating skew, male trait variance, or male trait distribution skew would serve as good tests of the models. Critics, however, could argue that the assumption of no concurrent selection on female trait values is often violated, that natural selection alone can lead to sexual dimorphism, or that interspecific differences in body size can generate differential dimorphism through the allometric processes without invoking any sexual selection. Any of these factors might complicate comparative tests. In fact, comparative methods have proved to be reasonably robust and generally support the predictions of intrasexual selection theory. Arak (1988b) took concurrent selection on female body size of anurans into account and still demonstrated strong links between sexual dimorphism and skew in male mating success. Even when allometric factors, diet and predation pressures are controlled for, clear correlations can be demonstrated between mating skew on the one hand and male traits such as larger body size, longer canines, thicker or more more elaborate horns and antlers, and presence of dermal shields on the other (Clutton-Brock & Harvey, 1977; Harvey et al., 1978; Packer, 1983; Arak, 1988b; Jarman, 1988).

7.3 Intersexual selection

The peacock's tail has become the paradigm case for a character that *must* have evolved through female choice or, more generally, through intersexual selection (Petrie *et al.* 1990). How could bizarre male characters evolve through female choice? For Darwin, there did not seem to be a problem. He knew of many extravagant male characters that obviously impaired survival, and the answer was that they must have evolved because females preferred to mate with males bearing such characters. In his own words '(t)he case of the male Argus pheasant is eminently interesting because it affords good evidence that the most refined beauty may serve as a charm for the female, and for no other purpose' (Darwin, 1871, Pt. II, p. 92). But why should female Argus pheasants have evolved the particular tastes they have? Why, for example, do they not prefer to mate with purely white males?

7.3.1 *Models of male trait evolution and female preference*

In a landmark paper aptly entitled 'The Evolution of Sexual Preference' Fisher (1915) tackled the question that Darwin had failed to pose. Fisher's answer involved both natural and sexual selection. He imagined a situation in which a phenotypic trait only rarely present in the population, say a long tail or runny nose, became favoured by natural selection. If there were additive genetic variance for the trait, females that chose mates with longer than average tails or runnier than average noses, would produce fitter offspring. But, the male offspring would not only have a survival advantage, they would also have a mating advantage. Fisher claimed that the mating advantage could then take the trait value beyond its natural selection optimum by what has come to be called 'the runaway process'. Why should that be so?

THE FISHERIAN MODEL

Consider a situation in which a given value of a female preference character (p) gives females a preference for mating with males that have a particular trait value (t), but preference is not absolute so that the probability of choosing a mate with a different trait value drops off with the extent to which the trait deviates from the preferred value. If the population contains individuals with a variety of values for both p and t (say that both are normally distributed), females with particular values of p will tend to mate with males having similar values of t. If both female preference and male trait values are under genetic control, genes for the male trait will tend to be found in the same individuals as the genes for its choice, resulting in a genetic co-variance between female preference and male trait.

If the mean values for p and t are the same, neither will change if there is no natural selection differential between individuals with different values of p and t. Now consider what will happen if, through drift perhaps, the mean value for p in the population differs from that of t: those males containing the value of a trait which is under-represented by choice genes in the population of females will leave fewer offspring than those males having the value of trait that is over-presented by choice genes. As a consequence, the mean value t will change. But the mean value of p will also change in the same direction because those males which failed to mate will have contained preference genes inherited from their mothers, and those preference genes will tend to match the trait genes.

For example, if the mean value of t is less than p in the population, males with low values of t may not get mates and their non-expressed preference genes, which will tend to be for small p, will be lost from the population together with their trait genes. If the variance in both p and t is maintained (say by recurrent mutation), and the genetic structure of the population is such that the trajectory of evolutionary change is not towards matching values of p and t, then the male trait and female preference will evolve indefinitely in a runaway process. Eventually, as Fisher pointed out, the population must reach some level of t at which a natural selection disadvantage kicks in and the runaway process stops.

Many assumptions go into the above scenario. For example, populations are assumed to be infinite, males can mate with any number of females, there is no viability cost to female choice, and recurrent mutation maintains the genetic variance in both male trait and female choice. Fisher never presented a formal model of the runaway process, which was left to O'Donald (1962; 1967), Lande (1980; 1981), Kirkpatrick (1982) and Seger (1985). These models were important because they were the first formal models of sexual selection through female choice. They showed that with certain rules of female choice, even if the male trait was costly in viability terms, runaway could occur. They also showed that for each strength of female preference there was a matching development of the male trait at which sexual selection for and natural selection against the trait were balanced. This means that any strength of female preference is a possible equilibrium value; these all fall on the now notorious 'line of equilibria' (Fig. 7.3). The slope of this line is set by the relative strength of the female preference and the viability costs. Note that a balance between mating benefits of exaggerated traits and associated viability costs is common to both female choice and intrasexual models. However, the important difference between the two is that in female choice models, the force determining male fertility, mating preference, is dragged along each time male trait distributions change, building a feedback into the process that is absent from intrasexual selection.

Both Darwin and Fisher also attempted to explain the pronounced colour and pattern dimorphism in many monogamous birds. The tails of monogamous male quetzals (*Pharomachrus* spp.) are much longer than those of many polygynous species (Fig. 7.4). Since males in monogamous birds usually have only a single partner, is it not impossible for sexual selec-

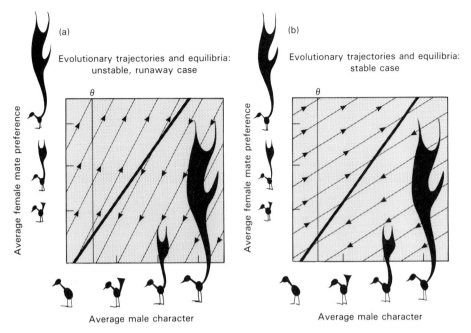

Fig. 7.3 Females choose mates on the basis of tail size. The light lines depict possible trajectories allowed by the underlying genetics for the joint evolution of male tail size and female preferences, and the arrows indicate the direction taken. The male tail size that is optimal for viability is indicated by the symbol θ. The heavier line connects the equilibrium states at which the male's tail length and the female's mating preference cease to undergo evolutionary change (other than by random drift along the line). Whether the evolutionary process leads to unstable (a) or stable (b) equilibria depends on genetic parameters and the strength of viability selection (from Harvey & Arnold, 1982).

tion, particularly by female choice, to operate? Darwin rose to the task and proposed that intersexual selection in monogamous birds could operate (i) if females differ in fecundity, and (ii) if the more fecund females are ready to breed earliest and thus have first choice of mates. Given any female preference for certain male traits, these conditions will lead to higher reproductive success for preferred males as well as the genetic co-variance between female choice and male trait which drives Fisher's process. Fisher gave a numerical example of Darwin's idea which was formally modelled by O'Donald (1980) who found that the proposed conditions can indeed lead to a classical Fisherian process. (An alternative scenario is that females from so-called monogamous species frequently produce offspring that are fathered by extra-pair copulation from attractive males (Møller, 1988a; Houtman, 1991).) A major difference between monogamous and polygynous contexts is that the line of equilibria in the former case is asymptotic: after a certain deviation away from the male ecological optimum, further increases are unlikely regardless of how extreme the female preference becomes (Kirkpatrick *et al.*, 1990). This is a consequence of differences in

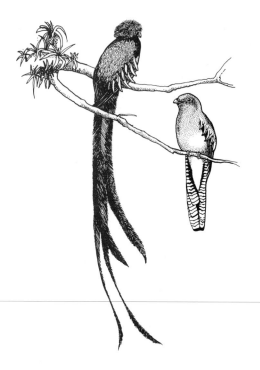

Fig. 7.4 Male and female Resplendent Quetzals, *Pharomachrus mocinno*. The species is reportedly monogamous, but is markedly sexually dimorphic (from Skutch, 1983).

reproductive success among monogamous males being limited by the amount of variation in female fecundity. In polygyny, there is in principle no end to how many mates a male may have. This result helps explain the common perception (beginning with Darwin) that monogamous species are generally less sexually dimorphic than polygynous ones.

But what selected for female preferences in the first place, particularly in species without paternal care? Fisher, as we have mentioned, suggested that genes for preference could invade a random mating population if the male trait was favoured by natural selection. Kirkpatrick (1987a,b) mentions a number of additional mechanisms. Where females already have specific 'search images' or 'releasers' associated with functions other than mate choice (e.g. foraging or maternal care), males with traits that mimic these releasing stimuli are more likely to catch the attention of females. Ryan *et al.* (1990) describe a possible example: basal papillar tuning in the ear of female frogs of the species *Physalaemus coloradorum* may have selected for the evolution of a low frequency male 'chuck call' (but see Gardner (1990) and Pomiankowski and Guilford (1990) who question the evidence). Alternatively, females may evolve preferences for particular male traits to avoid interspecific hybridization. These can also constitute the starting point for a Fisherian process. A number of other mechanisms can be envisaged.

Could the runaway process, or something akin to it, be responsible for the evolution of the peacock's tail? One major factor left out of many models is the possible cost to females of exercising choice. Such costs might be realized if traits preferred by females in males are also expressed in females. Alternatively, there might be immediate costs resulting from the extra effort or risk a female suffers during mate choice. Fisher (1930) and Lande (1980; 1987) showed that exaggeration of a male trait due to inter-sexual selection can occur even if females express the trait at some cost to their own viability. However, selection favours the diminution of the trait's expression in females, which will ultimately return to the ecological optimum (Lande, 1980; Lande & Arnold, 1985). Darwin had noted that the female versions of exaggerated male traits are often rudimentary (e.g. the crest on a peahen).

Immediate costs to females of choice have a more profound effect on the Fisherian models outlined earlier. If there is a cost associated with females choosing males that bear characters which have no benefit to the female or her offspring, the only stable evolutionary equilibrium for the population occurs when female fitness is maximized and therefore females would not be expected to exercise choice (Kirkpatrick, 1985; 1987a,b; Pomiankowski, 1987a,b; 1988). However, as Kirkpatrick (1987a,b) points out, there are many likely contexts in which the exhibition of certain traits or behaviours costly to males will *reduce* the costs to females of finding mates. The females of many vertebrate lekking species forage and roost solitarily and have very large home ranges (Bradbury, 1981; see Chapter 9), while many anurans are active only at night when visibility is low. In such cases, males which adopt bright coloration or loud calls make themselves conspicuous and so obtain more matings, as well as reducing the average time and energy costs of mate search to those females who prefer them. Whether choosy females pay more or less than non-choosy females will clearly depend on the trait chosen and on the ecology of the species considered.

Costly female choice appears to lessen the odds on Fisher's runaway process alone being the explanation of exaggerated traits like the peacock's tail. But one way in which costly choice genes might become established is for mutations on the male trait to be biased: if mutation tends to take the trait away from its natural selection optimum, costly female choice is selected for (Iwasa et al., submitted). Biased mutation is likely, especially in exaggerated or complex traits, because random changes to an organism's genotype will more frequently reduce rather than improve performance. Such bias has recently been documented in *Drosophila* bristle number mutations induced by transposable elements (Lai & Mackay, 1990). It is not yet clear how large an effect such biased mutations have on the evolution of costly female choice.

HANDICAP MODELS

What if costs to female choice are substantial and biased mutation explana-tions seem inappropriate? How could we then explain the extravagant characters seen in many species in which female choice seems to play a major role? A third viability trait lies at the heart of many recent investigations of sexual selection models which might help answer these questions. The models derive originally from a suggestion made by Zahavi (1975) that exaggeration of a male trait will be selected for if it advertises viability by handicapping the survival chances of its possessor (see also section 1.4). By choosing handicapped males as mates, females get viability genes for their offspring. Three types of handicap model have been considered (Maynard Smith, 1985) which differ in their predictions concerning the exaggeration of male traits. It is assumed that males differ in intrinsic viability and that investment in a handicap reduces adult male viability still further.

The first case is what Maynard Smith called the 'Zahavi handicap' model: all males invest in the handicap equally, but only those with high viability can survive the cost of that investment. The handicap thus acts as a 'survival filter' for males with the higher viability genes. The second type of model is the 'condition-dependent handicap' model (deriving from West-Eberhard, 1979): males invest in handicaps to optimize their survival and mating success so that good quality males still survive better than poor quality males despite having larger handicaps (Zeh & Zeh, 1988). The size of the handicap is thus a good index of intrinsic male viability. The third type of model is the 'revealing handicap': the expression of similar investments in handicaps is dependent on intrinsic male viability. Handicap values again are good indices of intrinsic male viability.

To distinguish between the condition-dependent handicap and the revealing handicap, consider the trait to be a long tail, with females preferring to mate with males that have long tails. Under the condition-dependent model, males would grow a tail of the length that maximized lifetime reproductive success, optimizing the balance between lowered viability and attractiveness to females. Under the revealing handicap model, all males with the long tail genes would attempt to grow a long tail, but the length of the tails they actually produced would reveal their intrinsic viabilities. Underlying many suggested examples of revealing handicaps is the idea that contributions to viability are observed directly by the chooser: mates might be chosen on the basis of the colour of their blood which is revealed at the cost of leaving bare patches of skin exposed without a cover of fur or feathers, and so reducing crypsis. With condition-dependent handicaps, some signal is often considered to stand in for viability: only mates with high viability could survive after amputating their digits, but the number of digits they decide to amputate would depend on their viability. It seems that the model Zahavi (1977) may have had in mind was of the conditional handicap, rather than the so-called Zahavi handicap.

Early models of the handicap principle considered only the spread of a single viability allele and, once this had spread to fixation, there was no longer any naturally selected advantage in choosing handicapped males (Davis & O'Donald, 1976; Maynard Smith, 1976; 1978; Kirkpatrick, 1986). More recent models which lead to the spread and maintenance of handicaps within populations assume that substantial heritability for viability exists in natural populations. One often repeated concern of the relevance of such models is the claim that insufficient additive genetic variance for components of fitness is likely to exist in natural populations. Is this likely to be the case? After reviewing the theory and data, Charlesworth (1987, p.21) concluded that there is little evidence for additive variance in fitness beyond that maintained by mutation pressure, but that which exists 'may be sufficient to generate a pressure of selection in favor of mate choice'. Genetic variance in viability may also be maintained within populations if there are continuing sources of environmental deterioration, so that alleles which used to increase viability become ones that decrease viability. An example is parasite–host coevolution, where host resistance alleles quickly become obsolete. Alternatively, recurrent deleterious mutation (a type of biased mutation) may maintain genetic variation in viability.

With discrete genetic models, handicap conditions can produce either of two equilibria: exaggerated trait value and costly female preference, or non-exaggerated trait value and random mating (Pomiankowski, 1988). Initial frequencies of the preference genes determine which equilibrium is reached. If the initial frequency of female preference lies above the Fisherian line of equilibria, the system will go to fixation of the male trait and the female preference. If the preference lies sufficiently below the Fisherian line that subsequent Fisherian coevolution of trait and preference will not hit the line, then the system goes to low trait value and random mating. Trajectories that begin below the line but then strike it are deflected up, reverse direction, and then go to fixation of the handicap.

Evolutionary trajectories in discrete models which increase both preference and handicap invariably fix the handicap before the preference reaches fixation. In the 'Zahavi handicap' model, females choose males on the basis of the trait, irrespective of viability. Once the male handicap is fixed, all males are identical and there is no longer any reason to choose. Selection then acts against the costly female preference and the system collapses back to random mating. With condition-dependent handicaps and revealing handicaps the situation is different. As Pomiankowski (1988, p. 177) summarizes the situation: 'with the other two types of handicap when the ornament is close to fixation . . ., females can still pick those males with currently favoured 'viability' alleles: only males with high viability are ornamented with the *condition-dependent handicap,* and males reveal whether they have high or low viability with the *revealing handicap*'. This causes both male ornament and female preference to go to fixation. These results have been recently confirmed using quantitative genetic models (Iwasa *et al.*, submitted).

Some caveats about theoretical treatments of female choice are appropriate here. Because intersexual selection has been proved difficult to model, various simplifying assumptions are invariably included. Discrete genetic models (for example, O'Donald, 1962; 1967; 1980; Maynard Smith,

1976; 1985; Andersson, 1982a; 1986; Kirkpatrick, 1982; Seger, 1985; Pomiankowski, 1987a,b; 1988; Tomlinson & O'Donald, 1989) have the advantage that important components such as genetic co-variance can be estimated as they evolve. They have the disadvantage that traits, preferences and investments cannot be selected continuously ('runaway' here means a run to fixation of the male trait), and that continuous elaboration of a trait must be viewed as a sequence of presumably independent single locus fixations. Quantitative genetic models (for example, Lande, 1980; 1981; Heisler, 1984; 1985; Arnold, 1985; Boake, 1985; Kirkpatrick, 1985; 1986; 1987a,b; Lande & Arnold, 1985; Lande & Kirkpatrick, 1988) appear to provide greater realism and permit the examination of continuous choice, selection and investment functions. They have the disadvantage that the modelling of the evolution of genetic co-variances between male traits and female preferences is currently intractable mathematically and hence these parameters are usually assumed to be constant. Several authors have questioned this assumption (O'Donald, 1983; Pomiankowski, 1988; Turelli, 1988), and recent evidence suggests that it may not be valid in many situations (Cohan & Hoffman, 1989; Wilkinson *et al.*, 1990). Despite these differences in approach, it is reassuring that most of the models of Fisherian and handicap processes appear to give qualitatively similar results. However, the votes are not yet all in and more sophisticated models may reveal new surprises.

7.3.2 *Observational and experimental studies*

Recent years have seen important advances in empirical work on the evolution of exaggerated characters by mate choice. While no single study can answer all the questions, taken together they signify major advances in our understanding of the processes involved. Key questions that have begun to be tackled include: (i) Do animals choose mates on the basis of extravagant characters? (ii) Is mate choice costly? (iii) Are extravagant characters handicaps to survival? (iv) Do extravagant characters reveal a mate's intrinsic viability either directly or as revealing handicaps and, if so, have they evolved for that reason? (v) Is there evidence that female choice enhances offspring viability? We cannot review all the work that has been done in each area, but we can provide illustrative examples for each question.

DO ANIMALS CHOOSE MATES ON THE BASIS OF EXTRAVAGANT CHARACTERS?

Few researchers doubt that female choice of males for paternal care or access to defended resources occurs in nature. A focus for competing theories of intersexual selection has been the study of lek mating. Males on leks provide only gametes (and perhaps inadvertently disease) for females, and hence the demonstration that particular male traits are maintained by sexual selection might be more easily made in such species. The first step is to identify the traits in lek males that the females are choosing by seeking significant correlations between male traits and male mating success. Unfortunately, this is not as simple as it seems. Some male traits, even on leks, will function in intrasexual competition and not female choice. These must be identified as such. Since the remaining male traits are likely to be correlated, all traits must be measured simultaneously on a number of males to determine inter-trait correlations. To obtain any power in the resulting multiple regressions, sample sizes must be large and the more male traits that are considered, the larger those samples must be. In many lekking species such as birds of paradise or intermediate-sized grouse, the requisite number of males for even a moderate number of traits soon exceeds typical lek size. Pooling data from several leks requires appropriate controls to ensure that dynamics on leks are not heterogeneous.

Sage grouse have leks of 30–150 males and between one third and one half of these males mate at least once each year. Gibson and Bradbury (1985) found that attendance at the lek, display rate and certain components of the display sounds were significantly correlated with male mating success after correcting for correlations with other traits such as territory location, anatomical measures and seasonal effects. Subsequent work on two adjacent leks indicates (Gibson *et al.*, unpublished data), that the relative significance of the different traits varies among both years and leks. This finding suggests either that the measured traits are not the ones used by females but only correlated with the true cues, or that females use a number of cues whose relative importance can vary annually. Multiple cues for female choice are likely to evolve in complicated ways if handicaps are involved (Tomlinson & O'Donald, 1989). Direct experimental manipulation of the acoustic cues is obviously needed to clarify what is going on in sage grouse.

Höglund (1989a) experimentally enlarged the size of the white tips on tail feathers of male great snipe. Control males were painted with dark liquid paper on existing dark feather spots. Of 17 matings observed, 16 were to males with experimentally enlarged white spots. The authors could find no other male traits which differed significantly between successful and unsuccessful males. However, since 10 of the matings were achieved by one male, the possibility remains that this male would have been successful anyway. Similarly, Borgia (1985) has shown that the number and colour of decorations on the bowers of male satin bowerbirds are significantly correlated with mating success. In one year, he divided 22 males randomly into two groups, removed all but a few leaves from the bowers of males in one group, and left unaltered the bowers of males in the other group. Of 15 recorded copulations in that year, only three copulations were obtained by males in the group with altered bowers, whereas 12 copulations were obtained by males with normal bowers.

More successful manipulations of physical male cues in intersexual selection have involved alteration of male tail length in non-lekking birds. Male African long-tailed widow birds (*Euplectes progne*) are black with red winged epaulets, tails measuring about 50 cm, whereas the females are mottled brown with short tails. Andersson (1982b; see also Moller, 1988a; 1989a;) experimentally shortened tails of some males and used the feathers to extend the tails of other males. The species is polygynous with males defending grassy sites within which females build nests. After the experiment, more new nests were found on the territories of the males with extended tails. As a control to this experiment, a further two groups of males were examined: some were unharmed and others had their tails cut off and stuck back on again. There was no difference in the number of females attracted by control males. Therefore, in this species, it seems that females do prefer males with longer tails. Additional work by Andersson suggested that tail length is not important in inter-male combat among these widow birds.

The most experimentally accessible male advertisement cues are the calls of anurans. Playbacks of synthetic and natural calls have been undertaken in both laboratory and field conditions with similar results. Despite this accessibility, the characterization of female choice of males in anurans has generated no clear pattern. In some species, and for some components of male calls (e.g. dominant frequency, pulse rate), females exhibit strong preferences for a modal value which is similar to that found in natural male populations. Such preferences appear to be more closely related to avoidance of hybridization than to runaway sexual selection. In other species, and for other components (e.g. call rates, call durations and lower frequency values), females prefer values that are very different from current male means and may even exceed those present in natural populations (Gerhardt, 1982; 1988; Sullivan, 1983; Robertson, 1986; Wells & Taigen, 1986; Gerhardt & Doherty, 1988; Ryan, 1988). Although such preferences might fuel runaway sexual selection, their expression is often suppressed by overriding effects of male proximity (Arak, 1983; 1988a), male call overlap (Dyson & Passmore, 1988) or convergence of male calls (Lopez *et al.*, 1988). The degree to which female preferences play important roles in anuran sexual selection remains open.

IS MATE CHOICE COSTLY TO FEMALES?

We noted earlier that females choice for males with particular preferences does not necessarily increase the costs of finding mates over that for females not exercising such a preference. In some cases (e.g. solitary forest birds or nocturnal anurans), female preferences for active male display might reduce such costs allowing classical Fisherian intersexual selection to operate. However, the benefit of this mechanism is the same whether a female finds one or several males. Once males are easy to find, perhaps because they are conspicuously aggregated and noisy, is there any reason why a female should exert any further effort to discriminate between them? The issue is critical because if choice among males can be shown to occur at some substantial fitness costs to females, then some compensatory benefits to females of that choice must be invoked (Pomiankowski, 1987; 1988). Where males undertake paternal care or provide defended or captured resources for females, the possible benefits of such choice are easy to imagine, However, on leks where males provide only gametes, costly female choice among males would be explicable (given current models) only if females were attempting to obtain viability genes via some handicap mechanism.

It is well known that females which mate on leks or at anuran choruses may spend a considerable amount of time visiting many of the males before mating (for example, Selous, 1927; Kruijt & Hogan, 1967; Ryan, 1985; Robertson, 1986). In some tropical birds, it is not clear that this results in a detectable increase in energetic or time costs or in risk. For example, female Lawes' birds of paradise often stop by male display courts as part of their normal foraging rounds months before they are ready to mate. They may only stay minutes, but cumulatively over several months they manage to visit most males in an area (Pruett-Jones, 1985). In a number of tropical lek species, females select mates early in the season and tend to return to the same male (presumably with no choice costs) for subsequent re-nestings that season (Lill, 1976; Trail & Adams, 1989). Female choice is made easier in classical lek species in which males are closely aggregated, allowing females to contrast and visit a large number in a short time. Except where males harass females, discrimination between males at classical leks must surely be easier for females than contrasting the same number of solitary males. However, is it more costly to compare several males at a lek rather than pick one randomly? If there is no real cost, then the patterns of trait exaggeration we see in these species can be explained by a simple Fisherian mechanism. If, however, the costs are significant, then it is likely that some handicap mechanism is operating. There is clearly a need to assess these costs, but we know of no study to date which has done so.

ARE EXTRAVAGANT CHARACTERS COSTLY TO MALES EXHIBITING THEM?

There is little doubt that extravagant characters can increase male predation risks. Brightly coloured male guppies are much more prone to predation risks in the wild (Endler, 1983). Male crickets are parasitized by flies which use their mating calls to locate them (Cade, 1979), and the carnivorous bat, *Trachops cirrhosus*, homes in on the advertisement calls of male Tungara frogs (Ryan *et al.*, 1982). Male sage grouse are particularly vulnerable to attack by golden eagles while displaying on leks (see Bradbury *et al.*, 1989). Circumstantial evidence, such as post-reproductive moult to more cryptic plumage by adult male birds, the more cryptic plumage of subadult males, and the fact that secondary sexual characters do not develop until reproductive age in many species, also suggests that secondary sexual characters are costly to males (Darwin, 1871; 1874).

A second possible cost to males is increased mortality as a result of high energetic investments in territorial defence, display to females, or production of secondary sexual traits. There are few available measures of the energetic costs of producing secondary sexual characters, much less the effect of such costs on viability. Recent work on display costs in lek and chorusing animals has shown that the energetic costs of display are very substantial (see below).

There are important theoretical reasons why such relationships should be examined in the field. If a simple Fisherian system is operating, one expects to find a negative selection gradient of viability on the exaggeration of the critical male traits which is balanced by a similar but opposite-signed gradient of male reproductive success on the trait. A failure to find a significant viability trait gradient could mean that intersexual selection is operating via a handicap mechanism and that males have adjusted investments to produce similar overall viabilities. Before accepting this conclusion one would also have to be sure that the relevant viability costs were not being incurred at some other stage in the life cycle and that sample sizes were large enough to avoid a type II statistical error. Perhaps this is another issue in which experimental manipulations may prove to be of more value than field observations.

DO EXTRAVAGANT CHARACTERS REVEAL A MALE'S VIABILITY AND, IF SO, HAVE THEY EVOLVED FOR THAT REASON?

A common conclusion in observational studies of lek and choruses is that male behaviour, not male morphology, is the major correlate of mating success (Bradbury & Gibson, 1983; Pruett-Jones,1985; Kruijt & de Vos, 1988; Höglund, 1989b; Trail & Adams, 1989; but see Petrie et al., 1990). Are the energetic costs of display performance likely to be sufficiently taxing on males that the resulting performance is a good index of male condition? Is display performance a handicap? Several studies suggest that the costs of display are so large in some species that they dominate male energy budgets, and that physiological differences in sustained display, not short-term maximal rates, determine differences in performance and thus mating success (Garson & Hunter, 1979; Taigen & Wells,1985; Gottlander, 1987; Klump & Gerhardt, 1987; Reid, 1987; Höglund, 1989c; Vehrencamp et al., 1989; Wells & Taigen, 1989). In sage grouse, for example, display accounts for 40–50% of the overall energy budgets for the more active and successful males (Vehrencamp et al., 1989). Although all males lose weight during the display season, active males lose less weight than inactive ones. Despite efforts to find a relationship, there appears to be no effect on survival of different patterns of lek attendance and display in these populations. At first glance, these results are highly suggestive of sexual selection via handicaps. However, they also fit an unrelated evolutionarily stable strategy model by Parker (1982) which does not invoke the coevolution of viability and sexual traits and does not even include Fisherian sexual selection. On the contrary, Parker's model requires only that males differ phenotypically (not necessarily genetically) in the costs suffered for a given investment. Although the fact that display effort is costly and differs among males in many species with exaggerated characters suggest that handicaps may be involved, more sophisticated tests are needed to discriminate among alternative interpretations (see Ryan, 1988).

In 1982 Hamilton and Zuk published a paper which suggested that animal coloration and displays, including song and mating calls, are often, in effect, revealing handicaps. Their argument was that animals are subjected to attack by debilitating parasites, that the resistance to the parasites is heritable, and that only resistant animals can fully express secondary sexual characters. As a consequence, mate choice based on the expression of secondary sexual characters results in the production of offspring that are likely to be more viable because they are resistant to parasites.

It is too soon to judge the hypothesis but a considerable amount of work has recently gone into testing its predictions. We shall mention the cross-species correlational evidence later, but here we are concerned with intraspecific tests. The major problem with testing the hypothesis is that the most convincing intraspecific evidence will have to be experimental. To show, for example, that the most brightly coloured birds both attract mates and have low parasite loads leaves many potentially confounding variables unexamined, for example subordinate birds may lose condition, become duller in colour, and then be more easily infected by a variety of parasite species. Read (1988) provided a review of the observational evidence available at that time, which he considered to be sketchy, circumstantial and sometimes contradictory, Even carefully designed experiments may not be sufficient. If randomly selected male birds are divided into groups, some of which are infected with parasites, and the parasitized birds then become duller and are eschewed by females, this may result from the parasitized birds investing less into body condition because nutrients are being used to sustain their parasites. Starved birds might similarly lose condition. This is not such a fine point as it may seem: the Hamilton–Zuk theory depends critically on heritable resistance to parasites being detected via the visual handicap, yet birds may simply be choosing mates on the basis of condition which might normally be influenced by other heritable factors than resistance to parasites. Read (1990) discusses other difficulties associated with interpreting experimental tests of the hypothesis.

It is important when assessing the Hamilton–Zuk theory to distinguish between mate choice based on choosing a good parent and mate choice based on ensuring fit offspring. For example, in species with parental care, it will pay the female to choose a mate that is likely to provide good parental care: parasitized males may be debilitated. What is more they may transmit their parasites, including ectoparasites such as lice, to their offspring and to the female. Under such circumstances, even in the absence of heritable disease resistance, it would pay females to choose parasite-free mates. Testing the theory is not even straightforward in species lacking parental care. For example, the physical contact necessary to copulate with a parasite-ridden mate could transmit parasites between individuals.

One of the best pieces of experimental evidence for the existence of revealing handicaps, although not necessarily the strict Hamilton–Zuk version, is Milinski and Bakker's (1990) 'green light' experiment with sticklebacks. Twenty-four male sticklebacks were scored for intensity of red breeding coloration. Each male was placed in its own aquarium. Pairs of aquaria, containing males 1 and 2, 3 and 4, and so on in ascending colour score rank, were placed in front of tanks each containing a receptive female stickleback. The females tended to adopt the head-up courtship posture to the brighter of the two males. Male brightness was also correlated with male condition, measured using the standard fisheries 'condition factor': (weight/(length)$^{2.76}$). This meant that the females preferred to mate with bright males that were in good condition.

Did the females use colour as a cue? The experiment was repeated in green light, conditions under which the authors judged that females should be almost unable to judge differences in the intensity of red coloration. The intensity of the light was such that feeding rates on *Daphnia* matched those under white light. This time, the males were allowed to build nests and each pair of males was subjected to choice by three or four females presented separately under the green light and another three or four females under white light. Although the males' courtship behaviour did not seem to differ under the two sorts of light, the females presented under white light tended to choose the brighter male but the green light females showed no detectable preference correlated with male coloration. Indeed, it seemed that the females were using colour as a cue in mate choice.

Why should the females use colour as a cue? Milinski and Bakker suggested that parasite infection might influence male coloration. Accordingly, they infected the brighter male of each pair with the ciliate *Ichthyophthirius multifiliis*, which causes the common fish disease known as white-spot. Cysts caused by the disease dropped off after a few days, but the coloration and condition of the infected males dropped to the level of the duller male of the pair. Now, under white light, the females prior preference for what had been the brighter male vanished. Under green light, the females' lack of preference between the males remained unchanged. Unfortunately, however, there was no significant interaction effect between female preference under white versus green light and female preference before and after parasitism. This means that there is no formal statistical support for a parasite-dependent change in preference based on coloration. Nevertheless, as Milinski and Bakker concluded, their results do imply 'that the female detected the prior parasitization of the males by their decreased intensity of breeding coloration, which is a necessary condition for coloration to be judged as a revealing handicap'.

There are a number of conclusions to be drawn from this experiment. First, female sticklebacks tend to choose mates that are in good condition. Second, females seem to be using male coloration rather than condition as a proximate cue in mate choice. The females may have been selected to choose mates that are effective at egg-guarding but, even if the females are selecting mates on the basis of their likely ability at parental care, as long as there is additive genetic variance for parasite resistance in the stickleback population the females will simultaneously be seeking mates that will pass on a measure of parasite resistance to their offspring.

Whether it ultimately turns out to be correct or not, and whether parasites are important under some other guise than originally envisioned, Hamilton and Zuk's (1982) paper has been of enormous heuristic value. Unfortunately, it is difficult to conceive of experiments that would demonstrate that a costly trait does not reveal directly the viability of its host. Just as it is possible that some selective force has been overlooked when genetic drift is held responsible for differentiation between two populations, so it is always possible that the expression of a character does reveal directly something yet undetected about an animal's health.

IS THERE EVIDENCE THAT FEMALE CHOICE ENHANCES OFFSPRING VIABILITY?

It might seem that a direct study on the relationship between female choice for costly male characters and offspring viability would be the best way to demonstrate a role for handicaps in sexual selection. Simply showing that female choice enhances offspring viability is a necessary, but not sufficient, part of such an exercise. For example, female mice prefer not to mate with males carrying the offspring-lethal *t* allele, and can use olfactory cues to discriminate between male genotypes (Lenington, 1983). Female choice clearly has viability consequences, but there is no evidence that male odours used by females carry any viability costs to males expressing them. On the other hand, females of many species avoid mating with males of other species, probably because non-viable hybrids would result. In many groups, such as ducks, it is the male which exhibits the distinctive and conspicuous plumage permitting species-specific identification and thus potentially higher predation risk. Is there direct evidence that female choice among conspecific males results in a greater number of offspring, or of more viable offspring?

Partridge (1980) demonstrated that *Drosophila melanogaster* females exposed to several males produced more offspring than females mated randomly with individual males. Attempts to replicate this study with a number of *Drosophila* species by Schaeffer *et al.* (1984) failed to show any effect on female choice. However, Taylor *et al.* (1987) were able to generate a similar effect using *Drosophila melanogaster.* Partridge and Taylor *et al.* emphasized that their experiments did not distinguish between the effects of female choice and those of male intrasexual interactions. Similar results were found by Crocker and Day (1987) in the seaweed fly *Coelopa frigida*: the fertility of females and the survival of their progeny were both higher when several males were made available for mating. Working with flour beetles, Boake (1986) found no correlation between the ability of males to attract mates and the viability of their offspring. Female field crickets *Gryllus bimaculatus* are known to mate preferentially with larger males (Simmons, 1986). However, when Simmons (1987) examined the consequences of choice the results were not so clear-cut. Some females were provided with a single large male, some with a single small male, others with a group of males from which to choose a mate. The offspring from the latter group of females had a higher survival rate and developed more rapidly from those of the non-choice groups. Perhaps females use a combination of traits in male choice, of which size is one. Or perhaps those females given a choice of males mated several times and derived extra nutrients for their offspring from additional male spermatophores?

As Simmons' work demonstrates, even where correlations between preference, male traits and offspring production are found, the right interpretation for the results may prove difficult to nail down. One of the most bizarre examples of such correlations is provided by Burley's fascinating work on zebra finches. Burley observed that captive male and female zebra finches both prefer mates with particular leg band colours (Burley, 1981). Males prefer mates with black bands and find those with light blue bands unattractive, whereas the equivalent colours for females are red and light green respectively. In a pair of reasonably long-term experiments, Burley banded the males in one aviary and the females in the other with random colours, leaving one sex unbanded in each aviary (Burley, 1985; 1986a,b,c). For both sexes, attractive birds left about twice as many offspring as unattractive birds. Part of this was due to the fact that attractive birds tended to attract higher quality mates that were physically able to raise more offspring (as predicted by Darwin). Since this was a laboratory study with no predators present, the degree to which bright leg bands are costly in nature cannot be evaluated. However, it is of considerable interest that even in the laboratory there were significant differences in adult survival rate depending on the colour of the leg band: unexpectedly, lifespan for attractive birds was twice as long as for unattractive birds. One reason for the differences in longevity is that the mates of attractive birds invested more heavily in the offspring than did mates of unattractive birds, reducing reproductive stress on the attractive birds.

In summary, the results to date linking female choice for male traits to offspring viability are sparse and any patterns still unclear. Much can clearly be done in experimental situations and perhaps some critical manipulations can be undertaken in the field.

7.3.3 Comparative approaches

Section 2.3 mentioned how, for mammals, comparative evidence could effectively be marshalled to support the argument that weaponry has evolved by intrasexual selection in mammals: those species with less well-developed weapons, be they canine teeth, antlers or horns, were those in which intersexual selection would be likely to be least intense. We have seen in the previous sections how particular examples of extravagent characters that are likely to have evolved by mate choice have been studied by experiment. How then, does cross-taxonomic diversity accord with a mate selection explanation for the evolution of extravagant characters? Several such studies have been done on birds and, taken together, they provide something of a disappointment.

LEK BREEDING AND PLUMAGE DIMORPHISM

Perhaps one of the most obvious patterns to be expected is that lekking species of birds are, on the whole, more sexually dimorphic than non-lekking species, with the males being both larger and more colourful than the females (Darwin, 1871; Hingston, 1933). The most straightforward way to test the idea is to compare dimorphism in closely related species that lek with those that do not lek, and to examine whether lekking has evolved to become associated with dimorphism more often than would be expected by chance. Lek breeding in birds is confined to 11 families (Tetraonidae, Otidae, Scolopacidae, Psitacidae, Trochilidae, Cotingidae, Pipridae, Tyrannidae, Pycnonotidae, Paradisaeidae and Ploecidae), 10 of which contain species that lek and species that do not. In a careful phylogenetically-based analysis, Högland (1989b) could find no significant association between the evolution of lekking and the evolution of either size or plumage dimorphism. In particular, lekking species are often sexually monomorphic for both size and plumage. However, all lekking species that display on the ground are dimorphic in body size whereas some that perform aerial or arboreal displays are monomorphic in size. Similar results were obtained in an earlier survey by Payne (1984). Högland suggests lekking may have promoted body size and plumage dimorphism in some species and acoustic display in others. This latter suggestion echoes the claim that, among bower building birds, those with the more elaborate bowers are the least showy (Gilliard, 1969).

BIRD COLORATION, BIRD SONG, SEXUAL SELECTION AND PARASITES

We have already mentioned Hamilton and Zuk's (1982) theory that many animal displays and ornaments have evolved by mate selection for parasite resistance. One of the reasons for taking their theory seriously was the comparative evidence that, across bird species, their scores for male brightness (a), female brightness (b), male song, (c), male showiness (a+c) and bisexual showiness (a+b+c) show weak but highly significant increases with recorded parasite load. The parasite species concerned were various haematozoa and microfilarial worms. However, more recent analyses call these conclusions into question.

Read and Weary (1990) found that the original song correlations were confounded by phylogenetic associations, while Harvey *et al.* (1991) found no correlations with parasite load using the song complexity figures of Hartshorne (1973). In a careful analysis using sonogram data, Read and Weary reported that within taxa there are no consistent relationships between haematozoa prevalence and song duration, intersong interval, song continuity, song rate, song versatility, and song and syllable repertoire size for 131 species of European and North American passerines.

Read and Harvey (1989a) questioned whether the prevalence of blood haematozoa is correlated with bird coloration when independently derived plumage colour scores were used instead of Hamilton and Zuk's original scores. A subsequent exchange between Zuk (1989), Read and Harvey (1989b) and Hamilton and Zuk (1989) suggested that, given the low correlation coefficients involved and the possibility of confounding ecological variables, comparative tests using these types of colour scores are unlikely to resolve the issue.

One interesting point to emerge from the above exchange is whether, under Hamilton and Zuk's original theory, female choice of resistant males would be expected to increase parasite resistance in a host population. Hamilton and Zuk assumed that it would not and therefore predicted a positive correlation between colour and parasite load. If, however, parasite resistance was increased in brightly coloured species, we could expect a negative correlation or no correlation at all! As Read and Harvey (1989b, p. 105) put it: '(t)he strength of natural selection against the evolution of bright coloration can be expected to differ among species, thus leading to different solutions to the conflict between the opposing forces of natural selection for dull coloration and sexual selection for bright coloration. Females of bright species will, according to the Hamilton and Zuk theory, more accurately assess resistance among males. Thus resistant males will gain a greater mating advantage in brighter species, which will consequently harbour a lower parasite load. We know of no data which bear on the issue of whether female choice increases resistance in the host population.'

Baker and Parker (1979) discussed an alternative theory for the evolution of bird coloration, which suggested that more colourful birds are likely to be less profitable prey because bright coloration has evolved as an honest advertisement of escape ability, and that variation in the coloration of adult birds can be explained in terms of the selective pressures generated through predation risk. In support of their theory, Baker and Parker produced a detailed analysis of coloration in 516 species of Western Palearctic birds which showed that diurnally active birds are more conspicuous than nocturnally active birds, gregarious birds are more conspicuous than solitary birds, and concealed nesters are more conspicuous than exposed nesters. Kerbs (1979) pointed out that all these correlations could be interpreted as consistent with having arisen by sexual selection through female choice, while Harvey and Partridge (1982) pointed out that *post hoc* explanations could be made for why each of those same categories might be subject to higher parasite infections. Again, comparative tests of selection theory involving mate choice have proved ineffective.

7.4 Sperm competition

As we emphasized at the beginning of this chapter, Darwin's primary interest in defining sexual selection was to explain variation in organs that give certain individuals advantages 'over other individuals of the same sex and species, in exclusive relation to reproduction'. Sexual selection is normally thought of in terms of competition for access to mates, but males can successfully mate females and still not father the offspring. In particular, when females mate with more than one male there is likely to be competition among the sperm for success at fertilizing the available eggs (Parker, 1970). Over the past decade there has been a steady accumulation of evidence that the morphology of an animal's genitalia and associated structures may vary according to whether sperm competition is likely to occur or not.

In a series of papers, Short (1977; 1979; 1981) suggested that some differences in testis size among the great apes might be related to sperm competition. For example, male chimpanzees weigh about one quarter of the weight of male gorillas, yet their testes are about four times heavier. It is well established that female gorillas mate with only one male during a given oestrus, whereas female chimpanzees will mate with several males. Could it be that chimpanzees' testes are so large because of selection to produce more sperm per ejaculate so as to have more tickets in the lottery, each having a roughly equal chance of fertilizing the receptive egg? Short's idea claims generality—primates in which females mate with more than one male during a single oestrus should have larger testes. However, testes are used for the production of both testosterone and sperm and, accordingly, we might expect larger bodied species to have larger testes in order to maintain threshold levels of testosterone in the blood and to produce enough sperm to counter the dilution effect of larger female reproductive tracts. Accordingly, body size needs to be controlled for in comparative studies that test the generality of Short's idea. It turns out that across primates as a whole, those species in which females live in multi-male societies are the ones in which males have larger testes relative to their body size (Harcourt *et al.*, 1981; Harvey & Harcourt, 1984). Furthermore, for their body sizes, primates living in multi-male societies also have higher sperm production rates, possess larger sperm reserves in terms of number of ejaculate equivalents, and produce larger volumes of ejaculate that not only contain more sperm, but also a higher proportion of motile sperm, ranging from a meagre 7% in the monogamous gibbons (*Hylobates* spp.) to about 90% in multi-male group macaques (*Macaca* spp.) (see Møller, 1988b). Casting the taxonomic net more widely, similar patterns have been found in other mammals and in birds (Clutton-Brock *et al.*, 1982; Kenagy & Tromulak, 1986; Møller 1988b; 1988c; 1989b; for a review see Harvey & May, 1989).

It is important to emphasize that there is a distinction between charac-

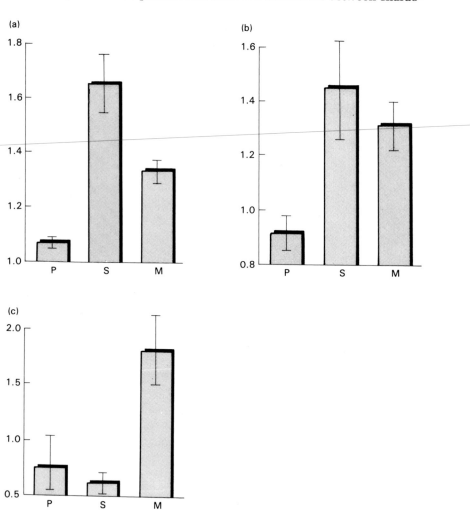

Fig. 7.5 (a) Body size dimorphism (adult male divided by adult female weight); (b) relative canine size (a measure of canine dimorphism: Harvey *et al.*, 1978); and (c) relative testes size (a measure of testes size after body size effects have been removed: Harcourt *et al.*, 1981) for primate genera with different breeding systems. Sample size for P, S and M respectively (a) 12, 9, 14; (b) 4, 7, 9; (c) 4, 7, 8. Bars indicate one standard error in each direction from the mean (from Clutton-Brock & Harvey, 1984) P, pair-living or monogamous; S, single male; M, multi-male.

ters that have evolved by sexual selection for access to mates and those that have evolved by post-mating sexual selection: polygynous species are not necessarily subjected to sperm competition. For example, males from species of primates where a single adult typically defends a group of sexually receptive females tend to have relatively large canine teeth to fight off other males, but relatively small testes since sperm competition is absent (Fig. 7.5; Harvey & Harcourt, 1984). The presence of sperm competition might even provide an explanation for sexual swellings which advertise oestrus in female primates that typically live in multi-male groups. It could be that such swellings promote copulation by more than a single male. If there is genetic variance for success at sperm competition, there will be selection for females to choose males whose sons are likely to be successful at sperm competition. One way to achieve that end is to test putative fathers against each other (Harvey & Bennett, 1985).

An echo of the above argument for the evolution of sexual swellings in primates is Eberhard's (1985) claim that enormous diversity in the morphology of intromittant organs of even closely related species (Fig. 7.6) has evolved as a consequence of female choice. Among species in which more than one male is likely to copulate with a single female, any advantage a male has in making his sperm more likely to fertilize the eggs will be favoured by natural selection. However, it will also be to a female's advantage to have her eggs fertilized by a male whose male offspring are effective at getting eggs fertilized. As a consequence, Eberhard (1985, pp. 70–71) argues '(t)hose males that are mechanically superior or are better stimulators are favoured. . . A female could discriminate among males' genitalia on the basis of their ability to fit mechanically into hers, or through other sensations occurring in her genital region'. Note that this hypothesis allows for both intrasexual competition (e.g. mechanical superiority) and Fisherian exaggeration (e.g. better stimulators). Eberhard considers other explanations for the evolution of diverse male intromittant organs and finds little evidence to support them in other than a minor role. In accord with his hypothesis is the observation that penile morphology is particularly diverse among species in which sperm competition is likely, such as primates with multi-male breeding systems (Dixson, 1987).

7.5 Sexual selection in plants: a concept in search of examples?

We have seen that Darwin explained the evolution of sexually selected traits as a consequence of the advantages they provide their bearers in competition with others of the same sex and species in exclusive relation to reproduction. Such a definition of sexual selection is not easily applicable to those 88% of flowering plant species for which the organs of both sexes are contained on each individual, which we describe here as 'bisexual species'. Nevertheless, the potential for sexual selection is clearly present. Brighter and larger flowers, for instance, may attract more insects to disperse pollen and might therefore be favoured by male–male competition. On the other hand, they may attract more insects to donate pollen, thus allowing ovules the choice of pollen from different males. Another example was given by Haldane (1932, p. 123) who pointed out that 'a higher plant is at the mercy of its pollen grains. A gene which greatly accelerates pollen tube growth will spread through a species even if it causes moderately disadvantageous changes in the adult plant'. What might provide a useful operational definition for sexual selection in bisexual species?

Charlesworth *et al.* (1987, p. 318) suggest that, for bisexual plants, sexual selection can be defined as 'competition between individuals for

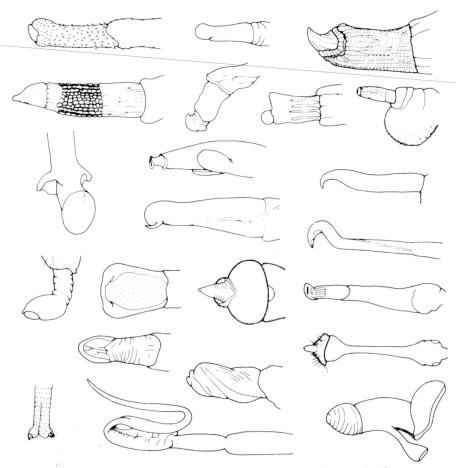

Fig. 7.6 'A sampler of mammalian penes (all flaccid, drawn to different scales) from groups in which penis morphology is generally poorly known and has not been used to make taxonomic distinctions. Those in the top two rows are all primates. The forms are varied and complex' (from Eberhard, 1985).

mates caused by traits which yield unequal fitness gains through male and female function, where the selective value of a trait is not due to its effects on survivorship of the sporophyte (offspring) itself'. Such a definition may, of course, also be applied to hermaphroditic animals. Comfortingly, from this definition, drop the two usual factors associated with sexual selection: male–male competition and female choice. Success at male–male competition or female choice can be realized through increased attraction to pollinators by producing, for example, more nectar or a better floral display. Similarly, Haldane's pollen competition scenario can be seen as an example of male–male conflict. Female choice is not likely to be so important as in mobile animals although, of course, flowers with female function can attract several pollen bearers thus providing the opportunity for discrimination among the pollen of different males, or even the selective abortion of offspring fathered by particular males.

When it comes down to identifying examples of characters that have evolved by sexual selection for male–male competition or female choice in flowering plants, the position is not so clear-cut. Consider the possibilities that we have already outlined. Plants need pollinators to disseminate their own pollen and also to obtain foreign pollen. Do more attractive bisexual flowers increase fitness through the male or the female line? One useful way of analysing the forces at work is to extend sex allocation theory (Charnov, 1982) to include attractiveness to pollinators. In addition to investment in the production of ovules and pollen, investment in attractiveness can then be partitioned into its influence on male and female fertility (Charnov & Bull, 1986). Charlesworth and Charlesworth (cited in Charlesworth et al., 1987) considered a flowering plant in which investment into female function is estimated by the number of ovules produced and into male function by pollen output. Female fitness depends on the proportion of non-selfed ovules that are fertilized, which will be a function of attractiveness to pollinators, as well as on the number of ovules produced. Similarly, male fitness will depend on the proportion of pollen grains that are exported to fertilize successfully ovules of other plants, which will also be a function of attractiveness to pollinators, as well as on the number of pollen grains produced. The question is how does increasing the attractiveness to pollinators influence fitness through male versus female function? When female fitness is strongly influenced by the production of expensive seeds, increased attractiveness to pollinators increases male fitness much more than female fitness. However, if seed initialization increases seed output then the Charlesworths' evolutionarily stable strategy model suggests that the influence of attractiveness on male and female fitness can be of similar magnitude.

Bell (1985) considered the male versus female functions of flowers, and came to the conclusion that the bisexual 'flower is primarily a male organ, in the sense that the bulk of allocation to secondary floral structures is designed to procure the export of pollen rather than the fertilization of ovules' (Bell, 1985, p. 223). Bell marshals both comparative and experimental evidence to support his conclusion. Using a variety of species with bisexual flowers, he finds that (1) insects visit larger flowers more frequently, and experimental reduction of floral biomass causes a proportional reduction in the frequency of visits by pollinators; (2) removal of attractive structures has no effect on the number of seeds set per fruit; (3) larger flowers may disperse a greater fraction of their pollen per unit time, and the removal of flowers from inflorescences causes a decline in the quantity of pollen exported per remaining flower. Furthermore, a comparative survey of sexually dimorphic plants shows that male flowers are generally larger than female flowers, male inflorescences bear more flowers, and male plants bear more inflorescences. The exceptions to male flowers being larger than female flowers occur in wind-pollinated species, where male flowers and female flowers tend to be the same size, which is what we might expect when male flowers are not competing for pollinators. Bell's study and other less extensive surveys provide exceptions to almost every generalization made; the question of resource allocation to the sexes by flowers is one that requires much more work.

What of pollen competition as envisaged by Haldane? Is there any evidence that pollen which succeeds at pollen competition produces less viable offspring? This would be a plant's equivalent of the meiotically driven t-alleles of mice. There seems no doubt that Haldane was right in his supposition that such conflicts could happen, and Wright (1969) has even shown that if genetic variants at a locus influence the fitness of haploid and diploid phases in different directions, as Haldane suggested, genetic variation can be maintained in a population. That is, if alleles causing higher rates of pollen tube growth cause greater reduction in fitness of the resulting offspring, a variety of alleles with a range of effects can be maintained by selection in a population. Despite the support of theory, experiments to determine the existence or magnitude of such effects have suffered from a lack of appropriate controls (see Charlesworth et al., 1987). For example, under conditions of pollen competition, it has often been found that the resulting offspring are more viable rather than less viable, the opposite of that predicted from Wright's model. However, pollen competition is induced by providing ovules with pollen from different sources without controlling for total pollen load size, yet it is known that large amounts of pollen added to a particular flower may cause the receiving plant to respond by investing more heavily in the offspring of that flower.

The next possibility is female choice among the pollen from different males. This certainly occurs but often through incompatibility systems: females reject pollen of similar genotype at incompatibility loci. Self-incompatibility systems probably evolved as inbreeding avoidance mechanisms (Charlesworth & Charlesworth, 1979) and can be profitably analysed from that perspective. Invoking sexual selection provides no additional explanatory power. Sutherland (1986) has pointed to the high fruit-set of dioecious species which suggests that, unless multiple pollination is common, female choice is not important (Charlesworth *et al.*, 1987).

Finally, what of fruit abortion? Although it has been suggested that this is an example of mate selection occurring after fertilization (Stephenson & Bertin, 1983; Willson & Burley, 1983), Queller (1987) points out that there is no evidence that choice is made on the basis of male characters used to increase mating success. Accordingly, Queller suggests that fruit abortion is best viewed simply in terms of choice by the female of which of her zygotes to invest in further.

In short, sexual selection may well be important in the evolution of floral displays and in pollen competition, but few suitable analyses and experiments to assess its importance as an evolutionary force producing particular phenotypes have yet been performed. Nevertheless, the basic conceptual foundations have been laid. We agree with Charlesworth *et al.* (1987, p. 335) that it does not seem particularly useful at present to include inbreeding depression and the evolution of genetically based self-incompatibility systems under the heading of sexual selection.

8: Parental investment

Tim Clutton-Brock and Charles Godfray

8.1 Introduction

As discussed in Chapter 2, all organisms face two fundamental decisions about reproduction. First, they must decide how much of the resources available to them should be spent on reproduction instead of on their own continued growth or survival. Second, they must decide how to divide the resources that they allocate to reproduction among their offspring. This chapter reviews theories concerning the distribution of parental resources among offspring, concentrating on six topics: measures of the costs and benefits of parental care; trade-offs between the size and number of offspring; the extent of parental care and the involvement of males and females; variation in parental care in relation to its potential benefits to offspring and its likely costs to parents; the influence of conflicts between parents and their offspring on patterns of parental care; and the allocation of resources to producing sons and daughters.

Parental behaviour can be measured at three different levels and it is important to distinguish clearly between them. First, we can describe the form, frequency or duration of parental care, for example the number of times a parent stops feeding to look for predators. Second, we can measure the parent's expenditure of energy (or other resources) on caring for its offspring, either in absolute terms or relative to some estimate of the total energetic resources available to the parent. And third, we can measure the costs of parental care to the parent's future fitness or residual reproductive value.

As Williams (1966) and Trivers (1972) have emphasized, the evolution of parental care will depend on the costs of care to the parent's residual reproductive value, as well as on its benefits to the fitness of offspring or other relatives. Unfortunately, fitness costs and benefits are notoriously difficult to measure (see below) and, as a result, measures of time or energy expended on care are commonly used instead. However, there will be many circumstances where time or energy costs do not reflect fitness costs and consequently provide an unreliable basis for predicting patterns of parental care (Trivers, 1972; Baylis, 1981; Clutton-Brock, 1991).

To distinguish between different measures of parental care we use three different terms: parental care, parental expenditure and parental investment.

1 *Parental care* describes any form of parental behaviour which appears likely to increase the fitness of a parent's offspring. This descriptive term

234

carries no implications about costs in terms of energy or fitness. It is useful to recognize two contrasting categories of care, although many forms may lie between these extremes: *depreciable care*, such as the provision of food, where the benefits of parental expenditure per offspring decline as brood size increases; and *non-depreciable* care, such as parental vigilance, where benefits per offspring do not decline with increasing brood size (see Altmann *et al.*, 1977; Lazarus & Inglis, 1978; 1986; Wittenberger, 1979). We have adopted Altmann *et al.*'s terms since an unfortunate terminology has developed. While Wittenberger (1979) refers to care of the first kind as 'shareable' and to care of the second kind as 'non-shareable', Lazarus and Inglis (1978; 1986) refer to the first as 'unshared' and the second as 'shared'!

2   *Parental expenditure* denotes the expenditure of parental resources (including time and energy) on parental care of one or more offspring. This may be expressed as a proportion of the parent's resources expended on one or more of its offspring, in which case we refer to it as relative parental expenditure.

3   *Parental investment* refers to the extent to which parental care of one or more offspring reduces the parent's residual reproductive value. Trivers (1972) originally excluded the costs of parental care to the parent's subsequent mating success from his definition of parental investment. However, there is no reason to distinguish between the costs of parental investment to the parent's subsequent ability to care for, versus produce, young, and today parental investment generally applies to the costs of parental care to any aspect of the parent's residual reproductive value (see Alexander & Borgia, 1979; Gwynne, 1984; Thornhill & Gwynne, 1986). We use parental investment to refer to the fitness costs of parental care of *individual* offspring while the total costs of caring for all progeny are designated *parental effort* which, together with *mating effort* is a part of the organism's reproductive effort (Low, 1978; Alexander & Borgia, 1979).

Not surprisingly, there are often problems in deciding what should and what should not be included in parental care. In particular, male expenditure on territorial defence or nuptial gifts can be hard to classify since they may both increase mating success (and thus be regarded as a form of mating effort) and improve the survival or quality of offspring (Zeh & Smith, 1985).

8.2   Measuring the costs and benefits of parental care

All theoretical predictions concerning the distribution or extent of parental investment rely on estimates of the fitness benefits and costs of parental care. There is extensive evidence that parental care affects the survival and reproductive success of offspring. For example, where males guard egg masses, experimental removal of the guarding male often leads to a substantial reduction in egg survival (Eberhard, 1975; Forester, 1979;

Dominey, 1981; Simon, 1983). Egg size and neonatal weight are also important: in birds, positive relationships between egg size and chick survival are common (Parsons, 1970; Lundberg & Vaisanen, 1979; Ankney, 1980; Galbraith, 1988) while in mammals, juvenile survival is often closely related to birth weight or early growth (Allden, 1970; Thorne et al., 1976; Guinness et al., 1978). There is, however, a danger that positive correlations between egg or neonate size and juvenile survival occur because both are related to some aspect of the mother's environment (Clutton-Brock, 1991).

In longer-lived animals where generations overlap, parental care commonly extends beyond the point at which offspring obtain their own food (Silk, 1983; Harcourt & Stewart, 1987). For example, in vervet monkeys, mothers help to protect their adolescent and adult offspring from competition with older or more dominant conspecifics, and females with mothers still present in the group have higher reproductive success than those without (Fairbanks & McGuire, 1986).

Since the benefits of parental care are often measured in terms of the number of young fledged or weaned, it is important to emphasize that parental care can also have substantial effects on the survival of offspring after fledging or weaning (Wolf et al., 1988) as well as on their breeding success if they survive. In red deer, for example, the birth weight of calves predicts their size as adults as well as the size and survival of calves that they produce over the rest of their lifespan (Albon & Clutton-Brock, 1988). In hamsters, depriving females of food during the first 50 days of life not only affects their fecundity and the weight and sex ratio of offspring produced over the rest of their lifespan, but also influences the birth weight and sex ratio of pups produced by their daughters, themselves raised with ad libitum access to food supplies (Huck et al., 1986; 1987). Presumably, these effects occur because early food deprivation has a permanent effect on a female's capacity for parental expenditure which, in turn, affects her daughter's capacity. One important implication of these results is that measures of weaning or fledging success may often seriously underestimate the true benefits of parental expenditure and that conventional measures of reproductive value may often be misleading.

The costs of care to parental fitness are also often substantial. Increased expenditure per offspring almost inevitably reduces the number of offspring that the parent can rear. Some of the clearest evidence for this effect comes from selection experiments on domestic chickens which show that (artificial) selection for large egg size reduces the rate of egg production while selection for laying rate reduces egg size (Nordskog et al., 1974; Nordskog, 1977).

The costs of egg care can be large, too. In ectotherms where one adult guards the eggs, the guarding parent commonly reduces food intake during parental care and may cease feeding altogether (Townsend, 1986). In some species, this period of starvation reduces the parent's reproductive performance in the following season (Hairston, 1983) and may also retard the parent's growth, thus affecting reproductive success in subsequent breeding attempts (Gross & Sargent, 1985). The guarding parent may also incur increased predation risks while, in species where males care for eggs or young, this may reduce their access to other mating partners (Clutton-Brock, 1991).

In endotherms, the energetic costs of egg care include the costs of maintaining egg temperature. In birds, estimates of the amount of heat transferred to the eggs ranges from 10% to 30% of basal metabolic rate (BMR) in passerines (King, 1973), increasing in smaller species with high egg weight to body weight ratios and in species with relatively large clutch sizes (Ricklefs, 1974; Coleman & Whittall, 1988). Where parents feed their young after hatching or birth, energy costs are typically high, exceeding those of egg production, incubation or gestation (Robbins, 1983). In birds, the energetic costs of feeding young rise to a peak of around four times BMR while the costs of lactation in mammals commonly lie between 2.5 and 5 times BMR—a level similar to the cost of heavy labour in humans (Drent & Daan, 1980; Oftedal, 1985). It is consequently unsurprising that the costs of feeding young to the parents' fitness are generally higher than those of gamete production, incubation of gestation (Clutton-Brock et al., 1989).

Finally, the costs of parental care after fledging or weaning can be substantial, in some cases because offspring compete with their parents or with subsequent sibs for access to resources. In red-necked wallabies, for example, mothers that associate closely with their weaned offspring are more likely to fail to rear their next joey (Johnson, 1985; 1987), while in several other social mammals the breeding success of females declines as the number of their resident daughters increases (van Schaik, 1983).

Measurements of the costs and benefits of parental care are beset with the methodological problems associated with studying trade-offs between different components of fitness, which were reviewed in Chapter 2. Much data on parental care is correlational which may be misleading, especially where parents differ in 'quality'. Although an increasing number of experimental studies is being performed (see Chapter 2), further work is required to provide a firm empirical basis to theories concerning adaptive variation in parental expenditure.

8.3 The trade-off between the size and number of offspring

Perhaps the most fundamental type of parental investment is in the resources required to provision the egg. However, because resources are finite, increasing the investment per egg leads to the production of fewer eggs. A large theoretical literature has grown up around the study of this trade-off, although empirical work in this area is chiefly observational, largely because of the difficulty in experimentally manipulating egg size.

8.3.1 The Smith–Fretwell model

The first quantitative model of the trade-offs between the size and the number of offspring was from Smith and Fretwell (1974) and this has formed the basis of nearly all subsequent work. Smith and Fretwell asked how a parent should distribute a fixed amount of resources (M) amongst an indefinite number of young. They had in mind a mother whose only investment in her offspring lay in provisioning the egg so that the only trade-off involved was between the size and number of eggs. Their approach assumes that the way in which parental resources are distributed within particular breeding attempts does not affect the risk of the whole clutch being lost or the parent's subsequent survival or breeding success.

Suppose the mother produces eggs of size s which survive to maturity with probability $k \cdot f(s)$. The probability of survival has two components, one ($f(s)$) influenced by parental investment and a second component (k) which summarizes all the other mortality risks that are unaffected by egg size. The number of eggs produced by the mother is simply M/s, the total resources available divided by the amount invested in each individual. The fitness of the mother is thus $(M/s) \cdot k \cdot f(s)$, the number of offspring multiplied by their own individual fitness. The optimal egg size is found by maximizing fitness with respect to investment per egg and is $f(s)/f'(s)$ where the prime refers to the derivative.

The solution to this problem is of a marginal value form (a ratio of a function and its derivative: see also Chapter 4) and can also be derived graphically (Fig. 8.1a). The optimal offspring size is defined by the point at which a straight line from the origin is tangent to the offspring fitness curve ($f'(s)$); at this point the parent maximizes its fitness returns per unit of investment (see also Chapter 4). Note that if the fitness of an offspring rises with increasing size but on a decelerating curve (Fig. 8.1b), a mother should produce the maximum number of the smallest viable individuals. Alternatively, if the offspring fitness curve always accelerates (Fig. 8.1c), all resources should be invested in a single large offspring. Unfortunately, the shape of the fitness curves in Fig. 8.1 remains largely hypothetical because of the problems associated with experimentally manipulating egg size.

The Smith and Fretwell model reveals a number of interesting properties about the trade-off between size and number of offspring (see also Parker & Macnair, 1978; Parker & Begon, 1986; Lloyd, 1987; Winkler & Walin, 1987). First, optimal egg size is not influenced by total parental investment in reproduction: total resources (M) do not appear in the expression for optimal egg size. The reason for this rather paradoxical result is that an implicit assumption is made that the fitness of an individual is unaffected by clutch size after the initial provisioning of the eggs. In many animals, egg size and clutch size will both affect fitness and will evolve together (for example, Parker & Begon, 1986).

Second, optimal egg size is uninfluenced by investment-independent mortality (k). Thus, at least in this simple model, high levels of 'random'

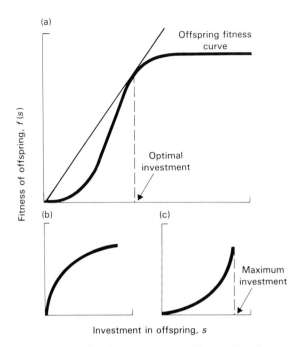

Fig. 8.1 A graphical representation of the Smith and Fretwell model. (a) The curve describes the relationship between offspring fitness and the resources invested in an offspring. The parent obtains the maximum returns on her investment at the point where a line from the origin is tangent to the offspring fitness curve. (b) When the offspring fitness curve is always decelerating, the parent is selected to produce the smallest possible offspring. (c) When the offspring fitness curve is always accelerating, a parent will allocate all resources to a single offspring.

mortality, perhaps associated with harsh environments, have no effect on optimal egg size. However, the shape of the relationship between fitness and egg size (Fig. 8.1) may differ between harsh and benign environments and so lead to different predictions for egg size. In general, harsh environments or intense competition with conspecifics are likely to raise the benefits of increased size to offspring and hence the optimal size for the parent to produce (Ito, 1980; McGinley et al., 1987). Both interspecific and intraspecific comparisons support an association between egg size and the severity of the environment faced by emerging juveniles. Among marine benthic invertebrates, the proportion of species producing large eggs and larvae that feed little on external food sources increases with latitude and depth (Thorson, 1950; Vance, 1973a; 1973b). In fish, species breeding in fresh water where environmental conditions are harsher and predators and parasites are more abundant produce larger eggs than those breeding in the sea (Wootton, 1979).

Within a number of species, egg or neonate size increases and clutch size declines as environmental conditions deteriorate (Bagenal, 1969; Capinera & Barbosa, 1977; Fraser, 1980). For example, in the terrestrial isopod *Armadillium vulgare*, females produce few, large offspring when conditions are unfavourable (Brody & Lawlor, 1984). Winter-breeding *Gammarus* (a freshwater amphipod) produce fewer and larger eggs than summer-breeding ones (Kolding & Fenchel, 1981). Similarly, the Atlantic herring, *Clupea harengus*, produce few, large eggs at the beginning and end of the spawning season when conditions are relatively unfavourable and a larger number of smaller eggs during the peak of spawning (Blaxter, 1969). Although phenotypic correlations of the sort discussed here must be treated with caution, they do provide at least strong circumstantial evidence for the influence of the environment on the trade-off between offspring size and number.

8.3.2 *Selection for variable offspring size*

The Smith and Fretwell model suggests that there should be a single optimal size for offspring, although this may change with environmental conditions or differ between parents. In constrast, some recent papers have argued that in heterogeneous or unpredictable environments, selection should favour genotypes that produce a range of propagule sizes, either because differently sized offspring are at an advantage in different situations, or because variable offspring size reduces the chance that the genotype will disappear from the population altogether (Janzen, 1977; Capinera, 1979). These arguments are similar to those which suggest that organisms should lay many small clutches, rather than a few large clutches, in a variable environment to 'spread the risk'. As Seger and Brockman (1987) have recently shown, verbal arguments of this type that rely on spatial rather than temporal variability, although temptingly plausible, are very difficult to justify by formal models.

McGinley *et al.* (1987) investigated the size–number trade-off in the face of different types of environmental heterogeneity. They have shown that spatial heterogeneity alone would not normally select for variable offspring size. They were able to find some circumstances where spatial heterogeneity led to selection for variable offspring size, but these conditions were quite restrictive: the environment must be divided into microhabitats in which optimal offspring size differs; there must be strong density dependence within microhabitats; and parents must be able to direct young of the appropriate size to the right microhabitat. They also investigated temporal variability, a process that had previously been shown to be capable of selecting for variability in life history parameters (Kaplan & Cooper, 1984). Again, they were able to demonstrate selection for variable offspring size, but concluded that in real systems this selection would be weak or absent. More theoretical work is needed to investigate a wider class of models but as yet there seems little support for the suggestion that environmental variability selects for variable offspring size. McGinley *et al.* reviewed variation in seed size and suggested that it results from a variety of environmental effects and developmental constraints.

Selection may, however, favour variation in offspring size for other reasons. Where parents cannot afford to rear their whole brood, they may increase their fitness by selectively reducing brood size (Lack, 1954; O'Connor, 1978). As we describe below, there is little evidence that they actively intervene to cause mortality although they not uncommonly permit some offspring to monopolize resources, leading to selective mortality of smaller juveniles (Mock, 1984). Parents may accelerate this process (thereby maximizing the gains from brood reduction) by creating size hierarchies among their young, either by varying egg size within clutches (Slagsvold *et al.*, 1984) or by ensuring that their offspring hatch asynchronously (Lack, 1954). Although both variation in egg size and hatching asynchrony may evolve for other reasons (reviewed by Lessells & Avery, 1989; Slagsvold & Lifjeld, 1989), there is at least some evidence suggesting that both facilitate brood reduction. In blackbirds, asynchronous broods produce more young than synchronous broods when food availability is low, but the provision of extra food removes this effect (Magrath, 1989). Similarly, in herring gulls, where third eggs are usually lighter than first or second eggs, experimental removal of the second egg causes parents to increase the relative size of their third egg (Parsons, 1976).

8.4 Distribution of parental care among species

The distribution of parental care raises two fundamental questions. First, why do some species show elaborate and protracted care of eggs or offspring while others (in some cases, closely related species) do not. And second, why are males responsible for care in some species, females in others, and both sexes in a few? We discuss the answers to these two questions in turn. As Chapter 9 describes, many differences in parental care are closely related to variation in mating systems.

Like large offspring size, parental care might be expected in animals where juveniles face physically harsh environments, heavy predation or intense competition with conspecifics. This appears to be the case, both among invertebrates and in fish and amphibia (Clutton-Brock, 1991). Although no systematic analysis of its distribution has yet been attempted, parental care in insects is thought to be commonest in species that live in physically harsh or biotically dangerous habitats such as the intertidal zone or in species whose young depend on rich but scattered and ephemeral resources, such as dung, dead wood or carrion (Wilson, 1971). Among teleost fishes, parental care of eggs or young occurs in only 16% of all families breeding in saltwater, but in 57% of families breeding in freshwater, where environmental conditions are commonly more unpredictable and predation rates are often higher (Baylis, 1981; Gross & Shine, 1981). In amphibia, parental care is associated with terrestrial breeding (Salthe, 1969; Wells, 1981; Nussbaum, 1985) and may be necessary to prevent desiccation or mould growth or to deter invertebrate predators (Wells, 1981).

Table 8.1 Maynard Smith's (1977) parental care game.

PAY-OFF MATRIX (MALE PAY-OFF | FEMALE PAY-OFF)

	Female guards	Female defects
Male guards	$vP_2 \mid vP_2$	$VP_1 \mid VP_1$
Male defects	$vP_1(1+p) \mid vP_1$	$VP_0(1+p) \mid VP_0$

EVOLUTIONARILY STABLE STRATEGIES

1 *Both sexes desert*
This requires that: (i) $vP_2 > VP_1$, the number of eggs laid by a non-caring female multiplied by their survival exceeds the number laid by a caring female multiplied by their survival, or the female will care, and that (ii) $P_2 > P_1(1+p)$, the survival of eggs that are not cared for by either parent multiplied by the number of matings achieved by a non-caring male $(1+p)$ cannot exceed the survival of eggs under uniparental care, or the male will care.

2 *Female deserts and male cares ('stickleback')*
This requires that: (i) $VP_1 > vP_2$, the number of eggs laid by a caring female multiplied by egg survival under uniparental care must exceed the number laid by a caring female multiplied by egg survival under biparental care, or the female will care, and that (ii) $P_1 > P_0(1+p)$, egg survival under uniparental care must exceed survival of uncared for eggs multiplied by the number of matings that a non-caring male can achieve, or the male will desert.

3 *Female cares and male deserts ('duck')*
This requires that: (i) $vP_1 > VP_0$, the number of eggs laid by a caring female multiplied by egg survival under uniparental care must exceed the number of eggs laid by a non-caring female multiplied by the survival of eggs that are not cared for, or the female will desert, and that (ii) $P_1(1+p) > P_2$, the mating success of a non-caring male multiplied by egg survival under uniparental care must exceed egg survival under biparental care, or the male will care.

4 *Both partners care*
This requires that: (i) $VP_0 > vP_1$, the number of eggs laid by a caring female multiplied by egg survival under biparental care must exceed the number laid by a non-caring female multiplied by egg survival under uniparental care, or the female will desert, and that (ii) $P_0(1+p) > P_1$, the number of eggs that survive under biparental care must exceed egg survival under uniparental care multiplied by the mating success of the non-caring male, or the male will desert.

8.4.1 Game theory models: who cares?

Some of the most important questions about parental investment are those concerning the evolution of male and female care. Since the evolution of care by one sex is likely to reduce its benefits to the other sex (Chase, 1980), game theory models provide the appropriate framework for analysing the conditions under which uniparental male care ('stickleback'), uniparental female care ('duck') and biparental care are likely to evolve (Maynard Smith, 1977; 1978; 1982; Vehrencamp & Bradbury, 1984). Maynard Smith's well-known model (1977) assumes that there are discrete breeding seasons; that a female's expenditure on egg laying and on parental care limits the number of young that she can produce; and that care by males and females has the same effects on offspring survival.

Let P_0 be the probability of survival of eggs that are not cared for by either parent, P_1 be the survival of eggs cared for by one parent and P_2 be the survival of eggs cared for by both parents, such that $P_2 > P_1 > P_0$. Suppose that a male who deserts has a chance p of mating again while a female who deserts after egg laying produces V eggs compared with v for a caring female. The pay-off matrix for this game and the four evolutionarily stable strategies (ESSs) that it generates are shown in Table 8.1.

If breeding seasons are continuous so that each individual breeds many times and two parents are less than twice as effective as one at ensuring that their egg survive, either uniparental male or uniparental female care can evolve from no-care (Maynard Smith, 1977). Whether male or female care evolves depends upon biological factors affecting the relative costs and benefits of care to each sex as well as the antecedent condition. Where the initial condition is biparental care, and two parents are less than twice as efficient at egg care as one, males may be more likely to desert than females where the absence of males affects egg survival less than the absence of females. It two parents are *more* than twice as good as one at caring for their eggs, the only possible ESSs are desertion by both sexes or biparental care (Maynard Smith, 1977).

Unfortunately, tests of Maynard Smith's model in contemporary populations may not tell us much about the initial evolution of parental care. This is because the behaviour of males and females and the development of eggs and juveniles are likely to become coadapted, following the initial evolution of parental care, with the result that manipulation of contemporary breeding systems may provide little indication of the initial pay-offs of care to males and females. For example, following the initial evolution of parental care, the development of young at hatching may change, and P_0 may consequently fall to zero. By removing *both* parents, we may be able to show that $P_2 \gg P_0$, but the original difference may have been far smaller. In species where one sex normally cares for eggs or young, the non-caring sex may gradually lose the capacity to provide effective care so that the experimental removal of one parent never leads to the assumption of care by the other and is usually followed by the death of the entire brood. Similarly, after the initial evolution of biparental provisioning, juvenile growth rates may increase so that successful rearing requires biparental care and P_1 fall to zero.

The most relevant empirical approach to investigating the evolution of male and female care is probably to examine the costs and benefits of care to contemporary species where either sex may care for the young (for example, Blumer, 1986). However, even here, the biology of one sex may have become coadapted to inequalities in investment between the sexes. As a result, explanations of the initial evolution and current distribution of male and female care rely on attempts to reconstruct its likely costs and benefits under particular antecedent conditions in the past and are necessarily speculative.

8.4.2 Influence of gamete size on who cares

Trivers (1972) initially suggested that females should usually be the care-giving sex because the costs of producing ova exceed those of sperm, committing deserted females to continue investing to avoid wasting their reproductive effort, an argument nicknamed the 'Concorde fallacy' by Dawkins and Carlisle (1976). As Dawkins and Carlisle pointed out, parents of both sexes should decide whether or not to prolong investment on the basis of its likely net benefits in the future (see also Maynard Smith, 1977). One reason why female care predominates in endotherms is that the large size of eggs (or of neonates in viviparous species) lowers their potential rate of reproduction relative to the rate that can be achieved by males. Where parental care inhibits further reproduction until it is complete, it usually has potentially higher costs to males than to females, with the result that uni-parental female care is more likely to evolve than uniparental male care. An additional reason is that most endotherms show internal fertilization, permitting the spatial separation of males and females after copulation.

The importance of the relative rates of reproduction in determining the evolution of parental care was first emphasized by Baylis (1978; 1981). Baylis (1981) suggested that the faster rates of gamete production by male fish lead to the evolution of predominant *male* care because they cause male fitness to be limited by access to mating partners, with the result that males compete for resources (e.g. nest sites) that will attract females. His model is most relevant to cases where parental care evolves from a non-caring ancestor and assumes that unit costs of parental care to males are low in many fish species because males can simultaneously care for mul-tiple clutches and continue to attract females—a situation rarely found in endotherms.

The contrasting consequences of anisogamy in ectotherms and endo-therms emphasize that there is no simple relationship between anisogamy and the evolution of parental care. Instead, differences in gamete size and in the rate of gamete production are one of a variety of factors that can influence the relative costs of parental care to the two sexes. They are most likely to be important where uniparental care evolves from an antecedent state of no care. Where either or both sexes already care for their young, their potential rates of reproduction may be limited by factors other than gamete size. In mammals, for example, it is the involvement of females in gestation rather than the difference between ovum and sperm size that is largely responsible for the difference in potential reproductive rate between the sexes. Similarly, in some pipefish, the reproductive rate of males is limited by the number of eggs they can brood in their pouches rather than by their rate of gamete production (see Berglund *et al.*, 1986).

8.4.3 *Uniparental care in fish*

The distribution of uniparental male and uniparental female care in teleost fishes provides some good examples of the way in which the costs and benefits of care affect the involvement of males and females. In most teleosts showing external fertilization, parental care only involves males (Fig. 8.2).

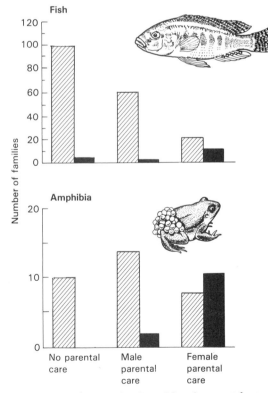

Fig. 8.2 Occurrence of male and female parental care (including biparental care) among 181 families of teleosts and 45 of amphibia in relation to mode of fertilization (from Gross & Shine, 1981). (▨) External fertilizers; (■) internal fertilizers. Families may appear in more than one category.

In the majority of these groups, male care probably evolved directly from a non-caring ancestor (Gross & Sargent, 1985). Although other ideas have been proposed (Trivers, 1972; Dawkins & Carlisle, 1976), the most likely explanation of the evolution of male (as against female) care in most of these species is that the costs of care to males are relatively low (Williams, 1975; Borgia, 1979; Baylis, 1981; Gross & Shine, 1981; Gross & Sargent, 1985). Males can guard large numbers of eggs, contributed by several females. For example, in some darters, nests commonly contain 2000 or more eggs while females lay around 150 eggs at a time (Gale & Deutsch, 1985). Moreover, in several species, females are attracted by the presence of previous eggs and males prefer to defend nests with fertilized eggs already in them (Unger & Sargent, 1988).

The fitness costs of energy expenditure on guarding may also be lower to males than females because the energetic costs of spawning are generally lower to males (Loiselle & Barlow, 1978; Perrrone & Zaret, 1979; Blumer, 1986) or because male size has little effect on mating success while fecundity is disproportionately high in larger females so that any effects of guarding on growth have higher costs to females (Gross & Sargent, 1985; but see Shine, 1988). Either of these differences could explain why in some monogamous fishes, including the anemone fish *Amphiprion*, males are responsible for virtually all egg tending (Fricke, 1974).

Uniparental *female* care in externally fertilizing fishes appears to be associated either with circumstances where the costs of care to females are unusually low or where the benefits are unusually high (Clutton-Brock, 1991). Uniparental female care is often associated with short breeding seasons or with semelparity, both of which are likely to reduce the costs of egg guarding to females (Perrone & Zaret, 1979; Gross & Sargent, 1985). For example, a close association between female egg guarding and short breeding seasons is found among the blennies, where egg guarding is confined to northern species where females typically lay all their eggs in a single clutch (Qasim, 1956; Thresher, 1984). In other species, female guarding is apparently associated with situations where the benefits of care to females are unusually high (Clutton-Brock, 1991). For example, female egg guarding occurs in a number of externally fertilizing species where males defend territories used simultaneously by several spawning females who defend smaller territories within that of the male (Fryer & Iles, 1972; Fricke, 1980). Here, either persistent egg cannibalism or competition between females for spawning sites could favour the evolution of female territoriality and uniparental female care, either from a non-caring ancestor or from uniparental male care (Clutton-Brock, 1991).

8.4.4 *Stable parental expenditure by males and females*

To account for the evolution of uniparental care from antecedent states where either or both sexes are involved in care, we need to understand both the relative costs of care to males and females and the likely response of their partners to changes in their level of expenditure (see above). Houston and Davies (1985) showed that where offspring fitness increases as an asymptotic function of total parental expenditure once parental expenditure exceeds some threshold level, each parent should respond to increases in care by its partner by reducing its own expenditure (Fig. 8.3). Conversely parents would be expected to respond to reductions in care by their mates by *increasing* their own expenditure. An ESS will be reached when the male's expenditure E_m is the male's best value given that the female expends E_f and where E_f is the female's best value if the male expends E_m (a

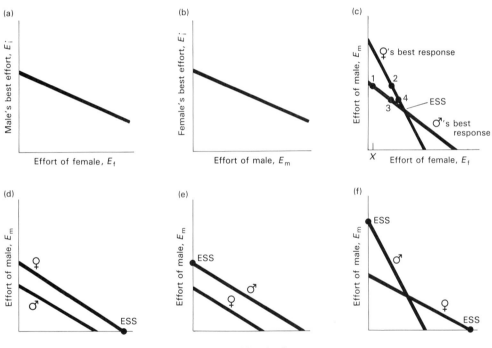

Fig. 8.3 Optimal parental expenditure for males and females plotted against the level of expenditure by the other sex (from Houston & Davies, 1985). (a) Optimal male effort plotted against female effort; (b) optimal female effort plotted against male effort. As both slopes are < -1, when plotted on the same graph (c) they predict an evolutionarily stable state (ESS) where both partners care. To see this, suppose the female expends X, then the male's best response is to invest at the level marked 1 on the graph. The female's best response to this level of male investment is found at point 2; the male responds at point 3, the female at point 4, and so on until the ESS is attained. In the ESS, it does not pay either individual to change its effort. Three other possible outcomes were suggested by Chase (1980). In (d), the female curve lies completely above that of the male. Here the ESS is for the female to do all the work. In (e), the male curve lies above that of the female. Here, the ESS is for the male to do all the work. In (f), although the curves intersect, the intersection is unstable and is not an ESS: both 'all female work', or 'all male work' are ESSs.

Nash equilibrium). If curves for E_m plotted against E_f and E_f plotted on E_m ('reaction' curves) intersect and each has a slope of less than -1, the intersection point should be an ESS for both partners (see Fig. 8.3c). In such cases, parents should respond to reductions in care by their partners by increasing their own effort while not fully compensating for the reduction so that total expenditure is reduced. Where two partners are making alternative investments in parental care, their expenditures are likely to converge on the ESS through a sequence of smaller and smaller changes.

If reaction curves do not intersect or the slope of either is greater than -1, other outcomes are possible (Chase, 1980; Winkler, 1987). If the female's reaction curve is completely above that of the male (Fig. 8.3d), then the male should desert and the female's best strategy would be to take over care of the young. Similarly, if the male's reaction curve is above that of the female (Fig. 8.3e), then the male should be responsible for parental care. Finally, if reaction curves have slopes greater than -1 at the region of intersection (Fig. 8.3f), the intersection point is unstable. Under these conditions, if one parent reduces its expenditure, the response of the other parent should be to *increase* its own expenditure by a larger amount, more than compensating for the reduction. Reactions then get larger and larger until one parent is responsible for all expenditure (Houston & Davies, 1985). Which parent this will be depends on the initial conditions.

The responses of partners to changes in the level of investment by their mates vary widely, which may help to account for the different effects of removing one partner. In some cases, increased assistance by the male has little or no effect on the level of parental care provided by its mate (Slagsvold & Lifjeld, 1988; Whittingham, 1989). In others, the remaining parent compensates partly or fully for the reduction in input by its mate, leading to negative correlations between the expenditure of partners (Martin, 1974; Weatherhead, 1979; Smith *et al.*, 1982; Breitwisch, 1988; Wanless *et al.*, 1988; Wolf *et al.*, 1988; 1989). Most experimental demonstrations of this effect involve the removal of one mate and have the disadvantage that the consequences of mate removal may be different from those of changes in investment by a parent that continues to help, especially if the level of expenditure is determined by a 'bargaining' process. One recent study of starlings (*Sturnus vulgarus*) avoids this problem by varying the rate of feeding by partners by attaching weights to their tails (Wright & Cuthill, 1989; Fig. 8.4). Weighted partners (either males or females) reduce their feeding rate and their unweighted mates compensate for this effect by increasing their own rate, although they do not do so fully. Finally, positive correlations between levels of investment have been found in a few cases. For example, great tits increase the intensity of their mobbing response to an (artificial) predator in relation to the intensity of their mate's behaviour (Regelmann & Curio, 1986).

As Chase (1980) predicted, similar effects of the level of input by other parents occur among cooperative and communal breeders. For example, in

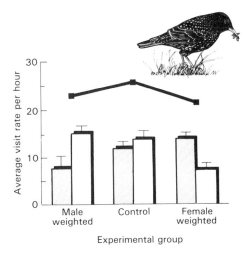

Fig. 8.4 Compensation in expenditure on feeding young by male and female starlings, *Sturnus vulgaris*, when their mates are artificially weighted, thereby reducing their feeding rate (from Wright & Cuthill, 1989). Bars show the average visit rate per hour by male (□) and female (□) parents when their mates are artificially weighted. Points above the histograms document the combined visit rate, showing that although mates do compensate for reductions in the number of visits by their partners, they do not do so fully. (■——■) Total visit rate.

the dunnock *Prunella modularis*, some females are polyandrously mated to two males, an α male who is responsible for most of the mating and a β male who mates less often (Davies, 1985). β males generally expend less effort on the broods than α males and their expenditure on care is related to the frequency with which they copulate with the female (Burke *et al.*, 1989). Hatchwell and Davies (1990) show that the provisioning efforts of both males and females are sensitive to the efforts of others. For example, females increase their provisioning rate when they have no male help and β males increase their rate when α males are removed from trios. As in the model (see Fig. 8.3c), these 'reactions' were not sufficient to compensate fully for the experimental losses.

In some species where both sexes care for the young, one sex may desert before parental care is complete (for example, Beissinger, 1987; Ezaki, 1988). This situation is likely to lead to conflicts of interest over who will care for the young unless biparental care is the best strategy for both sexes. The benefits of desertion will depend on whether the deserted partner takes over all care of the brood or deserts in its turn (Dawkins, 1976; Lazarus, 1990). A parent will be particularly likely to desert its mate at times when the potential pay-offs of desertion are high, when reductions in care have relatively little effect on the fitness of its brood, or when its partner is least likely to desert in response (Lazarus, 1989).

Houston and Davies' (1985) model begs the question of why biparental care is so common among birds. The best evidence of the benefits of male care in biparental species is provided by experimental removal of males (see also Chapter 9). In many altricial birds, the removal of males reduces the growth rate of chicks, the survival of nestlings, or the survival of fledglings before independence. For example, in dark-eyed juncos *Junco hyemalis*, experimental removal of males within 2 days of broods hatching had little effect on the survival of chicks to fledging but, after fledging, differences in chick survival increased (Wolf *et al.*, 1988). The effects of male removal are more variable in birds with precocial young (see Erckmann, 1983; Martin & Cooke, 1987).

8.5 Conflicts between parents and offspring

8.5.1 *Theory*

There are a number of conflicts that may arise during parental care. The young may demand more resources than the parent is willing to provide or may disagree with their brood-mates over the distribution of parental care, and, in species with biparental care, the parents may disagree over their relative investment in the young. These conflicts are best understood using Hamilton's (1964) theory of kin selection and, in particular, Trivers' (1974) extension of the theory to parent/offspring conflict. The basis of kin selection is Hamilton's rule (see also Chapter 1); a trait will be favoured if:

(Relatedness) (Benefits (costs) to others)−(Costs (benefits) to actor) > 0

Consider a mother providing care for a single offspring. Her optimum investment will be determined by the trade-off between the increase in her young's fitness as more resources are allocated to it and the reduction in her future reproductive success brought about by high investment in the current breeding season: the mother balances present offspring against future offspring. The amount of resources demanded by the young will also be influenced by this trade-off: the offspring has a genetic interest in the future reproductive success of its mother. However, whereas the mother is identically related to all its progeny, the offspring values investment in itself twice as much as investment in future siblings (assuming they are full sibs with relatedness of 1/2, the offspring would value itself four times as much as half sibs with relatedness of 1/4). There is thus a conflict of interest between the mother and her offspring, with the offspring demanding more resources than the parent is willing to give.

Conflict over the amount of parental investment will affect many aspects of the parent–young relationship (Trivers, 1974). When the young are being fed, they will tend to demand more food than the parent is willing to supply. As they grow older, there will be disagreement over the time of termination of parental care, the young wishing to prolong care beyond the parental optimum. Parental aggression against their young towards the end of the period of parental care is characteristic of many birds and mammals and is probably a direct result of parent/offspring conflict over weaning (Trivers, 1974; Lazarus & Inglis, 1986; Stamps et al., 1985).

Trivers' argument shows that parent/offspring conflict will exist but does not predict how much conflict will be found in nature. This problem was addressed in a series of papers by Parker and Macnair (1978; 1979; Macnair & Parker, 1978; 1979; Parker, 1985). They first asked what levels of greed will be observed if a mother completely acquiesces to her offspring's demands. Consider a mother investing in a single offspring and define her optimal investment as one unit of parental care. If the offspring demands m units of parental care it increases its fitness from $f(1)$ to $f(m)$ but at a cost in terms of its mother's future reproductive success. Parker and Macnair (1978; Macnair & Parker, 1979) solved for the ESS value of m using a population genetic argument, although a more general result can be obtained by an application of Hamilton's rule. Suppose all the offspring in the population are demanding m, if all alternative strategies, 'demand $m+\Delta$', are at a disadvantage by Hamilton's rule, then m is an ESS.

The benefit to an individual through switching from m to $m+\Delta$ is an increase in fitness: $f(m+\Delta)-f(m)$. The cost to the individual is a reduction in the number of future siblings: the Δ extra units of investment directed towards the individual could have been used by the parent to produce an extra Δ/m offspring, each with fitness $f(m)$. Substituting the costs and benefits into Hamilton's rule we obtain:

$$f(m+\Delta)-f(m) > r\frac{\Delta}{m}f(m)$$

where r is the coefficient of relatedness between siblings. Dividing both sides by Δ gives:

$$\frac{f(m+\Delta)-f(m)}{\Delta} > r\frac{f(m)}{m}$$

In an ESS, the costs exactly match the benefits and this inequality becomes an equality. Provided that offspring fitness shows a diminishing return with increased investment, the benefits will be greatest when Δ is small. As $\Delta \to 0$, the condition for the equality of costs and benefits becomes:

$$f'(m) = r\frac{f(m)}{m}$$

The left-hand side of this equation can be interpreted as the marginal gains from extra solicitation while the right-hand side represents lost future siblings. For a parent, equally related to all offspring, or for the offspring of an asexual species, the ESS level of solicitation is found by setting $r = 1$: $f'(m) = f(m)/m$. This is the same result as for the trade-off between egg size and egg number analysed in section 8.3. Here, by definition, the ESS condition is satisfied at $m = 1$. If the parent is monogamous with all her offspring full siblings, the young devalues future offspring by $r = \frac{1}{2}$ and the ESS condition is met at

$$f'(m) = \frac{f(m)}{2m}$$

If the parent has a different mate each breeding season, the young devalues future offspring by $r = \frac{1}{4}$ and the ESS condition is met at

$$f'(m) = \frac{f(m)}{4m}$$

The argument above considered the trade-off between the fitness of a current offspring and the number of its future siblings. Another source of conflict is sibling competition within a brood. Suppose, for example, that the total available resources for a brood are constant, but that the amount obtained by an individual depends on its level of solicitation relative to the other members of the brood: for a bird, the level of solicitation might be the noise of its begging (Macnair & Parker, 1979; Harper, 1986). Where there are no costs to solicitation, all individuals will be selected to solicit at the maximum possible rate. However, if increased solicitation incurs costs, selection will lead to more modest levels of solicitation. Note, the parental optimum is no solicitation, so there will always be some parent/offspring conflict. The exact result depends critically on whether the costs of solicitation are borne by the individual alone (for example, energetic costs) or by the whole brood (for example, if high levels of solicitation attract predators), and also on the presence of phenotypic asymmetries such as size hierarchies within the brood (H.C.J. Godfray & G.A. Parker, unpublished).

How might a parent retaliate to the extra demands of its offspring? In many, although certainly not all, cases the parent has complete discretion as to how resources are allocated. The parent thus has the opportunity to enforce its own optimum and to ignore the demands of its offspring. However, costs may be incurred through ignoring an offspring's demands: an unsatisfied offspring may solicit loudly and weaken itself, as well as attracting predators. In addition, as Dawkins (1976) pointed out, offspring solicitation may also carry important information for a parent (for example, that the offspring is relatively undernourished), information that is costly for the parent to ignore. Thus total parental indifference to offspring solicitation is unlikely to evolve.

Parker and Macnair (1979; Parker, 1985; for a different approach see Harper, 1986) model a number of cases of parental retaliation. The results of their studies are difficult to summarize as they depend critically on a number of assumptions about the nature of the parental response, about how the costs of solicitation are borne by the brood, and on the social structure of the species involved. In some cases, parent 'wins' or offspring 'wins' are ESSs (i.e. uninvasable by alternative parent or offspring strategies) while in other cases no ESS is possible—gene frequencies cycle indefinitely. However, the commonest solution, and probably the most likely result in nature, is that the overall ESS is a compromise between the parent and offspring optima. For example, consider the case of interbrood conflict discussed in detail above. If solicitation incurs costs to the individual and if the parent partially meets the demands of those offspring soliciting at a relatively high level, the ESS is

$$f'(m) = \frac{1}{2}(1+r)\frac{f(m)}{m}$$

(this is a slightly more general version of the results in Parker and Macnair (1979) obtained by solving Hamilton's rule simultaneously for parent and offspring). The joint ESS is intermediate between the offspring and parent optima.

The work of Parker, Macnair and others has helped resolve some of the early criticisms of the concept of parent–offspring conflict. For example, Alexander (1974) argued that a gene for conflict expressed in the young could never spread as it would reduce the reproductive success of the same individuals when they grew up and became parents themselves. As Dawkins (1976) pointed out, this argument was based on an artificial asymmetry between offspring and parents (the argument could be rephrased in favour of the offspring always winning) while explicit population genetic models show the conditions under which parent–offspring conflict can evolve. Alexander's second argument, that the parent holds the whip hand and so will always win may in some circumstances be true (see also Chapter 12). However, other results are possible, especially when a parental policy of ignoring offspring demands has other costs.

While theory suggests that conflicts of interest within the family are probably common and are likely to have important effects on parental investment and tactics, it is seldom possible to compare directly optimal levels of investment for offspring or their parents, or to measure the influence of offspring behaviour on levels of investment. However, the study of apparently extreme manifestations of within-family conflict, such as siblicide, allows at least qualitative tests of theory.

8.5.2 Siblicide

One of the most extreme outcomes of sibling competition is siblicide: when one offspring actually kills another. Siblicide has been extensively studied in birds where it is known from a number of groups including large raptors, gannets and boobies, herons and egrets, and skuas (jaegers). In some species of eagle, the parent normally lays two eggs, although the successful development of both offspring has hardly ever been observed (Meyburg, 1974; Garnett, 1978). The most probable explanation for the almost obligate siblicide of these species is that the mother lays a second egg to insure against infertility of a single egg (Stinson, 1979). Facultative siblicide may be an adaptation to variable food supply, the parent rearing the complete brood in years of plenty while brood reduction occurs in years of scarcity (Lack, 1966; Mock, 1984).

Parents and young will conflict over the exact conditions for brood reduction. Obviously, the victim will usually object to its own demise, but normally in birds the victim is a runt and is unable to influence events. A potentially siblicidal individual will be more willing to kill a sibling than its parents would wish as it values the redirection of resources from a sibling to itself more highly than do its parents. Parent–offspring conflict is predicted to be greatest in small broods because in small broods more of the redirected resources accrue directly to the siblicidal chick, rather than being spread around a large number of surviving offspring. O'Connor (1978), in a survey of siblicide in birds, found some evidence for this prediction.

The assumption in studies of bird siblicide that there is an identifiable victim, the runt, is very important for the outcome of parent–offspring competition (Godfray & Harper, 1990). Siblicide also occurs in many parasitoid wasps but in this case there is no runt. If a female lays two or more eggs into a host, the larvae fight with each other until only a single individual remains. The larval behaviour effectively prevents a parent increasing her clutch size and accounts for a dichotomy between solitary parasitoids (only one larva develops per host) and gregarious parasitoids (several larvae develop per host). In order for a female wasp to increase her clutch size, selection has to act simultaneously on her offspring to show tolerance for siblings. Genetic models show that it is extremely difficult for a rare gene for tolerance to spread, chiefly because, when rare, the gene for tolerance will frequently find itself in mixed sibships with a siblicidal individual and thus get eliminated (Godfray, 1987). These models predict that many solitary species will be found in hosts large enough to support a gregarious brood and there is evidence for this in at least one group of parasitoids (le Mesurier, 1987). The models also predict that the conditions for the spread of tolerance will be different to those for the spread of siblicide: once evolved, the solitary life style with fighting larvae will be very difficult to lose and thus it has some of the properties of an absorbing state.

8.6 Sex allocation and sex ratio manipulation

Our discussion of parental investment has so far largely ignored the sex of the offspring. In many circumstances this inattention can be excused by appealing to Fisher's (1930) argument which explains why equal sex ratios are so widespread. Fisher argued that if the population sex ratio is female biased, any gene that led its bearer to produce a preponderance of sons would be selected because these sons would mate with, on average, more than one female: when the population sex ratio is female biased, sons are a more efficient vehicle than daughters for transmitting genes to future generations. However, if the population sex ratio is male biased, sons will mate with, on average, less than one female and a gene that promotes female-biased sex ratios will be favoured. This frequency-dependent selection ceases only when the population sex ratio is at equality, which is thus an evolutionarily stable state. This result applies to the primary sex ratio, the sex ratio at the end of parental investment, and the prediction is unaffected by later differential mortality. A corollary of Fisher's argument is that at sex ratio equilibrium, a parent should be indifferent as to whether he or she invests in sons or daughters: the rate of return (measured in units of parental fitness) is identical for investment in either sex.

Fisher's argument makes a number of assumptions which, when violated, lead to the prediction of unequal sex ratios. The study of sex ratios, and sex allocation in general, has been one of the most profitable areas of evolutionary ecology and has spawned a huge literature, reviewed by Charnov (1982). Here we shall concentrate on three topics that are particularly relevant to the division of the parent's total investment between the sexes: (i) unequal costs of sons and daughters; (ii) environmental sex determination; and (iii) interactions between relatives.

8.6.1 Unequal costs of sons and daughters

If sons and daughters are equally costly to produce, Fisher's argument predicts an equal sex ratio. In fact, Fisher's result is more general and covers cases where one sex is more expensive to produce than the other. When this occurs, the ESS is to invest equally in the two sexes so that the sex ratio will be biased in favour of the cheaper sex. To understand this result, suppose the sex ratio is at equality but that sons cost more to produce than daughters. The reproductive success of sons and daughters will be equal but the gain in fitness through sons is achieved at greater cost than that through daughters: natural selection will thus switch investment towards daughters. When overall investment in the two sexes is the same, the increased cost of sons is exactly counterbalanced by their increased reproductive performance as the rarer sex.

A curious example of the consequences of equal allocation in the sexes comes from a group of parasitoid wasps called heteronomous aphelinids (Godfray & Waage, 1990). Female wasps search for two different types of host, one suitable for male eggs and the other for female eggs. In some species, males develop exclusively as parasitoids of moth eggs and females exclusively as parasitoids of homopteran nymphs. It seems likely that for most ovipositing wasps, the limiting investment in reproduction is the time spent searching for hosts. The wasp should thus allocate equal search time to looking for hosts suitable for males and those suitable for females so that the observed sex ratio will depend critically on the relative abundance and ease of discovery of the two host types. Heteronomous aphelinids often show highly biased sex ratios that are influenced by the relative abundance of host types (Hunter, 1989).

In dimorphic vertebrates, male juveniles commonly show higher differential mortality during the period of parental care compared with females (Fisher, 1930; Clutton-Brock et al., 1985). If sons and daughters are equally expensive to produce, but if sons often die before the termination of parental care, equal investment in both sexes will lead to a male-biased sex ratio at conception. However, by the end of parental care, and after the differential mortality affecting sons has acted, the sex ratio should now be biased towards females (Fisher, 1930). A second complication that arises in vertebrates (and in many other groups) is the constraint imposed by the sex determination mechanism. If sex determination is controlled by independently segregating sex chromosomes (such as the XY/XX system in mammals; for description of other mechanisms see Bull, 1983) the sex ratio at conception may be ineluctably fixed at $1:1$. Manipulation of the sex ratio is then difficult without selective abortion, which carries its own costs, although a parent is still able to channel parental investment subsequent to conception towards a favoured sex. Although the degree of bias in the birth sex ratio is related to the likely degree of sex differential mortality in some groups, this is not so in all cases (Clutton-Brock, 1991). One explanation may be that, in many of these species, the instantaneous costs of rearing males exceed those of rearing females (Clutton-Brock et al., 1981).

8.6.2 Environmental sex determination

Fisher's theory assumes that although the environment may be variable, sons and daughters suffer or benefit in equal measure from this variation. We use the word environment here in a broad sense, incorporating both changes in the abiotic environment as well as changes brought about by differences in maternal condition and numbers of competitors.

Trivers and Willard (1973) first explored the consequences of relaxing this assumption. Suppose females vary in condition and that this variation is reflected in the quality of the young they rear: for example, females in good condition might rear particularly large young. Now suppose that young of one sex, say males, gain more from being large than do young of the other sex: the reason for this might be that males have to compete with each other for mates and individuals that are large at weaning have an advantage. Trivers and Willard argued that, under these conditions, females in *relatively* good condition should rear sons and those in *relatively* poor condition should rear daughters.

Trivers and Willard's suggestion was supported by a series of population genetic models by Charnov (1979) and Bull (1981) who showed that a wide range of phenomena could be explained by this argument, which they termed environmental sex determination (also called conditional sex expression; Frank, 1987). Among the environmental gradients that have been suggested to influence the fitness of male and female offspring are maternal condition (mammals: Trivers & Willard, 1973), host size (parasitoid wasps: Charnov et al., 1981), the number of competitors in a host (nematodes: Charnov, 1982), temperature (reptiles: Charnov & Bull, 1989) and photoperiod (amphipod crustaceans: Adams et al., 1987). With minor modifications, the same theory applies to sex change in sequential hermaphrodites (Charnov, 1979). In all cases, a threshold in the environmental gradient is predicted above which females should produce the sex that can gain most from better conditions and below which they should produce the other sex (for complicated relationships between fitness and the gradient there may even be a number of thresholds). A second and strong prediction of the theory is that the position of the threshold should be relative and not absolute. In other words, the position of the threshold should depend on the frequency distribution of different environmental conditions.

Until recently, the consequences of environmental sex determination for the population sex ratio have been unclear and known only for special cases (Bull, 1981; Charnov, 1982). Frank and Swingland (1988) and Charnov and Bull (1989) have now proved that the population sex ratio will always be biased towards the sex produced at the poorer end of the environmental gradient. The actual sex ratio may be very difficult to calculate and may depend on a knowledge of the exact relationship between sex-specific fitness and the environmental gradient, but the direction of the sex ratio bias is clear.

Some of the best examples of environmental sex determination come from those parasitoid wasps that lay a single egg per host. Here, the environmental gradient is host size, which correlates very closely with the size of the wasp that eventually emerges from the host. It is argued that females benefit strongly from being large because female size and fecundity are closely correlated; males also benefit from being large but to a lesser extent because mating ability is only weakly correlated with size. As predicted by theory, many species of wasps lay female eggs in large hosts and male eggs in small hosts (Charnov et al., 1981; King, 1987): indeed, this pattern was first noted by Chewyreuv in 1913. However, a sharp threshold host size is seldom found, possibly because the wasp and the experimenter perceive subtly different gradients. Several investigators have presented wasps with combinations of either large and medium-sized hosts or medium and small hosts (Charnov et al., 1981; Jones, 1982). The prediction is that males should be placed in medium hosts when they are the smaller of the pair while female eggs should be laid in medium hosts when they are the larger of a pair. The results are rather mixed with some species showing a response to relative host size (Fig. 8.5) whilst other species show a response only to absolute host size. It is possible that animals with a response to absolute host density live in an environment where the frequency distribution of host types is constant and have thus never been exposed to selection to recognize and respond to changes in the distribution of host types.

Many reptiles bury their eggs in the ground and the sex of the offspring depends on the temperatures experienced during development (for example, all crocodilians, many turtles and at least a few lizards; Bull & Charnov, 1989). A number of different patterns occur: males may develop at cool temperatures; they may develop at warm temperatures; or males may develop at intermediate temperatures with females developing under hot or cold conditions. The only theory that can, at present, explain this pattern is environmental sex determination with temperature the gradient along which offspring fitness varies in a sex-specific manner (Charnov & Bull, 1977). Having said that, one of the greatest problems at the moment in research on sex allocation is to explain how temperature affects sex-specific fitnesses; currently there is no satisfactory explanation for this (Bull & Charnov, 1989).

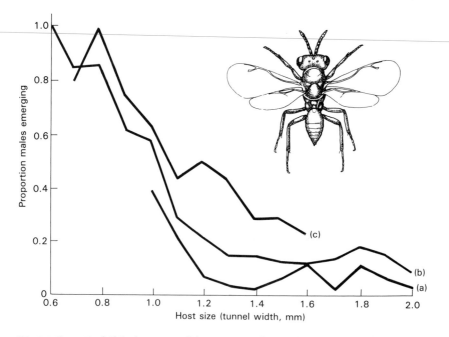

Fig. 8.5 Sex ratio shift in the petromalid wasp, *Lariophagus distinguendus* (from Charnov *et al.*, 1981). The proportion of males emerging is plotted against the size of hosts (weevil larvae) which were offered in sequence. Curve (b) resulted when females wasps were presented sequentially with 20 hosts of a single size. Curves (a) and (c) resulted when wasps were presented with the same sized hosts in alternating sequence with hosts differing in size by 0.4 mm. In curve (c), the host of interest was the smaller of the two and in curve (a) it was the larger of the two. For example, when hosts of 1.4 mm were offered alone, they gave a sex ratio of 15% male (curve (b)). When they were offered alternately with 1.8 mm hosts, they gave a sex ratio of 30% male (curve (c)) while when offered with 1.0 mm hosts this dropped to 2% male (curve (a)). This shows that female *Lariophagus* respond to relative host size.

There is now good evidence for the effects of environmental sex determination in several species of mammals, although these effects are strongly modulated by chromosomal sex determination (Clutton-Brock & Albon, 1982). In red deer, where there is great variance in male reproductive success, differences in dominance rank between hinds affect the reproductive success of their sons more than that of their daughters, suggesting that high ranking hinds should produce male calves and low ranking hinds female calves (Fig. 8.6). Data collected over 20 years from the red deer population on Rhum show that high ranking females consistently bias their sex ratio towards male calves while subordinate hinds produce an excess of daughters (Clutton-Brock et al., 1984). The mechanism by which the hinds manipulate sex ratio is not yet known.

Non-random sex allocation has been recorded in a wide variety of other mammals (Clutton-Brock & Iason, 1986), although frequently it is impossible to disentangle adaptive trends in the sex ratio from sex-differential mortality arising from sex differences in metabolism or growth rate. Studies of environmental sex determination in mammals are hampered by the lack of a well-developed theory that incorporates: (1) the constraints

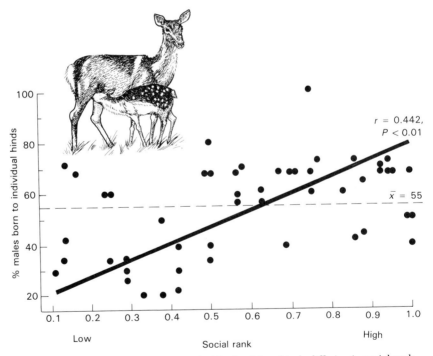

Fig. 8.6 Birth sex ratios produced by individual red deer hinds differing in social rank over their lifespans (from Clutton-Brock et al., 1986). Measures of maternal rank were based on the ratio of animals which the subject threatened or displaced to animals which threatened or displaced it weighted by the identity of the animals displaced. Values of this ratio range from 0.1 (low ranking) to 1.0 (high ranking). The figure shows that high ranking females tend to have sons.

due to chromosomal sex determination; (2) the costs of sex-ratio manipula-tion; (3) the small litter size of most mammals; and (4) differences in the costs of rearing sons and daughters. The latter can be very important; for example, red deer hinds that produced male calves in one year calved later the following year, and were almost twice as likely to be barren as hinds that reared female calves (Clutton-Brock et al., 1981). Important steps in the development of this theory have been taken by Maynard Smith (1980), Charnov (1982) and Frank (1987), but as yet only qualitative predictions can be made about mammalian sex ratio biases.

8.6.3 Interactions between relatives

A third important class of phenomena that lead to a biased sex ratio can be united under the heading of interactions between relatives. Considering in-teractions between siblings, Fisher (1930) implicitly assumed that the fitness of an individual was unaffected by the sexual composition of its sib-lings. If siblings of one sex compete for a resource, but those of the other sex do not compete (or suffer less competition), a parent will be selected to bias her sex ratio towards the sex that competes the less (Taylor, 1981). This selection pressure occurs because of a reduction in the efficiency of the competing sex as a means of transmitting genes to future generations.

One of the best studied examples of this type of process is local mate competition (Hamilton, 1967). If siblings tend to mate with each other then sons will compete for mates whereas daughters, the limiting sex, will not experience competition. A female-biased sex ratio is thus expected. In the limit, when mating takes place exclusively among members of a brood, the mother will be selected to lay just enough sons to fertilize all her daughters. Local mate competition has become a testing ground for the development of theoretical frameworks for explaining sex ratio biases. These different approaches fall into two main categories, those based around individual and kin selection models (Hamilton, 1967; Taylor, 1981; Grafen, 1984; Nunney, 1985) and those based around interdemic group selection and hierarchical selection models (Hamilton, 1979; Colwell, 1981; Wilson & Colwell, 1981; Frank, 1985; 1986). All theories give the same results but vary in their usefulness in analysing particular variants of the original problem and also in the ease with which they can be translated into plaus-ible verbal metaphors.

Many examples of biased sex ratios arising from local mate competition occur in haplodiploid insects and mites (see also Chapter 12). Highly female-biased sex ratios are not uncommon in these groups and are almost always associated with high levels of inbreeding (Hamilton, 1967; Waage, 1986; Griffiths & Godfray, 1988). There is also good evidence for facultative adjustment of the sex ratio in response to changes in the extent of local mate competition (Werren, 1980; 1983; Waage & Lane, 1984; but see also a criti-cal paper by Orzack, 1986). Some of the best experimental evidence comes from fig wasps. Fig wasps are fig tree mutualists which both fertilize and lay their eggs in the flowers inside figs. The young develop inside the growing fig and, when they become adult, mate among themselves inside the fig. Only females disperse to search for new breeding sites. The number of females that locate and enter a fig determines the magnitude of the local mate competition experienced by their sons in the next generation. Several studies have shown that the sex ratio produced by females becomes more male biased as the number of colonizers increases, as predicted by theory (Hamilton, 1979; Frank, 1985; Herre, 1985). Because fig wasps are haplodiploid, the predicted sex ratio in a fig depends not only on the number of wasps that colonize the fig but also on the level of inbreeding in the population, which is determined by the average frequency of sibling mating. Inbreeding in haplodiploids increases the coefficient of relatedness of a mother to her daughters relative to that of a mother to her sons. The increased relatedness to daughters results in selection for a further bias of the sex ratio towards females. In an elegant study, Herre (1987) estimated the magnitude of inbreeding in different fig wasp populations and showed that when the main effects of local mate competition were removed the residual variation in sex ratio was correlated with the level of inbreeding (Fig. 8.7).

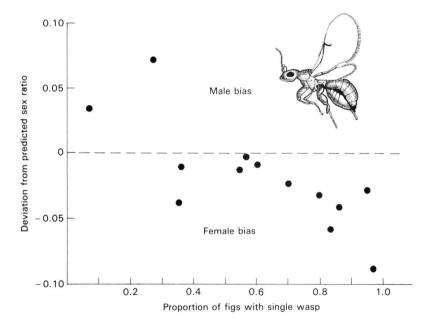

Fig. 8.7 Sex ratio and inbreeding in fig wasps. The optimal sex ratio in a fig depends on the number of females colonizing a fig. The figure plots deviations from the predicted sex ratio in figs colonized by two females for 13 species of Panamanian fig wasps as a function of the frequency of single-foundress broods in the population. The frequency of single-foundress broods correlates with the amount of inbreeding in the population. Sex ratios are more female biased when inbreeding is high, as predicted by theory (from Herre, 1987).

A second important case of competition between relatives is local resource competition (Clark, 1978). In many mammals, males disperse further than females from the site where they were born. If individuals compete with their neighbours for resources such as food, breeding sites or territories, there will be a greater probability of competition between female siblings than between male siblings. In addition, daughters will tend to compete with their mothers. These interactions lead to the prediction of a male-biased investment ratio. For example, Johnson (1988) found that more male-biased sex ratios were found in primate genera where females experienced the greatest intensity of competition (measured by the influence of group size on reproductive success). Furthermore, in an analysis which controlled for home range overlap, sex ratios were more male biased in genera where females did not disperse than in genera where males dispersed or where there was no sexual asymmetry in dispersal.

The opposite of competition is cooperation and this may also influence sex ratios, although in the opposite direction to competition. Thus if siblings of one sex cooperate with each other, or assist their parents in rearing future offspring, the sex ratio should be biased in favour of that sex (Malcolm & Marten, 1982). The red-cockaded woodpecker, *Picoides borealis*, has a male-biased sex ratio, and a possible explanation is that males of this species frequently assist their parents in rearing future young (Gowaty & Lennartz, 1985; Emlen *et al.*, 1986; Lessells & Avery, 1987). Similarly, in African wild dogs, *Lycaon pictus*, sons remain in their natal group and commonly help their mother to raise subsequent litters (Malcolm & Marten, 1982). As would be expected, birth sex ratios are consistently male-biased (see also Chapter 11).

8.7 Conclusions

Three developments would help to promote our understanding of the evolution of parental care in the future. First, we need to restrict the use of active verbs—invest, allocate, manipulate—to cases where there is some evidence of active discrimination by the parent. For example, it is misleading to refer to differential juvenile mortality as sex ratio manipulation unless there is evidence that the parent is responsible for the difference in survival. Similarly, we should not refer to partial mortality of broods as brood reduction unless there is evidence that parents (or young) are directly responsible for the death of some of the chicks. To do so begs the question that we are asking. Second, we should avoid testing the predictions of theoretical models of adaptive strategies with demographic data alone and should place considerable emphasis on the importance of understanding the causes of variation in the growth and survival of juveniles. Finally, we should not allow the dazzling perfection of some parental adaptations to lead us to assume that all aspects of parental behaviour are equally finely adapted or to blunt our interest in other factors which affect the development and survival of offspring.

9: Mating systems

Nicholas B. Davies

9.1 Introduction

The term mating system refers to the way in which individuals obtain mates, including a description of how mates are aquired, how many mates, the characteristics of pair bonds and patterns of parental care by each sex (Emlen & Oring, 1977). The variety in nature appears, at first sight, bewildering (see Figs 9.2 and 9.6), with differences not only between species but also within populations of the same species. What general principles can help us to understand this diversity? Beginning with Emlen and Oring's (1977) influential paper, a fruitful approach has been to view mating systems as outcomes of the behaviour of individuals competing to maximize their reproductive success. The diagrams in Figs 9.2 and 9.6 indicate how easy it might be to change from one mating system to another by changes in (i) male or female dispersion; or (ii) patterns of desertion by either sex.

Theoretical treatments of these topics have been inspired by two important themes.

DISPERSION

Females often need just one or a few matings to fertilize all their eggs. Where males provide only sperm (most insects and mammals), female reproductive success is therefore limited not by access to males but by access to resources; the more resources a female can gather the more offspring she can produce. A male, however, has the potential to father offspring at a faster rate than a female can produce them. For males, therefore, reproductive success is limited by access to females; the more females a male can mate with the greater his reproductive success (Bateman, 1948; Trivers, 1972). Thus it is males who tend to search for females, initiate courtship and fight for mates rather than vice versa. Resource limitation of female reproductive success and mate limitation of male reproductive success is nicely illustrated by some sex-changing fish where individuals become males only when large enough to monopolize several mates. Futhermore, when individuals change sex from female to male the amount of time spent feeding declines and more time is devoted to competition for mates (Hoffman, 1983).

This difference in the factors limiting the reproduction of the two sexes suggests that female dispersion should primarily be influenced by resources while male dispersion should primarily be influenced by female dispersion (Bradbury & Vehrencamp, 1977; Emlen & Oring, 1977; Fig. 9.1). Where males provide parental care (birds, some fish) this simple scheme is complicated by the fact that males become a resource important for female reproductive success, so female dispersion is influenced not only by resources (e.g. food, nest sites) but also by males.

PARENTAL CARE

The most important idea here is that mating systems should not be thought of as cooperative ventures where males and females raise offspring in perfect harmony, but rather that each individual should behave to maximize its own success even if this is at the expense of its mate (Trivers, 1972). We therefore need to consider, for each sex, the costs and benefits of caring versus desertion (Maynard Smith, 1977; Vehrencamp & Bradbury, 1984).

In this chapter I discuss how these two themes can help us to understand mating systems. I begin by considering cases where males do not provide parental care, where the simple scheme in Fig. 9.1. is most likely to apply. Note that males may compete for females directly ('female defence') or may anticipate how resources influence female dispersion and compete for resource rich sites ('resource defence'). In the second half of the chapter (section 9.5 onwards) I then consider the additional complication of male care.

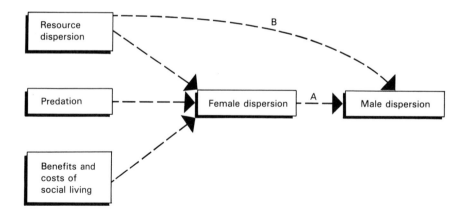

Fig. 9.1 Because female reproductive success tends to be limited by resources, whereas male reproductive success tends to be limited by access to females, female dispersion is expected to depend primarily on resource dispersion (modified by predation and benefits and costs of social living), while male dispersion is expected to depend primarily on female dispersion. Males may compete for females directly (A) or indirectly (B), by anticipating how resources influence female dispersion and competing for resource-rich sites.

9.2 Mating systems in mammals

9.2.1 Comparative studies

Comparative studies suggest that males typically impose themselves on females and that different mating systems arise as a consequence of differences in female dispersion (Wittenberger, 1980; Rubenstein & Wrangham, 1986; Clutton-Brock, 1989). The summary in Fig. 9.2 is based on educated guesses about whether female ranges or female groups are likely to be defendable by males. This broad comparative survey suggests that variation in male mating behaviour is attributable to four characteristics of females: (i) the extent to which female reproductive rate can be increased by male assistance with offspring care; (ii) the size of female ranges; (iii) the size and stability of female groups; and (iv) the density and distribution of females in space (Clutton-Brock, 1989). A male is better able to monopolize several females when male care with offspring is not needed, when females have small ranges or when they are in small stable groups. On the other hand, where females are solitary in large ranges, or occur in large or unstable groups, or where the female density is high, it may be uneconomic for males to defend several females at once and so they are more likely to be monogamous or monopolize individual females in succession.

The economic defendability of a female group may be influenced by both the size of the group and the temporal dispersion of receptive females. For example, among cercopithecine primates unimale harems occur when female group size is small, while several males associate with larger female groups (Andelman, 1986).

Where single males defend harems, harem size varies in relation to female distribution. For example, in seals which breed on pack ice, where suitable breeding habitat is superabundant, females are widely dispersed and males defend single females or small groups. In land-breeding species, by contrast, females may aggregate on particular safe beaches and males may be able to defend much larger harems (Le Boeuf, 1978). The temporal availability of receptive females also influences male defence. For example, in elephant seals *Mirounga angustirostris* (Le Boeuf, 1974) and red deer *Cervus elaphus* (Clutton-Brock *et al.*, 1982) males defend temporary harems in the short breeding season. In gelada baboons *Theropithecus gelada* (Dunbar, 1984) and Burchell's zebra *Equus burchelli* (Rubenstein, 1986), where breeding seasons are longer, harems may be defended throughout the year.

9.2.2 Measuring paternity

How well do these observations of male associations with females reflect male reproductive success? Several recent studies have measured paternity by analysis of polymorphic blood proteins as genetic markers. In the small

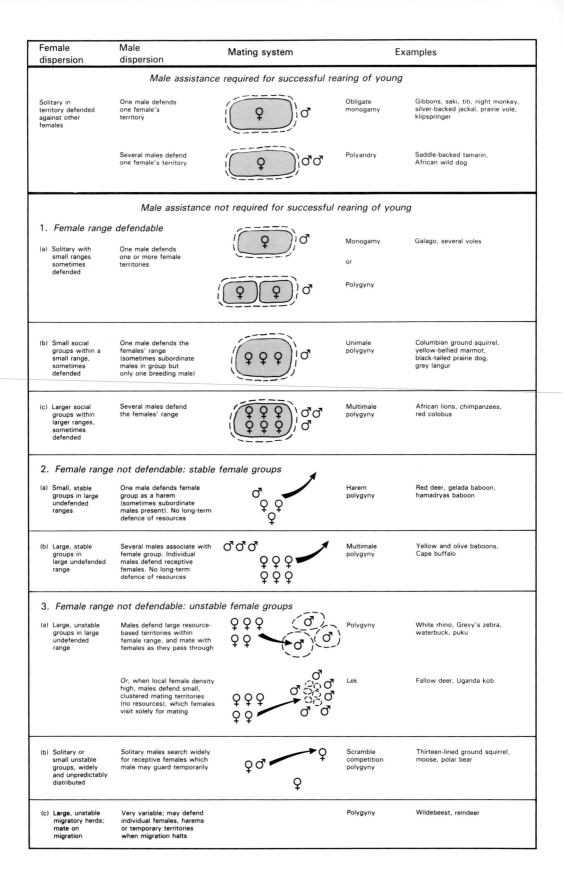

Female dispersion	Male dispersion	Mating system	Examples
Male assistance required for successful rearing of young			
Solitary in territory defended against other females	One male defends one female's territory	Obligate monogamy	Gibbons, saki, titi, night monkey, silver-backed jackal, prairie vole, klipspringer
	Several males defend one female's territory	Polyandry	Saddle-backed tamarin, African wild dog
Male assistance not required for successful rearing of young			
1. Female range defendable			
(a) Solitary with small ranges sometimes defended	One male defends one or more female territories	Monogamy or Polygyny	Galago, several voles
(b) Small social groups within a small range, sometimes defended	One male defends the females' range (sometimes subordinate males in group but only one breeding male)	Unimale polygyny	Columbian ground squirrel, yellow-bellied marmot, black-tailed prairie dog, grey langur
(c) Larger social groups within larger ranges, sometimes defended	Several males defend the females' range	Multimale polygyny	African lions, chimpanzees, red colobus
2. Female range not defendable: stable female groups			
(a) Small, stable groups in large undefended ranges	One male defends female group as a harem (sometimes subordinate males present). No long-term defence of resources	Harem polygyny	Red deer, gelada baboon, hamadryas baboon
(b) Large, stable groups in large undefended range	Several males associate with female group. Individual males defend receptive females. No long-term defence of resources	Multimale polygyny	Yellow and olive baboons, Cape buffalo
3. Female range not defendable: unstable female groups			
(a) Large, unstable groups in large undefended range	Males defend large resource-based territories within female range, and mate with females as they pass through	Polygyny	White rhino, Grevy's zebra, waterbuck, puku
	Or, when local female density high, males defend small, clustered mating territories (no resources), which females visit solely for mating	Lek	Fallow deer, Uganda kob
(b) Solitary or small unstable groups, widely and unpredictably distributed	Solitary males search widely for receptive females which male may guard temporarily	Scramble competition polygyny	Thirteen-lined ground squirrel, moose, polar bear
(c) Large, unstable migratory herds; mate on migration	Very variable; may defend individual females, harems or temporary territories when migration halts	Polygyny	Wildebeest, reindeer

rodent *Peromyscus polionotus* trapping showed that a male and a female remained together for their entire lives, suggesting long-term monogamy. Paternity analysis confirmed that all litters were sired by one male and that females remained with the same male for consecutive litters (Foltz, 1981).

Behavioural studies of black-tailed prairie dogs *Cynomys ludovicianus* (Hoogland, 1981) and yellow-bellied marmots *Marmota flaviventris* (Armitage, 1986) show that one adult male defends a group of females throughout the year, protects them and the young from predators, and chases off intruding males. Paternity analysis confirmed that in both species resident males sire the vast majority of the young (Schwartz & Armitage, 1980; Foltz & Hoogland, 1981).

In species where females regularly copulate with several males, differences in mating system not only reflect differences in female dispersion but perhaps also differences in sperm competition, as shown by field studies of three species of ground squirrel. In Belding's ground squirrel *Spermophilus beldingi*, females live in aggregations and males fight for dominance in areas of high female density, defending small mating territories. Fighting is severe and the heaviest territorial males gain the most matings (Sherman & Morton, 1984). A female typically mates with from three to five males and most litters (55–78%) are multiply sired (Hanken & Sherman, 1981). In the field, a female's first mate sires on average 60% of the litter, the second mate 30% and subsequent mates 10%. This apparent first-male advantage, together with the high density of receptive females, may explain why males do not guard females after mating but leave to compete for further mates as soon as copulation is over (Sherman, 1989).

In the 13-lined ground squirrel *S. tridecemlineatus* most females also mate with more than one male and in the field there is again apparent first-male advantage with the first male to copulate siring on average 75% of the offspring (Foltz & Schwagmeyer, 1988). As in *S. beldingi*, males do not guard females after mating but leave immediately and resume search for futher females. The mating system is different, however, because females are scattered at low densities. Males wander widely, aggression is very mild, and the premium is on competitive mate searching with a male's success determined by his ability to find females rather than his fighting ability. Individual males may mate with up to five females, the most successful males being those with the largest mate-searching ranges (Schwagmeyer, 1988).

Female Idaho ground squirrels *S. brunneus* are also widely dispersed, but unlike the other two species, a male guards a female for several hours after mating and chases off other males until the female is no longer receptive. Sometimes guarding males are ousted by replacement males who

---

Fig. 9.2 (Opposite) Mating systems in mammals (based on review and information in Clutton-Brock, 1989).

copulate with the female and then likewise guard her. In this species, paternity analysis revealed that pups born to females guarded by one male were all sired by that male, while litters of females guarded by several males were often multiply sired with most pups sired by the male who guarded last or longest (Sherman, 1989). The apparent absence of first-male advantage, and the wide dispersion of females, which limits opportunities for polygyny, may place a premium on mate guarding after copulation to protect paternity.

At first sight, these field observations seem to suggest that there may be differences between species in how mating order and frequency influence which male fertilizes the eggs. This would raise the intriguing possibility that females could influence the type of male–male competiton by evolving particular mechanisms for utilizing sperm. However, controlled experiments are needed to confirm order effects. Laboratory experiments with other rodents have shown that there is a period of peak fertilizability around the time of ovulation, when competition for matings is expected to be most intense. Within the same species, there can be first or second male advantage, or no order effect, depending on the timing of copulation in relation to this peak and also on the interval between matings (Dewsbury, 1984; 1988; Huck *et al.*, 1989).

9.2.3 *Experimental studies*

These comparative studies support the idea that female dispersion influences the mating system by affecting the way in which males can monopolize mates. Comparative data also suggest that female dispersion is correlated with resource dispersion and predator pressure, the first link in the chain in Fig. 9.1 (for example, Jarman, 1974; Clutton–Brock & Harvey, 1978; Wrangham, 1980a). However, we still do not have a convincing explanation for animal dispersion because it is so difficult to measure resource distribution and predation in the field. Qualitative descriptions of resources, such as 'patchy' and 'abundant', are of limited use because they mean different things in different studies. The most direct evidence for the two links in Fig. 9.1 comes from experiments in which resource distribution and female distribution have been manipulated.

Ostfeld (1986) used radiotelemetry to study space use by California voles *Microtus californicus*. Females were non-territorial, with overlapping ranges, and sometimes shared nests. Males defended territories containing from one to five female ranges. When extra food was provided, female ranges decreased in size but male ranges did not. This suggests that female dispersion is primarily influenced by food, but male dispersion is not. Ims (1987) showed that the dispersion of female grey-sided voles *Clethrionomys rufocanus* was also influenced by food. When food was provided in abundance at particular sites, female ranges became smaller and overlapped in the resource-rich areas. While these experiments are interesting, Hixon (1987) pointed out that the effects of provisioning food on territory size and polygyny can be variable depending on whether one or both sexes defend territories and on how the food influences intruder pressure.

The most direct test of the scheme in Fig. 9.1 is by Ims (1988) who showed experimentally for *C. rufocanus* that female dispersion influences male dispersion but not vice versa. He introduced a small population on to a little wooded island in S.E. Norway. In one experiment, females were kept individually in small cages and their positions were moved each day to simulate movement about a home range. When females were spaced out, free-ranging males (tracked by radiotelemetry) became dispersed, overlapping their ranges with the female ranges. When females were clumped, by placing cages close together, male dispersion changed and they aggregated on the female clumps. By contrast, when males were kept in individual cages the dispersion of free-ranging females was not affected by experimental changes in male dispersion. Thus, this study supports the idea that the causal links are from resources to female dispersion and then from female dispersion to male dispersion (Fig. 9.1).

9.3 Female dispersion and mating systems

Some other animal groups where males do not provide parental care are now briefly reviewed to show how female dispersion influences the mating system.

9.3.1 Lizards

In insectivorous species, females sometimes defend resource-based territories and males then compete to defend as many female territories as they can (Stamps, 1983). In species with small female ranges, males are more likely to be polygynous whereas defence of one female (monogamy) is more likely when female ranges are large (Fig. 9.3a). Similar trends are found within species, with males competing to defend exclusive territories containing females and larger males defending larger territories and so enjoying greater polygyny (Fig. 9.3b; see also Manzur & Fuentes, 1979).

Indirect evidence that female territory size is related to food distribution while male territory size is related to mate acquisition comes from Schoener and Schoener's (1982) study of *Anolis sagrei*. Adult males have much larger ranges than adult females—more than twice as large as expected from relative energetic requirements. Only in adult males does territory size consistently increase with body size, which reflects fighting ability. Female territory size, by contrast, sometimes declines with body size, perhaps because larger females can obtain the most resource-rich sites.

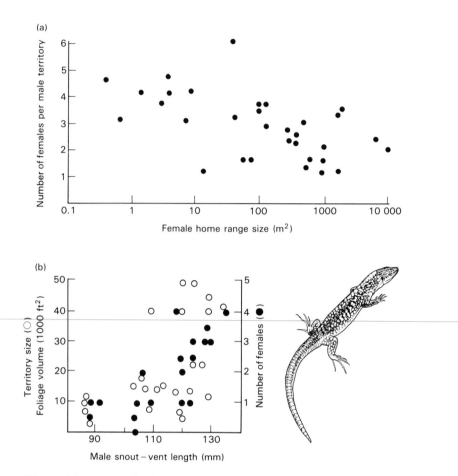

Fig. 9.3 Dispersion and polygyny in lizards. (a) In territorial species, males can defend more females when female home range is small. Each point refers to a different species (from Stamps, 1983). (b) In *Anolis garmani* larger males defend larger territories and so monopolize more females (from Trivers, 1976).

9.3.2 *Frogs and toads*

Two types of male–male competition can be recognized depending on the spatial and temporal distribution of receptive females (Wells, 1977).

EXPLOSIVE BREEDERS

In some species (e.g. *Bufo bufo*: Davies & Halliday, 1979; *Rana sylvatica*: Howard, 1980) females breed synchronously and spawn together in clumps. Synchronous breeding may sometimes be caused by seasonal availability of suitable spawning sites or it may be favoured to decrease predation on adults or eggs and tadpoles, through 'selfish herd' effects or other advantages of grouping. The sudden mass arrival of females at the pond within a few days leads to high densities of searching males and scramble competition for mates. Males often attempt to capture females on their way to the spawning sites and then remain in copula guarding the female until she has laid. Because of the short season there is limited opportunity to gain a second mate. In explosive breeders there is little scope for mate choice; in their frenzied search for females males grab at any passing object including other males, fish, sticks or even human fingers! In fights for females, larger males can generally displace smaller males. Höglund (1989) has shown experimentally that in *Bufo bufo* male–male competition is more intense (larger males are relatively more successful) when sex ratios are more male biased or when the time from pairing to spawning is increased.

PROLONGED BREEDERS

In other species females arrive at the breeding site a few at a time over a period of several weeks or months. Here, active searching for females is presumably unprofitable and males instead use vocalizations to attract mates (e.g. *Bufo calamita*: Arak, 1988; *Rana catesbeiana*: Howard, 1978). Males arrive at the breeding site first and compete to defend the best oviposition sites, which will attract the most females (e.g. *R. catesbeiana*). Thus the strongest competitors may gain several females in succession.

This mating system, known as 'resource defence polygyny', is really just a variation of the scheme in Fig. 9.1, where the males anticipate how resources influence female dispersion and wait for females in resource-rich sites. In some species females visit several males before spawning. Experiments are needed to test whether this represents resource choice or mate choice. In many cases females may use male vocalizations simply to lead them to a suitable breeding site. In some dendrobatid frogs males care for the eggs or tadpoles, which may be an extension of male defence of good oviposition sites. By remaining at a rich site, a male may care for eggs already laid while continuing to attract additional females (Wells, 1977).

9.3.3 Fish: wrasses

The contrasting mating systems of two coral reef wrasses (Labridae) show the influence of female dispersion in relation to resources (Robertson & Hoffman, 1977). The cleaner fish *Labroides dimidiatus* is a specialized feeder which removes parasites and other material from the body surface of various host fish, which return predictably to particular sites to be cleaned. Females remain permanently attached to particular sites, which they defend against other females. Males, which are larger, compete to defend female territories with the largest males typically defending up to six females.

The blueheaded wrasse *Thalassoma bifasciatum*, by contrast, is a generalist feeder taking zooplankton and benthic animal material as well as ectoparasites from the body surface of other fish. With food more widely dispersed, females have large overlapping ranges and sometimes form temporary aggregations at rich feeding sites. Males, therefore, are probably unable to exert permanent control over a group of females. Females do, however, spawn in favoured sites on the downcurrent edges of a reef where it may be safest to lay. Individual females return almost every day to a particular site where they lay eggs for a minute or so before leaving. The eggs are pelagic and there is no parental care. Males compete to defend territories at these preferred sites and the largest males gain the most mates. Warner (1987) showed that females were choosing a particular site rather than particular male characteristics. When a successful male was removed, the same females continued to visit the site even though a new male had taken over. Furthermore, when the new male was a previous neighbour, his females continued to visit his old site and did not follow him, even when his new territory was just a few metres away. Particular site preference may be maintained by cultural tradition, some sites being used daily for many years. When Warner (1988) removed entire populations from a reef and replaced them with new individuals, new sites were used and thereafter maintained.

To assess the roles of males and females in determining spawning sites, Warner (1990) removed either all breeding males or all females from local isolated populations and replaced them with fish from other populations. Where males were replaced, most of the spawning sites remained the same. By contrast, where females were replaced there were marked changes in sites used, even though some males continued to defend and display at the original sites. This neat experiment shows that females play the major role in choice of spawning site and that males simply compete to defend sites which females prefer. Like the experiment by Ims (1988), it provides strong support for the scheme in Fig. 9.1.

9.3.4 Insects

How males compete for mates depends not only on female dispersion but also on sperm competition (Thornhill & Alcock, 1983). Where females mate just once in their lives, males search for emerging virgin females and compete to be first there. If females emerge in abundance from particular sites then males may gather there in swarms and there is scramble competition for mates (e.g. bibionid fly *Plecia nearctica*: Thornhill, 1980). When females emerge from small defendable patches, the males may defend territories. For example, male eumenid wasps *Epsilon* defend clusters of brood cells from which virgin females will emerge (Smith & Alcock, 1980). Likewise male bees *Centris pallida* may defend small territories over the ground where female emergence sites are clumped, although where females are more scattered males do not defend territories but patrol over wide areas (Alcock *et al.*, 1977b).

By contrast, where females mate multiply, there is usually last-male sperm precedence and males often wait at oviposition sites where they compete for laying females and guard them during laying to protect their paternity (e.g. dungflies *Scatophaga stercoraria*: Parker, 1970; dragonflies and damselflies, Odonata: Waage, 1984). The influence of female dispersion and male density on the mating system is nicely shown by comparative studies of Odonata. When population density is low, males show no site attachment and patrol over large areas in search of females (e.g. *Aeschna cyanea*: Kaiser, 1985). At higher male densities, some species change from patrolling to territorial defence (e.g. *Pachydiplax longipennis*: Sherman, 1983). Models suggest that it may pay males to restrict their patrolling to regular territories at high male densities to minimize costly interactions with other males (Poethke & Kaiser, 1987; Ubukata, 1987). Where suitable oviposition sites are limited (e.g. particular shore lines or patches of vegetation) competition for territories is most intense and males defending these sites gain most mates (resource defence polygyny, *Plathemis lydia*, Koenig & Albano, 1985; *Erythemis simplicicollis*: McVey, 1988). Territorial males often guard females while they lay by hovering over them. Such 'non-contact' guarding frees the male to mate with other females should they arrive during laying (e.g. *Calopteryx maculata*: Alcock, 1979). When there is competition for limited sites, excluded males may adopt 'satellite' behaviour, hiding away inside defended territories (e.g. *Libellula quadrimaculata*: Convey, 1989a). At very high male densities, there is no territorial defence but rather scramble competition for mates, with males guarding females more intensely, often 'in tandem', while they lay (e.g. *Coenagrion puella*: Banks & Thompson, 1985). In *Sympetrum sanguineum* males remain in tandem for longer during egg laying when interference from other males is higher (Convey, 1989b).

As an alternative to waiting at oviposition sites, males may defend feeding areas. Male megachilid bees *Anthidium* defend patches of flowers and copulate with females who come to feed (Alcock *et al.*, 1977a). Male desert grasshoppers *Ligurotettix coquilletti* defend territories on bushes where females feed, with greatest male densities on bushes with better quality foliage. When males were removed to create equal male density on all bushes, females still accumulated at the same favoured sites, showing that females chose bushes in relation to resources rather than males (Shelly *et al.*, 1987). Male scorpionflies *Panorpa* compete to defend dead arthropods which attract females. Males then permit females to feed in exchange for sex. Thornhill (1981) showed by experiment that prey availability determined the degree of polygyny. When prey were scarce, a few successful males enjoyed high mating success whereas when prey were abundant variance in male mating success declined.

9.3.5 *Conclusion*

These comparative surveys and experiments support the scheme in Fig. 9.1, with mating systems determined largely by how resources influence female dispersion in space and time, which then determines the way in which males can best compete for mates. I now discuss a form of male competition which, at first sight, does not obviously fit into this scheme.

9.4 **Leks**

In all the examples discussed so far, males compete for females either directly, by defending the females themselves, or indirectly, by defending resources to which females are attracted. In some cases, by contrast, males defend small territories which contain no resources, often aggregate into groups and put all their effort into self-advertisement with visual, acoustic or olfactory displays. Females visit males solely for mating and males provide no parental care. In these mating systems, known as leks, females often visit several males before copulating and appear to be selective in their choice of mate. Male mating success is strongly skewed with the majority of matings performed by a small proportion of the males on the lek (Bradbury & Gibson, 1983).

Leks have been reported for seven species of mammals: walrus, hammerheaded bat and five ungulates, namely Uganda kob, white-eared kob, Kafue lechwe, topi and fallow deer (Clutton-Brock *et al.*, 1988); and some 35 species of birds including three shorebirds, six grouse, four hummingbirds, two cotingas, eight manakins, eight birds of paradise, the kakapo and great bustard (Oring, 1982). This breeding system is thus not common ($< 0.2\%$ of mammal species and $< 0.5\%$ of bird species). Similar mating systems occur in some frogs (e.g. treefrogs: Wells, 1977) and some insects (Thornhill & Alcock, 1983), where females visit groups of displaying males, choose a mate and then lay eggs away from the display site.

9.4.1 *Ecological conditions*

It has been suggested that leks occur when males are unable to defend economically either the females themselves or the resources they require (Bradbury, 1977; Emlen & Oring, 1977). This may arise where females exploit widely dispersed resources and so have large, undefendable ranges or because high population density, and thus high rates of interference between males, precludes economic female or resource defence. Interspecific comparisons support this view. In both antelope and grouse (Fig. 9.4) males defend smaller territories, and are more likely to be aggregated on leks, when females have large home ranges. In some species of digger-wasps *Philanthus* spp., males defend territories containing female nest sites where suitable sites allow clumping of nest burrows. However, in those species where nests are widely dispersed, males do not defend resource-based territories but display instead on leks, each male defending a small site containing no resources, which females visit to mate (Gwynne, 1980).

Variations within species provide further evidence. In Uganda kob, topi and fallow deer, males lek at high population density but defend harems or resource-based territories at low density where defence is presumably more economic (Clutton-Brock *et al.*, 1988). Some lekking bird species also show changes in male dispersion when female dispersion changes. For example, male buff-breasted sandpipers *Tringa subruficollis* display on leks at the start of the breeding season when females move about in groups over large ranges. Once females disperse and settle down on nest sites, males disperse too and defend territories centred on female nest sites,

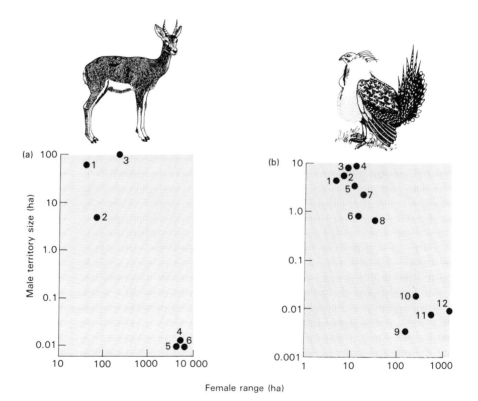

Fig. 9.4 Males defend smaller territories when female ranges are large. (a) Antelope (subfamily Reduncinae; data from Rosser, 1987; Clutton-Brock, 1989). 1, *Redunca arundinum*; 2, *Kobus vardoni*; 3, *Kobus defassa*; 4, *Kobus lechwe kafuensis*; 5, *Kobus kob thomasi*; 6, *Kobus kob leucotis*. (b) Grouse (family Tetranonidae; data from Bradbury *et al.*, 1986). 1, *Lagopus mutus*; 2, *Lagopus lagopus*; 3, *Lagopus leucurus*; 4, *Tetrastes bonasia*; 5, *Bonasa umbellus*; 6, *Dendragapus canadensis*; 7, *Dendragapus obscurus*; 8, *Tetrao urogallus*; 9, *Tympanuchus pallidicinctus*; 10, *Tympanuchus cupido*; 11, *Tympanuchus phasianellus*; 12. *Centrocercus urophasianus*.

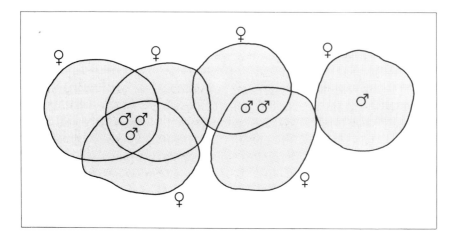

Fig. 9.5 An illustration of Bradbury's 'hotspot' model. Where females have large, overlapping ranges, males can maximize their encounters with females by settling in the overlap zones (hotspots). If males settle in an 'ideal free' manner, then the proportion of males in an area should match the proportion of females, so that average male mating success is the same in different areas. In this simple example, six males have settled on six female ranges. Note that once a male has settled in part of a female's range, he devalues the whole of that female's range for other males, so male groups are often expected to be separated by a distance of at least the diameter of one female's range.

perhaps hoping to fertilize replacement clutches should the nesting attempt fail (Pruett Jones, 1988).

If, as these data suggest, males lek when females cannot be monopolized economically, what causes the males to aggregate? Three main hypotheses have been proposed (Bradbury & Gibson, 1983). They are not necessarily mutually exclusive.

9.4.2 Males aggregate on 'hotspots'

Male aggregation may be explained by the familiar scheme in Fig. 9.1, with males clustering in areas where female encounter rate is particularly high (hotspots). For example, in some populations of topi *Damiliscus korrigum* males defend small clustered territories in areas of short grassland where groups of females prefer to rest because of improved predator detection (Gosling, 1986). Hotspots can also occur when females are solitary, but have large overlapping ranges (Fig. 9.5). Male settlement to maximize encounter rate with females can lead to aggregation in the overlap regions (Bradbury *et al.*, 1986).

The most convincing evidence for this last scenario comes from a detailed study of a bird of paradise *Parotia lawesii* (Pruett Jones & Pruett Jones, 1990). Males defend small courts on the ground, 0.5–20 m^2 in area, below perches where females come to observe their displays. Some males are solitary, while others are aggregated into small groups (2–14 males) with nearest-neighbour distances varying from 2 to 500 m. Radiotracking shows that females have large overlapping ranges with, on average, the courts of 17 males in each female range. The variation in male dispersion correlates with female dispersion; male aggregations are in regions where several female ranges overlap while solitary males tend to be sited on the ranges of females who are isolated, as shown in Fig. 9.5. The mating success of solitary and grouped males does not differ, as would be predicted if males settled in an 'ideal free' manner in relation to female dispersion. Finally, when males are removed from leks, female ranges remain unchanged. This supports the view that males settle in relation to female dispersion rather than vice versa (S.G. Pruett Jones, unpublished data).

In sage grouse *Centrocercus urophasianus*, between 20 and 200 males aggregate on leks in traditional sites on open grassy meadows. Radiotracking shows that females have very large, overlapping ranges and, although female ranges were not measured in sufficient detail to assess whether leks were sited in range overlap 'hotspots', two pieces of indirect evidence support the view that males settle in relation to female dispersion (Bradbury *et al.*, 1989). First, the siting of the leks varies as female dispersion varies. Early in the season, males display on leks near the females' wintering ranges but later these are abandoned in favour of other lek sites as females disperse to their nesting areas. This suggests that males follow changes in female dispersion rather than vice versa. Second, the average number of female visits per male is constant across leks of different sizes, suggesting that males distribute among leks in an 'ideal free' manner.

9.4.3 Males aggregate around 'hotshots'

Males may gain from 'stimulus pooling'; by displaying together, they may provide a greater attraction for females and draw in mates from a larger distance. Bradbury (1981) criticized the idea that the increase in signal distance could account for the benefits of aggregation on the grounds that communal displays were unlikely to be sufficiently effective to increase the number of females per male. However, if some males had particularly effective displays ('hotshots') it could pay poorer signallers to cluster around them to parasitize their attractiveness or disrupt their displays (Beehler & Foster, 1988).

Arak's (1988) study of natterjack toads *Bufo calamita* provides some support for this hypothesis. Larger males have louder calls which experiments showed could attract females from greater distances. Small males settle as satellites next to a larger caller when the caller's signal is more than twice as attractive as their own, and attempt to intercept females as they arrive. Whether such parasitism could give rise to very large aggregations, as found in sage grouse, depends on the degree of variation in male attractiveness and the pay-offs to parasitic males.

9.4.4 Female choice for particular sites or male aggregations themselves

Another possibility is that males aggregate because of female preference for particular sites or for clustered males *per se*. This may be seen as a refinement of the hotspot model, to explain why males aggregate in areas where females are most likely to be encountered, or it can be regarded as an alternative model to explain cases where leks are sited away from areas where females normally range (for instance, as occurs in capercaillie *Tetrao urogallus*; Wegge & Rolstad, 1986). Females may prefer particular sites because they are the safest place to mate or because this is an indirect way of forcing males to cluster. Females prefer to visit clusters of males because this facilitates mate choice (good genes) or because this provides a safe (from predators or male harassment) or quick mating (Alexander, 1975; Wrangham, 1980b).

Among insects, leks often occur in species where females are widely dispersed, either because their resources are widely dispersed or simply because the species is rare (Thornhill & Alcock, 1983). Leks are often sited at distinctive places, such as a clearing in a forest (orchid bees *Eulaema meriana*), on exposed vegetation (*Drosophila*), over prominent trees or bushes (mosquitoes *Aedes*) or on hilltops (butterflies, e.g. *Papilio zelicaon*). Males may prefer these sites because they provide good vantage points for spotting passing females (hotspots). Alternatively, Parker (1978) suggests that favoured encounter sites could evolve if females benefit from easy location of males to facilitate quick mating. Hilltops or prominent vegetation may be favoured mating sites not because of any location in relation to resources but simply as obvious landmarks for encountering otherwise widely dispersed mates.

9.4.5 Distinguishing the hypotheses

There is some evidence in favour of each of the three hypotheses for male aggregations (see above). Different hypotheses may apply to different cases and some leks may be explained by a mixture of all three. The main problem in testing the hotspot model is the difficulty of measuring female ranges, which are very large in most lekking species. However, although the hotspot model may explain lek dispersion on a coarse scale, males are often aggregated far more closely than expected from simple settlement in female range overlap regions (Bradbury *et al.*, 1986). Males may cluster to reduce predation (as in some frogs; Ryan *et al.*, 1981). However, in many bird leks, predation is rarely observed and may not be an important selective pressure favouring male aggregation.

There are two main sources of evidence against the 'hotshot' model as a general explanation for male aggregation. The first is that females appear, at least in some cases, to favour particular sites on a lek rather than particular 'hotshot' males. Two experiments suggest this. First, Rippin and Boag

(1974) sequentially 'removed' (shot) males from the centre of a sharp-tailed grouse *Pedioecetes phasianellus* lek. The vacancies created were quickly filled in an orderly and predictable manner, first by other central males, then by surrounding males, and finally by peripheral males. In the second experiment, Lill (1974) studied white bearded manakins *Manacus m. trinitatis* in Trinidad, where males displayed in aggregations of from six to 50 individuals. The males on a lek were spaced just a metre or less apart, each defending a small area of bare earth and some small saplings where they displayed. At one lek, Lill removed the top and the third-ranked males in terms of mating success. Within minutes their courts were occupied by other, less successful males. The most interesting result was that the same females continued to visit these sites for matings, even though the males had changed (as in Warner's wrasse; see section 9.3.3). Lill showed that the faithfulness of a female to a particular site was the same on these experimental territories as on control territories where the male was not removed.

Both results suggest that there is something about the site which influences female preference. The 'hotshot' model would predict that when the most successful male is removed, the next most preferred male (i.e. the next hotshot in line) would remain on his territory and rely on his attractiveness to gain the females. The male aggregation would then be predicted to rearrange around this new hotshot, rather than for replacement to occur on particular sites.

The second piece of evidence against the 'hotshot' model comes from an experiment on a fallow deer *Dama dama* lek. Clutton-Brock *et al.*, (1989) covered the territories of the most successful males with black polythene, so forcing them to change site. Even though these males set up new display sites several hundred metres away, they remained favoured by females. In this case, therefore, females were apparently choosing particular males rather than particular sites. The 'hotshot' model predicts that the other, less successful, males should have followed the movements of these attractive males to set up a lek at the new site. However, most remained on their former territories at the old lek. Thus the hotshot model may not explain male aggregations on leks even where females are choosing particular males rather than particular sites.

Distinguishing female preference for site versus male phenotype is difficult if the best males gain the best sites. Some studies have come to opposite conclusions, even with the same species. In sage grouse leks Wiley (1973) concluded that site characteristics (a central territory) were most important, while Gibson and Bradbury (1985) suggested that female preferences were most strongly correlated with differences in male display. In fallow deer, the experiment above suggests that male characteristics influence female choice, while in another study Appollonio *et al.* (1990) found changes in an individual male's mating success when he switched site on a lek, which correlated with differences in female preference for those sites.

The problem of showing the influence of male display is exacerbated by the fact that a male may display more intensively the more females that visit his site, so increased display could be a consequence of high mating success, not a cause. Gibson (1989) supplemented the natural display of a territorial male sage grouse with the tape-recorded vocalizations of another, reproductively successful, male. These broadcasts increased the number of females visiting the territory showing that display can influence female visitation independently of site.

In conclusion it seems likely that both site and male characteristics influence female choice within a lek. However, we still do not know to what extent lek formation by males can be explained by either the female choice (section 9.4.4) or the hotspot model (section 9.4.2), or a mixture of the two. Males are likely to settle in relation to female dispersion (causal links as in Fig. 9.1). Once they do, this may then modify female dispersion because male aggregations influence costs and benefits of mating for females (i.e. add an arrow from male to female dispersion in Fig. 9.1). As a challenge to field-workers, convincing evidence that the factors giving rise to a lek mating system have been understood, would be to choose a species with a variable mating system which includes lekking, and experimentally manipulate resources or females to show exactly how their dispersion leads to male aggregation.

9.5 Influence of male parental care

Where males provide parental care the males themselves become a resource which may influence female dispersion. In addition, the pay-offs of different mating systems to either sex depend not only on the potential to monopolize mates but also on the costs of desertion. In this section I illustrate this mainly with reference to birds (Fig. 9.6), where males commonly provide parental care, with some data from fish and mammal species where males care.

9.5.1 Monogamy: costs and benefits of desertion

Lack (1968) suggested that monogamy is the predominant mating system in birds (90% of species) because 'each male and each female will, on average, leave most descendants if they share in raising a brood'. This hypothesis certainly seems to explain obligate monogamy in many seabirds and shorebirds, where male and female share incubation and chick feeding and where the death or removal of one partner leads to complete breeding failure (Oring, 1982). In Bewick's swans *Cygnus columbianus bewickii*, males and females pair for life and both care for the cygnets. Breeding success not only increases with age, it also increases with duration of the

Fig. 9.6 (Opposite) Mating systems in birds. Female territories (solid lines); male territories (dashed lines); nests (●).

Mating system	Male and female dispersion		Parental care
Monogamy: (one male forms pair bond with one female)	*Resource defence* Many passerines, shorebirds and seabirds	Multipurpose territory defended by male or both sexes, or nest site only defended (seabirds)	Male and female
	Mate defence Some passerines, many ducks and geese	Male guards female and may defend nest site	Male and female (e.g. finches), or female only (many ducks)
Polygyny: (one male forms pair bond with several females simultaneously)	*Resource defence* Some passerines, and shorebirds	Male defends large multipurpose territory within which several females defend exclusive nest sites or smaller territories	Male and female, or mainly female
	Some passerines (Icteridae, Ploceidae)	Male defends clumped nest sites	Male and female, or mainly female
	Polyterritoriality (Pied flycatcher)	Male defends separate nest sites or territories, often several hundred metres apart, with one female at each site	Male and female, or mainly female
	Harem defence Ring-necked pheasant	Male defends group of females who nest solitarily	Female
Polyandry: (one female forms pair bond with several males simultaneously)	*Cooperative* Galapagos hawk, Tasmanian native hen	Several males defend one female's territory	All the males may help the female
	Resource defence Jacanas Spotted sandpiper	Female defends large multipurpose territory within which several males defend smaller, exclusive territories	Mainly or exclusively male
Polygynandry: (several males form pair bonds with several females simultaneously)	Dunnock	Several males defend a territory, within which several females may each defend smaller, exclusive territories, or share the whole territory and nest communally	All the males may help all the females
	Acorn woodpecker		
Promiscuity: (no pair bond — female and male meet briefly to copulate)	*Resource defence* Orange-rumped honeyguide, some hummingbirds	Males defend territory containing food and copulates with females who visit to feed	Female
	Display site defence Bowerbirds, manakins some birds of paradise, grouse and shorebirds, kakapo	Males defend small display site, containing no resources, which females visit solely to copulate. Males may be dispersed, loosely aggregated ('exploded leks') or densely aggregated ('leks')	Female
Sequential polygamy: (male or female form pair bond with one mate and then desert to find one or more mates in succession)	*Sequential polygyny* Woodcock	Male guards one female and then, after clutch completion, deserts to find another female. No territorial defence	Female
	Sequential polyandry Phalaropes Dotterel	Female guards one male and then, after clutch completion, deserts to find another male. No territorial defence	Male
	Sequential polygyny and sequential polyandry Temminck's stint	Female lays clutch in one male's territory, which male cares for, then deserts to lay for one or more other males. Males may gain a second female, who incubates the clutch herself if the male is already incubating a first female's clutch	Male or female
	Simultaneous polygyny and sequential polyandry Rhea Tinamou	Group of females lay in a communal nest, cared for by a male, and then desert to lay in the nests of other males	Male

pair bond; when one partner dies, the remaining partner suffers a temporary decrease in reproductive success when it pairs with a new mate. Lifelong monogamy is thus beneficial for both sexes (Scott, 1988). The importance of biparental care may also explain the occasional homosexual female pairs which form in gull colonies when there is a shortage of males (Conover & Hunt, 1984). One gull cannot rear young alone because other gulls are likely to destroy the eggs when a single parent leaves to feed, so cooperation between two females can bring benefits to both, in the same way that cooperation is favoured in male–female pairs.

Lack also used his hypothesis to explain the predominance of monogamy in passerine birds. He extended Crook's (1964) pioneering comparative work with weaverbirds (Ploceidae) to show that monogamy was the rule for every insectivorous passerine subfamily whereas about a quarter of seed-eating and frugivorous subfamilies were polygynous (Lack, 1968). The explanation proposed was that insects are hard to find, so two parents are needed to raise the brood (hence monogamy), while seeds and fruit are more easily exploited, enabling females to raise a brood alone, thus freeing the male to desert and monopolize several mates (hence polygyny).

Recent removal experiments have tested the importance of male care in passerines where both sexes commonly rear the young. When males are removed, females increase their care (e.g. provisioning rate) but usually not sufficiently to compensate fully for the loss of the male (Houston & Davies, 1985; Wright & Cuthill, 1989; Hatchwell & Davies, 1990). In 11 out of 16 passerine species, male removal during the early nestling period led to a significant decrease in the number of young surviving to fledging (Wolf *et al.*, 1988; Bart & Tornes, 1989). Only three studies have followed survival to independence, necessary for a reliable measure of success. In song sparrows *Melospiza melodia* male removal caused success to decrease to 51% of that of pair-fed broods (Smith *et al.*, 1982), and in seaside sparrows *Ammodramus maritimus* and dark eyed juncos *Junco hyemalis* the figures were about 66% and 38% respectively (Greenlaw & Post, 1985; Wolf *et al.*, 1988). For some species removals suggest that male help is more important when food is scarce (for example Lyon *et al.*, 1987) or earlier in the nestling period when young chicks need brooding and so females are less able to increase their provisioning to compensate for male removal (Burley, 1980; Sasvari, 1986).

These experiments show clearly that male help can increase reproductive success, but it is not essential. If male desertion reduces productivity to a fraction $1/x$ of a pair-fed brood, then provided a male can gain more than x females, desertion will be the more profitable option. Even if success is reduced to a half or less and a male can gain only two females, polygyny will still pay provided the male helps provision at least one of the broods. As predicted, male passerines readily desert to gain extra females if given the chance, for example by removal of neighbouring males (Smith *et al.*, 1982), and they often help to provision either one of their female's broods full-time, or several females' broods part-time. Occasional polygyny has been reported in 39% of 122 well-studied European passerines (Møller, 1986).

In some species with precocial young, where the chicks are not fed but simply brooded and protected, male removal causes little if any decrease in reproductive success (e.g. willow ptarmigan *Lagopus lagopus*: Martin & Cooke, 1987; lesser snow goose *Chen caerulescens*: Martin et al., 1985). Here a male's continued presence during incubation and chick care may largely represent mate-guarding of the female as a resource for a replacement clutch if the brood fails. In willow ptarmigan, male removals led to increased polygyny (Hannon, 1984).

These experiments suggest that the predominance of monogamy in many birds arises not, as Lack proposed, because each sex has the greatest success with monogamy but because of the limited opportunities for polygyny. Two obvious constraints are: (i) strong competition among males may make it difficult for a male to gain another female; and (ii) females are likely to suffer in polygyny through the loss of male help, so females should be aggressive to other females to decrease the chance that their partners gain another mate.

If one partner can raise some young alone then the way is open for one sex to desert and attempt to monopolize further mates. Why is it usually the male who deserts? Two likely explanations are, first, that males have the opportunity to desert earlier, for example while females are completing the clutch, thus leaving their partner in a 'cruel bind' (Trivers, 1972) and, second, that males may have more to gain from desertion because they can potentially fertilize eggs at a faster rate than a female can lay them. Thus, in most birds, males desert females and polygyny is far commoner than polyandry (Oring, 1982). Some studies, however, suggest sexual conflict over opportunities to desert first. In the penduline tit *Remiz pendulinus* (Persson & Öhrström, 1989) and the Florida snail kite *Rostrhamus sociabilis* (Beissinger & Snyder, 1987), either sex may desert leaving the other to care for the brood. Which sex deserts seems to depend on male versus female opportunities to gain further mates. In the kites, desertion is more frequent when feeding conditions are good and so the remaining partner is better able to raise the young unaided.

Keenleyside (1983) has performed removal experiments to test the benefits of biparental care in the cichlid fish *Herotilapia multispinosa*, which is often monogamous. Females tend to do more parental care and female removal led to greater brood loss from predation than male removal. Adult sex ratios were experimentally biased to give either sex the chance to desert and gain extra mates. With a surplus of males, females did not desert but remained to care for their current brood. With a surplus of females, males frequently deserted but tended to wait until their current brood had reached the free-swimming fry stage, when they were less vulnerable to predation. Thus in both birds and fish there is evidence that the costs and benefits of desertion influence a male's decision of whether to leave to monopolize further mates.

9.5.2 Sperm competition and egg dumping in monogamous birds

Even where males and females are unable to gain extra mates, individuals can increase their reproductive success by more subtle means. In most monogamous bird species, males adopt a 'mixed reproductive strategy', not only guarding their own female and helping her to raise a brood, but also attempting to copulate with other females, especially those of neighbouring males (Birkhead, 1987). Experiments with genetic plumage markers have shown that such extra-pair copulations can be successful in fertilizing eggs (Burns *et al.*, 1980; Birkhead *et al.*, 1988). Males attempt to protect their paternity either by guarding their females closely during the fertile period or, in cases where this is not possible, because either male or female feeds alone while the other guards the nest (raptors, colonial sea-birds), by frequent copulation to swamp the sperm of rivals (Birkhead, 1988).

Extra-pair paternity may be commoner in colonial species or poly-gynous species, where opportunities for extra-pair matings are greater because females are clumped or more often left unguarded. Too few studies have been done to test this, and estimates vary widely. For example, DNA fingerprinting showed that all 120 young from 19 broods of willow warbler *Phylloscopus trochilus* ($<$ 5% of males are polygynous) and all 56 young from 13 broods of wood warbler *P. sibilatrix* (23% of males polygynous) were sired by the resident male (Gyllensten *et al.*, 1990). Likewise, in a population of dunnocks *Prunella modularis*, a bird with a variable mating system (see section 9.5.4), DNA fingerprinting revealed that all but one of 133 young (45 broods) were fathered by a resident male; the exception was fathered by a neighbour (Burke *et al.*, 1989). By contrast, high levels of extra-pair paternity ($>$ 14%) have been estimated, using polymorphic enzymes, for two species which have occasional polygyny (Indigo bunting *Passerina cyanea*: Westneat, 1987; white crowned sparrow *Zonotrichia leucophrys*: Sherman & Morton, 1988). Still higher estimates (about 24% of extra-pair paternity) have been made, using morphological heritability, for the colonial, monogamous swallow *Hirundo rustica* (Møller, 1987a) and the regularly polygynous pied flycatcher *Ficedula hypoleuca* (Alatalo *et al.*, 1984).

In some cases females may accept extra-pair copulations simply because it is costly to resist them. In other cases, however, females may actively solicit matings from males other than their mates. In dunnocks *Prunella modularis* a female may gain the help of a second male in the care of her offspring if she distributes copulations between two males (see section 9.5.4). In some species females encourage extra-pair copulations from dominant males (Smith, 1988) or males with attractive plumage (Møller, 1988a). Unlike the dunnocks, they do not thereby gain another male's help with parental care, so in these cases there is the intriguing possibility that females are shopping for good genes. It is possible that females also suffer a cost from engaging in extra-pair copulations, namely reduced parental investment from a mate that is aware of being cuckolded (Trivers, 1972). Males should prefer to invest more in broods with higher paternity than broods with lower paternity. The current evidence on this is equivocal. Møller (1988b) suggested that male swallows *Hirundo rustica* reduced parental care if their partners were subjected to frequent sexual chases by other males and if other males were experimentally given increased access to the female during the mating period (by temporary removal of the paired male). However, the interpretation of his results is confounded by brood reduction in experimental nests. Other studies have failed to show a significant reduction in paternal care in response to female promiscuity (Morton, 1987; Westneat, 1988). Further experimental work is needed.

Females can also increase their success by laying in the nests of neighbours in addition to raising a brood themselves. The proportion of nests parasitized by conspecifics has been estimated as 5–46% for various populations of starlings *Sturnus vulgaris* and 3–31% for swallows *Hirundo* spp. (Andersson, 1984; Brown, 1984; Møller, 1987b). The message for field-workers is clear: studies need to obtain good measures of maternity and paternity before concluding that the number of young raised by a pair gives a good measure of male and female reproductive success (e.g. see Birkhead *et al.*, 1990). In many cases the term monogamy may well describe the parental care shared by one male and one female but not their genetic contribution to future generations (Mock, 1983; Gowaty, 1985).

9.5.3 *Polygyny*

If monogamy in birds often occurs because males are unable to gain another female, rather than because it pays males to remain faithful to one mate, what permits regular polygyny in some species? Polygyny in birds usually arises through males monopolizing females indirectly, through the control of scarce resources such as food or nest sites. Where these are patchily distributed, males able to defend the best patches can gain the most mates (resource defence polygyny; Fig. 9.6).

Various female settlement patterns can lead to polygyny (Table 9.1). When males contribute little or nothing to parental care (some Ploceidae and Icteridae), females may suffer no cost from mating polygynously and so may settle at random in suitable habitat, or nest close together in safe sites where they may benefit from cooperative nest defence. In these cases females may be largely indifferent to the mating system that emerges, which is determined simply by a male's ability to monopolize females. If a small number of males is able to control all the suitable nest sites, then high degrees of polygyny may occur (Wittenberger, 1976).

In other circumstances, however, females may suffer costs from polygyny through having to share either the resources a male controls or his contribution to parental care. Females may be forced to accept these costs if a fraction of the males controls all the suitable breeding habitat, their choice being 'accept polygyny' versus 'forego breeding' (Fig. 9.7a). If, however, most of the males are able to defend breeding territories, and if there is variation in territory quality, then a female's choice may be 'settle on a good territory with an already mated male, i.e. choose polygyny' versus 'settle on a poor territory with an unmated male, i.e. choose monogamy' (Fig. 9.7b). Here, a female may gain compensation for the costs of polygyny by choosing a high quality breeding situation and female choice plays an important role in causing polygyny (Verner & Willson, 1966; Orians, 1969).

Most studies have focused on this last hypothesis, the 'polygyny threshold model', but showing that females make the best choice is difficult unless a great deal is known about the costs of sharing and the choices available to individual females (Davies, 1989). Futhermore, females may not always be free to settle where they choose; if the first female suffers from the arrival of a second female it will pay the first female to try to prevent her from settling. Future studies should put more emphasis on testing between

Table 9.1 Hypotheses for how female settlement patterns can lead to polygyny in passerine birds. Possible examples are given, although in most cases further experiments are needed to test between the hypotheses (partly after Searcy & Yasukawa, 1989).

No cost of polygyny to females

1   *No benefit.* Females settle at random in suitable habitat. Female reproductive success does not increase with harem size. 'Neutral mate choice hypothesis' (Lightbody & Weatherhead, 1988), e.g. yellow-headed blackbirds

2   *Benefit.* Females settle together. Female reproductive success increases with harem size, e.g. due to cooperative nest defence. 'Cooperative female choice' (Altmann *et al.*, 1977), e.g. yellow-rumped cacique (Robinson, 1986)

Cost of polygyny to females

1   *Females accept cost* because:
 (a) No unmated males available, e.g. marsh wren (Leonard & Picman, 1987) or unmated males hard to find, e.g. pied flycatcher (Stenmark *et al.*, 1988)
 (b) Unable to distinguish unmated versus mated males (deception), e.g. pied flycatcher (Alatalo *et al.*, 1981)
 (c) Unable to drive off other females/monopolize male, e.g. dunnock (Davies & Houston, 1986), song sparrow (Arcese, 1989)

2   *Females choose polygyny* because cost compensated by benefits. 'Polygyny threshold model' (Verner & Willson, 1966; Orians, 1969):
 (a) Access to better breeding situation (food, nest sites), e.g. lark bunting (Pleszczynska, 1978), great reed warbler (Ezaki, 1990)
 (b) Access to better genes ('sexy son' hypothesis) (Weatherhead & Robertson, 1979)

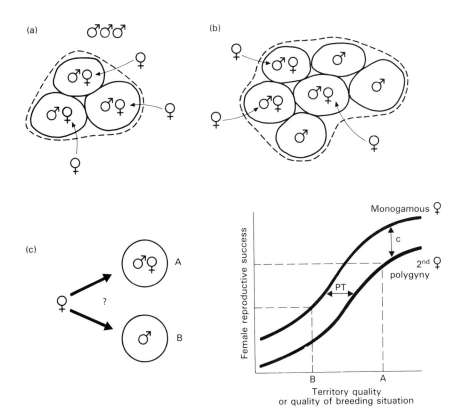

Fig. 9.7 Polygyny in birds. (a) Where a small proportion of the males can monopolize all the suitable breeding sites (enclosed by dashed line) females may be forced to accept polygyny. (b) Where most males can defend breeding sites, females may choose to settle polygynously if some males defend better sites than others (the Orians–Verner–Willson 'polygyny threshold model'). For example, in (c) a female has the choice of settling with an unmated male on a poor quality territory B, or with an already mated male on a good quality territory A. Female reproductive success increases with territory quality. There is a cost C of sharing with another female, so the curve for a secondary female in polygyny lies below that for a monogamous female. Provided the difference in territory quality exceeds PT (the polygyny threshold), a female does better by choosing to settle with an already mated male on territory A rather than with an unmated male on territory B.

the alternative hypotheses in Table 9.1 (Searcy & Yasukawa, 1989). Different hypotheses may apply to different populations of the same species. For example, in some populations of red-winged blackbirds *Agelaius phoeniceus* males provide help with parental care, while in others they do not (Orians, 1980). In some populations experiments have shown that the addition of food to male territories increases polygyny (Ewald & Rohwer, 1982). In others, experiments show that heavy predation favours nesting aggregations due to the advantages of cooperative nest defence (Picman *et al.*, 1988). Thus the costs and benefits of polygyny to females may vary even within species. Different hypotheses in Table 9.1 may even apply to different individual females within a population. For example, first settling (primary) females may have greater choices available and may not suffer polygyny costs (for example, if males help to feed the broods of first females), while second settling (secondary) females may have more limited choice and may suffer costs. Three questions need particular attention.

1 IS THERE A COST OF POLYGYNY TO FEMALES?

Most studies have looked for costs by comparing female reproductive success in harems of different sizes (for example, Orians, 1980; Lightbody & Weatherhead, 1987), but observations of natural variation are confounded by other variables such as female quality or territory quality. The only experiment to test whether polygyny is costly for females failed to show a cost; reproductive success of female red-winged blackbirds was not affected by a reduction in harem size (Searcy, 1988).

2 WHAT ARE THE CHOICES FACING INDIVIDUAL FEMALES?

More detailed observations are needed on female settlement patterns. In the pied flycatcher *Ficedula hypoleuca*, females who mate with already paired males are left to raise their brood alone (males usually help only their first females) and so raise fewer young. Why, then, do they settle polygynously? One possiblity is that they are deceived into doing so because male polyterritoriality (Fig. 9.6) prevents the females from distinguishing mated from unmated males (Alatalo *et al.*, 1981). An alternative is that females are not deceived but accept polygyny costs simply because this is their best option when unmated males are hard to find (Stenmark *et al.*, 1988). According to this view, polyterritoriality by males is not to aid female deception but to decrease the chance that aggression from the first female will prevent the second female from settling. Testing between these hypotheses requires detailed observations on how females sample males and territories, and measurements of the profitability of their alternative options.

Alatalo *et al.* (1991) have recently performed a clever experiment to test between these explanations. By erecting nest boxes in careful sequence, they arranged for neighbouring boxes, less than 100 m apart, to be occupied by an unmated male and a mated male, whose first female was incubating a clutch in another box 100–300 m away. Boxes were put up at random sites, so there was no difference in territory quality between mated and unmated males. In this situation females could clearly sample both males (some were seen to do so) and the songs of both could be heard from either nest site. In 20 such paired choices, nine females settled with the unmated male and 11 with the mated male—clearly no difference. Furthermore, the females who chose the mated males raised significantly fewer young than those who later chose the unmated males they had rejected.

This result supports the deception hypothesis; females did not discriminate between mated and unmated males even when they had a simultaneous choice between them, and even though it would have paid them to make a choice.

In the polyterritorial wood warbler *Phylloscopus sibilatrix*, by contrast, Temrin (1989) showed that females who settled as secondary females had similar or even greater reproductive success than those who later paired with the unpaired males they had rejected. This suggests that in this species females may choose polygyny because it is the best mating option available at the time they settle. Clearly, then, different hypotheses may apply to different cases, not only between species but perhaps within a species.

3 DO FEMALES CHOOSE TERRITORIES OR MALES?

In some cases female reproductive success is influenced most strongly by territory quality (Searcy, 1979), and in others by male parental care (Davies, 1986). Where both are important, it may be difficult to distinguish their effects if the best quality males gain the best quality territories. In pied flycatchers, for example, males arrive on the breeding grounds before females. When females settle, there is a correlation between the order in which they choose territories and the order previously chosen by the males. In an ingenious experiment, Alatalo *et al.* (1986) forced males to settle randomly by putting up nest boxes one at a time at randomly chosen sites. This treatment destroyed the correlation between male and female settlement order, suggesting that it arises normally because males and females each independently settle in order of territory quality, not because earlier arriving males themselves are preferred by females.

9.5.4 *Cooperative polyandry*

In some birds two or more males may share one female (polyandry) or several females (polygynandry) and all the males may help to raise the young (Fig. 9.6). Cooperative polyandry is likely to be of benefit to females because more young can be raised with more male help (Fig. 9.8a). However, the benefit to males will depend on whether the increase in offspring production offsets the cost of shared paternity. Figure 9.8b suggests that male cooperation is more likely to occur in harsh environments where one male and a female have difficulty in raising young (see also Gowaty, 1981). This is illustrated by two contrasting cases of polyandry, in dunnocks (passerine bird) and saddle-backed Tamarins (primate).

In dunnocks *Prunella modularis* there is a variable mating system which reflects the different outcomes of sexual conflict (Fig. 9.9). Females have the greatest reproductive success with cooperative polyandry, where they have the help of two (unrelated) males, less success with monogamy, where they have one male's help, and least success with polygyny, where

they have to share one male's help. For males, however, the increased production of trio-fed broods does not compensate them for shared paternity
(assessed by DNA fingerprinting). Males do better with monogamy, where
they have sole paternity of a smaller brood, and best of all with polygyny,
where they have two females rearing broods (Davies & Houston, 1986;

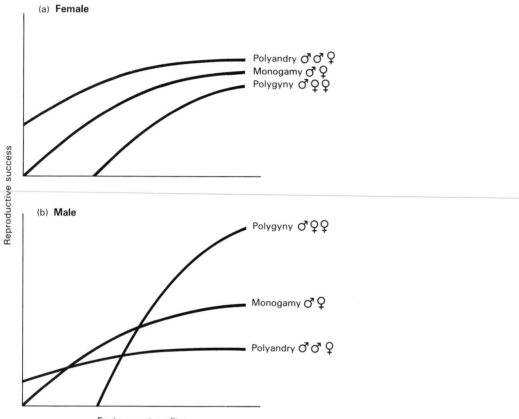

Fig. 9.8 A graphical illustration of sexual conflict. (a) A female's reproductive success per
brood is greatest with polyandry, where she has the full-time help of two males, less in
monogamy, where she has full-time help of one male, and least in polygyny, where she has
only part-time help of one male. For each mating system, success increases with
environmental quality. The differences between the curves are greatest in poor
environments, where increased male help is of greater benefit. In very rich environments
the curves converge as resources become superabundant in relation to a brood's needs. (b)
A male's reproductive success, calculated from the curves in (a). The monogamy curve is
the same as in (a). The polygyny curve is two times the female polygyny curve because a
male's success is the sum of the two females in polygyny. The polyandry curve is half the
female polyandry curve, assuming that the two males share paternity equally. In rich
environments male success follows the order polygyny > monogamy > polyandry, the
exact reverse of the order for female success; an example of such sexual conflict occurs in
dunnocks. In poor environments, however, male success (like female success) may be
greatest with polyandry and so there may be less conflict.

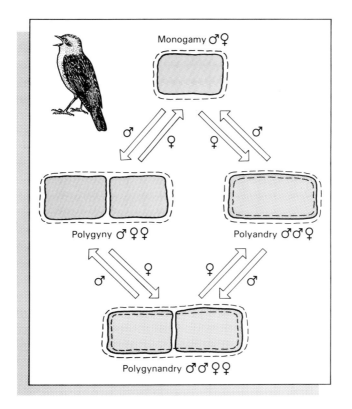

Fig. 9.9 Sexual conflict in dunnocks. Female territories (solid lines) are exclusive and may be defended by one or more (usually two) males (dashed lines). Arrows indicate the directions in which a male and female behaviour encourage changes in the mating system (from Davies, 1989).

Burke *et al.*, 1989). Thus the situation is as in the right hand part of Fig. 9.8, with male–female conflict.

These conflicts in reproductive pay-offs give rise to intense behavioural conflict, with each sex attempting to increase its number of mates while restricting their partner's access to additional mates (Davies & Lundberg, 1984; Davies, 1985). For example, monogamous males attempt to drive off second males. If they are unsuccessful they attempt to prevent the second male from copulating with the female, or at least to limit his access. Females, by contrast, attempt to escape the dominant (α) male's guarding and actively solicit matings from the second (β) male. This makes good sense because males provide parental care only if they copulate with the female, copulations being a good predictor of paternity. If a female is able to get two males to copulate with her she gains the help of both in care for her offspring.

Which individuals gain their preferred options varies depending on competitive ability and on the choices available. For example, a female prefers cooperative polyandry to monogamy, an α male prefers monogamy to cooperative polyandry, while a β male prefers cooperative polyandry to not breeding at all. Sometimes the conflicts reach a 'stalemate'; a male is unable to evict a second male and a female is unable to evict a second female, so the outcome is polygynandry with two males sharing two females (Fig. 9.9). This example makes the general point that there are often conflicts of interest within and between the sexes, so individuals will not always end up with their preferred options. Thus mating systems may not easily be explained by the choices of one party in the system.

Saddle-backed tamarins *Saguinus fuscicollis* are neotropical primates which also have a variable mating system including cooperative polyandry, monogamy and occasional polygyny (Terborgh & Goldizen, 1985; Goldizen, 1987; 1989). In polyandry, the males are probably unrelated. Both copulate with the female and both help to look after the infants. An increased number of carers seems to be very important for successful reproduction. Lone pairs rarely attempt to raise young unless they have previous offspring, which act as helpers. In a monogamous pair without helpers, it probably pays both male and female to allow a second male to join them to gain his help with parental care. In striking contrast to dunnocks, polyandrous males are rarely aggressive to each other and rarely disrupt each others' copulations. Further data on reproductive success are needed but the situation is probably as in the left hand part of Fig. 9.8, with no male–female conflict.

9.5.5 Sequential polygamy: female desertion

In most birds, if one sex deserts it is usually the male (see above). In shorebirds (Charadrii), by contrast, although most species are monogamous with biparental care, if one sex deserts it is often the female, leaving the male to care alone (Fig. 9.6). Females may share incubation but desert soon after hatching (Lenington, 1984), perhaps because in some species successful incubation requires both partners but brood care can be done by one (in most shorebirds the young feed themselves and require only brooding and protection). In these cases females may desert because, having laid the eggs, they are more stressed energetically and less able to provide care (Ashkenazie & Safriel, 1979). Some species may mate with the same partners in successive years, so the increased survival of the female may benefit the male too (Oring & Lank, 1984).

In other species (see Fig. 9.6 for examples), however, females desert immediately after clutch completion and search for another mate (sequential polyandry). Shorebirds are characterized by their small clutch size which never normally exceeds four eggs (Maclean, 1972). The four large eggs fit snugly together and experiments suggest that they represent an incubation limit because adding an extra egg reduces hatching success (Hills, 1980). Thus a clutch of four may represent a local adaptive peak. It may not be possible to evolve an increase in clutch size without first incurring a decrease in fitness associated with reduced egg size. If shorebirds are indeed 'stuck' with a maximum clutch of four, selection may particularly favour female desertion because, with a fixed clutch size, the only way females can increase their reproductive output if conditions become more favourable is to lay more clutches (Erckmann, 1983).

9.5.6 Sex role reversal

In some shorebirds, females are the larger sex, males normally do all the parental care, and females compete for males by defending large territories (Fig. 9.6; resource defence polyandry). Invasion of temperate (spotted sandpiper *Actitis macularia*: Lank *et al.*, 1985) and tropical regions (*Jacana* spp.: Jenni & Collier, 1972), with longer potential breeding seasons, may give females of these species the opportunity to lay more clutches per year than their Arctic ancestors. Given an ancestral condition where males provide care and clutch size is limited, more intense female competition for males may have led to sex role reversal.

In the Arctic-breeding phalaropes, too, females are the larger sex and are more brightly coloured, but they defend individual males rather than territories (Reynolds, 1987). Individuals may wander widely and large numbers congregate at local pools with superabundant food. These conditions may preclude economic territory defence and favour mate guarding instead. However less than 10% of females gain a second mate which raises the question of why they do not rear a second brood themselves. Perhaps a short breeding season coupled with long migration favours early departure to wintering grounds to increase survival chances (Myers, 1981).

One other family of birds, the button quails (Turnicidae), which has sex role reversal (larger and more brightly coloured females) and sequential polyandry, also has a small clutch size (two to four eggs). However, their eggs are remarkably small (Lack, 1968), which suggests that, unlike shorebirds, clutch size may not be fixed by an incubation limit. Thus the 'adaptive peak' explanation for the evolution of successive polyandry may not apply unless some other factor limits clutch size in this family.

9.6 Conclusion

The scheme 'resources influence female dispersion influences male dispersion' (Fig. 9.1) is supported by comparative studies of species where males do not provide parental care. Evidence for the second link in the scheme is stronger than for the first, because it is easier to measure female dispersion than resource dispersion. More experiments would be useful, involving manipulation of resource or female dispersion to test directly how these influence the mating system which emerges.

Where males provide care, comparative studies and removal experiments show how mating systems vary not only in relation to resource dispersion but also the costs and benefits of desertion by either sex. More studies of species with variable mating systems would help to identify the conditions leading to one outcome rather than another. It is important to focus on the reproductive pay-offs of alternative options facing particular individuals and to recognize conflicts of interest. A key question to ask is 'under what circumstances will particular individuals be likely to achieve their preferred option, despite the conflicting preferences of others?'

More measurements are needed of paternity and maternity (for example by DNA fingerprinting) to show how behaviour maps on to reproductive success. Different mating systems may emerge depending on the details of sperm competition (last versus first male advantage; potency of extra-pair matings). Females may be able to control males by how often or where they mate and by particular mechanisms of sperm utilization. Males, in turn, may control females by, for example, varying their investment in parental care depending on female fidelity. The outcome of such evolutionary games is likely to have an important influence on which mating system emerges. Future progress will depend as much on good field studies as on advances in theory.

PART 4

COOPERATION AND CONFLICT

Introduction

The paradox of the evolution of cooperation, especially altruism, has been a central theme in neo-Darwinian theory in the past 25 years. The issue is simply stated: how can the process of natural selection, which by its very nature should favour self-interest, promote traits that cause their bearers to sacrifice their own survival and reproduction whilst increasing that of others. Since the mid-1960s, four not necessarily mutually exclusive ideas have been put forward to account for apparent altruism: mutualism, reciprocity, kin selection and parental manipulation.

The most fertile testing ground for these ideas has been the study of cooperative breeding in birds and mammals and its analogue, eusociality, in hymenopteran insects. The difference between vertebrates and insects is one of degree. In both taxa individuals appear to forego their own reproductive opportunities to help others to produce offspring. In insects, but not in vertebrates, the helpers have evolved morphological and physiological specializations associated with helping, such as defensive weapons, food gathering equipment and sterility. With the discovery of naked mole-rat colonies, even this distinction is becoming blurred: mole-rats show incipient caste specialization and permanently non-reproductive individuals, rendering them close to the eusocial insects in terms of social organization.

Conflicts of interest are closely intertwined with cooperation, because even though cooperating individuals may be driven by ecological and genetic circumstances to share common goals, the balance of costs and benefits is usually delicate enough for conflict to show through. One of the most startling discoveries of the past 20 years is the extent to which even the well organized monolithic nests of ants and bees are riddled with non-cooperative behaviour.

Emlen's review of helping in birds and mammals draws two major conclusions: (i) crucial to the understanding of helping is to recognize that ecological constraints give rise to delayed dispersal (juveniles stay at home). These constraints may mean that alternatives to staying at home are especially poor (no, or poor quality, empty habitat) or that staying at home is especially good (benefits of an exceptionally good territory or large group in which to live); the genetic gains from helping are twofold: self improvement (practice at breeding, increased survival, etc.) and kinship—helping to raise close relatives. In about half a dozen species, the relative importance of these two has been assessed and in all but two cases, both factors seem to be necessary to maintain helping at a selective advantage.

The conclusions reached by Seger in many ways parallel those of

Emlen, but the story for eusocial insects is probably more complex. The taxonomic distribution of eusociality among insects suggests that no single factor will account for its origin. For example, most origins of eusociality occurred in the aculeate hymenoptera, but most aculeates are not eusocial. In other words, one has to think in terms of constellations of factors that predispose a group to develop eusociality.

The most famous 'predisposing factor', immortalized by W.D. Hamilton, is of course haplodiploidy. One effect of haplodiploidy (or more strictly the development of males from unfertilized eggs) is that it allows a female to combine the benefits of helping with that of laying her own (male) eggs. The more widely discussed consequence is that it produces a pattern of relatedness in which, all other things being equal, females benefit more in genetic terms from rearing reproductive sisters than from rearing reproductive sons or daughters. This simplified version of the role of haplodiploidy conceals many complications. For one thing, the benefits of rearing closely related sisters only accrues if the workers in a nest produce a sex ratio in their reproductive siblings that is female biased relative to the rest of the population (to see why, read Seger's chapter!). Seger discusses a number of ways in which, through spatial or temporal variation in production of the two sexes, this situation may arise as a preadaptation for eusociality in primitively social insects.

As already pointed out, haplodiploidy, whilst it may be a predisposing factor, is not enough on its own to cause the evolution of eusociality. Other factors, equivalent to Emlen's ecological 'constraints' are discussed by Seger, especially the construction of an elaborate nest, which shifts the balance against setting out to breed alone and in favour of joining an existing breeding unit, and the fact that aculeate hymenoptera have a sting which they use in nest defence. This again favours being a member of a large group. A further factor which might influence the balance of costs and benefits of eusociality is the recent discovery by Shykoff and Schmid-Hempel (1991) that among bumble bees genetically more homogeneous colonies are more susceptible to interindividual transmission of an intestinal trypanosome parasite.

The sex ratio of reproductives is important not only in the evolutionary origin of eusociality but also in the analysis of conflicts of interest between workers and queen once eusociality has been established. Trivers and Hare's (1976) classic finding that workers manipulate the ratio of investment in their favour and against the queen has been substantiated in a number of more recent studies. Now that it has been shown that social hymenoptera are able to recognize their kin using olfactory cues, the possibility of more subtle discrimination by workers is suggested; this remains to be studied.

The final chapter of this section discusses cooperation and conflict in the context of communication. It is now generally accepted that animal communication is not the purely cooperative venture seen by early ethologists, but that conflicts of interest between signaller and receiver are important in understanding the design of signals. On the one hand signallers are selected to alter the behaviour of receivers in a way that benefits the actor. On the other hand receivers are selected to respond only to 'reliable' or 'honest' signals. Some models suggest that the equilibrium outcome of this coevolutionary interaction is honest signalling. A question for the future is to establish to what extent this is true and whether, as Zahavi has suggested, signal design can be understood in terms of the logical link between the signal and the quality of the actor that is being signalled.

10: Evolution of cooperative breeding in birds and mammals

Stephen T. Emlen

10.1 Introduction

The term cooperative breeding describes situations in which adult individuals in addition to the genetic parents regularly aid in the rearing of young. Such breeding systems have been described in approximately 220 species of birds (Brown, 1987) and 120 species of mammals (Riedman, 1982). Helpers (also called auxiliaries) at the nest were first described by Skutch in 1935. Skutch's observations raised a number of intriguing questions. Who were these extra adults? Why were they not breeding on their own? And why were they raising another's young? It was not until the mid-1960s, however, with the advent of modern behavioural ecology, that widespread attention began to focus on cooperatively breeding species. In particular, Hamilton's (1964) theory of inclusive fitness provided a theoretical construct for explaining how individuals might benefit from engaging in phenotypically costly behaviours. Trivers' (1974) later model of parent-offspring conflict stressed the need to examine helping from the separate perspectives of the different participants. Many workers recognized that cooperatively breeding species provided an excellent arena for testing emerging ideas concerning the evolution of altruism, as well as the existence and resolution of conflicts of interest among members of social groups.

The 1970s saw the initiation of a number of long-term field studies aimed at measuring the fitness consequences of helping from the perspective of both donor (helper) and recipient (breeder) of the aid. Many of those studies are now coming to fruition, as evidenced by the recent publication of a compilation of 18 contributions on cooperatively breeding birds (Stacey & Koenig, 1990).

In the last edition of this volume, I outlined what I perceived to be an emerging conceptual framework for understanding the evolution of cooperative breeding (Fig. 10.1). For heuristic reasons, I subdivided the issue into two questions: First, 'Why don't helpers breed on their own?' I concluded that the predominant reason was that they could not, that some constraint existed that made it difficult for them to breed independently. Most cooperatively breeding groups formed when grown offspring delayed dispersal and remained with their parents.

The second question became, 'Why do such individuals become helpers?' Specifically, what fitness benefits accrue to helpers that might offset the costs of engaging in alloparental care? Although a number of

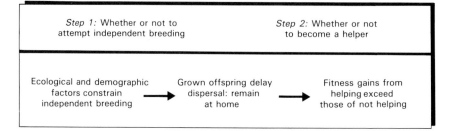

Fig. 10.1 The evolution of helping as a two-step process.

hypotheses had been advanced, insufficient empirical data were available at that time to justify any general conclusions concerning their relative importance.

In the 6 years since the publication of the second edition, new hypotheses have been offered regarding both of the questions raised above. A surge of new demographic and behavioural data on helper species has become available. In the following pages, I again raise the same two heuristic questions, incorporating the new empirical data as well as new challenges and additions to the theoretical framework.

10.2 Diversity of cooperative breeding systems

Skutch's first published account in 1935 of cooperative breeding described extra adults provisioning young in brown jays, banded-backed wrens and bush tits. We now know that helpers in these species are younger individuals that delay dispersal, and remain with and aid their parents in the rearing of a later brood. This simple helper-at-the-nest/den system is the most common form of cooperative breeding in both birds and mammals.

The Florida scrub jay provides a good example (Woolfenden, 1975; Woolfenden & Fitzpatrick, 1984; Fitzpatrick & Woolfenden, 1988; Mc-Gowan & Woolfenden, 1989). The basic social unit of these birds consists of a monogamous breeding pair together with some of their young from the previous one or more years. These mature offspring assist in group defence of the natal territory. During the breeding season, they provide food for the nestlings and play a major role in tending and guarding the nest when others in the group are off foraging. After the young fledge from the nest, all group members continue to provide food and protection to the fledglings during the latter's transition to full independence.

The silver-backed jackal provides a mammalian analogue. Moehlman (1979; 1983; 1986), studying this species on the plains of Tanzania, found that roughly one third of the young remained with their parents through the following breeding season. Groups of from three to five adults were formed, and the non-breeding members served as helpers in rearing the next year's litter of pups. Such help took the form of provisioning the nursing female, regurgitating food to the young, and guarding the pups when the parents were absent from the den. All members of jackal groups also play with and groom the young as well as helping teach them to hunt.

Not all cooperative breeders have such a simple social organization. In some species, groups become larger, extended rather than nuclear families predominate, and individuals from several overlapping generations may be involved. The term 'plural breeder' (Brown, 1978) is used when more than one pair within a social unit breeds simultaneously.

Among grey-breasted jays in Arizona, extended family groups of from eight to 18 individuals live on permanent, all purpose territories (Brown & Brown, 1980; 1981; 1990). Several monogamous pairs within each family may reproduce and non-breeding group members act as helpers, dividing their provisioning and defence among the active nests. After the young fledge, they intermingle and are fed indiscriminately by all group members.

Banded mongooses are plural breeding mammals that live in packs of four to 40 individuals and exhibit helping at the den (Neal, 1970; Rood, 1974; 1978; 1980). Several females in each pack are reproductives, and they produce their litters synchronously. As a result, the young are the same age and they suckle indiscriminately from any lactating female. Most members of the pack bring food (insects) to the young and take turns in guarding the den. Rood (1974) found that three quarters of the guarding was performed by adult males and that lactating females never acted in this capacity. This partial division of labour freed the breeding females to spend more time foraging and replenishing the nutrient reserves needed during lactation.

Still other cooperative breeders are highly gregarious. White-fronted bee-eaters roost and nest in large colonies of from 50 to 300 birds. Each colony is composed of a number of smaller extended family groups called clans within which helping behaviours occur (Emlen, 1981; 1990; Hegner & Emlen, 1987; Emlen & Wrege, 1988; 1991). Unlike grey-breasted jays, bee-eater helpers attach themselves to a single nest. They participate in virtually all phases of breeding activity, including excavating and defending the nest, allofeeding the breeding female, incubating the eggs, feeding the nestlings, and caring for the fledglings after they leave the nest. Should a nesting attempt fail, individuals often join up as helpers at another on-going nest of the clan. Such 'redirected' helping results in adult bee-eaters shifting back and forth between breeder and helper status repeatedly during their lifetimes.

Breeding in the examples mentioned above is restricted to mono-gamous pairs. But cooperative breeders exhibit a variety of mating arrange-ments. Polyandry (where a female mates with multiple males) and polygynandry (in which multiple males and females mate with one another) both occur.

Acorn woodpeckers in coastal California live in groups consisting of from one to four breeding males, one or two breeding females, and up to eight non-breeding auxiliaries (generally grown offspring from previous years; Koenig & Mumme, 1987; Koenig & Stacey, 1990). During breeding, only a single nest is tended, and most or all group members help to incubate the eggs and feed and defend the young. Pair bonds appear to be absent and several males engage in sexual activity with the nesting female(s) (Stacey, 1979b; Joste et al., 1985; Mumme et al., 1985). In roughly one quarter of the groups studied in California, two females (usually sisters) became reproductives and laid eggs communally in the same nest (Koenig et al., 1983; Mumme et al., 1983). Because females copulate with multiple males, and males will copulate with both females (when females nest jointly), the mating system is polygynandrous.

Polyandrous mating is found in cooperatively breeding callitrichid primates. Among saddle-backed tamarins, reproductive groups consist of two or more adult males and a single breeding female. The 'extra' male can be a grown offspring that remains as a non-breeder with its parents, or an unrelated adult that shares reproductive access to the female (Terborgh & Goldizen, 1985; Sussman & Garber, 1987; Goldizen & Terborgh, 1989). Females virtually always give birth to twins, and the 'helping' consists primarily of carrying the two young as the group moves through the forest canopy. All group members also jointly defend both the territory and the young. Virtually no pairs without helpers attempt to reproduce, and Goldizen (1987) has speculated that the high costs of lactation and infant carrying require a minimal group size of three participants.

Naked mole-rats provide a final mammalian example (Jarvis, 1981; Sherman et al., 1990). These rodents live in large subterranean colonies of from 40 to 300 animals. Each colony is composed of one breeding female, one to three breeding males, and large numbers of non-breeding helpers. The reproductive female mates polyandrously with the breeding males. The breeders perform the majority of direct care of the pups. Non-breeders of both sexes defend and maintain the colony's extensive tunnel system (up to 3 km in length). Further, such non-breeders show a size-dependent polyethism (Lacey & Sherman, 1990). Smaller individuals build the communal nest, carry food to it, and keep tunnels free from debris, while larger individuals defend the colony against snakes and foreign colonies, and excavate new tunnels. Colonies of naked mole-rats form via retention of grown offspring. Recent molecular data indicate that matings are consanguineous, resulting in the members of any given colony being highly inbred (Reeve et al., 1990).

The social organization of naked mole-rat colonies, with their large number of non-reproductives, their single female breeder, and their incipient divison of labour, is the closest parallel to eusocial insect societies known among vertebrates.

Even this review does not fully cover the breadth of social organizations found in cooperative breeders. There can be no question that helping systems have evolved independently a number of times, in a large number of different taxonomic groups. The challenge for the behavioral ecologist is to find pattern in this diversity, to search for common ecological and demo-graphic patterns that might explain the evolution of cooperative breeding across such a broad array of species.

10.3 Do helpers really help?

Before addressing the two basic quesions raised in the introduction, it is first necessary to ask whether helpers really do help. Is it possible that such behaviours have been misnamed as a result of anthropomorphic inter-pretations on the part of researchers? Or, as has recently been argued, is helping merely an unselected consequence of non-dispersal of grown young (Jamieson, 1986; 1989; Jamieson & Craig, 1987)? Does helping have important fitness consequences for recipient breeders, or does it not?

There are three major ways in which helpers might increase the fitness of breeders. First, they might increase the success of individual repro-ductive attempts. This could occur through enhancement of the breeder's clutch or litter size, increased detection and deterrence of predators, and/or increased provisioning of the young. Second, helpers might reduce the 'workload' of the breeders, enabling them to initiate more reproductive attempts per season. And third, the presence of helpers could enhance breeder survival.

10.3.1 Increasing the success of breeding attempts

CORRELATIONAL APPROACHES

The simplest way to assess the importance of helpers on the success of in-dividual reproductive attempts is to compare the production of young from dens/nests tended by the breeding parents alone with that from dens/nests tended by three or more adults. When more than one helper may be present, one examines the regression of reproductive success upon group size. Many simple correlational analyses of this sort show a seemingly posi-tive influence of helpers (for recent reviews see Brown, 1987; Koenig & Mumme, 1990).

Figure 10.2 shows three examples. Similar trends have been reported in a broad range of additional cooperative species (e.g. dwarf mongoose: Rood, 1990; red fox: Macdonald, 1979; brown hyaena: Owens & Owens, 1984; purple gallinule: Hunter, 1985; moorhen: Gibbons, 1987; Tasmanian native hen: Ridpath, 1972; hoatzin: Strahl & Schmitz, 1990; pied kingfisher: Reyer, 1980; red-throated bee-eater: Dyer & Fry, 1980; stripe-backed wren: Rabenold, 1984; bicoloured wren: Austad & Rabenold, 1985; Florida scrub jay: Woolfenden, 1975).

Univariate analyses of this type provide a very weak basis for making inferences about helping behaviour, however. If productivity was higher on better quality territories, one result could be the retention of larger

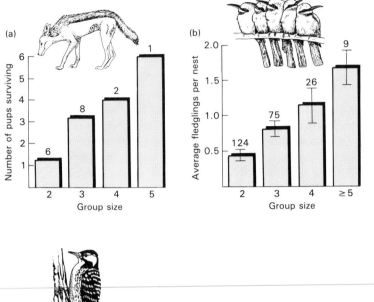

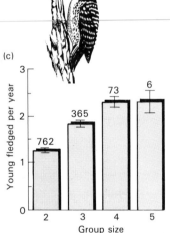

(c)

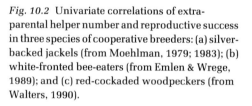

Fig. 10.2 Univariate correlations of extra-parental helper number and reproductive success in three species of cooperative breeders: (a) silver-backed jackels (from Moehlman, 1979; 1983); (b) white-fronted bee-eaters (from Emlen & Wrege, 1989); and (c) red-cockaded woodpeckers (from Walters, 1990).

groups on such territories. Similarly, if breeding success was influenced by the age or past experience of the breeders, larger groups could form on the territories of such experienced breeders. In either case, the presence of helpers would be an epiphenomenon of successful breeding in the past, and positive correlations of the type shown in Fig. 10.2 could be due to territory quality or breeder experience rather than the actions of helpers *per se*.

One way of separating these variables and determining the relative importance of each is through the use of multivariate analyses. The importance of helpers in predicting reproductive success can then be examined while statistically controlling the effects of other variables such as territory quality or breeder experience.

In white-fronted bee-eaters, the linear trend of increasing fledging success with increasing helper number was maintained when environmental conditions, as well as age and experience of the breeding pair, were treated as co-variates (Emlen & Wrege, 1991). The actions of helpers have a major impact in increasing fledging success in this species.

Among red-cockaded woodpeckers, however, much of the trend seen in Fig. 10.2c disappeared when Walters (1990; Walters *et al.*, 1988) included the age of the breeders in a multivariate analysis. In another study of this species, Lennartz *et al.* (1987) found that territory quality was positively correlated with both group size and reproductive success. When Walters (1990) compared pairs breeding on the same territory with and without helpers (in different years), no influence of helpers was detected. It thus appears that helpers do *not* enhance breeding success. This example demonstrates the difficulty of interpreting helper effects without also considering the influence of confounding variables.

EXPERIMENTAL REMOVAL OF HELPERS

A more powerful way of examining the influence of helpers on reproductive success involves removing them from breeding units and directly comparing the success of natural sized and artificially reduced groups. To date, three such experiments have been performed (Table 10.1). Brown *et al.* (1982) removed helpers in grey-crowned babblers. These birds live in groups of from two to 13 individuals. Only one pair in each group breeds; the remaining individuals behave as helpers (Councilman, 1977). Twenty groups, roughly matched for initial helper number and territory quality, were selected for study. In nine experimental groups, helper number was reduced to one. In 11 control groups, helper numbers ranged from four to six. The subsequent success of nests with their full complement of helpers was nearly three times that of nests tended by the artificially depleted groups.

Mumme (in preparation) removed all helpers from 16 randomly selected groups of Florida scrub jays immediately before the onset of nest-

Table 10.1 Results from experiments in which helpers were artificially removed from breeding groups.

Species	Controls			Experimental			References
	Group size	RS*	*n*	Group size	RS*	*n*	
Grey-crowned babblers	7	2.4	11	3	0.8	9	Brown *et al.* (1982)
Florida scrub jays	3.8	1.62 + 0.29	21	2	0.56 ± 0.18	16	R. Mumme (personal communication)
Moorhens	4.2	2.20 ± 0.21	10	2	2.50 ± 0.17	10	Leonard *et al.* (1989)

*Mean ± standard error; RS, reproductive success (number of young).

ing. The reproductive success of 21 control groups (averaging 1.8 helpers) was three times that of the experimentals. Woolfenden and Fitzpatrick (1984) had previously shown through correlated analyses that more young fledged from nests with helpers than without, and had suggested that this helper effect was due to enhanced detection and deterrence of predators. Mumme confirmed this by demonstrating that much of his observed difference in reproductive success was due to reduced predation losses in control (helpers present) compared with experimental (helpers absent) groups.

The third experiment was performed on moorhens, a species in which young from first broods often remain and assist their parents with the second brood. Such juvenile helpers feed their younger siblings and join in territory and predator defence. Gibbons (1987) had reported that groups with helpers had greater success than did simple pairs in rearing second broods. But when Leonard *et al.* (1989) experimentally removed such juvenile helpers from 10 groups, they found no difference from controls in subsequent reproductive success of the breeding pair. They concluded that the helper effect originally reported (from the same population) must have been due to confounding variables of territory and/or breeder quality (Leonard *et al.*, 1989).

There is a strong need for more multi variate analyses of, and for additional removal experiments with, helper species. On the basis of the available evidence, however, it appears that helpers have a moderate effect in enhancing breeding success in a large number of species, and a strong effect in a few. There are a number of species, however, in which no influence of helpers on the success of individual reproduction attempts has been found (for example, Bekoff & Wells, 1982;) Zacj & Ligon, 1985a, b; Nias 1986; Bednarz, 1987; Koenig & Mumme, 1987; Leonard *et al.*, 1989; Sydeman, 1989; Marzluff & Balda, 1989).

10.3.2 *Increasing the number of breedings per season*

Helpers can also reduce the workload associated with reproduction, thereby enabling breeders to re-nest or produce a second litter more rapidly.

Among grey-crowned babblers, the presence of many helpers does not lead to an increase in total provisioning rate (Brown *et al.*, 1978). But by sharing the provisioning task each individual makes fewer foraging trips (Fig. 10.3). Similar findings have been reported for many other avian cooperative breeders (e.g. hoatzin: Strahl & Schmitz, 1990; moorhen: Eden, 1987; Gibbons, 1987; pied kingfisher: Reyer, 1980; green woodhoopoe: Ligon & Ligon, 1978a; stripe-backed wren: Rabenold, 1984; bicoloured wren: Austad & Rabenold, 1985; splendid fairy wren: Rowley, 1981; bell miner: Clarke, 1989; Galapagos mockingbird: Curry, 1988b; white-browed sparrow weaver: Lewis, 1981; Florida scrub jay: Stallcup & Woolfenden,

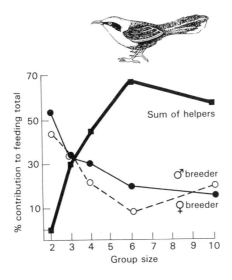

Fig. 10.3 The relative contributions of different group members to the feeding of nestlings in the grey-crowned babbler (from Brown *et al.*, 1978). Note the decrease in the workload of the breeders as helper number increases.

1978; Beechey jay: Raitt *et al.*, 1984).

When young babblers fledge from the nest, they are tended by the breeding male and helpers, thus freeing the female breeder to initiate re-nesting. Brown and Brown (1981) found that groups with many helpers re-nested more rapidly and reared more broods per season than did groups with few helpers.

Fairy wrens provide additional examples of this phenomenon. The presence of helpers has only a minor influence on the number of young fledged per brood in these species (Nias, 1986; Russell & Rowley, 1988). Helpers do release the female from *fledgling* care, however, and such females re-nest earlier and produce significantly more broods per year (Rowley, 1965; Russell & Rowley, 1988).

10.3.3 Enhancing breeder survivorship

Helpers can also influence the likelihood of survival of the breeders. Accurate data on survivorship are difficult to obtain because they require long-term study of large populations of recognizable individuals. Woolfenden and Fitzpatrick (1984) were the first to report that breeder survival was higher in Florida scrub jay groups with helpers. Similar findings have now been reported for dwarf mongooses (Rood, 1990), pied kingfishers (Reyer, 1984), acorn woodpeckers (Koenig & Mumme, 1987), bicoloured wrens (Austad & Rabenold, 1986), splendid fairy wrens (Russell & Rowley, 1988) and Galapagos mockingbirds (Curry, 1988b), and I expect that many more species will be added to this list as more long-term studies come to completion. Longevity is proving to be a major component of lifetime fitness estimates in birds and mammals (Clutton-Brock, 1988), and even a small increase in annual survivorship can have a large influence on lifetime productivity (Woolfenden & Fitzpatrick, 1984; Mumme *et al.*, 1989). The magnitude of reported helper effects on breeder survivorship range from 8% to 39%. This raises the possibility that, for some species, enhanced survivorship may prove to be the most important breeder benefit gained from having helpers.

In conclusion, the presence of helpers typically does enhance the fitness of the recipient breeder. The magnitude of this benefit, as well as the way in which it occurs, varies between species. Although it would be erroneous to conclude that helpers *always* enhance breeder fitness, we can answer the question raised at the beginning of this section in the affirmative. Helpers generally *do* help and the breeders are the obvious beneficiaries. But what of the helpers themselves? Are they also benefiting or would they do better to leave their groups and attempt to breed independently?

10.4 Why don't helpers become independent breeders?

10.4.1 Routes to sociality

There are two primary routes by which social groups form among organisms. In the first, individuals aggregate because of inherent advantages of group living. The most often cited advantages are increased alertness and defence against predators and increased capabilities for detecting and harvesting food resources that are difficult to locate (Alexander, 1974; Hoogland & Sherman, 1976; Bertram, 1978). Both lead to enhanced survivorship of the participants. By this route, the average fitness of individual group members $(\overline{W})$ will increase as some function of increasing group size (k) up to some optimum size, and decrease thereafter (see Fig. 10.4a).

In the second route, offspring delay dispersal and remain with their natal group because of various constraining factors that restrict their option of dispersing and breeding on their own. Consider a species in which grouping leads to a net decrease in per capita fitness compared with solitary breeding (shown as the solid line in Fig. 10.4b). But consider further that openings for breeding are scarce and competition for them is severe. The intrinsic benefit of independent breeding must be devalued by the probability of dispersing and becoming established successfully, here denoted as s. When s is low, a 'hump' is created in the fitness curve and we could speak of individuals being 'forced' to remain in their natal groups (dashed line in Fig. 10.4b).

The magnitude of the constraint, s, for any given species may vary across different geographical areas and across years. Such changes in s lead

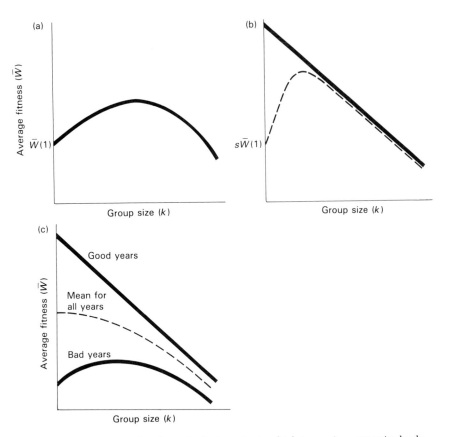

Fig. 10.4 (a) Fitness curve for a hypothetical species in which increasing group size leads to a direct benefit to group members; $\overline{W}(1)$ denotes the fitness of a solitary individual. (b) Fitness curve for a hypothetical species in which increasing group size leads to a decrease in average individual fitness (solid line), but the possibility of leaving the group is severely constrained (the dashed line, where s denotes the probability of successfully dispersing and becoming established independently). (c) Fitness curves for a hypothetical species in which grouping is disadvantageous but the intensity of constraining factors limiting dispersal varies greatly across years. 'Good' years denote times when constraining factors are minimal; 'bad' years denote times when constraints are severe.

to predictable differences in the size and complexity of the social group-ings, with concomitant changes in the potential for development of helping behaviour. Fitness curves for a hypothetical species residing in a variable environment where the severity of constraining factors changes from year to year, are shown in Fig. 10.4c. In 'good' years all individuals will breed independently; in 'bad' years, many individuals will postpone dispersal, groups will develop, and the potential for helping interactions will be present.

   In the overwhelming majority of cooperatively breeding birds and mammals, groups form as a result of delayed dispersal of grown young. The retention of such individuals within their natal units creates social groupings comprised of genetic family members. It is within these family units that helping behaviours have proliferated.

Part of the challenge for behavioural ecologists is to identify the reasons for delayed dispersal among cooperative breeders. Ultimately, this depends upon the relative fitness pay-offs of the opposing dispersal strategies of staying at home versus dispersing early. Many authors have modelled this trade-off (Brown, 1978; 1987; Vehrencamp, 1979; Emlen, 1982a,b; Stacey, 1982; Woolfenden & Fitzpatrick, 1984; Stacey & Ligon, 1987; Walters *et al.*, in press) and attention has been focused on the search for ecological or demographic features of cooperative breeders that might tip the balance in favour of delayed dispersal and retention in the parental unit.

10.4.2 The habitat saturation hypothesis for delayed dispersal

The most widespread hypothesis to explain delayed dispersal is that of territory limitation or habitat saturation. It was first proposed by Selander (1964) in his study of *Campylorhynchus* wrens and later elaborated by Brown (1974). The model has been refined several times since (for example, Stacey, 1979a; Koenig & Pitelka, 1981; Emlen, 1982a; Woolfenden & Fitzpatrick, 1984; Brown, 1987) and has been embraced by a large number of field-workers. In an earlier paper (Emlen, 1982a), I went so far as to state that the habitat saturation model had become the modis operandi for answering the question, 'Why don't helpers breed on their own?'

In its simplest form, the hypothesis states that the reason many helpers do not breed independently is that they cannot; they are constrained from becoming independent breeders by a shortage of suitable breeding territories. If occupancy of a territory is a prerequiste for independent breeding, then auxiliaries must wait until an established breeder dies or is displaced, and then compete for the resulting vacancy. Waiting is best done at home, on a territory of proven quality and in the company of familiar kin.

The demographic profile of a typical cooperative breeder includes high juvenile survivorship, high adult survivorship, and permanent residency on all-purpose territories (Brown, 1974). The result is the production of a considerable surplus of mature individuals relative to the number of territorial vacancies.

Shortages of suitable breeding openings have been stressed as causal factors underlying the formation of helping groups in hoatzins (Strahl & Schmitz, 1990), Tasmanian gallinules (Ridpath, 1972), New Zealand pukekos (Craig, 1979), red-cockaded and acorn woodpeckers (Walters *et al.*, 1988; Koenig & Mumme, 1987), green woodhoopoes (Ligon & Ligon, 1978a), Arabian, common and jungle babblers (Zahavi, 1974; Gaston, 1978a) and numerous species of jays (for example, Hardy, 1961; Brown, 1974; Woolfenden, 1975; Hardy *et al.*, 1981; Woolfenden & Fitzpatrick,

1984). Among mammals, the same type of constraint probably underlies the formation of cooperative groups in many carnivores (dwarf mongoose, red fox, silver-backed jackal, wild dog, brown hyaena: Macdonald & Moehlman, 1982), as well as in tamarins (Goldizen & Terborgh, 1989).

10.4.3 The benefits of philopatry hypothesis for delayed dispersal

The habitat saturation hypothesis stresses the costs associated with early dispersal. Recently, Stacey and Ligon (1987; personal communication) have offered a 'benefits of philopatry' hypothesis which emphasizes the *gains* of staying at home.

As evidence against the ubiquity of habitat saturation, they cite data from a number of cooperatively breeding birds in which seemingly suitable vacant territories are available, but remain unoccupied (Rabenold, 1984; Zack & Ligon, 1985b; Hegner & Emlen, 1987; Stacey & Ligon, 1987; Bednarz & Ligon, 1988). They suggest that when variance in the quality of available territories is large, selection will favour individuals that delay dispersal and wait for a *high quality* opening rather than accepting the first vacancy available.

Remaining at home could increase an auxiliary's future chance of gaining a high quality breeding site in at least two ways. First, continued residency can lead to inheritance of the natal territory itself. Inheritance as a strategy for achieving breeding status is a widespread phenomenon among both cooperatively breeding birds and mammals (see section 10.5.2). Second, residency may provide a competitive edge in contests for openings that arise nearby. Recent studies suggest that contestants from neighbouring groups fill a disproportionate share of such vacancies (Zack & Rabenold, 1989; Walters *et al.*, in press). This competitive advantage may be due to a resident's earlier detection of, and arrival at, the contested vacancy. Whatever the proximate explanation, any asymmetry of 'resident neighbour wins' would enhance the benefits of staying home.

10.4.4 Reconciling the differences

Stacey and Ligon (1987) have stimulated considerable rethinking of habitat saturation ideas. However, I suggest that the differences between the two hypotheses are more semantic than real. When the benefits of philopatry and habitat saturation hypotheses are formulated in mathematical terms, they reduce to the same inequality. Both describe conditions that lead to group formation via the second route discussed in section 10.4.1, namely the retention of grown offspring in the natal unit. In each, the driving factor underlying delayed dispersal is the existence of strong competition for a limited number of *suitable* territory vacancies.

A shortage of breeding territories by itself, however, is not a sufficient explanation for delayed dispersal and the creation of family groups. Many territorial organisms face such shortages, yet in only a small fraction do grown young remain on the natal area. More typically, young disperse shortly after attaining sexual maturity. Successful dispersers obtain territories of their own and become independent breeders. Unsuccessful individuals become 'floaters', living in marginal habitats or moving among interstitial spaces between occupied territories until a suitable vacancy can be found (for example, Smith, 1978). We must ask why young of cooperative breeders do not become floaters.

Koenig and Pitelka (1981) proposed that an additional factor was important in explaining natal retention. Not only must optimal habitat be limited, but marginal habitat must be rare. When this is the case, a maturing individual is constrained from either establishing itself as an independent breeder in the optimal habitat, or successfully surviving as a floater in a more marginal area.

Their graphical model is redrawn in Fig. 10.5 which plots the fitness of breeding occupants against the quality of habitat occupied. Koenig and Pitelka (1981) argued that retention of young would be most common when the 'habitat gradient slope' was steep (curve A). In the absence of marginal habitat, dispersers cannot become floaters, and young will remain on their natal area until a vacancy within the optimal habitat occurs.

Koenig and Pitelka (1981) initially phrased their model in terms of different *habitat* types. But the abscissa could just as easily be defined in terms of *territory quality* (cf. Koenig & Mumme, 1987, p. 62; Walters *et al.,* in press). Curve A would then have a steep 'territory quality gradient' and would describe situations where all high quality territories were occupied and most available vacancies were on territories of much lower quality. In this broadened form, Fig. 10.5 accommodates many of Stacey and Ligon's criticisms.

What might cause a steep gradient in the quality of breeding territories? Four factors deserve mention.

1   Suitable breeding habitat may actually be limited. This would be true for species occupying habitats that are physically restricted or relict in distribution. The Florida scrub jay, which is dependent upon isolated patches of fire-maintained oak scrub habitat, is a good example (Woolfenden & Fitzpatrick, 1984).

2   The quality of a territory may be strongly dependent upon some critical resource which is distributed unevenly among territories. Green woodhoopoes which rely upon scarce tree cavities for roosting and nesting provide an example (Ligon & Ligon, 1988).

3   The occupants may extensively modify their habitat, thereby magnifying differences between occupied and unoccupied areas. Acorn woodpeckers, which store food in huge granaries constructed, provisioned and defended by the group, provide one example (Koenig & Mumme, 1987). Bannertailed kangaroo rats (*Dipodomys spectabilis*) which construct

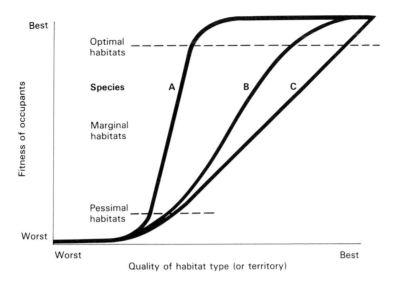

Fig. 10.5 Breeder fitness plotted as a function of habitat quality for three hypothetical species of organism (modified from Koenig & Pitelka, 1981).

elaborate mounds essential for security from predation and shelter from climatic extremes, provide another (Waser, 1988).

4 In species where the presence of helpers is critical for successful breeding, the quality of a territory can be determined by what Brown (1987, p. 80) has termed its 'labour force'. In stripe-backed wrens, unassisted pairs are rarely successful and the number of young reared increases sixfold in groups with two or more helpers (Rabenold, 1984). Similarly, successful reproduction in saddle-backed tamarins is largely restricted to groups of three or more adults (Goldizen, 1987).

The effect of each of these factors is to heighten the *difference* between good and poor quality territories. As the territory quality gradient steepens, 'suitable' (high quality) vacancies become scarcer and competition for them increases. This restricts early dispersal while simultaneously magnifying the advantages of staying at home.

10.4.5 A generalized constraints model of delayed dispersal

A shortage of high quality territories is not the only factor that can tip the balance between dispersing early and remaining at home. Three additional factors influence the fitness pay-off associated with early departure: (i) the mortality risk associated with dispersal itself; (ii) the probability of securing a mate; and (iii) the likelihood of successfully reproducing once 'established' (Table 10.2). Because each can lead to the retention of grown offspring, I have urged that they be combined under the conceptual

Table 10.2 Factors influencing dispersal from natal groups.

Risk of dispersal	Probability of obtaining a suitable territory	Probability of obtaining a mate	Probability of successfully reproducing once established	Resulting behaviour
Low	High	High	High	Disperse and breed independently
High	Low	Low	Low	Remain in natal group as an auxiliary

umbrella of 'constraint' on the option of independent breeding (Emlen, 1982a).

SHORTAGE OF MATES

In many species of cooperative breeders, there is a skew in the adult sex ratio. Among birds, an excess of males has been reported in Tasmanian native hens, several species of bee-eaters, kookaburras, pied kingfishers, red-cockaded woodpeckers, Australian noisy and bell miners, superb and splendid wrens, Galapagos mockingbirds, and pinyon jays. Among mammals, a strong excess of males also typifies wild dogs. The reason for such skewing is poorly understood (see Emlen *et al.*, 1986), but its effect is to increase competition for mates. When the skew is sufficiently great, a significant proportion of individuals will be unable to obtain mates, thus creating a *demographic constraint* on the option of becoming an independent breeder. Mate shortage has been suggested as a causal factor favouring delayed dispersal and helping Tasmanian native hens (Maynard Smith & Ridpath, 1972), splendid and suberb blue fairy wrens (Rowley, 1965; 1981; but see Rowley & Russell, 1990), pied kingfishers (Reyer, 1984), bell miners (Clarke, 1989) and pinyon jays (Marzluff & Balda, 1990).

PROHIBITIVE COSTS OF INDEPENDENT REPRODUCTION

Many cooperative breeders inhabit areas that are subject to large scale, between-year fluctuations in environmental conditions. I have argued elsewhere (Emlen, 1982a) that such fluctuations can create the functional equivalents of breeding openings and closures. As environmental conditions change from year to year, so too does the degree of difficulty associated with successful independent breeding. In benign seasons, abundant food and cover decrease the costs of breeding independently. In harsher seasons, the costs associated with such reproductive ventures increase, eventually reaching prohibitive levels. As conditions deteriorate, breeding options become more constrained. The result may be an inability of inexperienced or unassisted pairs to reproduce successfully during severe years. Because older, established pairs will be more experienced and will have a larger resident workforce of potential helpers available to them, this constraint will hit younger individuals the hardest. The predicted outcome is delayed reproduction and continued retention in the parental group. This process may be important in species such as the pied kingfisher (Reyer, 1980), white-fronted bee-eater (Emlen, 1982a,b), Galapagos mockingbird (Curry, 1988b) and African wild dog (Frame *et al.*, 1979).

10.4.6 Tests of the constraints model

SHORTAGES OF TERRITORY VACANCIES

A large number of studies of cooperatively breeding birds and mammals have concluded that suitable territories are in short supply (see section 10.4.2). One way of documenting this shortage is to contrast the numbers of individuals reaching sexual maturity with the number of territories becoming vacant each year. Woolfenden and Fitzpatrick (1984) have provided such data for the Florida scrub jay. As described previously, this species occupies isolated islands of relict oak scrub habitat. The number of occupied territories (= breeding pairs) remained remarkably constant over a period of 10 years, whereas the number of non-breeding helpers and juveniles fluctuated greatly. All high quality areas were constantly occupied, suggesting that potential breeders were always waiting to fill vacancies.

Similar evidence has recently been documented for a number of other cooperative breeders (Stacey & Koenig, 1990).

COMPETITION OVER TERRITORY VACANCIES

If suitable territorial openings are scarce, we expect intense competition to fill them. Such competition has often been observed. When a breeder dies, same-sex auxiliaries from nearby territories often converge on the 'vacated' territory within hours. There generally follows a period of intense challanges and aggressive contests which may last for several days. Contesting auxiliaries that fail to secure the breeding slot return to the security of their natal territories where they continue the waiting game. These 'power struggles', as they have been called, are vividly described by Koenig (1981), Hannon *et al.* (1985) and Zack and Rabenold (1989).

Several workers have artificially removed breeders and then quantified the ensuing competition for the vacancy. In each study, the intensity of competition (measured as number of contesting groups and duration of the struggle) was related to the 'quality' of the vacancy. In white-browed sparrow weavers (Lewis, 1981; 1982), the number of intruders was greatest in contests for territories with large amounts of preferred feeding cover. Such cover, in turn, was positively correlated with breeding success. Among acorn woodpeckers (Hannon *et al.*, 1985), competition was the greatest for breeding openings on territories containing large and well-stocked acorn granaries. Work in both California and New Mexico has shown that granary size is an important determinant of reproductive success in this species (Koenig & Mumme, 1987; Stacey & Ligon, 1987). Finally, in stripe-backed wrens, contests were most extreme over openings on territories with a large resident 'labour force' of non-breeding helpers (Zack & Rabenold, 1989). Previous analyses of nesting success have shown an over-whelming effect of group size in these wrens, with a minimum of four birds being necessary for successful reproduction (Rabenold, 1984).

PREDICTING THE FREQUENCY OF RETENTION

The constraints model predicts that the frequency of occurrence of retained auxiliaries will vary directly with (i) the magnitude of the constraint on be-coming a breeder; and (ii) the degree of difficulty of successful breeding, once established. Two examples drawn from acorn woodpeckers and white-fronted bee-eaters are shown in Fig. 10.6.

EXPERIMENTAL TESTS

The strongest test of the constraints model would involve experimentally creating breeding vacancies and noting whether helpers leave their natal groups in order to fill them. Two such experiments have recently been performed.

Pruett-Jones and Lewis (1990) captured and removed breeding male superb blue fairy wrens from an area in southeastern Australia. In this species helpers are predominantly male, and both territory and mate (female) shortages have been stressed as probable constraints to indepen-dent breeding (Rowley, 1965; Emlen, 1984). Pruett-Jones and Lewis (1990) report that of 32 male helpers in the study population prior to the manipu-lation, 31 dispersed from their groups and became breeders on the experi-mentally vacated territories. These individuals paired with the resident female breeder and typically initiated nesting within 2 weeks. Helpers did not disperse into vacated territories in the absence of a female (when both members of the original breeding pair had been removed). They did, however, fill such vacancies if the original female was first reintroduced onto the territory. The results confirm that shortages of both territories and females limit breeding opportunities in this species.

Walters *et al.* (in preparation) performed an analogous experiment with red-cockaded woodpeckers. In this species, breeding has been hypothesized to be constrained by a shortage of suitable roosting and nest-

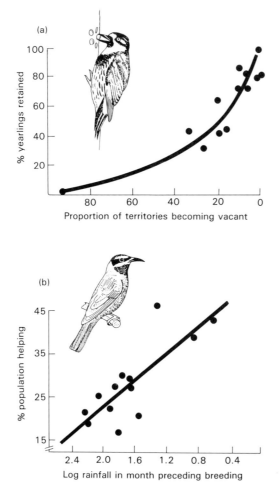

Fig. 10.6 (a) Retention of yearling acorn woodpeckers plotted as a function of the severity of territorial shortages (from Emlen, 1984). (b) Incidence of non-breeding helpers in white-fronted bee-eaters plotted as a function of increasing harshness of environmental conditions (from Emlen, 1982a). Survival of nestlings depends on insect food, which in turn depends on rainfall. Drier conditions are harsher for breeding.

ing cavities (Walters *et al.*, in press). Such cavities are excavated in living pine trees and their construction represents a considerable time and energy investment (each cavity taking 10 months to 3 years to complete; Jackson *et al.*, 1979; J.R. Walters, personal communication). In this experiment, artificial cavities were constructed at 20 previously unoccupied areas. Twenty additional unoccupied areas, matched for other vegetative parameters, served as controls. In the following breeding season, 18 of the 20 experimental sites were occupied and nesting occurred at six; of the controls, all remained vacant. By creating new clusters of cavities the researchers were able to change the dispersal decisions of helpers.

These experiments provide the strongest evidence to date of the link between delayed dispersal and specific constraining factors.

10.5 Why do helpers help?

The retention of grown offspring in their natal groups is the usual precurser to the expression of alloparental care. Retention by itself, however, does not provide an adequate explanation for helping. We must ask why retained auxiliaries help in rearing offspring that are not their own.

Behavioural ecologists generally assume that the behaviours they are studying are adaptive in the sense that the fitness consequences of engaging in the activity are positive (the benefits exceed the costs). They then devise and test different hypotheses of the *ways* in which the behaviour might enhance the fitness of the performer. In contrast, Jamieson and Craig (Jamieson, 1986; 1989; Jamieson & Craig, 1987) argue that helping in many instances is not adaptive. Instead they suggest that helping might be pleiotropically linked to parental care itself. Because selection for decreased helping would then result in decreased parental care, helping would not be eliminated and would be 'carried along' by positive selection for the latter (see also Halliday & Arnold, 1987). Emlen *et al.* (1991) have argued against this genetic corollary mechanism, on the basis that selection would favour any genetic variant that broke this linkage, in which case helping could be subject to selection independent of parental care. If helping carried a fitness cost, it could then be reduced or eliminated. If it conferred a benefit, it could be modified in ways that enhanced the magnitude of such benefit.

Numerous reports exist of organisms infrequently (and often mistakenly) provisioning young that are not their own offspring (reviewed in Brown, 1987). For many such 'helpers' alloparenting is undoubtedly non-adaptive. But for species in which helping is a well-developed and commonly expressed behaviour, the evidence demonstrates that helpers generally do enhance their inclusive fitness through their actions. Emlen *et al.* (1991) and Koenig and Mumme (1990) have recently summarized the data on this question. Both conclude that while non-adaptive hypotheses should always be included among the possible explanations of helping behaviour, there are very few documented cases of persistent non-adaptive helping among either avian or mammalian cooperative breeders.

How then might helping benefit the helper? Nine different mechanisms have been proposed which can be arranged into four basic categories. Through its actions, a helper may: (i) enhance its probability of survival; (ii) enhance its likelihood of becoming a breeder in the future; (iii) increase its fecundity when it does become a breeder; and (iv) increase the production of non-descendent kin.

It is useful to place these into the conceptual framework of Brown (1980) who partitioned fitness benefits along two axes: direct–indirect and present–future. Direct fitness gains are those accruing to the helper or its

Table 10.4 Adaptive hypotheses to account for the current expression of helping (modified from Emlen & Wrege, 1989).

Hypothesis	Hypothesized benefit to the helper	Hypothesized fitness effect
1	Enhanced survivorship (group size effects)	Helping leads to increased production of young, thereby augmenting group size; larger group size enhances the helper's survivorship through improved group vigilance, antipredator behaviour or intergroup competitive abilities
2	Enhanced survivorship (resource effects)	Helping is a 'payment' to breeders to be allowed to retain access to the physical and social resources of the natal group
3	Enhanced future probability of breeding	(For species in which breeding is constrained by a shortage of territories.) Through helping, the size of the social group is enlarged; the group then expands the size of its territory, which enables the helper to 'bud off' a portion of the natal area as its own breeding territory
4	Enhanced future probability of breeding	(For species in which breeding is constrained by a shortage of territories.) Helping leads to the formation of social coalitions. Later such coalitions may disperse and compete for vacant breeding territories as a group. If groups out-compete solitary individuals, then helping can enhance the donor's later probability of gaining a breeding territory
5	Enhanced future probability of breeding	(For species in which breeding is constrained by a shortage of mates.) Through its actions, the helper demonstrates its parental abilities, thus enhancing its probability of being chosen as a mate by the original breeding female in a subsequent year
6	Increased reproductive success	Helping provides experience at parenting which translates into higher reproductive success when the helper becomes a breeder
7	Increased reproductive success	Helping increases the likelihood that the helper will gain the services of the original recipients as its own helpers in the future; this might occur if the young gain more by remaining with their natal group (e.g. hypotheses 1–5 above) than by dispersing
8	Increased production of non-descendent kin	Helping increases the survival of the breeders (who are assumed to be kin), thereby increasing the probability of their being alive to reproduce later in the same or in future seasons
9	Increased production of non-descendent kin	The actions of the helper increase the production of young (assumed to be kin) who will effectively increase the helper's inclusive fitness through their own reproductive efforts

own offspring; indirect benefits accrue through the increased production of the helper's non-descendent kin. Present or immediate benefits are those achieved during the current breeding; future benefits are those where the pay-off is delayed. Of the four categories listed above, the first three provide future direct gains to the helper; in the fourth, the benefit is indirect and may have immediate as well as future components.

A listing of all currently proposed mechanisms of helper benefit is presented in Table 10.4. The specific hypotheses have recently been reviewed by Emlen and Wrege (1989), from which much of the following discussion is drawn.

10.5.1 Enhanced survivorship?

According to hypothesis 1, the chance of surviving to become a future breeder is increased through helping. The argument goes as follows: helping increases the number of young fledged, which augments the size of the helper's group. If vigilance or other antipredator behaviour improves with increasing group size, then survivorship for all group members, including the helper, will be enhanced following a successful breeding attempt (Brown, 1987; Stacey & Ligon, 1987). Two critical predictions of this hypothesis are that helping augments group size and that individual survival is higher in large, than in small, groups. Data from several species of avian and mammalian cooperative breeders support these predictions.

Gaston (1978b) also proposed that helping might lead to higher survivorship, but through a somewhat different mechanism (hypothesis 2). He suggested that helping was a 'payment' made to the breeders to be allowed to remain on the natal territory. Individuals that did not help would be expelled from the home area. This hypothesis is based on the assumption that it is safer to remain in the familiar area of home rather than to risk early dispersal. For helping to be an evolutionarily stable strategy under either hypothesis 1 or 2, helpers must somehow benefit more than non-helpers. Gaston's hypothesis provides such a mechanism.

The payment hypothesis predicts that non-breeding individuals that do not help will be expelled from their natal territories. Although data are available from only a small number of species, no compelling evidence of differential treatment of helpers versus non-helpers, or of expulsion of the latter, has been reported (for example, Hegner & Emlen, 1987). More data are required from species in which non-helping auxiliaries are common before this hypothesis can be properly tested.

10.5.2 Increased probability of future breeding?

In this category of hypotheses, helping leads to an increased likelihood of becoming a breeder in the future. The mechanisms by which this occurs depend upon the constraints operating on the species in question. Three specific hypotheses have been proposed. The first two are especially relevant to species in which breeding is constrained by a shortage of suitable territories; the third is relevant to species in which the constraint is a shortage of mates.

TERRITORIAL INHERITANCE

Consider a species in which breeding vacancies on high quality territories are in short supply. One route to becoming a breeder is to remain at home and wait to inherit the natal territory. This is a widespread practice among cooperative breeders, having been described in mongooses (Rood, 1990), jackals (Moehlman, 1979; 1983), red foxes (Macdonald, 1979; von Schantz, 1981), wolves (Mech, 1970) and lions (Bertram, 1975) as well as hoatzins (Strahl & Schmitz, 1990), woodpeckers (Koenig & Mumme, 1987), woodhoopoes (Ligon, 1981), babblers (Zahavi, 1974; Gaston, 1978b) and jays (Brown, 1974; Woolfenden & Fitzpatrick, 1978; 1984).

According to hypothesis 3, helping enhances the reproductive success of the breeders, leading to an increase in the size of the family unit. Larger groups outcompete smaller ones at territorial boundaries, and the size of the natal territory increases. This, in turn, increases the likelihood that the former helper will be able to 'bud off' a portion of the enlarged parental territory for itself. Among Florida scrub jays, in which this process has been studied intensively, 48% of males become breeders through acquiring part or all of the parental territory itself (Woolfenden & Fitzpatrick, 1986).

When several helpers are members of the same cooperative unit, a dominance hierarchy normally forms with the older, most dominant, member being first to inherit the territory. Consequently, the magnitude of this advantage to helping will depend upon the severity of the territorial shortage, the average residency time required to gain inheritance, and the number of more dominant competitors 'in line' to inherit (Gaston, 1978a; Ligon, 1981; Wiley & Rabenold, 1984).

The inheritance hypothesis (as an explanation for helping) assumes that the act of providing alloparental care is critical to the future acquisition of the natal area. This assumption has not yet been tested, so it remains unknown whether the benefit of territory inheritance is linked to helping *per se,* or is a passive effect of delayed dispersal alone. Ironically, although a great deal of evidence is consistent with the hypothesis, and a large number of species exhibit the phenomenon of territorial budding and inheritance, virtually no data exist on the *relative* success of helping versus non-helping auxilliaries or of major versus minor contributing helpers in obtaining breeding slots on the natal area.

COALITION FORMATION

When competition for reproductive vacancies is intense, auxiliaries may gain by dispersing together. Among many cooperative breeders, non-

reproductives typically use the natal territory as a base from which to pros-
pect and challenge for vacancies that arise nearby. In a number of species,
auxiliaries from the same natal unit form coalitions that challenge for such
vacancies as a group.

Among lions, most female young remain and reproduce in their natal
groups and female lineages endure for many generations (Schaller, 1972;
Bertram, 1975; Packer *et al.,* 1988). Males, however, disperse from the natal
pride when approximately three years old, and wander nomadically until
they are able to take over another pride by challenging and defeating the
resident male(s). Lionesses in a pride frequently come into oestrus simultaneously, with the effect that several litters of cubs are born synchronously. The females then communally nurse the cubs. When the young
males reach dispersal age, they emigrate together, forming small coalitions
of from one to seven individuals. Data collected in the Serengeti over a
period of 12 years have shown that both the probability of pride take-over,
and the tenure time of control of a pride, are proportional to male coalition
size. Figure 10.7 graphs the combined effect of these two benefits of group
dispersal on male lifetime reproductive success.

Similar coalitions form in many species of cooperatively breeding
birds. Unisexual groups of differently aged green woodhoopoes explore,
locate and fight for breeding vacancies together (Ligon & Ligon, 1978a).
Typically, the older members of these coalitions helped to raise the
younger ones, and the Ligons (1978b) hypothesized that one advantage of

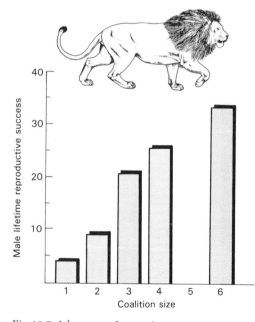

Fig. 10.7 Advantage of group dispersal in lions. Expected lifetime reproductive success of
males in different sized coalitions. For discussion see text (from Packer *et al.,* 1988).

helping was that it cemented a tight social bond between helper and helped. When the older, more dominant, auxiliary emigrates and attempts to secure a territorial position, it 'takes along' several younger, more sub-ordinate, members that it helped to rear in previous breeding seasons. As with the lions, these coalitions are better able to secure breeding openings than are solitary individuals.

Analagous cases in which former helpers and the young they helped to rear disperse together have been reported in dwarf mongooses (Rood, 1990), Tasmanian native hens (Ridpath, 1972), acorn woodpeckers (Koenig, 1981; Hannon et al., 1985), Arabian and jungle babblers (Zahavi, 1974; Gaston, 1978a), yellow-billed shrikes (Grimes, 1980) and white-browed sparrow weavers (Lewis, 1982).

The coalition formation hypothesis makes two predictions: first that larger coalitions have a competitive advantage over smaller ones; and second that the act of helping is important to the formation of such coali-tions. Size of the competing unit has been found to be a significant pre-dictor of the outcome of territoral contests in acorn woodpeckers as well as in woodhoopoes and lions. Few data are available regarding the second prediction. To the degree that social bonds formed during helping are important in the development of dispersal coalitions, we can conclude that helping in one season can increase a helper's breeding prospects in future years.

FUTURE MATE ACQUISITION

In many species of cooperative breeders, there is an imbalance in the adult sex ratio (see section 10.4.5.). The resulting shortage of potential mates creates a demographic constraint on members of the limited sex in be-coming established as breeders. Under such situations, Reyer (1980) prop-osed that helping could lead to the formation of social bonds that enhanced the likelihood of the helper pairing with the opposite sex breeder in the future.

The mate acquisition hypothesis was based on studies of pied kingfishers (Reyer, 1980; 1984; 1986). Two types of helpers occur in this species: primary helpers, which are grown sons that remain with and help their parent(s) in the subsequent breeding season, and secondary helpers, which are unrelated males that join breeding groups after young have hatched. These secondary helpers are initially repelled by the breeders but are gradually accepted if food conditions are poor. When a breeding male dies, the female typically pairs with her former secondary helper. Fully 48% of secondary males obtained mates the following year in this manner. In contrast, incest was avoided and mothers never paired with their prim-ary helper sons.

Additional support for this hypothesis comes from bell miners. These Australian honeyeaters live in groups of from three to 20 individuals among which from one to four pairs breed simultaneously. Virtually all birds provision multiple nests (Clarke, 1984; 1989). Clarke (1989, p. 303) states that 'the shortage of breeding vacancies appears more acute for breeding-age males than females'. In each case when a breeder male disappeared ($n = 5$), the widowed female preferentially paired with the unmated male helper that had contributed the most aid at her previous nesting attempt.

10.5.3 Improved reproductive success as a breeder?

The third category of hypotheses is based on the assumption that experience gained by being a helper translates into improved reproductive success when the donor later becomes a breeder itself.

GAINING BREEDING EXPERIENCE

In this hypothesis, the crucial experience gained from helping consists of learning and/or practising the skills associated with caring for and rearing young. It is a well-known phenomenon among many higher vertebrates that inexperienced breeders are often less successful parents than experienced ones (for example, Clutton-Brock, 1988). If breeding requires the development of specialized skills, prior experience may be a *prerequisite* for successful reproduction (the 'skills' hypothesis of Brown, 1987). Alternatively, helping experience might simply improve future breeding success.

This hypothesis predicts that reproductive success of first-time breeders will increase as a result of prior helping activities. Individuals with helping experience will often be older than ones without, and may have inherited breeding vacancies with 'built in' helpers. Consequently, tests should control for the effects of age and number of helpers as well as territory quality.

Very few data are available. In Florida scrub jays, prior helping experience improved reproductive performance among males, but possible confounding variables were not fully controlled (Woolfenden & Fitzpatrick, 1984). Similiar trends of higher nesting success among novice breeders with past helping experience have recently been reported for female splendid fairy wrens (Rowley & Russell, 1990) and white-winged choughs (R.G. Heinsohn, personal communication). In contrast, no significant effect of prior helping was found in acorn woodpeckers or white-fronted bee-eaters (Koenig & Mumme, 1987; Emlen & Wrege, 1989). Despite the paucity of current data, I expect this benefit will emerge as being of at least minor importance in a large number of species.

GAINING FUTURE HELPERS

Through its helping, an auxiliary might also gain 'social experience' that later translates into a heightened probability of gaining helpers itself when it becomes a breeder. This would be beneficial for any species in which the presence of helpers enhanced breeder fitness. According to this hypothesis, 'young . . . in the nest can be viewed as an essential resource that can be utilized by the current helpers for their own personal gain' (Ligon, 1981, p. 240; see also Carlisle & Zahavi, 1986). Ligon's argument is that social bonds formed during helping may increase the helper's probability of recruiting the former recipient as *its* helper in the future. Such recruitment has been termed generational mutualism by Brown (1983) and delayed reciprocity by Wiley and Rabenold (1984).

Cases of helpers recruiting young they previously helped to rear have been reported in Tasmanian native hens (Ridpath, 1972), acorn wood-peckers (Koenig, 1981), green woodhoopoes (Ligon & Ligon, 1978a,b), hel-mut shrikes (C. Vernon, cited in Brown, 1978), stripe-backed wrens (Rabenold, 1985), jungle and Arabian babblers (Gaston, 1978a; A. Zahavi, personal communication), Galapagos mockingbirds (Curry, in prepara-tion) and bell miners (Clarke, 1989). Helping patterns of this sort would be expected as a natural outcome of the demographic combination of high longevity and delayed dispersal. Consequently, a critical prediction of this hypothesis is that recruitment of the next generation of helpers be depen-dent upon prior helping associations. No such data currently are available. Ideal candidates for study would be plural breeders in which helpers have a choice of recipients. One could then test whether grown young *preferen-tially* aid individuals that previously had helped to rear them.

10.5.4 *Increased production of non-descendent kin*

The remaining category of hypotheses involves indirect benefits of helping. Helpers increase their own inclusive fitness when, through their actions, they increase the production of non-descendent kin (see section 1.3). This can occur in two ways. First, helpers might increase the likelihood that breeders (assumed to be relatives) survive to reproduce again in the future. Second, helpers might increase the success of the current breeding effort. The extra young attributable to the activities of the helper may themselves become helpers for kin (generally the original breeders) in the next year. In this way, helping to rear additional young may produce a future, as well as a present, indirect benefit. Methods of calculating both components of indirect fitness are discussed by Mumme *et al.* (1989).

These hypotheses make two predictions: (i) that the activities of helpers significantly increase the fitness of breeders; and (ii) that helpers, on aver-age, are closely related to such breeders. Failure to confirm *either* predic-tion would lead to the rejection of the hypotheses.

Evidence that helpers generally do enhance breeder fitness was pre-sented in section 10.3. And data on average relatedness indicate that helpers typically are younger kin of the breeders they aid. Published helper–breeder coefficients of relatedness range from 0.27 to 0.40. Indirect benefits of helping are thus expected to be commonplace.

White-fronted bee-eaters provide strong evidence of indirect selection for helping. Helpers have a major effect on reproductive success in this species and helpers are closely related to the young they help to rear (aver-age $r = 0.33$; Emlen & Wrege, 1988; 1989). The result is a large indirect fitness gain from helping. Bee-eater helpers also show nepotistic favouritism in their decisions of *whom* to aid. As a consequence of being plural breeders and living in extended family groups, potential helpers often have a large assortment of breeders of varying degrees of relatedness available as potential recipients of their aid. Emlen and Wrege (1988) found that helpers preferentially chose to aid their closest genetic relatives.

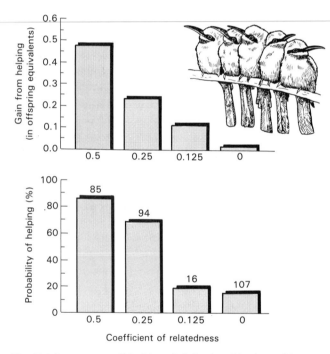

Fig. 10.8 Importance of kinship to helping in white-fronted bee-eaters. (a) The indirect fitness gain realized by a helper plotted as a function of the helper's relatedness to the nestlings it aids. This gain is measured in offspring equivalents and is calculated as the product of two terms: the average number of additional offspring successfully fledged as a result of a helper's activities, and the coefficient of relatedness between the nestlings and helper (sibling = 0.5, half-sibling = 0.25, etc.). (b) The probability that a potential helper (a non-breeding bee-eater with a recipient nest available within its group) becomes an actual helper plotted as a function of its coefficient of relatedness to the recipient nestlings. Numbers above histograms are sample sizes of potential helpers in each kin category.

Kinship was also a predictor of whether non-breeders would help *at all*. Almost half of the individuals that could have become helpers (those having active breeders in their family group) did not do so. Many of these non-helpers were unrelated mates that had paired into their partner's group and had no genetic link to the available recipients. Bee-eaters were most likely to become helpers when the kin benefits realized by helping were large (Fig. 10.8). Similar favouritism of helpers for closely related recipients has been reported among Galapagos mockingbirds (Curry, 1988a), bell miners (Clarke, 1984; 1989) and brown hyaenas (Owens & Owens, 1984).

To summarize, there are numerous ways in which helpers have been shown to benefit through their helping actions. In some the gain is direct, in others it is indirect. The effects of all benefits are additive. Helping thus no longer poses the major evolutionary paradox that it once appeared to do.

10.6 Roles of mutualism, reciprocity and kin selection

Much of the interest in cooperative breeding in the 1960s and 1970s stemmed from reports of individuals seemingly forgoing personal reproduction in order to help rear another's offspring. Such behaviour was assumed to be altruistic, and explanations were sought to account for its evolutionary origin and maintenance. Two major theories, reciprocity and kin selection, offered possible explanations for altruism, and it was inevitable that cooperative breeders would become a focal point for testing their importance.

We now recognize that individuals generally become helpers only when their options of personal reproduction are curtailed, and that helpers may accrue fitness benefits in several ways through their actions. It is thus useful to reassess the issue of altruism, as well as the applicability of reciprocity and kinship theory, in light of recent empirical findings.

10.6.1 Mutualism

Hamilton (1964) categorized social interactions into four types on the basis of the resulting consequences to the *direct* component of fitness of both donor (helper) and recipient (breeder): 'When a recipient benefits but at a cost to the donor, the interaction is considered altruistic. When the payoffs to the two parties are reversed, the behavior is selfish. And when both participants benefit, the interaction is termed mutualistic.' In the vast majority of cooperative breeders, helping should be considered mutualistic rather than altruistic, because both helper and breeder benefit directly from the actions of the former (see sections 10.3. and 10.5). Many of these benefits are not realized until future seasons, however, raising the question of whether they are dependent upon reciprocal interactions.

10.6.2 *Reciprocal altruism*

The idea of reciprocity as a mechanism for maintaining altruism was first developed by Trivers (1971). The basic idea was that an individual that engaged in a costly act today would be paid back for its actions in the future. To quote an early definition: '(1) one individual aids another (2) in anticipation that the recipient will return the favor, (3) benefiting the actor at some time in the future' (Waltz, 1981).

The idea of reciprocal altruism quickly generated considerable interest and scepticism: interest because such reciprocal interactions might be commonplace among highly social organisms; scepticism because the system seemed evolutionarily unstable and prone to cheating. If the original recipients of aid died or dispersed before repayment, or if they reneged on repaying, reciprocity would not persist (for example, West Eberhard, 1975; Dawkins, 1976). The theoretical plausibility of reciprocal altruism was strengthened when Axelrod and Hamilton (1981) showed, through game theory analyses, that reciprocity could be resistant to cheating provided there was a sufficiently high probability of repeated encounters between the same participants. And cheating ceased to be an insurmountable problem if the species possessed sufficient individual recognition and memory capabilities. Individuals then were predicted to maintain a knowledge of the behavioural attributes and past behavioural actions of other members of their group and to dispense their own behaviours accordingly.

Early workers (myself included, Emlen, 1981; 1984) felt that cooperatively breeding birds and mammals were logical candidates in which to expect reciprocal altruism. Populations are divided into small groups whose membership is highly stable; individuals are long lived; and individual recognition capabilities appear to be universal. Individuals thus have repeated opportunities for interaction with the same small group of known acquaintances.

Many of the direct benefits of helping listed in Table 10.4 are delayed benefits, occurring in future seasons. To the degree that these future benefits are provided by recipients of the original aid, should helping be considered as reciprocal altruism?

Probably not. As several authors have pointed out, early definitions of reciprocal altruism were incomplete (Brown, 1983; Connor, 1986; Koenig, 1987). For reciprocity to be important to the maintenance of any behaviour, there must be a fitness *cost* associated with performing the behaviour, both for the original actor and for the future reciprocator. In the absence of such costs, each action (the original helping and the later repayment) becomes mutualistic in itself and would occur in the absence of any reciprocal exchange. Consider a pied kingfisher secondary helper (see section 10.5.2). By helping, it increases its future probability of pairing with the current recipient of its help (the breeding female). This female, however, incurs no cost by later pairing with the male. One might even argue that the male has demonstrated his 'parental' abilities by helping in the previous season. By pairing with her former helper, the female is acting selfishly in her own best interest. Such cases are distinguished from true reciprocal altruism by most authors and have been called by such names as byproduct mutualism (Brown, 1983), return benefit altruism (Trivers, 1985) and pseudoreciprocity (Connor, 1986).

Koenig (1987, p. 75), in an excellent review, defined reciprocal altruism as requiring five elements: '(1) one individual aids another, (2) at some fitness cost to itself, (3) in anticipation that the recipient will choose to return the favor, (4) again at some fitness cost to the actor, and (5) benefiting the actor at some time in the future'.

Do helping interactions in cooperative breeders fit these more restrictive criteria? In a number of species, an individual that helps in one year often gains the services of the young it helped to rear as its own helpers in a later year (section 10.5.3). Ligon (1983) carefully reviewed such interactions and interpreted them as examples of reciprocity. Evidence is lacking, however, that the reciprocators incur a cost. In the best studied example (green woodhoopoes: Ligon & Ligon, 1978b), grown young that later 'reciprocate' as helpers may benefit directly through the formation of dispersal coalitions of their own, by inheriting the natal territory, or even indirectly by increasing the reproductive success of relatives.

Reciprocal exchanges of helping have been reported to occur within a single season in plural breeding species such as white-fronted bee-eaters, Galapagos mockingbirds, bell miners, and grey-breasted jays. In the first two cases, failed breeders commonly become 'redirected' helpers at other ongoing nests within their social group (Emlen, 1981; Curry, in preparation). In the latter, breeders simultanously help at nests other than their own. Again, although reciprocal exchanges do occur, no strong evidence has been presented that reciprocating helpers either incur a cost through their helping or that prior receipt of aid is a significant predictor of the choice of recipient of later repayment (Rabenold, 1985; Clarke, 1989; Curry, in preparation). I previously interpreted redirected exchanges of helping between adult bee-eaters as an example of reciprocity (Emlen, 1981). I now realize that through redirected helping such individuals recoup almost as much in indirect benefits as they would have gained in direct fitness had their own breeding attempt succeeded (Emlen & Wrege, 1989).

In short, cooperative breeders do *not* provide any unequivocal cases of reciprocal altruism. Many benefits of helping are deferred, and reciprocal exchanges of helping occur, but as yet there is no solid evidence that the reciprocator incurs a cost and thus is 'repaying' the orginal donor. Instead, each participant at each level of the exchange, appears to be acting in its own best interest. The exchanges described by Emlen (1981), Ligon (1983) and others are better considered as pseudoreciprocity (*sensu* Connor, 1986; see also Koenig, 1987).

Very few cases of reciprocal altruism have been documented for *any* animal society, despite considerable effort expended in looking (Taylor & McGuire, 1987). This raises the larger question of why such reciprocal interactions are so rare in species other than our own.

10.6.3 *Importance of kinship*

Considerable controversy continues to exist among ornithologists and mammalogists over the importance of indirect selection to both the origin and the current maintenance of helping behaviours (for reviews see Woolfenden & Fitzpatrick, 1984; Brown, 1987; Koenig & Mumme, 1987). In my opinion, the controversy has been unproductive. To oversimplify the extreme positions: in the vast majority of cooperative breeders, groups form by retention of grown young and helpers are therefore close genetic kin (typically offspring) of the breeders they aid. Whenever helping enhances the fitness of such recipient breeders (section 10.3), the helpers will gain an indirect benefit from their actions. One might (mistakenly) take the widespread occurrence of indirect benefits as sufficient evidence to conclude that kin selection is both necessary and sufficient to explain the evolution of helping behaviours. However, recent research is also documenting sizeable direct fitness benefits to helpers in an ever-growing number of species (section 10.5). At the other extreme, one might (equally mistakenly) interpret this evidence to mean that kin selection is unimportant to understanding helping behaviour.

To take this debate out of the realm of semantic argument and into the realm of quantitative biology, Vehrencamp (1979) derived an algebraic index which can be used to evaluate both the relative importance of direct and indirect benefits in the maintenance of a behavioural strategy, and whether the behaviour is 'altruistic' (involving a loss of direct fitness relative to an alternative strategy):

$$I_k = \frac{(W_{RA}-W_R)r_{ARy}}{(W_{AR}-W_A)r_{Ay}+(W_{RA}-W_R)r_{ARy}}$$

where $(W_{RA}-W_R)$ is the change in lifetime reproductive success of the recipient, R, when it is aided by a donor A; $(W_{AR}-W_A)$ is the change in lifetime reproductive success of the donor A, when it provides aid to R; r_{ARy} is the relatedness of A to R's young; and r_{Ay} is the relatedness of A to its own young.

The index is applicable only to situations in which the net change in inclusive fitness of A (i.e. the denominator) is positive. When I_k is greater than one, there is a net cost to A's personal reproduction and pure indirect selection is acting. When I_k is less than zero, then A is manipulating R at a net cost to R's personal reproduction. When I_k is between zero and one, both direct and indirect benefits are occurring, and the value of the kin index gives the proportion of the inclusive fitness gain that is due to the indirect component.

Estimates of I_k have been made for a number of species of cooperative breeders, comparing the alternative strategies of 'remaining at home and helping' versus 'dispersing and attempting to breed'. Most reported values lie between zero and one, implying that both direct and indirect benefits accrue to helpers. These include values of 0.24–0.57 for acorn woodpeckers in California (Koenig & Mumme, 1987), 0.24–0.48 for acorn woodpeckers in New Mexico (calculated from data in Stacey & Ligon, 1987), 0.35–0.57 for splendid fairy wrens in Australia (Rowley, 1981; Russell & Rowley, 1988), 0.55 for Florida scrub jays (Vehrencamp, 1979; but see Woolfenden & Fitzpatrick, 1984) and 0.51 for female lionesses (Vehrencamp, 1979).

The lowest value reported is 0.13, from secondary helpers among pied kingfishers. Recall that these secondary helpers are males that disperse from their natal family, then join and aid unrelated breeders (Reyer, 1980; 1984). Indirect benefits from helping are small because of the low relatedness between donor and recipient. However, the future direct gains from helping are large. This is because secondary helpers have a high probability of pairing with the original female breeder, thereby becoming breeders themselves in the following season.

Two studies have yielded I_k values greater than one. Interestingly, the first also comes from pied kingfishers. Primary helpers in this species are grown sons that remain with and help their parents in a succeeding breeding. Both types of helper increase fledgling production, but since primary helpers are closely related to the young being reared, they gain a large indirect benefit. However, by helping, they also *decrease* their future probability of becoming breeders themselves (Reyer, 1984); hence the I_k value > 1.0.

The second case comes from white-fronted bee-eaters, for which $I_k = 2.17$ (Emlen & Wrege, 1989). In this species no significant future direct benefit from helping could be measured. Instead, auxiliaries forfeited a current direct fitness gain that could have been realized by attempting to breed. As a consequence they incurred a loss in direct fitness by acting as helpers. However, helpers preferentially aided close relatives and their influence on nesting success was large. The resulting gain in indirect fitness realized through helping more than compensated for the loss in direct fitness.

White-fronted bee-eaters and primary helpers in pied kingfishers provide two cases where helping can be classified as altruistic in the Hamiltonian sense (Hamilton, 1964). In both instances, helping would not be maintained in the population except for the indirect fitness gained through the increased production of close kin.

A second line of evidence that indirect selection has been important in shaping helping behaviours comes from studies showing adaptive fine tuning or modification of helping in ways that increase the magnitude of the indirect benefit gained. In both white-fronted bee-eaters and Galapagos mockingbirds, auxiliaries exhibit discriminative nepotism; (i) being more likely to become helpers when the recipients are kin; (ii) suppressing helping where no kin recipients are available; and (iii) choosing to aid the most closely related recipients when multiple choices are present (Curry, 1988a,b; Emlen & Wrege, 1988). Among bell miners and brown hyaenas, helpers have been reported to modify their rates of provisioning, bringing more food to young to whom they are more closely related (Clarke, 1984; 1989; Owens & Owens, 1984). Finally, the two types of helper in pied kingfishers show predictable differences in their alloparenting behaviours (Reyer, 1986). Secondary helpers (which benefit directly by pairing with the breeding female) expended less energy in provisioning; they brought lower quality food items (a different species of fish), and fewer of them, than did the primary helpers. Further, secondary helpers fed such items to the breeding female (rather than the nestlings) more often than did primary helpers.

In summary, indirect selection appears to be a contributory factor in the current maintenance of helping in a large number of species. An indirect benefit will be present *whenever* helpers are genetic relatives of the breeders they aid and helping contributes to the fitness of the recipients. We have seen that most cooperatively breeding units are family groups formed through the retention of grown offspring with their parents (section 10.4). Further, helpers typically do enhance the fitness of the breeders they aid (section 10.3). Consequently, some level of indirect selection is expected to be nearly ubiquitous. However, the relative importance of indirect versus direct gains varies considerably across species. In two species, indirect selection appears to be essential for the expression of helping; in most species, its importance is more modest. To date, no cases have been reported where a kin benefit is totally absent ($I_k = 0$), but they may be forthcoming as more data are collected on species in which breeders seemingly gain no benefits from helpers, or if cooperative species are found in which groups form through the joining of unrelated individuals.

10.7 Concluding comment

Cooperative breeding systems have provided critical information for the development and testing of basic social evolution theory. Initially, researchers were drawn to them as a means of studying the evolution of altruism. More recently, workers have come to realize that most are better described as mutualistic. A basic conceptual framework for understanding the adaptiveness of helping is currently in place. It emphasizes ecological and demographic constraints on early dispersal and independent breeding. Once grown offspring remain at home with their parents, the task of explaining the actual expression of helping becomes less formidable. Retained auxiliaries generally benefit in several additive ways by engaging in alloparental activities.

Cooperative breeders continue to provide a rich supply of unanswered research questions. In closing, I briefly mention three areas of future work that I feel are especially promising.

First, the generality of the conceptual 'solution' to helping described here must be tested in other taxa. Virtually all of the quantitative data currently available come from avian studies. There is a need for more work from cooperatively breeding mammals as well as from various species of fish that also exhibit alloparental care (for example, Taborsky, 1984).

Second, the diversity of social systems found among cooperative breeders demands explanation. What factors determine whether a single male and female are the sole breeders in a cooperative unit, or whether several pairs breed simultaneously? And why do some species pair monogamously while in others reproductive access to the breeding female(s) is shared by most or all of the males present?

Finally, cooperative breeding systems are ideally suited for studies of the interplay of competition with cooperation. Conflicts of interest are expected between individuals in any social group, and conflicts within kin groups are especially valuable for testing social theory. How common is it for dominant individuals to coerce others into helping them, or into providing more help than would be provided otherwise? How often, and under what conditions, are helpers actually the genetic parents of some of the offspring they help to raise? Molecular testing using DNA fingerprinting techniques will answer questions of the incidence of extra-pair copulations by helper males and of brood parasitism by helper females. I anticipate that we will be in for some fascinating surprises.

Appendix 10.1 Common and scientific names of the species referred to in the text.

Family	Common name	Genus and species
Mammals		
Callitrichidae	Saddle-backed tamarin	*Saguinus oedipus*
Canidae	Coyote	*Canis latrans*
	Silver-backed jackel	*Canis mesomelas*
	Golden jackel	*Canis aureus*
	African wild dog	*Lycaon pictus*
Felidae	African lion	*Panthera leo*
Viverridae	Dwarf mongoose	*Helogale parvula*
	Banded mongoose	*Mungus mungo*
Hyaenidae	Brown hyaena	*Hyaena brunnea*
	Spotted hyaena	*Crocuta crocuta*
Bathyergidae	Naked mole-rat	*Heterocephalus glaber*
Birds		
Accipitridae	Harris' hawk	*Parabuteo unicinctus*
Opisthocomidae	Hoatzin	*Opisthocomus hoazin*
Rallidae	Pukeko	*Porphyrio porphyrio*
	Tasmanian native hen	*Tribonyx mortieri*
	Purple gallinule	*Gallinula martinica*
	Moorhen	*Gallinula chloropus*
Cuculidae	Groove-billed ani	*Crotophaga sulcirostris*
Alcedinidae	Pied kingfisher	*Ceryle rudis*
	Kookaburra	*Dacelo gigas*
Meropidae	White-fronted bee-eater	*Merops bullockoides*
	Red-throated bee-eater	*Merops bullocki*
	European bee-eater	*Merops apiaster*
Phoeniculidae	Green woodhoopoe	*Phoeniculus purpureus*
Picidae	Red-cockaded woodpecker	*Picoides borealis*
	Acorn woodpecker	*Melanerpes formicivorus*
Laniidae	Grey-backed fiscal shrike	*Lanius excubitorius*
	Yellow-billed shrike	*Corvinella corvina*
Troglodytidae	Stripe-backed wren	*Campylorhynchus nuchalis*
	Bicoloured wren	*Campylorhynchus griseus*
	Banded-backed wren	*Campylorhynchus zonatus*
Mimidae	Galapagos mockingbird	*Nesominus parvulus*

Family	Common name	Genus and species
Muscicapidae	Common babbler	*Turdoides caudatus*
	Arabian babbler	*Turdoides squamiceps*
	Jungle babbler	*Turdoides striatus*
	Grey-crowned babbler	*Pomatostomus temporalis*
	Splendid fairy wren	*Malurus splendens*
	Superb blue fairy wren	*Malurus cyaneus*
Sittidae	Pygmy nuthatch	*Sitta pygmaea*
Meliphagidae	Bell miner	*Manorina melanophrys*
Ploceidae	White-browed sparrow weaver	*Plocepasser mahali*
Grallinidae	White-winged chough	*Corcorax melanorhamphus*
Corvidae	Florida scrub jay	*Aphelocoma coerulescens*
	Grey-breasted jay	*Aphelocoma ultramarina*
	Brown jay	*Psilorhinus morio*
	Beechey jay	*Cyanocorax beecheii*
	Pinyon jay	*Gymnorhinus cyanocephalus*
Aegithelidae	Bush tit	*Psaltriparus minimus*

11: Cooperation and conflict in social insects

Jon Seger

11.1 Introduction

Social insects are diverse, abundant and ecologically important; in some terrestrial habitats they may constitute 10–40% of the animal biomass (for example, Hölldobler & Wilson, 1990). The hallmark of sociality and the key to its great success is the cooperative division of labour that occurs in all social species from the most primitive to the most advanced. This division of labour takes many forms, but is always in part *reproductive*; some individuals reproduce, while some (usually most) do not. A large ant or termite colony may contain 10^5–10^7 workers, and it may occupy a complex dwelling more than 10^9 times as massive as any of the individuals that built it (Fig. 11.1); yet the sexual reproductives produced by the colony are likely to be the offspring of just one or a few individuals.

Human beings have long been fascinated by the spectacular organization and seemingly purposeful intelligence of social insects. The honeybee, in particular, is a universal icon for industriousness. More generally, people tend to see idealizations of their own societies reflected in the societies of insects, as when speakers of English refer to the principal egg layer as the 'queen'. Social insects always appear prominently in discussions of the 'superorganism' concept, a venerable and recurring analogy that equates individuals within a society to the cells of a single organism (see Wilson & Sober, 1989). But the current scientific vision of insect sociality is neither royalist nor superorganic. Instead, it portrays an almost melodramatically volatile politics, in which individuals attempt to gain advantage by taking part in profoundly cooperative relationships, but none the less find themselves constantly tempted to cheat.

This chapter is intended to serve as a field guide to the logic of the modern view of insect social life, with its emphasis on the tension between cooperation and conflict. It does not give exhaustive or authoritative reviews of the subjects it touches on, many of which are fast-moving fields of research with large literatures. Instead it surveys the major theoretical ideas and empirical findings that have motivated recent work on insect sociality, and it indicates some directions that future studies are likely to take. Many of these topics have recently been reviewed by Andersson (1984b), Brian (1980), Brockmann (1984), Fletcher and Ross (1985), Gadagkar (1985), Gamboa *et al.* (1986), Hölldobler and Wilson (1990), Itô (1989), Myles and Nutting (1988), Nonacs (1986), Page and Breed (1987), Roubik (1989), Strassman (1989), Sudd and Franks (1987), Trivers (1985) and various authors in the volumes edited by Breed *et al.* (1982), Breed and Page (1989), Fletcher and Michener (1987), Hepper (1991), Hermann (1979; 1981; 1982a,b), Itô *et al.* (1987), Jeanne (1988), Pasteels and Deneubourg (1987), Ross and Matthews (1991), Trager (1988) and Watson *et al.* (1985).

The next section describes two fundamental problems that are posed by the existence of a reproductive division of labour among imperfectly related organisms, and it outlines the three evolutionary mechanisms most often invoked as possible solutions to these problems: worker altruism, parental manipulation and mutualism. Section 11.3 briefly discusses presocial behaviour in insects, and the problem of the restricted taxonomic distribution of true eusociality. Section 11.4 reviews the natural history of social behaviour in primitive and advanced eusocial insects, with emphasis on interactions that reveal conflict within colonies. Section 11.5 considers some special behavioural and physiological attributes of the Hymenoptera that may help to explain why most of the independent origins of eusociality have occurred in that order.

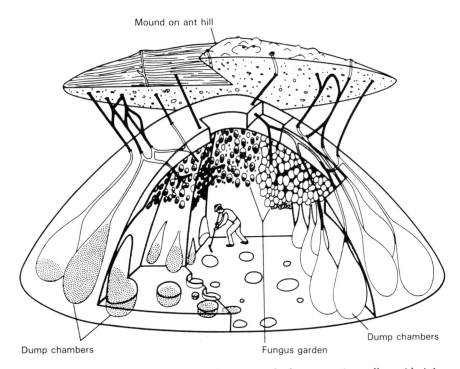

Fig. 11.1 Cutaway view of a nest of a South American leafcutter ant, *Atta vollenweideri*. A person with a rake (representing the excavator) is included for scale. The mound is composed of soil brought to the surface by the ants. Fungus is cultured on leaf fragments in the fungus garden chambers; exhausted substrate is then transferred to the dump chambers (from Hölldobler & Wilson, 1990; after J.C.M. Jonkman, in Weber, 1979).

11.2 Problems and possible solutions

Social insects pose two distinct problems for the theory of natural selection, and both were clearly identified by Darwin (1859, Chapter VII). These are, first, the evolution of worker sterility and, second, the evolution of morphologies and behaviours possessed only by the sterile castes. Today we worry mainly about the first problem, but to Darwin the second was more troubling.

Given his necessarily imprecise understanding of heredity, Darwin's tentative solution to the first problem was remarkably close to modern ideas based on kin selection: sterility might evolve if it proved sufficiently 'profitable to the community' of (implicitly) related individuals who were already enagaged in some form of cooperative reproduction. But the subsequent *modification* of worker morphology and behaviour seemed to raise a more serious difficulty because it required the accumulation of many small changes, rather than a single loss of function; it therefore seemed to pose a fundamental problem of evolutionary mechanics. To us the problem of modification seems almost trivial because we know about genes. Mutations that improve the performance of workers will be carried and transmitted by the reproductive members of their colonies; thus if sterile workers exist, they can easily diverge in structure and behaviour.

Darwin's relative complacency about the *origin* of sterility can be explained in part by his ignorance of the many cases in which workers are *not* unconditionally sterile. If 'absolute' sterility were almost universal (as Darwin apparently believed), then there would be less scope for self-interested worker behaviour than there is where workers can attempt to reproduce (Hamilton, 1972). Moreover, we know that even where workers are 'utterly' sterile, they may none the less take part in conflicts over the colony's production of male and female reproductives (Trivers, 1974; Trivers & Hare, 1976). This understanding has emerged from an integration of population-genetic theories of kin selection and sex-ratio evolution, together with a knowledge of the haplodiploid genetics of the Hymenoptera, and would therefore have remained beyond Darwin's reach, even had he known as much as we do about the natural history of worker behaviour.

Three kinds of processes have been proposed to explain how non-reproductive workers could evolve. These can be referred to as (i) kin-selected worker altruism; (ii) parental manipulation of offspring; and (iii) mutualism. The boundaries between these processes are fuzzy, and any two, or all three, could operate simultaneously; none the less, it is often important to distinguish between them because they can lead to different predictions and to different interpretations of the evidence, so they are presented here in fairly stark, simplified forms.

11.2.1 Worker altruism

Fisher (1930), Haldane (1932; 1955), Wright (1945) and others showed in outline how Mendelian genetics could explain the evolution of reproductive altruism. Hamilton (1963; 1964; 1972) pursued the idea to its logical conclusion, which we now know as the theory of *inclusive fitness* or kin selection (Maynard Smith, 1964; 1985c; see also section 1.3). A few years before Hamilton published the general theory, Williams and Williams (1957) analysed a genetic model for the initial evolution of sibling 'donorism', and discussed its relevance to the evolution of insect societies. Their paper did not uncover the simple relationship between benefit, cost and relatedness that would later come to be known as 'Hamilton's rule' (see Chapter 1) for the positive selection of altruism ($b/c \equiv k > 1/r$ or $br - c > 0$, where b is benefit to the recipient and c is cost to the actor), but it showed concretely how altruism might evolve in groups of close kin, and it argued that such a process could explain the emergence of insect societies. Hamilton's papers generalized these earlier ideas in a powerful and intuitive way, showing how the concept of inclusive fitness could account for both *altruism* (performing an act when b and c are positive) and *restraint from selfishness* (not performing an act when b and c are negative, in which case the actor would benefit while the recipient was harmed), for particular combinations of the benefit/cost ratio and the relatedness of actor and recipient (Fig. 11.2).

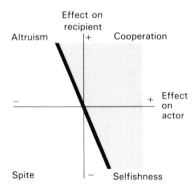

Fig. 11.2 Hamilton's fourfold classification of behaviour. The effect of an act on the fitness of the actor is represented on the horizontal axis, and the effect on the recipient is represented on the vertical axis. A positive effect is equivalent to an increase in lifetime production of successful offspring, and a negative effect to a decrease. Each quadrant is labelled with the kind of act that corresponds to its combination of signs of the two effects. For example, altruism corresponds to a *negative* effect on the actor (cost) and a *positive* effect on the recipient (benefit). The shaded region contains those acts positively favoured by selection for an actor related by ½ to the recipient; acts in the open region are opposed by selection. In general, the slope of the line dividing the hatched and open regions is $-1/r$. Note that the conditions for positive selection of an altruistic act (e.g. voluntary worker behaviour) are more restrictive than those for a selfish act (e.g. parental manipulation) whenever $r < 1$.

Coefficients of relatedness are unusual in the Hymenoptera owing to the haplodiploid genetic system that occurs throughout the order (and in some other arthropod taxa; for a review see Bell, 1982). Hymenopteran males develop from unfertilized eggs and are therefore haploid; if an egg is fertilized, it develops as a diploid female. Because males are haploid, they produce sperm cells that carry identical sets of chromosomes. A male's daughters are related by $r = 1$ with respect to the paternal half of their genotypes, and by the usual $r = \frac{1}{2}$ with respect to the maternal half, giving an overall relatedness of $r = \frac{3}{4}$. But a female shares only one parent (her mother) with a brother; her mother's meiotically produced eggs were *not* identical, so a female is related to her brother by $r = \frac{1}{4}$. As in diploid species, a female is related to her offspring (of both sexes) by $r = \frac{1}{2}$. (Definitions of relatedness and techniques for calculating it are explained more fully in Chapter 1 and in Grafen, 1985.)

Because hymenopteran females are more closely related to their full sisters than they are to their own offspring (Hamilton, 1964b), they might 'easily [evolve] an inclination to work in the maternal nest' rather than attempting to reproduce on their own, if 'the sex ratio or some ability to discriminate allows the worker to work mainly in rearing sisters' (Hamilton, 1972). If a worker adds as many additional offspring to her mother's net reproduction as she would have produced on her own, and if these additional siblings are mostly full sisters, then worker habits will be favoured by selection, at least when rare. But as workers become more common the ratio of investment will tend to become female biased, increasing the relative reproductive success of males (Fisher, 1930); when the ratio of investment reaches 1:1.5 (male:female), a female will do equally well to rear sons or sisters (Trivers & Hare, 1976). (At this investment ratio the value of sons is $\frac{1}{2} \times 1\frac{1}{2} = \frac{3}{4}$ relative to $\frac{3}{4} \times 1 = \frac{3}{4}$ for sisters, per unit invested.) This implies that in the earliest stages of the evolution of eusociality 'a polymorphism will naturally develop', in which eusocial nests 'will specialize in the production of female reproductives' while solitary and semisocial nests 'will specialize in the production of males' (Trivers & Hare, 1976). If this state of very primitive (mixed) sociality promotes the evolution of behaviours that increase the efficiency of eusocial nests, leading to the abandonment of solitary and semisocial reproduction, then the relatedness asymmetry caused by haplodiploidy will have played an important role in the *origin* of eusociality, even if it plays no direct role in its *maintenance*, once a state of complete eusociality has been reached.

To exploit the relatedness asymmetries most effectively, females must be able to control the sexes of their offspring. Hymenopteran females store sperm in an organ (the spermatheca) that connects to the oviduct, and they are able to determine whether a given egg is fertilized, just before oviposition. Many solitary species employ adaptive sex-ratio strategies (for example, Herre, 1987; Werren, 1987). Extremely complicated strategies (involving worker laying and modification of the colony sex ratio) may arise under eusociality; some of these are discussed below.

11.2.2 *Parental manipulation and parasitism*

Until the early 1970s the problem of insect sociality had been seen almost entirely as one of *voluntary* worker behaviour; the assumption (often implicit) was that during the early stages in the evolution of helping, a worker could choose either to remain a sterile helper or to leave the colony and attempt to reproduce directly. This assumption was pointed out and challenged by several authors, most notably by Alexander (1974) in an influential paper that argued the importance of *parental manipulation of progeny* as a general alternative to kin selection and reciprocity, in the evolution of many kinds of social behaviour.

Alexander devoted a large section of the paper to social insects, and he argued that 'the parents of sterile insects have made them so in their own interests, and have made them altruistic beyond the possibilities of kin selection as so far formulated'. A sterile worker's relative efficiency at increasing its parent's reproduction can be much lower under parental manipulation than it needs to be under voluntary helping (Fig. 11.2), and the problem of the initial spread of an allele for altruism is seemingly reduced, since an immediate advantage accrues to the parent that succeeds in manipulating its offspring to be altruistic. Michener and Brothers (1974) argued that workers in a primitively eusocial bee (*Lasioglossum zephyrum*) do in fact seem to be more 'oppressed' than 'altruistic', based on a detailed analysis of dominance interactions within colonies of this species.

Several theorists (for example, Charnov, 1978a; Craig, 1979) pointed out that a hymenopteran mother that shared her nest with an adult daughter need only substitute her own eggs for her daughter's (for example, while the daughter was foraging away from the nest) in order to reap a twofold reproductive advantage (by trading grandoffspring for offspring). Happily for the 'parasitized' daughter, she might suffer no fitness loss as a consequence, since she would, in effect, trade offspring for siblings. If her mother had mated with only one male, then the parasitized daughter would be on average as closely related to these substituted full siblings as she would have been to her own progeny, assuming her mother produced a 1 : 1 sex ratio. Thus the daughter's motivation to resist being exploited would in general be much weaker than her mother's motivation to exploit, so that the advantages of remaining in her established natal nest might easily outweigh the small (possibly zero) cost of being transformed (perhaps unknowingly) into a worker.

Stubblefield and Charnov (1986) showed that if several daughters were sharing the nest with their mother, each would be positively selected to help her mother exploit the others (thereby trading nieces and nephews for siblings, to which the daughter would be, on average, twice as closely related). As Stubblefield and Charnov admit, this mechanism makes the evolution of sociality seem almost too easy. It fails to explain 'why eusociality has arisen repeatedly in the haplodiploid Hymenoptera but only once in diploid insects' and why, within the Hymenoptera, the many origins are confined to a few families. This is a central problem to which we will return later. It leads to the idea that certain features of the biologies and life histories of aculeate (wasp-waisted, stinging) Hymenoptera must provide opportunities for helping (and exploitation) that are absent in most groups of insects, including the non-aculeate Hymenoptera and even some groups of aculeates. The features most likely to have been important include *advanced parental care in a permanent nest* and an *overlap of generations* that brings adult offspring into direct contact with their parents (for example, Evans, 1977).

11.2.3 Mutualism

In their pure forms, both of the previous explanations (altruism and exploitation) assume that 'workers' do not reproduce. But in many primitively social species (and in some advanced species) workers often lay unfertilized eggs that develop into males. This is usually viewed as a source of conflict, but it also suggests the possibility that, at least under certain circumstances, 'workers' may benefit directly from remaining in their mother's established nest, even in a subordinate role (for example, Michener, 1969; West Eberhard, 1978a,b). A daughter who remains in her mother's nest also stands to inherit it (and the role of principal egg layer) if her mother dies in mid-season; even if most 'workers' will never win this prize, it increases their expected fitness. (A similar idea is used to explain helpers at the nest in bird species where breeding territories are the critical limiting resource.) Thus in many cases an apparently sterile worker may be either a cryptic reproductive or a *hopeful reproductive* (West Eberhard, 1978a).

11.2.4 Summary

Each of these three processes 'takes the altruism out of altruism' in a different way: kin-selected voluntary helping increases a worker's inclusive fitness and thereby advances her genetic self-interest; a parentally manipulated worker is by definition exploited, not altruistic; and mutualistic associations are maintained by the expectation, on each female's part, that cooperation will pay off directly. Female behaviour may be influenced by more than one of these processes at any given time; their relative importance seems likely to vary both spatially (over a species' range) and temporally (over the course of a season, and over evolutionary time within a given lineage). For example, each partner in an initially mutualistic nesting association might increase her inclusive fitness by persuading the others to forage more and to reproduce less. From this point of view, cooperation and conflict are almost inseparable.

The problem of insect sociality is a larger, richer problem for us than it was for Darwin. Worker modification *per se* is no longer a fundamental dilemma, but many of its details seem increasingly puzzling; examples include the relationship between underlying developmental mechanisms and the morphologies and ergonomics of worker castes (for example, Oster & Wilson, 1978; Wheeler, 1986) and, most importantly in the present context, the problem of who wins conflicts and how they get away with it. The problem of origins is more troublesome for us than it was for Darwin, in part because we have a fuller appreciation of the potential conflicts of interest, and of the opportunities and constraints that may arise from particular details of a species' life history and ecology, and in part because we know much more about the taxonomic distribution of independent origins of eusociality, and about primitively social species with facultative worker behaviour (see West Eberhard, 1987).

11.3 Approaches to sociality

Many insect taxa show advanced forms of social behaviour other than reproductive altruism of the kind seen in termites and social Hymenoptera (Wilson, 1971; Eickwort, 1981). For example, extremely vicious and specialized forms of aggression occur among males in species of beetles, fig wasps, ants, bees (Fig. 11.3) and other insects that breed in confined spaces; in some cases the combatants are close relatives (Hamilton, 1979). This kind of unrestrained selfishness can be taken to establish one end of a spectrum that includes at its other end the 'soldier' morphs of some aphids (Aoki, 1987; Itô, 1989) and parasitoid hymenoptera (Cruz, 1981) that are often clonally related to the individuals they defend, and completely sterile.

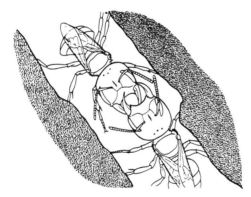

Fig. 11.3 Macrocephalic fighting males in an Australian halictid bee, *Lasioglossum (Chialictus) erythrurum*. Pugnacious, flightless males occur in at least two families of bees, in fig wasps and in several other orders of insects (reviewed by Hamilton, 1979). In some species all males are fighters, but in others (as in *L. erythrurum*) 'normal' fully winged males also occur, sometimes at high frequency (see Stuart *et al.*, 1987, for a recently discovered case of male dimorphism in ants). Ten or more females of *L. erythrurum* may share a single communal nest; most nests are also occupied by one macrocephalic male who mates with females within the nest and probably defends it against intruders such as ants (from Kukuk & Schwarz, 1988; drawing by Byron Alexander).

There are many species in which immature siblings cooperate, for example in larval group defences and mutualistic feeding. Communal nesting by adults also occurs in cockroaches, beetles, wasps, bees, spiders and other groups, and spectacular examples of parental care involving extended feeding and defence of immatures occur (leaving aside the solitary Hymenoptera) in roaches, crickets, bugs, beetles and several other insect groups (Eickwort, 1982; Tallamy & Wood, 1986). Many cases of advanced parental care involve a permanent nest; some involve both male and female parents; and some give rise to overlapping generations. These species seem to stand on the threshold. Why have none of them gone on to develop reproductive division of labour, thereby taking the final step to eusociality?

Although most species of solitary aculeate (wasp-waisted, stinging) Hymenoptera gather food for their offspring, direct contact between adults and immatures is relatively uncommon. A typical wasp or bee develops from egg to adult in a brood cell constructed by its mother, and is nourished entirely on food provided by its mother. But in most cases the food is 'mass provisioned' in a single bout of parental investment, immediately before or after the egg is laid in the cell. After provisioning the cell and laying the egg, the mother typically closes the cell (e.g. by backfilling the burrow leading to it) and moves on to construct and provision another cell. In species with one generation per year, the offspring usually overwinter in their brood cells (either as mature larvae or as adults), and therefore never see their mothers. Even in species with several generations per year, the mother is likely to have died before the first of her offspring emerge. There are almost certainly more independent origins of eusociality in the Halictidae ('sweat bees') than in any other insect family, and perhaps more than in all other insects combined. Most of the social species in this otherwise solitary family retain the primitive habit of mass-provisioning and then closing their brood cells. Thus contact between parents and their immature offspring is not an important precondition for the evolution of eusociality, even though is occurs in some primitively eusocial halictids and anthophorids, and is characteristic of the more highly derived eusocial lineages (e.g. ants, paper wasps, apid bees, termites). Likewise, serial (progressive) provisioning of the developing larva does not seem to strongly predispose a lineage to develop eusociality, since it is found in many solitary taxa, including some solitary Hymenoptera.

11.4 Origins and varieties of eusociality

Temporarily non-reproductive 'helpers at the nest' occur in many animal taxa, especially birds and canids (see Chapter 10). They are usually offspring of the individuals they help, and their presence usually increases the number of offspring produced by the family group (Trivers, 1985; Brown, 1987). Such species are at least temporarily eusocial, but the term is usually reserved for species in which colonies come to exhibit a more or less fixed division of reproductive labour, involving functionally sterile individuals that work on behalf of their parents (Wilson, 1975). Even by this strict definition, a few mammals such as the naked mole-rat (Jarvis, 1981; Sherman et al., 1990) and some spiders would seem to qualify as eusocial, but the overwhelming majority of eusocial species are either termites (Isoptera) or Hymenoptera.

Like the wood-eating cockroaches, to which they are related, termites specialize in digesting cellulose with the aid of a complex community of intestinal symbionts that must be obtained (in some cases after every molt) from other individuals; this fundamental adaptation requires some form of group living, but not eusociality. Termites are 'hemimetabolous' insects that pass through a series of larval instars more or less similar to the final adult instar; the larval instars are able to feed themselves and to do useful work within the colony. (Hymenoptera are 'holometabolous', with complete metamorphosis from a helpless, grub-like larval stage to a final adult stage.)

All species of termites are fully eusocial, which suggests that they represent a single origin of eusociality. Some have systems of morphologically and behaviourally differentiated worker castes that are as complex as those of advanced ants. However, the differences between termites and all eusocial Hymenoptera are in many ways as significant as the similarities. One important difference is that termite workers are both male and female, while in Hymenoptera all workers are female. The mating systems of termites and social Hymenoptera also differ in important ways. For example, the primary male and female reproductives ('king' and 'queen') typically remain together for life, whereas in social Hymenoptera males die shortly after mating and the queen (who may mate with several males) uses their stored sperm throughout a lifetime that may exceed 10 years. Also, termite colonies often have secondary and replacement reproductives that are the offspring of other colony members, and that mate within the nest (see Myles & Nutting, 1988). Dominance hierarchies among workers appear to be much less common among termites than among social Hymenoptera (for example, Wilson, 1971; Noirot & Pasteels, 1987). Primary polygyny (where several foundresses establish a new colony) is also relatively uncommon, occurring only within a few genera in the advanced family Termitidae, and even where it occurs it appears to be facultative rather than obligate (Thorne, 1982).

All other origins of insect eusociality have occurred within the Hymenoptera. The ants, like the termites, are entirely eusocial and presumably represent one origin. Social wasps represent a few origins, while the bees contain an undetermined but possibly large number of origins (Table 11.1). Given their diversity, it is not surprising that social Hymenoptera differ in many details of colony foundation and structure. For example, there may be one queen or several (at colony foundation or maturity); queens may mate with one or several males; workers may be completely and per-

Table 11.1 An informal and abbreviated taxonomy of the aculeate (stinging) Hymenoptera. Taxa containing eusocial species are shown in bold.

CHRYSIDOIDEA (cuckoo wasps)	S(P)		
APOIDEA (sphecid wasps, bees)			
Sphecidae (digger wasps, sand wasps, mud daubers)			
Pemphredoninae (mostly solitary, but eusociality in *Microstigmus*)	S	C	E
Sphecinae, Astatinae, Larrinae, Crabroninae, Nyssoninae, Philanthinae, etc.	S	C	
Colletidae (colletid bees)	S		
Stenotritidae (stenotritid bees)	S		
Andrenidae (andrenid bees)	S		
Oxaeidae (oxaeid bees)	S		
Halictidae (sweat bees)			
Halictinae	S	C	E
Nomiinae	S	C	
Dufoureinae	S		
Melittidae (melittid bees)	S		
Megachilidae (leaf-cutter, resin, mason bees)	S	C	
Fideliidae (fideliid bees)	S		
Anthophoridae (anthophorid bees)			
Nomadinae	S(P)		
Anthophorinae	S	C	
Xylocopinae (carpenter bees)			
Ceratinini (small carpenter bees)			
'Allodapine bees' (*Allodapa, Allodapula, Exoneura*)		C	E
Ceratina	S		
Xylocopinae (large carpenter bees)	S	C	
Apidae (apid bees)			
Euglossinae (orchid bees)	S	C	
Bombinae (bumblebees)	S(P)		E
Apinae (honeybees)			E
Meliponinae (stingless bees)			E
VESPOIDEA (vespoid wasps, ants)			
Tiphiidae, Sapygidae, Mutillidae (velvet ants), Sierolomorphidae, Rhopalosomatidae, Pompilidae (spider wasps), Bradynobaenidae, Scoliidae	S		
Vespidae (social wasps, potter wasps, etc.)			
Euparagiinae	S		
Masarinae (pollen wasps)	S	C	
Eumeninae (potter wasps)	S	C	
Stenogastrinae (hover wasps)			E
Polistinae (paper wasps)			E
Vespinae (hornets, yellowjackets)			E
Formicidae (ants)			E

S indicates that the group contains strictly solitary species, C indicates that some otherwise solitary species show communal (shared) nesting, and E indicates that eusociality occurs. S(P) indicates that all solitary species are cleptoparasites (brood parasites or 'cuckoos'); cleptoparasitism also occurs sporadically in many of the other taxa shown here. (Modified from Brockmann, 1984.)

manently or partially and facultatively sterile; and colonies may persist for one to many seasons. Furthermore, individual species may contain several different kinds of colonies, in some cases spanning the range from strictly solitary females working without helpers to fully eusocial family groups. Each of these variables affects the kinds of conflicts of interest that may arise within a colony, and the kinds of tactics that individuals involved in such conflicts may be able to employ to increase their inclusive fitness.

Early efforts to analyse hymenopteran sociality from a comparative point of view gave rise to an important distinction between the *subsocial* and *parasocial* routes of colony formation (Wheeler, 1923; Evans, 1958; Michener, 1969; Lin & Michener, 1972). In the subsocial route, a female cares for her own offspring which emerge and then, instead of departing, assist her in rearing additional offspring who are their siblings. In the parasocial route, females of the same generation cooperate in rearing each other's offspring; if some of them act as sterile workers, the colony is said to be *semisocial*; when the offspring emerge and begin working for one or all of the original foundresses, the colony becomes eusocial. The subsocial route leads somewhat more directly (in principle) to 'ideal' eusociality of the kind envisioned in most theoretical models. The parasocial route requires that some individuals rear brood to which they are less closely related than they would be to their own offspring, but it eventually reaches much the same state, in which offspring assist their parents (and possibly also some collateral relatives). Each of these scenarios can be interpreted in either of two ways: first, as a description of the process of colony foundation in contemporary eusocial species; and second, as a possible evolutionary sequence in which successive levels of presocial behaviour converge on full eusociality (but see Michener (1985) for a critique of the evolutionary interpretation).

11.4.1 Primitive eusociality

Species classified as 'primitively' eusocial tend to have small, short-lived colonies and little or no morphological differentiation between queens and workers. In temperate latitudes, colonies are typically established in the spring by overwintered foundresses who rear a first brood of female off-spring (sometimes with a few males). These first-generation female off-spring then take over the tasks of foraging and brood care. In some species, one or more additional broods of workers are produced; in all species, the final brood consists of reproductive females and males. After mating, the reproductive females seek shelter for their winter hibernation. No workers or males survive the winter. Even the simplest monogynous versions of this colony cycle contain plenty of room for conflict. For example, workers may disagree with the queen as to what the sex ratio of the final brood should be, and they may attempt to substitute their own eggs for hers, or for each others'.

The life cycles of many temperate bees in the family Halictidae follow essentially this course. But in some species, such as *Lasioglossum zephyrum*, several of the foundress's daughters mate and become full-fledged reproductives within the colony, while others act as workers (for example, Batra, 1964; 1966; 1968; Michener, 1982). The distinction between workers and reproductives is far from absolute, however, and it is determined and enforced largely through dominance interactions (for example, Michener & Brothers, 1974; Smith, 1987). Most halictid bees retain the habit, primitive for aculeates, of nesting in the ground. Michener and his colleagues have developed techniques for culturing *L. zephyrum* and other halictids in plexiglas-fronted 'sandwich' boxes that permit close observation of behaviour within the nest. This has made it possible to observe certain conflicts directly. For example, Batra (1964; 1968) saw females eat eggs that had recently been laid by nest-mates. Note that although colonies of *L. zephyrum* usually begin along the subsocial pathway (mother rears daughters who may then help her to rear more offspring), they can later become at least partly semisocial (females work for their sisters). In many species, mated females overwinter in their natal burrows; if several remain in the burrow the following spring, the colony may begin the year in a semisocial state, while the dispersing females each attempt to found new colonies (Knerer, 1983; Packer, 1986b). Although halictid colonies may frequently pass through semisocial states, these states are thought to be unstable and shortlived in most species, whose colonies usually end up being eusocial under a single queen (Michener, 1985).

However, in at least a few primitively social species, not all females take part in colony life. In *Halictus rubicundus*, many newly emerged females of every generation (including the first summer generation) appear to leave the nest, mate and enter hibernation, without ever having worked or reproduced; the following spring they emerge from hibernation and found new (initially solitary) nests (Yanega, 1988; 1989). Because some of *their* daughters will leave, mate and hibernate in the same way, the population is polymorphic for solitary and fully eusocial life cycles. This implies that females in many primitively eusocial species may have a wide range of life-history options open to them, at least on first emergence. But once a female commits herself to an established colony (say, with her mother and possibly some sisters), subsequent dominance interactions may force her into a subordinate role from which there is no escape.

Geographical clines in the level of sociality have been found within several halictid species. For example, the normally eusocial *Lasioglossum calceatum* has solitary populations at high altitudes in Japan, where its season is brief (Sakagami & Munakata, 1972). The best-studied case is that of *Halictus ligatus*, an abundant and widespread species that ranges from southern Canada into the neotropics. Northern populations are annual and eusocial, but subtropical and tropical populations remain active during all or most of the year, and the females within a nest seem to act more as a communal association of independently reproducing females than as a eusocial colony (Michener & Bennett, 1977; Packer, 1986a; Packer & Knerer, 1986; 1987). The loose organization of these communal nests appears to have created an excellent opportunity for *intraspecific brood parasitism*; some females do not forage, but instead roam from nest to nest placing their eggs in other females' brood cells (Packer, 1986c). *Interspecific* brood parasitism is common in bees, but this intraspecific form appears to be rare (leaving aside the intrigues that normally take place within colonies), although it could easily have escaped notice in many species.

Facultative sociality occurs in other groups of mainly solitary bees, particularly among small carpenter bees (Xylocopinae, family Anthophoridae) (Sakagami & Maeta, 1982; 1984; 1987; Michener, 1985; Schwarz, 1987). For example, most species of *Ceratina* are believed to be solitary, but 20% of natural twig nests contain two females in the univoltine *C. japonica*, and multifemale nests readily form in experimental populations where females outnumber twigs (Sakagami & Maeta, 1984; 1987). One female (usually the larger) tends to remain in the nest and to do most of the egg-laying, while the other does most of the foraging. Solitary and social nests are also found in a related multivoltine species, *C. okinawana*; here the two females sharing a nest often differ both in age and in ovarian development, with the older female (mother?) being the one whose ovaries are better developed (Sakagami & Maeta, 1987).

Even though every female is potentially able to reproduce, many or most never do so in most species of primitively social bees. Do these functionally sterile individuals work on behalf of close relatives? A great deal of behavioural evidence suggests that in most species they usually do (as discussed in the context of kin recognition, section 11.5.6), and this inference has recently been confirmed by genetic studies. Using enzyme electrophoresis, Crozier *et al.* (1987) assayed gene and genotype frequencies at five polymorphic loci in the Kansas population of *Lasioglossum zephyrum* that has long been the object of behavioural studies by Michener and his associates. The resulting data on genetic variation within and among nests and clusters of nests were then used in a regression procedure (Pamilo, 1984a) that estimates average coefficients of relatedness among different categories of individuals. The relatedness of nest-mates within subpopulations was very high; an estimate for the subpopulation with the largest sample of nests ($n = 20$) was $b = 0.82$, with a lower confidence limit of 0.64, implying that most of the females sampled were full sisters (i.e. daughters of a singly inseminated foundress). Kukuk (1989) found somewhat lower, but none the less substantial, levels of relatedness among females within colonies in central New York state, at two different times in the colony cycle (Wcislo, 1987).

The bees can be viewed as a group of sphecid wasps that switched from taking animal prey to taking vegetable 'prey' (pollen and nectar). The modern Sphecidae continue to hunt insects and other arthropods. They are a very large and diverse assemblage in which many forms of advanced parental care and communal nesting occur. Eusociality was unknown in the family until Matthews (1968a,b) discovered evidence of primitive eusociality in the neotropical pemphredonine *Microstigmus comes*. This tiny wasp builds nests of silk and plant fibres on the undersides of palm fronds. One to 18 females may be associated with a given nest, and there appears to be at least a modest division of reproductive labour (Matthews, 1968a,b; 1991). The average relatedness of female nest-mates is roughly 0.65, and in many nests the genotype of the heaviest individual is consistent with the hypothesis that she is the mother of all the others (Ross & Matthews, 1989a,b). Small, newly established nests sometimes contain several adult females whose average relatedness is lower (roughly 0.45), implying that foundress groups may often be mixtures of first and second degree relatives (Ross & Matthews, 1989b).

Microstigmus appears to represent the most phylogenetically isolated origin of eusociality in the Hymenoptera. Communal nesting and other forms of incipient sociality occur elsewhere in the Pemphredoninae (e.g. McCorquodale & Naumann, 1988; Matthews & Naumann, 1988; Matthews, 1991), and the subfamily is not as well studied as many others in the Sphecidae, so there is every reason to suppose that it may yield more surprises.

Wasps of the family Vespidae are more closely related to the ants than to the bees and sphecid wasps. One subfamily is strictly solitary (Euparagiinae), two are solitary with some communal species (Eumeninae, Masarinae) and three are entirely social (Stenogastrinae, Polistinae, Vespinae). The vespines (hornets, yellowjackets) show advanced eusociality, even though their colony cycles are usually annual at temperate latitudes (see Wilson, 1971). The polistines (paper wasps) comprise a large and diverse assemblage of somewhat more primitively social species. Many polistine species build open-faced aerial nests. Their behaviour can therefore be readily observed in undisturbed natural settings, and they are relatively amenable to various kinds of experimental manipulation. For these reasons they are among the best species in which to study the costs and benefits of primitive sociality (for example, West Eberhard, 1969; Noonan, 1981; Queller & Strassman, 1988; 1989; Strassman et al., 1988; Strassmann & Queller, 1989). The stenogastrines (hover wasps) are a small group, closely related to the Polistinae and Vespinae, that occurs only in south-east Asia. Many species appear to be facultatively social, but their behaviour is relatively poorly known (see Carpenter, 1988).

In temperate species of the worldwide genus *Polistes*, colonies are typically established by one or several related foundresses who overwintered as mated adults. These females rear a brood that consists mainly (but often not entirely) of females. Most of the first-generation females usually remain on their natal nest and act as workers but, as in primitively social bees, the distinction between workers and reproductives is mainly behavioural and is maintained through dominance interactions among females on the nest (for example, Jeanne, 1980; Fletcher & Ross, 1985). If the principal egg layer dies, her place is immediately taken by a high-ranking subordinate, whose ovaries may already be better developed than those of lower-ranking workers. Males and females produced at the end of the season mate, and in temperate regions the mated females are the only individuals that hibernate and survive the winter.

No polistines are known to be solitary, but some species of *Ropalidia* build aggregations of very small open-faced nests, which suggested to Richards (1978) that they might be more nearly communal than eusocial. Wenzel (1987) quantified several aspects of the nesting biology of *R. formosa* at various locations in Madagascar, and concluded that this species may be 'nearly solitary'. The mean number of adult females per nest is only slightly greater than one; the mean and maximum number of cells per nest are both smaller than for any other polistine that has been studied, and there are fewer cells at first emergence of both female and male offspring; helpers, where present, seem to make smaller contributions to the colony's productivity than in other polistines; and finally, nesting females are completely docile in the presence of human investigators, in striking contrast to the aggressive nest defence shown by most social polistines, including many species of *Ropalidia*.

Because there is often more than one egg layer, both at colony foundation and later in the colony cycle (see Itô, 1987), the relatedness of colony members is expected to be variable both within and among species of primitively social wasps. Strassman *et al.* (1989) used enzyme electrophoresis to estimate the intracolony relatedness of females in 15 populations of 14 species of *Polistes* and *Mischocyttarus* (a related tropical genus in the Polistinae). The estimates are fairly evenly distributed over a wide range (0.31–0.77, standard errors 0.04–0.16), with a mean of 0.54. Thus nest-mates appear to be substantially related in all species, but not, on average, more closely than they would be to their own offspring. In this respect primitively eusocial wasps and bees appear to be similar to each other, and different from more socially advanced groups.

11.4.2 Advanced eusociality

There are fundamental differences between primitive eusociality as seen, for example, in *Lasioglossum zephyrum* and *Halictus rubicundus*, and advanced eusociality as seen in many ants and termites, but these are the ends of a spectrum (or better, a multidimensional space) in which species occupy almost every conceivable intermediate position. Specialized, morphologically distinct, worker castes are a hallmark of advanced eusociality (Fig. 11.4), and they tend to be associated with high degrees of worker sterility and large, long-lived colonies. But not all advanced species exhibit these characteristics. What sets them apart is their more complete

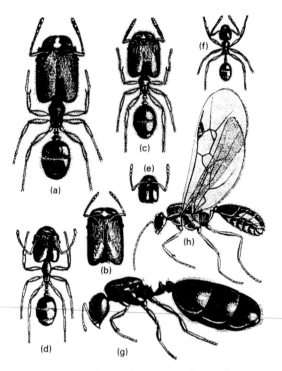

Fig. 11.4 Caste polymorphism in an advanced myrmicine ant, *Pheidole tepicana*. This famous drawing by W.M. Wheeler (1910) shows the queen (g), the male (h), and six worker subcastes, from the major worker (a), through several sizes of media workers (b–e), to the minor worker (f). Note that proportions (especially those of the head) change dramatically with size.

reliance on the division of reproductive labour. Queens may establish new colonies on their own, and workers may reproduce within the colony or form queenless satellite colonies, but no solitary individual could come close to successfully producing sexual offspring.

In most species of ants, new colonies are established by single queens who rear a first brood of workers on their own (*haplometrosis* or primary monogyny), but colony foundation by several queens (*pleometrosis* or primary polygyny) has evolved repeatedly in many different lineages (Höll-dobler & Wilson, 1977; 1990). Species that are initially monogynous usually remain monogynous, and colonies therefore develop along the classic subsocial pathway to full eusociality. In such species, workers are closely related to each other and to the reproductive brood they rear; the exact pattern of relatedness depends on the number of males with which the queen mated, and on whether any of the colony's male offspring are derived from worker-laid eggs. At least in some species the queen is effectively singly mated, and the relatedness of workers and female offspring approaches its theoretical upper limit of three quarters. Examples documented by enzyme electrophoresis include monogynous colonies of several species of *Rhytidoponera* (Ward, 1983a), monogynous colonies of the fire ant *Solenopsis invicta* (Ross & Fletcher, 1985) and the regularly monogynous species *Lasius niger* (van der Have *et al.*, 1988).

Many socially advanced wasps and bees (and most or all army ants) found new colonies by swarming (also called budding or fissioning), a process whereby one queen (in vespine wasps, most apine bees, and army ants) or several queens (in polistine wasps) depart with a portion of the parent colony's worker force. As a consequence, primary polygyny tends to involve related foundresses, as it does in primitively social wasps and bees. Many swarm-founding polistine wasps build elaborate nests and have large colony sizes, but like *Polistes* and *Mischocyttarus* they tend to show relatively little morphological caste differentiation. Colonies may have several to many active egg layers, and are therefore expected to show low to intermediate levels of intracolony relatedness. Queller *et al.* (1988) obtained relatedness estimates of 0.34 for *Polybia occidentalis*, 0.28 for *P. sericea*, and 0.11 for *Parachartergus colobopterus* (with standard errors of 0.05–0.07).

In most ants, pleometrotic associations typically involve unrelated foundresses (Ross & Fletcher, 1985; Hagen *et al.*, 1988; Rissing & Pollock, 1988; Strassman, 1989; Hölldobler & Wilson, 1990). In many of these species primary polygyny reduces to secondary monogyny when the first workers eclose, either because fighting erupts among the queens themselves, or because the workers evict all but one queen. However, queens of the desert leaf-cutting ant *Acromyrmex versicolor* apparantly continue to cooperate indefinitely (Rissing *et al.*, 1989), and in the honey ant *Myrmecocystus mimicus*, a few nests may still contain two or three queens as long as two years after founding, even though most queens are ejected by workers early in the colony's growth (Bartz & Hölldobler, 1982).

Pleometrosis involving unrelated foundresses would be expected to evolve if *n* foundresses are more than *n* times as likely as a single foundress is to succeed in establishing a new colony. If this condition is satisfied, then a female's expected fitness as part of a pleometrotic association is greater than it would be alone, even though she stands only a $1/n$ chance of becoming the colony's eventual queen. In *A. versicolor, M. mimicus* and other primarily polygynous ants, the main threat to incipient colonies appears to be brood raiding by other incipient colonies of the same species; thus a new colony is under strong pressure to quickly produce a large worker force, with which to sieze an early advantage in the tournament of reciprocal kidnapping. From each foundress's point of view, the best value of *n* should be that which maximizes the *per-foundress* probability of eventual success, not the *per-colony* probability of success. In laboratory studies of *M. mimicus*, Bartz and Hölldobler (1982) found that trios seemed to produce the largest numbers of workers per foundress, even though groups of five or six produced the largest absolute numbers of workers; in nature, most foundress groups contain from two to four females, in remarkable agreement with the optimal group size predicted by the behaviour of this species in the laboratory.

Some primarily polygynous ants remain polygynous and even recruit additional queens (as do a few primarily monogynous species which thereby become secondarily polygynous). Whether primarily or secondarily polygynous, colonies with several to many functional queens are expected to show relatively low average relatedness among nest-mates. This expectation seems to hold for the polygynous species whose genetic structures have been studied by enzyme electrophoresis (see Boomsma, 1988). For example, estimates of relatedness between nest-mate workers in polygynous species of *Formica* range from −0.02 to 0.42 (Pamilo & Varvio-Aho, 1979; Pamilo, 1981; 1982b). Similarly low estimates have been obtained for queenless nests in several species of *Rhytidoponera* (Ward, 1983a; Crozier *et al.*, 1984), for polygynous species of *Myrmecia* (Craig & Crozier, 1979) and *Myrmica* (Pearson, 1983), and for polygynous colonies of the fire ant *Solenopsis invicta* (Ross & Fletcher, 1985). Workers and brood may have low average coefficients of relatedness even where queens are closely related (e.g. 0.4 for nest-mate queens of *Leptothorax acervorum*; Douwes *et al.*, 1987).

Workers retain functional ovaries in many species of ants and socially advanced wasps and bees. Thus, even without being mated, workers are potentially able to produce males, which arise from unfertilized haploid eggs. Worker reproduction has been documented in every major subfamily of ants, but it tends to occur mainly in orphaned or otherwise queenless colonies, especially in the relatively advanced subfamilies Myrmicinae and Formicinae (reviewed by Bourke, 1988b; Choe, 1988; Hölldobler & Wilson, 1990). Worker reproduction also appears to be widespread in highly eusocial wasps and bees, but, again, it usually takes place at much reduced levels in queenright colonies (reviewed by Fletcher & Ross, 1985; Bourke, 1988b). Although queens appear to suppress worker reproduction substantially or entirely in most species, there are exceptions (for example, Cole, 1981; 1986; Franks & Scovell, 1983; Bourke, 1988a; Bourke *et al.*, 1988). *How* queens suppress worker reproduction is not understood for any species except the honeybee, either in a proximate (mechanistic) or an ultimate (evolutionary) sense, especially for species with large colony sizes, where direct intimidation of workers (as occurs in many primitively social species) would seem to be impossible. Various kinds of evidence suggest that pheromones are often involved (see Fletcher & Ross, 1985; Hölldobler & Wilson, 1990), but why workers should be inhibited by a mere 'signal' is still somewhat mysterious. Here Darwin's problem of worker modification returns to haunt us, in a thoroughly modern form.

A partial resolution of this paradox has recently been suggested by studies of attempted worker reproduction in the honeybee (*Apis mellifera*). A small fraction of worker honeybees have well-developed ovaries (one to a few per cent), but in studies using genetic markers, only a tiny fraction of the drones produced by a typical colony (about 0.1%) can be attributed to workers (Visscher, 1989). Honeybee queens may mate with as many as 20 different males on their nuptial flights (for example, Adams *et al.*, 1977; Page, 1986), so a typical colony contains many full-sister patrilines ($r = \frac{3}{4}$); because there are many different patrilines in the colony, most workers are related to each other as half-sisters ($r = \frac{1}{4}$). A worker therefore expects to be related to a worker-produced male by slightly more than $\frac{1}{4} \times \frac{1}{2} = \frac{1}{8}$, which is less than her relatedness to a queen-produced male ($\frac{1}{2} \times \frac{1}{2} = \frac{1}{4}$); she should therefore prefer to see the colony's male reproductives be produced by the queen. This chain of reasoning led Ratnieks (1988) to propose that if honeybee workers can distinguish worker-laid from queen-laid eggs, then they should 'police' the brood comb, destroying any worker-laid eggs they find. In a subsequent series of egg-introduction experiments, Ratnieks and Visscher (1989) showed that workers given arrays of worker-laid and queen-laid haploid eggs preferentially destroy the worker-laid eggs. This occurs even if both kinds of eggs are derived from colonies unrelated to the experimental colony, so direct assessment of kinship is not required.

One interesting implication of this finding is that worker suppression of worker reproduction may be very effective in species where the queen usually mates with several unrelated males, but not in species where the queen usually mates with one male. In the latter (*monandrous*) case, a worker is more closely related to her sister's male offspring ($\frac{3}{4} \times \frac{1}{2} = \frac{3}{8}$) than she is to her mother's male offspring ($r = \frac{1}{4}$). Of course in all cases a worker should prefer her own male offspring ($r = \frac{1}{2}$) to any others. But even if prevented from reproducing through worker–worker dominance interactions, she should prefer worker-laid to queen-laid male offspring in monandrous species. There are only a few species for which data exist on both multiplicity of mating and prevalence of worker reproduction, but they seem to satisfy this prediction. Bumblebees and stingless bees tend to be monandrous and to have relatively high levels of worker reproduction in the presence of queens; honeybees and yellowjackets tend to be polyandrous and to have low levels of worker reproduction, except when colonies are orphaned (see Plowright & Laverty, 1984; Fletcher & Ross, 1985; Ratniecks, 1988; Ratnieks & Visscher, 1989; Hölldobler & Wilson, 1990). If this pattern holds up, it could explain at least some of the variation in levels of worker reproduction seen in advanced monogynous species with relatively large colony sizes.

11.4.3 Summary

Some primitively eusocial species are morphologically, ecologically and behaviourally almost indistinguishable from their solitary relatives, but the most advanced are like nothing else on earth. Although advanced species are very far removed, in many ways, from their solitary and primitively social ancestors, they none the less continue to exhibit many ancient forms of reproductive competition. For example, worker production of males is almost as prevalent among advanced as among primitively eusocial Hymenoptera, and it seems to give rise to similar kinds of dominance hierarchies (for example, Cole, 1981; 1986; Franks & Scovell, 1983).

The term 'caste' denotes role assignments that are fixed from birth. As applied to social insects, the term also implies that the members of a given caste are so alike as to be interchangeable. But the more we learn about primitively social species, the more flexible and individualistic their behaviour seems to be (for example, West Eberhard, 1987). Even within the differentiated worker castes of advanced species, an amazing richness of individual behavioural variation is now being documented; these detailed pictures of colony life are undermining the image of the worker as a mere cog in a well-oiled machine (for example, Frumhoff & Baker, 1988; Jeanne, 1988; Peters & Crozier, 1988; Robinson & Page, 1988; Breed & Page, 1989; Crozier, 1989).

Species are also coming to seem more individual as more of them are studied in depth, and this growing awareness of species-level variation tends to undermine the idea that we need a monolithic 'explanation' for eusociality (one pathway, driven by a single overriding selective factor). However, as the number of detailed studies of individual species increases, so does the probability that significant and previously unknown patterns will emerge; one example is the apparent tendency of nest-mates to be relatively closely related in primitively social species and more variably related in advanced species. Such patterns will both stimulate and constrain future theoretical models for the origin and maintenance of different styles of eusociality.

11.5 Enabling mechanisms

The peculiar phylogenetic distribution of eusociality demands explanation, and it is also a pattern than can be used to understand the evolution of eusocial habits. We want to answer the question, 'Why did it happen in these lineages, and not in those?' There is no essential circularity involved in using our knowledge of phylogeny in this attempt, because sociality is only one character among many whose joint distribution we would like to explain.

To be useful, a theory for the evolution of eusociality must make predictions that could be contradicted by the actual pattern of associations. At least in their simplest forms, the existing theories tend to predict that if any aculeate Hymenoptera are eusocial, most should be. The reason why we cannot explain the distribution of eusociality with any precision is probably not that we have failed to identify some key ingredient, but rather that we do not understand in sufficient detail how the various (known) ingredients interact with each other. It does little good to ask, 'What do all eusocial lineages have in common?', because the character states shared by most eusocial lineages are likely to be shared with many solitary lineages as well. Instead, we need to ask, 'What *combinations* of factors set eusocial lineages apart from nearby lineages in which eusociality has not evolved?' Unfortunately, it may often be impossible to know many important characteristics of the ancestors of anciently eusocial groups such as termites and ants, but the phylogenetic approach should work well for social wasps (Carpenter, 1989) and especially for primitively eusocial bees, who have many living solitary relatives.

Because the great majority of independent origins of eusociality have occurred in the aculeate Hymenoptera, the question, 'What is special about the ancestors of social insects?' has so far tended to elicit answers that might equally well be responses to the question, 'What is special about aculeates?' This is a reasonable and necessary starting point but, for the reasons emphasized above, it is not sufficient because most aculeates remain solitary. The rest of this section briefly reviews some of these major factors that have been identified as being likely to favour the development of eusocial habits.

11.5.1 The nest

Almost all social insects live in permanent or semipermanent nests. The notable exceptions are the army ants, whose enormous colonies are constantly on the move. Legionary behaviour has arisen more than once in the ants and is a highly derived condition (Hölldobler & Wilson, 1990), so its occurrence does not contradict the hypothesis that nesting is a required preadaptation for the evolution of eusociality. If we ignore the termites for a moment, we can specify that the nest be a place to which food is brought, for the benefit of helpless offspring who cannot feed themselves. Contact between parents and immature offspring is not required (recall the primitively social halictid bees), but at least two adult generations must overlap.

There are several important implications of this style of nesting. First, the nest is a place where parents and offspring can easily meet each other; no ability to recognize kin is required, just the ability to recognize the nest. Second, there are significant ways in which an adult offspring can help its parent (or sibling) to reproduce, and it can begin productive work as soon as it emerges, without the delays involved in establishing an independent nest (Queller, 1989). For example, it can forage. If adults must provide everything an immature offspring eats, then the rate of reproduction is likely to be limited principally by the rate of provisioning. One adult can therefore directly increase another adult's reproductive rate. (In many insect groups such as flies and Lepidoptera, where the limiting factor is egg production, there is little a surrogate parent could do.) The nest also allows adults to cooperate in defending the brood, and this can be a great advantage if one guard is nearly as effective as two (for example, Lin, 1964). Third, nests are likely to be expensive, either because good sites are a limited resource, or because time and energy are spent in nest construction, or both. Thus an individual may find it profitable to join an established nest rather than attempting to initiate a new one. This immediately favours communal nesting, which is likely to be an important step on the way to eusociality, under almost any scenario.

There is evidence to support all three of these implications. Parents and offspring do meet in many (but not all) species of multivoltine solitary aculeates. For example, several generations of females may use the same burrow in wasps of the genus *Cerceris* (Evans & Hook, 1982a,b; 1986), and sisters associate communally even in nests away from their natal nest in *C. antipodes* (McCorquodale, 1988). Cleptoparasitic flies, wasps and bees are often a major cause of larval mortality in many species of solitary aculeates. Communal nesting can dramatically reduce the success of such parasites (for example, Abrams & Eickwort, 1981), especially where one female usually guards the nest entrance. The hypothesis that nests can be expensive is undoubtedly correct, especially for burrowing species that nest in firmly packed soils (see Evans & Hook, 1986). The mandibles of older individuals often show severe wear in such species, and burrow systems can be up to several metres long in arid regions where the best temperatures and humidities for larval development may be located at relatively great depths.

If communal nesting directly benefits the females involved, then there is no reason why nestmates need to be related. McCorquodale and Thomson (1988) recently found an instance of interspecific nest sharing in which a female of *Cerceris antipodes* was joined for more than two weeks by a female of *C. australis*. On excavation, the nest was found to contain many vacated cells of *C. antipodes* (one unworn female left the nest during the period of joint occupancy), six cells of *C. australis* containing immatures, and two cells probably of *C. australis*, but containing some prey species normally taken by *C. antipodes*. This seemingly bizarre incident suggests that 'factors such as cost of nest building and defense of the nest against nest parasites and conspecific females' may often favour joint nesting. Lin and Michener (1972), Evans (1977) and others have long advocated this idea.

Defence of the nest against conspecific females is a serious problem in some solitary species. The females of most solitary wasps and bees are usually faithful to a nest for at least several days at a time, and often for life. But apparently chronic patterns of nest switching, nest usurpation and joint provisioning occur in a number of sphecids, for example in the great golden digger *Sphex ichneumoneus* (Brockmann, 1979) and in several species of *Cerceris* (Alcock, 1975; Hook, 1987). Females in these species sometimes abandon nests for no apparent reason, to begin working at a nearby nest which may or may not be currently occupied by another female. When two females who are provisioning the same nest happen to meet they may fight, and in any case one evicts the other. The evicted female may then enter yet another existing nest, or dig a new one. However, new nests are sometimes dug by females who have not been evicted from the nest they previously occupied. During a lifetime of 2–4 weeks, a female may end up working at several or many different nests; a few females have been observed alternately provisioning two different nests on the same day.

The reasons for this strange behaviour are only partly understood. Stealing a nest may clearly be better than digging one, especially if it already contains some prey. But if most individuals attempted to steal nests, then few nests would be dug, and usurpation would be less profitable than digging. This suggests that the evolutionarily stable strategy may be a mixed one, in which each female switches randomly between episodes of digging and episodes of entering other nests. The data for *S. ichneumoneus* are in many ways consistent with this hypothesis (Brockmann & Dawkins, 1979; Brockmann *et al.*, 1979), given that females spontaneously abandon burrows at some finite rate, but there seems to be no convincing explanation for the unforced abandonment of burrows. Brockmann and Dawkins (1979) suggest that quantitative changes in the costs and benefits of digging, usurping and sharing (possibly caused by climatic or other long-term ecological changes) could cause systems like these to evolve into ones in which females jointly provision in a sustained and cooperative way, as in the communal species of *Cerceris* mentioned earlier.

In summary, there is every reason to believe that nesting is a supremely important preadaptation, without which eusociality is very unlikely to evolve. Although the ancestors of termites did not bring food to helpless immatures, they probably lived in family groups that excavated galleries in rotting wood, where defence of resources and protection of relatively vulnerable early instars might well have been important activities that could profitably be taken up by surrogate parents. However, although nesting may be a *necessary* condition, it is clearly not a *sufficient* condition for the evolution of eusociality, because most aculeate Hymenoptera (and many other nesting insects) remain thoroughly solitary. Some forms of nesting seem more likely to favour sociality than others; for example, twig nesting may tend to be unfavourable, as may those forms of nest sharing in which nest-mates tend to be unrelated.

11.5.2 The sting

The ovipositor of sawflies and parasitoid Hymenoptera was transformed, in the ancestor of aculeates, into a specialized sting apparatus. Most solitary aculeate wasps still use it for its original purpose, which is to paralyse their prey, although it is also sometimes used defensively against arthropod enemies. But in many social aculeates, the sting has evolved into a primarily defensive weapon which is often used against large vertebrate enemies. Large animals usually cannot afford to search for and process small insects where they occur singly, but where hundreds or thousands are concentrated in a nest, they may be very attractive.

Starr (1985) revived a suggestion made previously by various authors, to the effect that the sting is a uniquely effective deterrent against these large enemies, which would otherwise make advanced sociality impossible.

Kukuk *et al.* (1989) pointed out that invertebrate predators and parasites are far more damaging than vertebrates, for social and solitary insects alike. Moreover, sociality evolves 'in the context of small colonies' that are not yet subject to vertebrate predation. For example, ground-nesting halictids are never bothered by vertebrates. And some advanced ants and bees have even lost the sting. Starr (1989) replied that although vertebrate predation may be quantitatively minor in relation to invertebrate predation, it falls differentially on eusocial groups, including such primitively social genera as *Polistes*. He agreed that the origin of eusociality would not be much affected by pressure from vertebrate enemies, but argued that its subsequent elaboration might be substantially blocked without an effective and economical way to deter them. He also noted that 'eusociality [may result] from the confluence of several independent factors', the sting being one of these and perhaps more important than previously recognized.

11.5.3 *Arrhenotoky*

Parthenogenic male production is a consequence of haplodiploidy, and it appears to be primitive for the Hymenoptera as a whole. Haplodiploidy also gives rise to the unusual patterns of relatedness that have already been mentioned and to precise sex-ratio control, which will be discussed under the next subheading. But it may be worth emphasizing that the ability of unmated females to reproduce would seem to open up possibilities for novel strategies that involve both helping and direct reproduction. These possibilities are logically independent of the other consequences of haplodiploidy. Thus a hymenopteran female can emerge at a time of year when there may be very few adult males on the wing, without risking complete sterility. The ability to reproduce without mating can also be advantageous to solitary species, of course, and arrhenotoky occurs in many groups of arthropods other than Hymenoptera (see Bell, 1982).

11.5.4 *Asymmetrical relatedness and sex-ratio control*

The average reproductive success of male and female reproductives varies inversely with their relative abundances. For example, if there is twice as much investment in females as in males, then each female achieves only half as much reproductive success per unit invested. Because of this automatic devaluation of the overproduced sex, female-biased investment ratios do not increase the value of helping (relative to direct reproduction) for eusocial workers in the Hymenoptera, if all nests in the population produce the same sex ratio (for example, Craig, 1979; 1980b).

To exploit their greater relatedness to sisters than to offspring, workers must invest at ratios that are more female biased than the population as a whole. For example, if the population contains solitary and/or semisocial nests that produce relatively male-biased sex ratios (see section 11.2.1), and if females that work for their mothers can somehow arrange to work mainly on behalf of reproductive full sisters ($r = \frac{3}{4}$), then these non-reproductive working females may achieve greater inclusive fitness through their sisters than they would have achieved through their own offspring. Yanega (1989) observed just such a pattern in *Halictus rubicundus*, where a sample of parasocial nests (foundress not present) produced 78% males, and a sample of eusocial nests (foundress present) produced 45% males in the final generation of the year. Males are smaller than females in this species, so the ratios of investment were probably more female biased than the numerical sex ratios would indicate.

During the earliest stages in the evolution of eusociality, before workers are able to discriminate between male and female eggs or larvae, how might a population come to contain a mixture of male-specialist and female-specialist colonies? One possibility is that some females simply fail to mate and therefore produce all-male broods (Godfray & Grafen, 1988). Mated females are selected to produce compensating female biases that balance the ratio of investment in the population as a whole. A daughter that works for her mother will therefore expect to be related by more than one half to the reproductive sisters and brothers she rears (rather than being related by exactly one half to her own offspring). Working will therefore yield a higher inclusive fitness than reproducing independently, on the assumption that, as a worker, the daughter will add as many additional offspring to her mother's brood as she would have produced on her own. (The ratio of extra siblings reared to potential offspring forgone is often referred to as k; it is the benefit/cost ratio of Hamilton's rule; thus if $k > 1$ a worker is more efficient than a solitary female, while if $k < 1$ a worker is less efficient.) Under this model, a potential worker need not discriminate between immature male and female offspring, because *on average* a sibling is more valuable than an offspring. Thus helping may be favoured, even where workers are unable to direct their helping preferentially toward sisters, and even where k is somewhat less than one.

More generally, almost any mechanism that gives rise to 'split sex ratios' (Grafen, 1986) in a basically solitary species with overlapping generations may encourage the daughters of female-specialist mothers to work rather than attempting to reproduce on their own (Stubblefield & Charnov, 1986). The main requirement is that each newly emerged daughter have some clue as to whether her mother is a female specialist or a male specialist. Simply *being* a daughter is evidence that one's mother may be a female specialist, because most female offspring are produced by such mothers; in addition, some feature of the habitat, the nest or the mother herself may indicate that she is likely to produce an excess of fertilized eggs. For example, relatively large females might tend to produce an excess of female offspring, in species where females are larger than males (as they are in most aculeates), if large females are better able than small ones are to construct the large rearing cells or gather the large provision masses that are needed to produce a daughter of optimal size. Once daughters occasionally help, their presence in the nest could induce a mother to bias her sex ratio further, if helpers tend to cause more food to be provisioned per offspring, and if females gain more from being relatively large than do males (Frank & Crespi, 1989). Thus an initially modest difference between female- and male-biased mothers might quickly be amplified into a dimorphism of the kind envisioned by Trivers and Hare (1976), where solitary male-specialist and eusocial female-specialist nests coexist at stable frequencies, long before workers have evolved any means to control their colony's ratio of investment.

Sex-ratio differences *between* generations can also favour helping, and such differences are expected to arise under certain kinds of multivoltine life histories (several generations per year) with partial overlap of generations. For example, if there are two generations per year and males of the first generation commonly survive to mate with second-generation females (but second-generation males do not survive to mate with next year's first-generation females), then the sex ratios of the first- and second-generation offspring are expected to be male and female biased, respectively (Werren & Charnov, 1978). These biases arise because first-generation males have the opportunity to mate with two generations of females (which increases their value to the parents that produced them) while second-generation males mate with only one generation of females, and they also compete with surviving first-generation males.

If some females also contribute to two successive generations, their newly emerged daughters may have an opportunity to help them rear a biased sex ratio. Two different kinds of *partially bivoltine* life histories occur in the aculeate Hymenoptera, and they are expected to produce different patterns of alternating sex-ratio biases (Seger, 1983). Under the 'sphecid' or 'larval diapause' life history, both sexes overwinter as unmated immatures; the daughters of the overwintered females are members of the female-biased summer generation, and they contribute (either directly or as helpers) to the next overwintering generation, which is male biased. This life history tends to discourage helping, because the critical value of k is greater than one. But under the 'halictine' or 'female hibernation' life history, the overwintering generation consists of mated adult females; the daughters of these females belong to the male-biased generation, and they contribute (either directly or as helpers) to the female-biased generation that emerges and mates at the end of the season; these inseminated females go directly into hibernation, to emerge and reproduce in the following spring. This life history tends to encourage helping, because the critical value of k is less than one, even if all females contributing to the overwintering generation produce the same female-biased investment ratio.

Alternating sex-ratio biases of the predicted kind have been found in several species of bivoltine wasps and bees (for example, Seger, 1983; Tepedino & Parker, 1988; H.J. Brockmann & A. Grafen, unpublished data), but there is almost no direct evidence that males commonly survive to mate with two generations of females, which suggests that in many cases the alternating biases may have other causes. The predicted taxonomic associations between life history syndromes and origins of eusociality are also found, female hibernation being characteristic of temperate eusocial species and their close solitary relatives, and larval diapause being characteristic of many exclusively solitary taxa (Seger, 1983; Brockmann, 1984), but here, too, there are other possible explanations for the pattern (for example, Grafen, 1986; Stubblefield & Charnov, 1986).

As soon as facultative helping becomes established, for whatever reasons, selection will tend to improve the efficiency of cooperative breeding, and k may therefore increase to values well in excess of one. As k increases, the balance between solitary and eusocial nests will shift toward increased sociality. At sufficiently large values of k, a female will do best to remain a worker no matter what her colony's sex ratio, and the solitary life style will disappear. During this transition, foundresses may help the process along by making daughters who are slightly too small to be fully viable solitary females, and who therefore do best to remain and work in the natal nest (Craig, 1983). This form of parental manipulation should easily evolve in facultatively social species, because daughters are expected to make their decision (whether to leave or stay) on the basis of as much relevant information as possible, including their own expected fertility as solitary reproductives. It is possible that the basically solitary *Ceratina okinawana* (discussed earlier, in section 11.4.1) exhibits this subtle form of manipulation.

As the solitary life style disappears and colonies become large, two basic strategies emerge, by means of which workers can attempt to increase their inclusive fitnesses (Hamilton, 1972; Trivers & Hare, 1976). First, they can attempt to *monopolize male production*, by substituting sons ($r = \frac{1}{2}$) and nephews ($r = \frac{3}{8}$) for brothers ($r = \frac{1}{4}$); this pays off directly, and it has an additional indirect effect, similar to that of solitary male-specialist nests, since it induces queens to specialize voluntarily on female reproductives (Iwasa, 1981). Second, workers can attempt to *control the colony's investment ratio*; if they push it in the female direction, they may at least temporarily achieve the kind of relatedness advantage that their partially solitary ancestors achieved by exploiting opportunities such as those provided by split sex ratios. In both cases, the interests of workers are opposed to those of queens; this is obvious in the case of male production, but the conflict is somewhat more subtle in the case of pure sex-ratio manipulation, where queens produce all offspring.

Being equally related to her sons and daughters, a queen would do best to specialize on the sex that is currently under-represented in the population; if investment in females exceeds that in males, then a queen's inclusive fitness would be maximized if her colony produced nothing but males (her sons). She is indifferent as to her colony's sex ratio only when the population-wide investment ratio is balanced. But the workers see it differently. If their mother mated with only one male, then they are three times as closely related to their sisters as they are to their brothers, and they would do best to produce nothing but sisters, whenever the population-wide investment ratio is less biased than 1:3 (male:female). If the population were *more* biased than 1:3, they would actually do best to produce only brothers, because at such an investment ratio the relatively great fitness of males would more than offset the workers' relatively low relatedness to them. Workers are *indifferent* only when the population-wide investment ratio is exactly 1:3. This means that at all investment ratios between 1:1 and 1:3 the queen would prefer to make only males, and the workers would prefer to make only females. Qualitatively similar (but quantitatively different) zones of conflict arise even if workers produce some of the males, or if the queen had more than one mate, or if there is more than one queen (Trivers & Hare, 1976; for detailed analytical and population-genetic treatments of the model see, for example, Oster *et al.*, 1977; Benford, 1978; Charnov, 1978b; Macnair, 1978; Craig, 1980a,b; Uyenoyama & Bengtsson, 1981; Pamilo, 1982a; Bulmer, 1983b).

How are these conflicts resolved? If queens control the investment ratio, it should be 1:1 over the population as a whole (although it may vary among colonies), in the absence of some factor such as local mate competition that creates a biased equilibrium from a queen's point of view. If workers control the ratio, it should tend to be female biased, up to a limit of 1:3, depending on the extent of worker reproduction, multiple mating and polygyny. Trivers and Hare (1976) estimated investment ratios for 21 species of monogynous ants, and found a ratio remarkably close to 1:3, on average, but with considerable variation both among and within species (Fig. 11.5). Several polygynous species showed lower investment ratios, and two 'slave-making' species (where workers are stolen from other species) showed ratios near 1:1. These patterns are consistent with the hypothesis that workers have a significant degree of control in most species of ants, except in the slave-makers where the workers (being completely unrelated to the brood) are not under any selection to resist the queen's influence.

Alexander and Sherman (1977) proposed that the female biases might be caused by local mate competition (Hamilton, 1967), which occurs when males competing directly with each other for mates are closely related. This process gives rise to strong biases in various groups of solitary arthropods that mate before dispersal. Most monogynous ants appear on behavioural grounds to be outbreeding (for example, they often have synchronized nuptial flights involving colonies distributed over wide areas), and recent population-genetic studies of ants have consistently shown that these be-

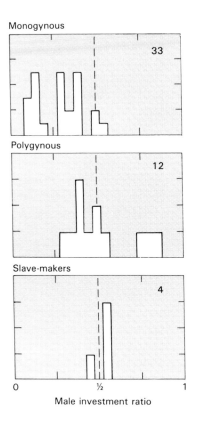

Monogynous

Polygynous

Slave-makers

0   ½   1

Male investment ratio

Fig. 11.5 Population-level investment ratios in various monogynous, polygynous and slave-making species of ants. Investment ratios are represented on the horizontal axis as the proportion invested in males (1:3 = ¼); they are estimated from the sex ratio and the dry weight ratio of males and females (Trivers and Hare, 1976). Data are from various sources, reviewed by Nonacs (1986); sample sizes (number of species) are shown at the upper right of each distribution. As predicted by Trivers and Hare (1976), the monogynous species tend to be strongly female biased, the polygynous species tend to be weakly female biased, and the slave-makers show no tendency to be biased in either direction (from Bull & Charnov, 1988).

haviours do indeed correlate with outbreeding (references cited above, in connection with relatedness in advanced eusocial Hymenoptera).

Investment ratios have now been estimated for many additional monogynous, polygynous and slave-making ants, and the combined data have been reanalysed in several different ways (Pamilo & Rosengren, 1983; Nonacs, 1986; Boomsma, 1988; 1989; Bull & Charnov, 1988; Bourke, 1989). Adult dry weights (used to estimate the costs of males and females) may tend to overestimate the costs of females, and thereby the extent of bias (Boomsma, 1989), but the basic pattern discovered by Trivers and Hare continues to hold up (Fig. 11.5), and many new details are emerging. For example, Ward (1983b) found that queenright nests in three *Rhytidoponera* species produced increasingly biased investment ratios (averaging 82%

female) as the proportion of queenless (relatively male-producing) colonies increased in their local populations.

In many species of ants there is a pronounced tendency for colonies to specialize either on male or female reproductives (reviewed by Nonacs, 1986). Boomsma and Grafen (1990) have proposed that these bimodal distributions of colony sex ratio may reflect the fact that for any given actor, the best sex-ratio strategy is often to invest entirely in one sex or the other (as explained above). If the workers' relatedness to females varies substantially among colonies, owing to variation in the number of times the queen mated (in monogynous species) or the number of queens (in polygynous species), then in some colonies (e.g. those where the queen mated once) workers will do best to specialize on female reproductives, while in others (e.g. those where the queen mated several times) workers will do best to specialize on males, since their average relatedness to female reproductives will be relatively low.

A pattern of investment and relatedness that seems to be consistent with this model was described for the monogynous ant *Lasius niger* by van der Have et al. (1988). Populations at three locations had different average ratios of investment, ranging from roughly 75% female to roughly 55% female; the population with the least biased investment ratio also showed the strongest tendency toward a split between male-specialist and female-specialist colonies (Fig. 11.6). The three populations also had different average levels of relatedness among female nest-mates, ranging from about 0.70 to about 0.55, and these relatedness estimates were ranked in the same order as the investment ratios, implying that investment was under worker control. The population with the lowest *average* relatedness (owing to the highest frequency of multiple mating) would be expected to have the highest *variance* of number of matings, and thus of relatedness; if so, workers in this population would be expected to produce extreme sex ratios relatively often (by the argument of Boomsma and Grafen), as in fact they did (Fig. 11.6). Several other mechanisms that could give rise to intercolony sex-ratio variation in eusocial Hymenoptera have been proposed by Frank (1987b), Herbers (1984), Nonacs (1986) and others.

In summary, the haplodiploid genetic system of the Hymenoptera sets up an interaction between relatedness asymmetries and sex ratios that may (i) encourage the initial evolution of helping, and (ii) give rise to conflicts over the sex ratio in fully eusocial species. The importance of relatedness asymmetries in the origin of helping has often been questioned, on the grounds that the first workers would have needed to evolve simultaneously both the working habit and the ability to manipulate the colony's investment ratio, since neither would be of use without the other. This problem now seems less severe than it once did, owing to the discovery of several plausible mechanisms that could generate favourable investment-ratio biases *before* helping had evolved. These ideas have stimulated increasing interest in the natural histories of solitary and primitively social wasps and

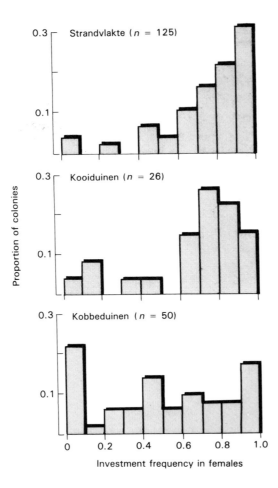

Fig. 11.6 Colony-level investment ratios in the monogynous ant *Lasius niger*, at three different locations on an island in the Dutch Wadden Sea. The estimated investment ratio is shown on the horizontal axis as the proportion invested in females (1:3 = ¾). Most queens in the Strandvlakte and Kooiduinen populations appear to have mated with only one male; genetic estimates of the relatedness of females within colonies are high (roughly 0.7). Most queens in the Kobbeduinen population appear to have mated with two or more males; relatedness within colonies is much lower (roughly 0.55). The average investment ratios are much more strongly female biased in the populations with predominant monandry than they are in the population with predominant polyandry, as expected, and the polyandrous population has a large proportion of colonies that specialize strongly on male or female reproductives (from van der Have *et al.*, 1988).

bees. There is also increasing interest in patterns of intercolony sex-ratio variation and in worker reproduction in advanced eusocial species. In both of these areas, studies that combine estimates of reproductive allocation with direct estimates of relatedness have been especially valuable.

11.5.5 Inbreeding and chromosomal translocations in termites

Termites are diploid and therefore do not exhibit the relatedness asymmetries caused by haplodiploidy in the Hymenoptera. But there are at least two ways in which patterns of relatedness favourable to the evolution of helping could arise under diploidy. One involves inbreeding, and one involves non-random segregation of chromosomes at meiosis in males.

Inbreeding increases the relatedness of family members and is therefore expected to promote altruism and to discourage selfishness (for example, Hamilton, 1972). But just as it increases an individual's relatedness to its siblings and their offspring, inbreeding also increases an individual's relatedness to its own offspring, and this tends to limit the extent to which inclusive fitness can be increased by helping, other things being equal. However, if *cycles* of inbreeding and outbreeding occur, such that new colonies are founded by primary reproductives (queen and king) that tend to be unrelated to each other but inbred, then first-generation offspring (who are outbred but relatively homogeneous) may be more closely related to each other than they would be to their own (outbred) offspring, and the members of subsequent inbred generations (produced by sib–sib or parent–offspring mating within the nest) may also be more closely related to their siblings than they would be to outbred offspring (Bartz, 1979; Pamilo, 1984b; Tyson, 1984; see Flesness, 1978). Many termite species produce supplementary reproductives that mate within the nest, and many of the winged reproductives (alates) produced late in a colony's life are therefore likely to be inbred; but primary reproductives often live long enough to directly produce significant numbers of alates, who therefore tend to be outbred (Myles & Nutting, 1988). As long as some primary reproductives are inbred an expected relatedness asymmetry will persist, but the magnitude of this asymmetry will decrease as the proportion of outbred alates increases. The termites have no close solitary relatives, so it is difficult to evaluate the likely importance of this effect during their origin.

The X chromosomes of species with heterogametic (XY) males are formally equivalent to a haplodiploid genetic system. Every one of a male's daughters inherits his only X, and full sisters are therefore related by three quarters at loci on their sex chromosomes. Similarly, every son inherits the same Y. (If recombination between X and Y is suppressed then the Y will tend to accumulate damaging mutations and may become largely inert genetically; for example, Nei, 1970; Charlesworth, 1978; Bull, 1983). In many species of termites the males carry several reciprocally translocated chromosomes that form rings (rather than bivalents) during meiosis (Syren & Luykx, 1977; 1981; Luykx & Syren, 1979; 1981). This implies that the Y is one of the translocated chromosomes, and that segregation from the ring is 'alternate', such that the entire set of translocated chromosomes passes together into a male-determining gamete. In effect, this entire set is a gigantic segmented Y chromosome, and its homologous (non-translocated) set is an X chromosome. In some species, half or more of the entire genome may be involved in this translocation complex and may therefore exhibit many of the relatedness patterns that arise under haplodiploidy. In principle, this could favour the evolution of increased altruism among the same-sex siblings (Lacy, 1980; 1984). However, no evidence of sex-biased interaction or association within colonies has been found (Luykx *et al.*, 1986; Hahn & Stuart, 1987), and any attempt at sex-ratio manipulation would place male and female reproductives into direct conflict, with respect to genes on chromosomes involved in the translocation complex (Leinaas, 1983). Also, the complexes are of variable size both within and among species, which implies that they may have arisen repeatedly within various termite lineages (Crozier & Luykx, 1985). Thus a 'haplodiploid analogy' seems unlikely to have played a significant role in the origin of termite sociality, but why it should exist so widely and variably remains a fascinating unsolved puzzle.

11.5.6 Kin recognition

The conditions under which an altruistic or selfish behaviour should be (or should not be) performed depend in part on the actor's relatedness to the recipient; it follows that there must be many circumstances under which eusocial insects could benefit from mechanisms that enabled them to estimate relatedness directly. Kin recognition in animals is a very rapidly developing field of study, and there have been several recent reviews devoted specifically to social insects (e.g. Gadagkar, 1985; Gamboa *et al.*, 1986; Breed & Bennett, 1987; Michener & Smith, 1987; Page & Breed, 1987; Carlin, 1988; 1989; Gamboa, 1988; Greenberg, 1988; see also section 12.3.2).

Many species of eusocial Hymenoptera are able to discriminate (i) nest-mates from non-nest-mates, and/or (ii) the kin of nest-mates from the kin of non-nest-mates. For example, in colonies of *Lasioglossum zephyrum*, one female usually stations herself at the nest entrance as a guard; a returning forager (or other bee) can enter the nest only if the guard gives way. Buckle and Greenberg (1981) set up colonies in which variable proportions of the worker force belonged to two unrelated sibships. Unfamiliar females from other colonies were then presented to the guards; some of these bees were related to the guard and her nest-mate sisters, if any, while others were related to the guard's unrelated nest-mates. Guards accepted novel bees that were related to any of their nest-mates, indicating that odours are genetically determined, but that discrimination is learned. However, in colonies where the guard was the only member of her sibship, she readily admitted her unrelated nest-mates' sisters (as usual), but she tended to reject her own unfamiliar sisters, indicating that she did not learn her own odour; her unrelated nest-mates *did* admit her sisters at nearly normal rates, indicating that an odour can be learned from a single unrelated bee in an otherwise related colony.

The pheromones involved seem likely to be mixtures of macrocyclic lactones and other similar compounds that are produced by the Dufour's gland (Smith *et al.*, 1985), in proportions that are highly variable but correlated among relatives (Smith & Wenzel, 1988). This system of pheromone production and recognition appears to play a role in nest marking (by females) and in mate recognition (by males), and is therefore likely to be present, at least in a rudimentary form, in many solitary bees (Michener & Smith, 1987).

Although the signals involved are not as well understood for other taxa as they are for *Lasioglossum*, the mechanism of learned recognition appears to be widely distributed. Some components of the signal often appear to be genetic and individual-specific, but others may be acquired from either the physical or the social environment (for example, Gamboa, 1988; Stuart, 1988; Crosland, 1989a,b). For purposes of defending a nest or a foraging territory (for example, Gordon, 1989), these distinctions should matter little and all available information should be used.

The ability to detect variation among one's nest-mates may allow females to make general assessments of average relatedness within the colony (Smith, 1987); such information could be useful in many contexts, for example in allocating investment to male and female reproductives. A major unresolved question concerns the extent to which individuals can distinguish among particular eggs, larvae or adults of different degrees of relatedness (e.g. full-sisters versus half-sisters). Such abilities could certainly be useful to a worker who possessed them, most notably and universally when investing in (or culling) immature reproductives, but also in food exchange with other workers, in the feeding (or culling) of multiple foundresses, and in forming a virgin queen's entourage, in species that reproduce by swarming (Breed & Bennett, 1987). However, to make such distinctions a worker would apparently need to compare herself directly to other individuals; this would require at least that she be able to learn her own odour. There is evidence to suggest that workers in some species may be able to distinguish kinship lineages within a colony under simplified experimental conditions (for example, Evers & Seeley, 1986; Getz & Smith, 1986; Carlin *et al.*, 1987; Frumhoff & Schneider, 1987; Hogendoorn & Velthuis, 1988), but it is not yet clear whether the observed biases, which tend to be weak (for example, Page & Erickson, 1986; Page *et al.*, 1989), are likely to produce significant inclusive-fitness benefits under natural conditions (for example, Page & Breed, 1987; Carlin, 1989; Carlin & Frumhoff, 1990).

11.6 Conclusion

Although social insects have invented a few new ecological niches, such as

fungus gardening, most species still compete directly with solitary pollinators, predators and scavengers. Their success must therefore flow largely
from the efficiencies made possible by a cooperative division of labour. Not
long ago this cooperation seemed almost entirely selfless; workers were
identical little automata, and their colony was a remorselessly efficient
machine, possesed of a single will. Have the discoveries of the last two
decades completely destroyed this vision of insect sociality? In some respects they clearly have. For example, workers and queens may waste time
and energy establishing dominance hierarchies, eating each others' eggs,
and even killing each other; workers may develop idiosyncratic habits, express genetic predispositions toward particular roles, establish breakaway
satellite colonies, or abandon social life altogether. These lapses from ideal
sociality are not confined to just a few taxa; different groups may specialize
in different forms of selfish disorder, but all seem to exhibit at least a few of
them.

It would be hard to invent a creature more charming and seemingly
peaceable than a sweat bee. Yet inside the nest—among family—dominance relationships may be established in pointedly unsubtle ways, as
Packer (1986b) observed in *Halictus ligatus*:

> Large individuals were observed to grab hold of smaller ones by the
> neck with their mandibles. The small females were then taken to the
> bottom of the burrow and repeatedly pummelled into the earth at the
> end or bashed from side to side against the burrow walls.

If Darwin had known about these behaviours, his treatment of insect sociality might have been even longer and more troubled than it was.

But even though life inside a colony may be more complicated and less
harmonious than anyone would have imagined until recently, it remains
fundamentally cooperative. Subordinated foundresses usually remain
with their dominant sister, and workers frequently die defending their nest-
mates. That these acts take place against a background of equivocation
makes them all the more remarkable. If female nest-mates were perfectly
related, the tension would disappear. The ant *Pristomyrmex pungens* appears to be entirely or almost entirely asexual (female parthenogenic or
thelytokous), but all individuals appear to function as reproductives early
in life and as foragers later in life, and there appear to be no dominance
hierarchies (Tsuji, 1988; 1990). This remarkable ant may therefore be considered communal rather than eusocial. Perhaps some kinds of cooperation evolve more easily in the presence of conflict than they do in its
absence.

12: Communication

David G.C. Harper

The elaborate songs of birds, such as the nightingale *Luscinia mega-rhynchos*, fascinate and entertain many people. Like other animal signals, they also raise many questions for ethologists. What information do they convey (section 12.3)? How might the signal have originated and why has it become more complex (section 12.4)? Can signals be used to deceive other animals (section 12.5)? What factors favour the use of song rather than smells or bright plumage (section 12.6)? Why are many signals, including nightingale song (Hultsch & Todt, 1989), highly variable (section 12.7)? Before examining these questions, I will introduce the participants involved in communication (section 12.1) and the technical concept of noise (12.2).

12.1 Signallers and receivers

Animals need to be good ethologists so that they can predict what other individuals are about to do. For example, an approaching rival may betray hostile intent by preparing its weaponry or reveal fear by hesitating. These cues can be used to select an appropriate response. Krebs and Dawkins (1984) suggested the term *mind-reading* to describe this ability to forecast other animals' behaviour.

Communication is what evolves if being mind-read is beneficial. An animal which is perceived as aggressive may drive mind-readers away without having to spend time and energy fighting. Aggressive animals should therefore make this particular act of mind-reading as easy as possible. They can do so by exaggerating the cues used by mind-readers. This process of exaggeration is call *ritualization* and produces what we recognize as *signals*. Signalling is adaptive because the signaller gains a net benefit from the responses of receivers, even though some responses may harm them. For example, male field crickets *Gryllus integer* call loudly at night and benefit by attracting fertile females. However, they also attract male rivals who approach silently to sneak copulations and parasitic flies *Calcocondmyia auditrix* which lay their eggs on calling males, eventually killing them. These *eavesdroppers* benefit at the signaller's expense, but do not do enough harm to outweigh the benefits of signalling (Cade, 1979). Other eavesdroppers exploit receivers; Mediterranean house geckos *Hemidactylus tursicus* intercept decorated crickets *Gryllodes supplicans* of both sexes as they approach calling males (Sakaluk & Belwood, 1984).

374

Receivers do not always benefit by responding to signals. Female bolas spiders (*Mastophora* sp.) from New Guinea lure male armyworm moths (*Spadoptera frugiperda*) within catching distance by emitting a pheromone that resembles the sexual attractant of female moths (Eberhard, 1977). The spider manipulates the male moths into using their own muscle power to propel them to their deaths. Dawkins and Krebs (1978) suggested that signals are the means by which signallers exploit the effectors, especially the muscles, of receivers. This emphasis on exploitation by signallers can be overdone: when receivers benefit by responding to a signal, they can equally be said to be exploiting the sense organs of the signaller (Krebs & Dawkins, 1984). It is therefore useful to distinguish between cases of *cooperative signalling,* in which receivers benefit (e.g. female field crickets), and of *non-cooperative signalling,* in which they do not (e.g. male armyworm moths).

Signals can evolve without being preceded directly by mind-reading. Some signals mimic cues that receivers use in other circumstances. Angler fish attract smaller fish into their mouths using elaborate lures, which mimic the cues used by the small fish to detect food (Wootton, 1990). Other signals may have evolved because they emphasized pre-existing signals. For example, male orang-utans *Pongo pygmaeus* hurl branches to the ground before calling to distant receivers (Galdikas, 1979), as if to attract their attention. Finally, some signals appear to have evolved from other signals. Among the bowerbirds of New Guinea, species with elaborate plumage tend to have simple bowers and vice versa (Brown, 1975; Diamond, 1988). Perhaps the risks of predation favoured a transfer of conspicuous signals from the males' bodies (an *intrinsic* signal) to the environment (an *extrinsic* signal).

The evolution of communication is constrained: mutant signals may be missed or misunderstood; mutant receivers may overlook or misinterpret signals. Some mutations influence both signalling and receiving roles: such *genetic coupling* would allow rapid evolutionary change in signals (Alexander, 1962). But if receivers respond most strongly to signals resembling the average they will impose stabilizing selection on signals, which may reduce that rate of evolution (Gerhardt, 1982; Butlin et al., 1985). However, receivers can have open-ended preferences for particular stimuli. Often this is because their sense organs respond more strongly as stimulus intensity increases (Hopkins, 1983). For example, female three-spine sticklebacks *Gasterosteus aculeatus* prefer to approach the larger of two dummy males, even if it is 25% longer than any natural male (Rowland, 1989). In some cases, receivers are not fussy about the position of the stimulus. Female zebra finches *Taenopygia guttata* prefer males with red rings on their legs (Burley, 1988), even though the only naturally red part of a male is his beak. Open-ended preferences among receivers will make it much easier for signallers to ritualize the subtle cues used by mind-readers into signals.

The coevolution of the signaller and receiver roles faces particular constraints in intraspecific signalling. These can be most clearly seen in intersexual communication. Selection for a signal in one sex may result in its
expression in the other sex due to *genetic correlation* (Muma & Weatherhead, 1989), unless the genes influencing the trait are either sex linked or
their expression is sex limited (Goodenough, 1984). Since sexually selected
signals are often energetically expensive (Ryan, 1988) and can increase the
risks of predation (Ryan, 1985), their expression in the receiver sex will
select against their elaboration in the signalling sex. Signaller–receiver
coevolution is also constrained when individuals can change roles, as in
agonistic displays, because the signals must be easily reversible (Hansen &
Rohwer, 1986).

12.2 Noise

One of the problems faced by signallers is that receivers are imperfect.
Receivers need to detect and recognize signals, a process which involves
coding the signal into an internal message. Often they also need to localize
the signaller in order to respond appropriately. Receivers can fail at any of
these stages. At the level of detection, they may miss some signals or
respond to irrelevant stimuli. It is impossible to minimize both missed
detections and false alarms. Factors that increase the frequency of these
mistakes are known as *noise*. From a signaller's point of view there are
three important sources of noise (Gerhardt, 1983; Wiley, 1983): (i) attenuation and degradation of signals through space or time; (ii) high levels of
competing stimuli; and (iii) receivers selected to avoid false alarms or to
resist exploitation.

Receivers usually have to use their sense organs to perform many tasks,
but a few sense organs detect only one signal. Male mosquitos *Aedes
aegypti* have the Johnston's organs on their antennae tuned so precisely to
the frequency of the wing-beats of mature females that they cannot hear
immature ones (Chapman, 1971). Specialized receptors reduce noise
because they can be optimized to detect one type of signal which is recognized at the point of reception. The drawback is the loss of flexibility.
Receivers will not evolve specialized receptors unless, like male
mosquitos, they benefit from responding to the signal. Signallers that are
harming their receivers will usually face increased noise, simply because
their signals have to be detected by generalist sense organs. More importantly, they face the increased noise due to selection on receivers to resist
exploitation.

12.3 Information

In the jargon of information theory (Shannon & Weaver, 1949), all signals
transmit information. Shannon information is the reduction in the uncer-

tainty of an observer about: (i) the signaller's future behaviour (*broadcast information*) e.g. a threat display may often be followed by an attack; and (ii) the receiver's future behaviour (*transmitted information*), e.g. a threat display may often be followed by retreat.

It is possible to use information theory to quantify the amount of information transmitted (Krebs & Dawkins, 1984). However, this has been of limited use in ethology because the values obtained depend critically on our ability to categorize signals in the same way as receivers (section 12.7.2).

A more interesting concept is that of semantic information. This focuses on the receiver, not the observer. Has the receiver acquired any information from the signaller?

12.3.1 *Information about the environment*

Alarm calls are a widespread example of signallers reporting on their external environment. Many small passerines give short (about 0.5 s) high-pitched (about 7 kHz) calls when they detect dangerous predators such as hawks. The similarity in the 'hawk-alarms' given by distantly related species has been suggested to reflect the convergent evolution of a call which is hard for predators to detect (Marler, 1955). Although red-tailed hawks *Buten jamaicensis* and great-horned owls *Bubo virgianus* seem to agree with humans that hawk alarms are hard to locate (Brown, 1982), pygmy owls *Glaucidium passerinum* and goshawks *Accipiter gentilis* find them no harder to locate than other calls (Shalter, 1978).

Vervet monkeys *Cercopithecus aethiops* give different alarm calls when they detect three different types of predator. Receivers dash into trees when they hear the 'leopard alarm', scan the sky after an 'eagle alarm' and scutinize the ground after a 'snake alarm' (Seyfarth *et al.*, 1980). They also recognize and respond differently to two alarm calls of the superb starling *Spreo superbus* (Cheney & Seyfarth, 1985). Young vervets have to learn to produce each alarm call in the right circumstances and to respond appropriately to conspecific and starling calls (Seyfarth & Cheney, 1986; Hauser, 1988).

12.3.2 *Information about identity*

Experienced entomologists need to hear just a few seconds of chirping to identify species of grasshoppers by their song. Indeed it is *so* easy for us to identify species using some signals, that we refuse to believe that the species differences evolved merely to signal species identity. But the baby needs saving from the bathwater; signalling species identity is an important function of communication (Halliday, 1983).

GROUPS

Members of social groups often share similar signals. For example, Australian magpies *Gymnorhina tibicen* share more songs within their cooperative groups than expected by chance, regardless of kinship (Brown *et al.*, 1988). However, this pattern may simply reflect similar learning opportunities and many cases of animals discriminating between group members and strangers (Halliday, 1983) are equally well explained by individual recognition (see below). Some animals, such as dwarf mongooses *Helogale undulata* (Rasa, 1973), mark other group members with chemical secretions. Their ability to recognize group members may be soley based on an ability to recognize their own signals. But some social animals do signal group membership. For example, black-capped chickadees *Parus atricapillus* modify features of their calls so that they converge on the common mean within their winter flock (Nowicki, 1989).

KIN

Previously encountered kin may be recognized as individuals (see below) and kin which live in groups may recognize novel kin using group 'labels' (Waldman *et al.*, 1988). However, some animals are able to discriminate between strangers on the basis of genetic similarity, which has been interpreted as evidence for direct signalling of kinship (Hepper, 1986; Fletcher & Michener, 1987; section 11.5.6). However, mechanisms for recognizing conspecifics or group members using genetic cues will produce this ability as a non-selected byproduct (Grafen, 1990c). Only one example demonstrates that the ability to discriminate genetic similarity provides an accurate estimate of relatedness and that it is used to make decisions about social behaviour (Grosberg & Quinn, 1986). The planktonic larvae of the ascidian *Botryllus schlosseri* aggregate and subsequently fuse with relatives. Fusion depends on a chemical signal: it occurs only if individuals share at least one allele at a histocompatibility locus. Recognition of close kin is probably important for larvae since unrelated individuals would be more likely to attempt to take over the reproductive role within a fused colony.

PAIRS

Some birds, such as twites *Acanthis flavirostris* (Marler & Mundinger, 1975), modify their vocalizations to resemble those of their mates. An obvious explanation is that this helps them to detect each other in flocks. However, a shared cell may serve other functions, especially in birds that synchronize their calls into complex duets. It has been suggested that duetting strengthens the pair bond (Farabaugh, 1982) or tests the likely fidelity of prospective mates (Wickler, 1980). But in canary-winged parakeets *Brotogeris versicolurus* duetting occurs during contests with other individuals and not during courtship. This suggests that duetting signals to rivals that they are facing a pair rather than a lone individual (Arrowood, 1988).

INDIVIDUALS

Individuals do not have to share a signal to recognize each other. Relatives and mates usually use individual differences in signals to recognize each other (Beer, 1970; Halliday, 1983). Many territorial animals can distinguish the acoustic signals of neighbours from those of strangers—fish (Myrberg & Riggio, 1985), amphibia (Davis, 1987), birds (Weary *et al.*, 1987) and mammals (Conner, 1985; Randall, 1989)—which will reduce the costs of maintaining established territory boundaries (Ydenberg *et al.*, 1988). Plumage variation allows turnstones *Arenaria interpres* to recognize individual flock members (Whitfield, 1986; 1988).

12.3.3 *Information about abilities*

CONTESTS

Contests could be settled in three ways:
1   *By an all-out fight.* The phenotypic characters determining the outcome of a fight are called resource-holding potential or RHP (Parker, 1974). High RHP is costly in contexts other than fighting; for example, being large may increase RHP, but will increase the costs of growth and maintenance.
2   *By a limited fight.* The winner will be the animal prepared to fight more intensely or for longer. This willingness has been called aggressiveness (Maynard Smith & Harper, 1988). Unlike high RHP, high aggressiveness is costly only in the context of fighting.
3   *By signals.* Contests would be settled by *assessment* or *conventional signals.* Assessment signals are biologically correlated with RHP and cannot be faked. For example, the spider *Agelenopsis aperta* assesses the mass of rivals from vibrations in their webs (Riechert, 1978). Conventional signals are those such as colour patterns that are not logically correlated with RHP.

FIGHT OR SIGNAL?

Fighting takes up time and energy, but these costs are usually trivial compared with those of injury and death, including the risk of predation (Hammerstein & Reichert, 1988). Whether it is better to settle a contest by signalling or fighting depends on the value of the resource, the costs of fighting, and the behaviour of opponents. Because the best thing to do depends on what your rival does, we need to use a game theory approach to seek the evolutionarily stable strategy (ESS). When the ESS is common, no alternative phenotype can invade the population.

The hawk–dove game (Maynard Smith, 1974) was developed to identify the ESS when animals can either signal or fight over a resource. Animals adopt one of two strategies, hawk and dove, but are otherwise identical. Hawks fight until they win or sustain enough injury to force them to retire. Doves try to settle disputes by signalling and retreat if attacked. A population of doves can be invaded by hawks because doves always lose to hawks. If the costs of fighting are less than the value of the contested resource, hawk is a pure ESS and animals should always fight. A more complex games model shows that when the resource is so valuable that a large fraction of expected future reproductive success is at stake, animals should fight to the death (Enquist & Leimar, 1990). However, if injury costs more than the resource value, doves can invade a population of hawks. This means that the ESS is a mixture of both strategies: play hawk with a probability p and dove with a probability $1-p$. This ESS could be reached in two ways. First, individuals could adopt a mixed strategy, fighting with probability p. Second, the population could become genetically polymorphic, with proportion p of animals always playing hawk and proportion $1-p$ always playing dove.

The hawk–dove game assumes that animals can differ only in strategy, but many games models have asked whether other differences between opponents can resolve contests without fighting. Three types of asymmetry have been considered (Maynard Smith & Parker, 1976):

1 *Pay-off asymmetries*. These cause differences in the net benefit of an ascalated fight (Hammerstein, 1981).

(a) *Asymmetries in RHP*. Opponents differ in fighting ability.

(b) *Asymmetries in resource value*. Food is worth more to a hungrier animal.

2 *Uncorrelated asymmetries*. Opponents differ in ways that do not cause differences in the net benefit of fighting. Each can then be said to adopt a particular role, e.g. owner or intruder.

As in the simple hawk–dove game, if the costs of injury are less than the value of the resource, animals should always play pure hawk, whatever asymmetries exist. When injury is more costly than the contested resource, games with asymmetries can have both 'commonsense' and 'paradoxical' solutions. In the case of pay-off asymmetries, paradoxical solutions (e.g. better fighter or hungrier animal loses) are only ESSs when the asymmetry is small. The 'commonsense' solution (e.g. better fighter or hungrier animal wins) is always an ESS, regardless of the size of asymmetry (Maynard Smith & Parker, 1976). Therefore, if we see a slightly larger or hungrier animal submit without a fight, we could be observing a 'paradoxical' ESS. But when we see animals defer to much smaller or much better-fed opponents we need to ask whether one type of asymmetry is outweighed by another (e.g. the smaller animal is hungrier or vice versa), whether the animals have perfect information about the asymmetry (Parker, 1984), and whether the contest is settled by uncorrelated asymmetries (Hammerstein, 1981).

Whenever fighting is costly, exchange of information about pay-off asymmetries is favoured. Contests should escalate gradually in a way that increases the probability that any asymmetry is detected before injury is risked (Parker & Rubenstein, 1981; Enquist & Leimar, 1983). As predicted, gradual escalation is a common feature of animal contests (Clutton-Brock et al., 1979).

In models of uncorrelated asymmetries, when the costs of injury are greater than the value of the resource there are always two pure ESSs. When the uncorrelated asymmetry is prior ownership of the resource, these are: (i) 'bourgeois': play hawk if owner, play dove if intruder; and (ii) 'anti-bourgeois': play dove if owner, play hawk if intruder.

Whichever ESS is adopted no fighting will occur and all contests will be settled conventionally (Maynard Smith & Parker, 1976). Intruders usually defer to owners (for example, Krebs, 1977; Davies, 1978), but in a few cases owners always lose (Burgess, 1976; J. Dawson in Dawkins & Krebs, 1978). The apparent rarity of anti-bourgeois behaviour suggests that the usual dominance of owners over intruders is the result of correlated asymmetries (Dawkins & Krebs, 1978). Asymmetries in RHP favouring owners could arise if owners often had to fight to defend the resource, if owners were hard to evict for physical reasons (e.g. male copulating with female; Parker, 1974), or if the owner could run off with the resource (Hamilton & Busse, 1982). Owners will benefit from asymmetries in resource value if individuals have to invest time to exploit a resource efficiently, such as learning about a territory (Krebs, 1982) or forming a pair bond with a mate (Yokel, 1989).

Respect for asymmetries is open to exploitation. Threatened male baboons may 'consort' with infertile females (Packer, 1979) or carry an unrelated infant (Packer, 1980). Both behaviours mimic situations in which they are favoured by a pay-off asymmetry and reduce the risk that they are attacked.

WHAT TYPE OF SIGNAL?

Assessment signals offer receivers useful information. Individuals with low RHP do not benefit by refusing to signal because receivers will underestimate their RHP (Krebs & Dawkins, 1984).

Conventionial signals have three potential advantages for signallers. First, they can provide privacy: eavesdroppers may find it harder to extract information. Second, patterns of even just two signals can carry very complex information (e.g. binary code in computers). Finally, conventional signals can be faked. So why should receivers respond to conventional signals?

The conventional signals that have received most attention are badges of status in birds (Whitfield, 1987). These are variable plumage features which influence the outcome of contests between birds of the same age and sex (Jarvi & Bakken, 1984). For example, male house sparrows *Passer domesticus* with larger black bibs on their throats tend to be dominant and experiments with dummies have shown that this is causal (Møller, 1987a,b). These birds defend better territories and are more likely to pair. Females implanted with oestradiol solicit more copulations from stuffed males with large bibs (Møller, 1988a). These results suggest that plumage variation signals individual quality and is important in both intersexual and intrasexual selection. If so, although the signal is conventional in the sense that it is not logically correlated with RHP, cheating is prevented because only some individuals can afford the costs of making and wearing badges of high status. For example, if male sparrows have their bibs enlarged by dyeing before being introduced into a captive flock, they do not become more dominant than controls and are repeatedly attacked by genuinely large-bibbed males (Møller, 1987b). However, badges are not strongly correlated, if at all, with components of RHP (Maynard Smith & Harper, 1988), which raises the possibility that they signal aggressive intent (section 12.2.4).

SIGNALS TO PREDATORS

Several signals once supposed to be warning signals may be directed at predators (Smythe, 1970; Baker & Parker, 1979). When approached by predators, Thomson's gazelles *Gazella thomsoni* give a display called stotting, in which they jump off the ground with their legs held stiff and straight. African hunting dogs *Lycaon pictus* chase those which stot at lower rates and are most likely to catch those which do not stot (Fitzgibbon & Fanshawe, 1988). As previously suggested (A. Zahavi in Dawkins, 1976), stotting appears to be an honest signal of a gazelle's ability to escape.

SIGNALS TO POTENTIAL MATES

These are discussed in Chapter 7. As with signals to rivals, it is important to distinguish signals that are necessarily correlated with particular attributes (e.g. handicaps) from those that are not (e.g. Fisherian traits).

12.3.4 *Information about intentions*

Both games theory (Caryl, 1979; 1982) and gene selection (Dawkins & Krebs, 1978) approaches suggest that signals will rarely provide information about intentions. First, signallers could lie. Second, signalling intentions is like laying cards on the table straight after a poker deal. One exception is that it will pay to signal retreat if the costs of continuing a contest become too high. Studies on Siamese fighting fish *Betta splendens* (Simpson, 1968; but see Bronstein, 1985) and the cichlid *Nannacara anomala* (Jakobsson *et al.*, 1979) found that winners could be identified only at the end of contests, supporting the idea that rivals are poker faced. Similarly, Caryl (1979) found that the displays of blue tits *Parus caeruleus* (Stokes, 1962), rose-breasted grosbeaks *Pheucticus ludovicianus* (Dunham, 1966) and great skuas *Stercorarius skua* (Andersson, 1976) were poor predictors of attack. Eleven different threat displays have been described in the great skua but none was consistently associated with attack (Paton & Caryl, 1986).

However, it may be naïve to expect a signal accurately to predict a single behaviour (Hinde, 1981). First, signals often occur in sequences that require more complex analysis. However, this cannot explain the great skua data because most interactions involved only one display. Second, signals may offer imprecise information ('I will attack or stay put, but not retreat'), conditional information ('I will retaliate if attacked') or may be attempts to gain information ('What will you do if I escalate briefly?'). The relationship between a signal and the signaller's subsequent behaviour could thus be very complex and we might do better to examine the responses of receivers to different signals. In the case of the great skua, receivers did not respond differently to different threat displays (Paton, 1986). This raises the question of why skuas have so many displays (section 12.7.1). A third hypothesis about the weak correlation between many signals and subsequent behaviour is that important variations between signals of the same type are being overlooked (section 12.7.2).

Contrary to expectation, some animals appear to signal aggressive intent. The fact that the eventual winner of contests can sometimes be identified from its behaviour very early in a contest (Turner & Huntingford, 1986; Glass & Huntingford, 1988; Franck & Ribowski, 1989) is inconclusive because the winner may be signalling a higher expected pay-off rather than an intention to win. Male African elephants *Loxodonta africana* periodically enter a state called musth, during which they are extremely aggressive towards other males and search for receptive females. They signal that they are in musth by producing copious glandular secretions, urine marking and vocalizing. Small musth males in poor condition can dominate larger males in good condition to whom they are normally subordinate, suggesting that the signal does not reflect asymmetries in RHP. Apparently small musth males can stop signalling very quickly and so avoid fights with large musth males (Poole, 1989). Similarly, rutting male Pere David's deer *Elaphurus davidianus* signal aggression by dilating their preorbital gland, but close it if they meet a dominant with a dilated gland (Wemmer *et al.*, 1983). In both cases, males appear to gain increased access to females by signalling aggression, but without paying the cost of fighting. Maynard Smith (1982) has argued that such behaviour could be stable only if the signal was irrevocable, although it is not necessary that males signalling aggression *always* pay costs when they meet each other. Infrequent but extreme costs would suffice. The signals of musth are so dramatic, including a vivid green penis, that it is hard to believe that small males always avoid fights with larger musth males. Elephant fights are rare, but are often fatal (Poole, 1989).

Badges of status in birds (section 12.3.3) may be signals of aggressive intent. A games theory model suggested that a polymorphism in aggression can be stable so long as the cost of a contest rises sufficiently steeply with aggression. Conventional signals could be used to resolve contests only if individuals signalling high aggression have to pay a cost when they meet each other (Maynard Smith & Harper, 1988). The observation that male house sparrows with experimentally enlarged badges were attacked by birds with naturally large badges (Møller, 1987b) is consistent with this prediction.

12.3.5 Information about nothing

Finally, communication may have nothing to do with semantic information, but simply be a method by which signallers manipulate receivers (Dawkins & Krebs, 1978; Krebs & Dawkins, 1984).

12.4 Ritualization

Any cues used by mind-readers can become ritualized. Comparative studies have suggested four main sources of raw material for signals (Hinde, 1970; Brown, 1975):

1   *Intention movements.* Before take off, many bird crouch, withdrawing their heads and raising their tails. Many bird displays involve these movements, suggesting that mind-readers once used them to predict when birds were about to fly.

2   *Motivational conflict.* When an animal is motivated to behave in two or more conflicting ways it may:

(a) *Alternate between different behaviours.* The zigzag dance of male sticklebacks may be the ritualized result of males tending to alternate between retreat and approach when meeting a gravid female.

(b) *Behave ambivalently.* Many threat signals appear to be made up of elements from both aggressive and fearful behaviour.

(c) *Perform a displacement activity.* Ducks sometimes preen when dithering between different behaviours. This suggests a reason why mind-readers have selected for mock preening in the courtship displays of many male ducks; in the past, males torn between sex and aggression may have given away their indecision to females by preening.

(d) *Redirect the behaviour.* Animals torn between attacking or fleeing from a rival sometimes attack an innocent third party. It has been suggested that the aerial displays of some seabirds have been derived from such redirected attacks.

3   *Autonomic responses.* The erect fur of a cat threatened by a dog and the

flushed face of its enraged owner may both be the ritualized products of changes in thermoregulation.

4 *Protective responses.* Primates prepare for a fight by partially closing their eyes and withdrawing the corners of their mouth. Different elements of these responses seem to have given rise to signals ranging from fear 'grins' in chimpanzees *Pan troglodytes* to the eyebrow lowering used by dominant baboons *Papio anubis* (Andrew, 1963).

12.4.1 Features of ritualized signals

Wiley (1983) has reviewed the features of ritualized signals.

REDUNDANCY

There are two basic types of redundancy. The first is repetition of the same signal: about 80% of the signals beamed from ground control to spacecraft are repeats. A second type of redundancy involves using more than one signal. We do this when we write a cheque in both words and figures.

CONSPICUOUSNESS

Signals can be made to stand out from the background by being of high intensity or by contrasting strongly. In order to appear contrasting to receivers, signals need to differ from those given by other signallers. Signals with opposite meanings, such as threat and appeasement, often differ dramatically in appearance (Darwin, 1872). The courtship displays of closely related species are often more divergent than their threat displays, presumably because mating with the wrong species is more costly than threatening them.

STEREOTYPY

Receivers need to classify the signals that they detect. Signallers can help receivers to do this by producing a small range of signals each with a very distinct form or *typical intensity*. Sometimes, it may pay signallers to produce signals resembling those of other species for this reason (section 12.5).

ALERTING COMPONENTS

Receivers may miss signals because they are doing something else. It therefore pays signallers to give a highly detectable warning that they are about to signal. Male orang-utans *Pongo pygmaeus* hurl branches to the ground before calling (Galdikas,1979). The introductory tonal elements of the song of rufous-sided towhees *Pipilo erythrophthalmus* have been shown experimentally to draw receivers attention to the trill that follows (Richards, 1981).

12.4.2 Selective advantages of ritualization

INCREASED RANGE

Ritualization, especially conspicuousness, can increase the range at which receivers can detect a signal. The louder calls of animals ranging from red-winged blackbirds *Agelaius phoeniceus* to blue monkeys *Cercopithecus mitis* can be detected by conspecifics up to about two home-range diameters away (Brenowitz, 1982; Brown & Schwagmeyer, 1984; Brown, 1989). Signals with this range will be detectable by a neighbour whenever it and the signaller are in their respective home ranges.

REDUCING AMBIGUITY

Ritualization may reduce the risk that receivers confuse one signal for another (Cullen, 1966). An obvious drawback is that ritualization reduces the information available to receivers. Producing small repertoires of signals given with typical intensity does so directly, while redundancy uses up time or body parts that could have been used to give additional signals. This loss of information is often regarded as the price paid by the signaller to overcome noise (Morris, 1957).

MANIPULATION

However, if signals benefit only the signaller, the most important source of noise is the 'sales resistance' of receivers (Dawkins & Krebs, 1978). Ritualized signals, which are used extensively by the advertising industry, may be the method used to overcome this noise. But, if they benefit from responding to a signal, receivers will be selected to reduce the noise faced by signallers. One method is to become more sensitive to the signal. Male silkworm moths *Bombyx mori* can respond to just one molecule of the female pheromone bombykol (Schneider, 1974). The females have to produce vast quantities of pheromone because males are few and far between. However, in other cooperative signalling systems there is little noise to overcome; signaller and receiver might be close together with few competing stimuli. In these cases signals will evolve to become as cheap as possible; they will be barely detectable and human observers may overlook most cooperative signalling.

HONESTY

Especially when signalling is non-cooperative, receivers should pay particular attention to those signals that reliably transfer semantic information. Deceit, like information, is not required for successful manipulation (Krebs & Davies, 1987) and honest signalling will often be the best strategy. Bluff by signallers can be countered in a variety of ways (section 12.5) and if honest signals are costly (e.g. section 12.3.3, 12.3.4) they may be impossible to mimic (Zahavi, 1975; 1977a,b; 1979; 1986; see also section 1.4). Many aspects of ritualization, such as conspicuousness and repetition, will increase the costs of signals. Zahavi (1979) also suggested that stereotypy provides a uniform background against which subtle differences between signallers are easier to detect and that repetition gives receivers more opportunities to glean information. The strut display of the male sage grouse *Centrocercus urophasianus* has often been quoted as an example of extreme stereotypy (e.g. Wiley, 1983). Ironically, male mating success is correlated with repetition rate and small differences in the calls given during the display (Gibson & Bradbury, 1985), as predicted by Zahavi (1977a).

RELATIVE IMPORTANCE

In some cases, one type of noise seems to have been particularly important for the ritualization of the signal. The begging calls of nestling birds seem far louder and more repetitive than needed to be detected by a parent (Krebs & Dawkins, 1984). For example, 90% of food deliveries to American white pelican chicks *Pelecanus erythrorhynchos* are accompanied by frantic convulsions, during which chicks often peck at and batter themselves (Cash & Evans, 1987). Parent–offspring interactions are likely to involve manipulative young and sales-resistant parents (Trivers, 1974), suggesting that the important type of noise faced by begging chicks is that due to the sales-resistance of their parents.

The type of noise that favours ritualization can change during the evolution of a signal (Krebs & Davies, 1987). Initially ritualization reduces ambiguity, but then manipulation may take over, selecting for discrimination by receivers. If the receivers become sufficiently discriminating, signals that do not transfer accurate information are predicted to fall into disuse. In many cases, manipulation will always involve honest signalling.

12.5 Deceit

Signallers could deceive receivers by giving qualitatively incorrect signals, like giving an alarm call when there is no danger, or by giving quantitatively misleading ones, like signalling an ability to fight for a long time when they have only enough energy to fight for seconds. One advantage of studying humans is that they can be told to lie. When they do so, they tend to wave their hands about, shrug, and talk in a high-pitched voice. Listeners appear to be able to use these cues to detect attempted deception (Ekman *et al.*, 1976; Streeter *et al.*, 1977). So, at least in our own species, mind-readers are skilled enough to detect some deceptions and it is possible that the need for such skills lies behind many of our mental abilities (Humphrey, 1976). How do other receivers respond?

12.5.1 Mimicry

Although the term was originally applied to palatable animals that gain protection by resembling distasteful or dangerous species (*models*) any signallers giving qualitatively incorrect signals can be viewed as Batesian mimics. They exploit receivers by giving signals that are also used in other contexts. The nestlings of viduine finches have the same complex colour patterns in their mouths as the young of the estildid finches which their mother brood-parasitized (Friedmann, 1960). Female grey partridges *Perdix perdix* pretend to have a broken wing when predators approach their young (Skutch, 1976). Several species of birds, including great tits *Parus major* in foraging flocks, sometimes give hawk alarms (section 12.3.1) in the absence of a predator. Receivers of their own and other species flee, allowing the signaller to grab food (Munn, 1986; Møller, 1988b). Batesian mimics will tend to make receivers cautious about responding to signals and can therefore harm the mimicked signallers.

Batesian mimicry is contrasted with Mullerian mimicry in which the receivers are not exploited. For example, many Hymenoptera benefit from sharing a yellow and black colour pattern because this makes it easier for receivers to learn that this is a warning signal. It is possible that birds with similar hawk alarms are Mullerian mimics, rather than independent solutions to the design problem of producing an unlocatable call (section 12.3.1). If signallers benefit from giving alarms because the predator fails to catch food and moves elsewhere, they will benefit from warning as many species as possible.

12.5.2 Responses to deceit

Receivers have several non-exclusive reponses to deceit.

IGNORE THE SIGNAL

This response is probably more common when receivers face exploitation by Batesian mimicry (qualitatively incorrect signals) because quantitatively misleading signals still contain some information. An exaggerated threat display is at least not an appeasement display.

If receivers cannot discriminate between a model and a Batesian mimic, they should ignore the signal if:

$$m > B/(B+C)$$

where m is the proportion of mimetic signals, B is the benefit obtained by responding to the genuine signal, and C is the cost of responding to a mimetic one. For a fox encountering an apparently crippled partridge, B is the profit gained by attacking a genuinely injured bird, while C combines the smaller profit gained by attacking a fit parent and the reduced probability of locating the brood.

Three conditions should favour Batesian mimicry:

1   Mimicry rare (*m* small): in cases involving interspecific communication, this can arise simply because the signalling species is kept rare by factors other than the availability of victims. For example, cuckoos *Cuculus canorus* are much rarer than their hosts (Sharrock, 1976). However, it is not always true: most birdwatchers (and foxes?) see more grey partridges pretending to have broken wings than genuinely injured birds.

2   Mimics do little harm (*C* small): great tits that leave a feeding site for a few seconds when they hear a false hawk alarm may not suffer much.

3   Real signal is very important (*B* large): great tits not responding to a genuine hawk alarm may be killed.

Exploitation by Batesian mimics is an obvious feature of interspecific communication. It may be less common between members of the same species, because there are fewer constraints on the frequency of mimicry (1 above). Sometimes mimicry is used only in certain conditions. The stomatopod crustacean *Conodactylus bredini* is very vulnerable to being evicted from its burrow when newly moulted because it lacks body armour and the use of its claws. Some individuals attempt to bluff intruders by giving a meral spread display which usually precedes attack. If the intruder ignores the signal, bluffers have to flee for their lives (Steger & Caldwell, 1983). This is probably why newly moulted owners are less likely to bluff if the intruder is larger (Adams & Caldwell, 1990).

DEVALUE THE SIGNAL

This penalizes honest signallers and can lead to increasing exaggeration by signallers until the costs of signalling match its benefits. At this point we might expect this signal to fall into disuse, because it pays signallers to adopt an energetically more efficient signal. However, they are largely dependent on their mind-readers: receivers have to respond to new cues for them to work as signals. As a result, costly signals may be the 'ghosts of manipulation past'.

PICK NEW CUES

Deceit selects for efficient mind-readers. However, receivers will eventually run out of cues, however subtle. As above, we may see very costly signals and very cautious receivers. Courtship displays are often remarkable for the ridiculous contortions of males and the apparent indifference of females (for example, Snow, 1958).

PROBE THE SIGNALLER

Receivers could demand to see evidence to back up signals. They might try to force suitors signalling high parental ability to go and catch food, or to force rivals signalling high aggression to behave aggressively. The draw-back is that this strategy will often be costly for receivers.

THIRD PARTIES AND DECEIT

Attempts to deceive some receivers may be limited by the responses of others. Young male elephant seals *Mirounga angustirostris* attempt to sneak copulations by entering the harems of territorial bulls and behaving like females. They are often detected only because females protest loudly when the sneaks attempt to mount (Cox & Le Boeuf, 1977).

12.6 Types of signal

The problems of overcoming noise (section 12.2) differ according to the type of energy used to signal with.

12.6.1 *Chemical signals*

Chemical receptors are probably the oldest sense organs and so chemical signals, or pheromones are believed to be the oldest type of signal. They can be released in three ways.

RELEASE INTO AIR OR WATER

When the air or water is stationary, the pheromone spreads by diffusion. This is a slow process, especially in water. Signallers can increase the area over which receivers can detect the pheromone (its active space) by pro-ducing more pheromone, releasing pheromone continuously rather than as a single puff, using a more volatile chemical, and by exploiting gentle movements in the air or water (sometimes generated by the signaller). If re-ceivers benefit by responding to the signal they will be selected to increase their sensitivity to the pheromone (Bossert & Wilson, 1963).

When pheromones are released into air or water, signallers can be hard to find because diffusion is unlikely to produce steep concentration gradients (Kennedy & Marsh, 1974). This problem is simplified if the medium is mov-ing since receivers can move up-current to help them locate the signaller. However, turbulence is likely to be greater in a moving medium and can make source localization very difficult (Gerhardt, 1983).

Signallers can reach more receivers by using a more volatile chemical. This may not be appropriate; for example, threat signals may be directed at one specific individual. Even if signallers would benefit from increasing their audience, increasing a chemical's volatility has two drawbacks. First, the easiest way of increasing volatility is to decrease molecular weight, which may reduce the specificity of the signal. A widely used solution to this problem is for the signaller to produce a cocktail of chemicals. Even when closely related species use the same pheromones they produce them in differing amounts (Roelofs, 1979). This helps receivers to discriminate between species, although with variable success (Shorey, 1976). The second problem with increasing volatility is that it reduces the fade-out time, the time taken for the active space to vanish (Bossert & Wilson, 1963).

DEPOSIT ON TO STATIONARY OBJECT

If molecules diffuse slowly, signallers can leave durable signals on stationary objects. The drawback is that pheromones of low volatility will be hard for receivers to detect. Signallers often use visual cues to reduce this problem. Many mammals scent-mark on conspicuous objects (Barette & Messier, 1980; Gosling, 1987; Smith *et al.*, 1989), while desert iguanas *Diposaurus dorsalis* use an ultraviolet marker to make pheromones easier for receivers to locate (Alberts, 1989).

DIRECT TRANSFER TO RECEIVER

Honeybee queens produce a pheromone, 'queen substance', which is dispersed around the colony by workers. Exposure to queen substance inhibits workers from rearing rebel queens and from laying eggs themselves (Seeley, 1985).

12.6.2 *Mechanical signals*

SOUNDS

Sound is a particular type of mechanical disturbance transmitted through air or water involving longitudinal vibrations of the medium. Sounds can be relatively easy to localize, are very flexible and are virtually instantaneous, so that they can convey a lot of information. Signallers can alter the range of sound signals by changing their volume or frequency. The main drawbacks are that sound signals can be energetically expensive (Ryan, 1988; Wells & Taigen, 1989) and that noise levels from other signallers and from abiotic sources can be very high. Sound signals can be distorted in many ways, including scattering and absorption by obstacles and microclimatic effects such as temperature gradients in air. Distortion can alter the amplitude, frequency and temporal patterning of signals. Moving vegetation is a source of especially complex distortion (Michelsen & Larson, 1983). Small animals face particular problems because they can efficiently emit only ultrasonic frequencies which suffer acute distortion (Michelsen *et al.*, 1982) and they therefore tend to use other types of mechanical signals (see below). On the other hand, large animals such as whales and elephants can efficiently use very low-pitched sounds to communicate over kilometres (Mobley *et al.*, 1988; Poole, *et al.*, 1988).

Sound travels about five times faster in water than in air (Popper & Coombs, 1980) and suffers much less attenuation (Hawkins, 1986). However, there are drawbacks to using sound under water. First, the similarity in acoustic impedance of water and body tissues makes it difficult to construct sensitive receptors. Second, sounds are hard to locate under water because the speed of their arrival means that they reach all parts of the animal almost simultaneously (Hopkins, 1983).

Many studies have looked at the factors that influence the distances over which sounds can be detected by receivers. This is a complex subject in which generalizations look increasingly sparse (Piercy & Embelton, 1977; Michelsen & Larson, 1983; Sorjonen, 1986; Boarman, 1990). One 'rule' is that sounds attenuate least in air at dawn and dusk and less by night than by day (Wiley & Richards, 1978). It is therefore not surprising that many sound producers call nocturnally or at dawn and dusk, although other factors are clearly involved (Mace, 1987; Cuthill & Macdonald, 1990).

The best time to call can also be influenced by other species. A Panamanian katydid *Neoconocephallus spiza* is inhibited from singing by three closely related species. In the few habitats where it lives alone, it sings by night. Elsewhere it sings by day, and removal of the other katydid species shows that this is due to their presence (Greenfield, 1988).

Males attempting to attract mates often form groups known as choruses. They may benefit from the increased amplitude and repetition of the signal, but suffer increased competition for females and broadcast time. Overlap between calls will often disrupt important features (Schwartz, 1987), but attempting to avoid all overlaps between calls would be self-defeating, since most males would be forced to remain silent. Computer simulation suggests that the best solution to this problem is to avoid overlaps with only one or two of the loudest (probably nearest) rival callers. This appears to be the tactic used by the tree frog *Eleutherodactylus coqui* (Brush & Narins, 1989). When males call sequentially, females sometimes prefer the first caller (Whitney & Krebs, 1975; Dyson & Passmore, 1988), but not always (Forester & Harrison, 1987). Experiments with the bushcricket *Tettigonia viridissima* demonstrate that male mating success is lowered by proximity to other callers, which explains why males are regularly dispersed within a chorus (Arak *et al.*, 1990).

OTHER MECHANICAL SIGNALS

All animals can use *tactile signals*. These can obviously be used only at very short ranges, but are easy to locate and act instantaneously.

Many small insects generate *contact vibrations*, transmitting low frequency vibrations through solid objects such as plants (Michelsen *et al.*, 1982). This is probably the least studied form of communication, but attenuation and distortion limit it to ranges of less than 2 m. Contact vibrations seem to convey little specfic information unless used together with other signals, especially pheromones (Markl, 1983).

Compression vibrations differ from sounds because they involve compression, rather than longitudinal, waves in the medium. They are efficient only if the signaller uses a radiator which is about one third of a wavelength in diameter. A 1 cm insect would therefore have to signal at above 10 kHz in air and above 50 kHz in water (Markl, 1983). These signals would be energetically expensive and would suffer extreme attenuation and distortion due to their high frequency. However, movements of the medium very close to the signaller can be effective short range signals and are not so expensive to create. Female *Drosophila* can detect the wing vibrations made by courting males in this way, and many fish use their lateral lines to detect the fin movements of signallers (Bleckmann, 1986).

Signallers can also set up vibrations at the boundary between two media. For example, *boundary vibrations* are produced by the water strider *Gerris remigis* in the form of ripples on the water surface. Males signal their sex by producing higher frequency surface waves than females (Wilcox, 1979).

12.6.3 *Electromagnetic signals*

LIGHT

Visual signals can be limited in range by dense vegetation or by low light levels. The former problem cannot be overcome, but some animals cope with darkness by producing light of their own (Lloyd, 1971; Marshall, 1979). The ease with which visual signals can be located is a major advantage, but also leaves signallers vulnerable to eavesdroppers. One way of overcoming this problem is to use visual signals that can be rapidly hidden, such as the epaulettes of red-winged blackbirds *Agelaius phoenicus* (Hansen & Rohwer, 1986).

ELECTRIC FIELDS

Fish of the families Mormyridae and Gymnotidae communicate using the discharge of an electric organ in their tail, which also help them to locate objects in their murky environments by electrolocation. This behaviour is best known in the Mormyrids, which use it extensively in contests and courtship (Hopkins, 1977). The Gymnotid *Hypopomus occidentalis* can discriminate between the sexes using features of the electric organ discharge (Shumway & Zelick, 1988), and there are characteristic differences between the discharges of dominant and subordinate males (Hagedorn & Zelick, 1989). Because electric fields attenuate rapidly in water, these signals are important only at short range, despite the great sensitivity of the receptors (Hopkins, 1983).

12.7 Signal variety

12.7.1 Why are there so many threat displays?

Many animals have several threat displays; 11 have been described for the great skua (Paton & Caryl, 1986). Tinbergen (1959) suggested that different displays indicate different levels of aggression. This raises the problem of why signallers ever indicate low aggression (Andersson, 1980). Enquist *et al.* (1985) found that the effectiveness of the threat displays of fulmars *Fulmarus glacialis* increases with their cost. Some displays are less likely to cause a rival to flee, but are unlikely to provoke a fight. Displays that are more likely to persuade a rival to retreat are also more likely to cause fighting. Perhaps fulmars select high cost–high benefit displays when the contested resource is valuable to them.

Tinbergen (1959) also suggested that different threat displays may be adapted for signalling at receivers in different locations. For example, the postures adopted by Sabine's gull *Xema sabini* while calling aggressively are related to the position of the opponent rather than the signaller's motivation. Similarly, Paton (1986) has suggested that some of the 11 threat displays described for great skuas are simply variants of one display caused by differences in the relative position of the signaller and receiver.

A third hypothesis to explain the diversity of threat displays is that the efficiency of a signal changes in a frequency-dependent way (Andersson, 1980). A new threat signal may be efficient because it reliably predicts attack. It therefore pays signallers to bluff, which will reduce the signal's efficiency, until it pays receivers to ignore it. If so, what limits the number of threat displays? One possibility is that there are only a certain number of features which predict attack well enough to become ritualized. Another is that receivers may find displays hard to categorize if there are too many of them. The efficiency hypothesis suggests that signals that are hard to bluff (e.g. assessment signals) or for which it is disadvantageous to bluff (e.g. 'I'm going to submit') should emerge slowly because their efficiency is not continually eroded. They ought also to be better predictors of subsequent behaviour than bluffable signals, and retreat does appear to be signalled more reliably than attack (Caryl, 1979; but see Paton & Caryl, 1986). The hypothesis also suggests that different threat displays should be about equally good at predicting attacks. Although this is not true for fulmars, it is true for some species (Andersson, 1980). For example three blue tit displays were followed by attack on 48%, 44% and 43% of occasions (Stokes, 1962).

12.7.2 Variable signals

One reason for signals appearing to be weak predictors of the signaller's future behaviour (section 12.3.4) could be that we are not dividing signals into enough categories. The acoustic features of many calls vary according to context and receivers are sometimes able to make use of this extra information (Green, 1975; Veen, 1987; Gouzoules & Gouzoules, 1989; Masataka, 1989). For example, juvenile rhesus macaques *Macaca mulatta* scream in many situations. When attacked by dominants they give 'noisy' screams, but when interacting with kin they give 'pulsed' screams. Mothers respond differently to playbacks of these two variants (Gouzoules *et al.*, 1984). Belted kingfishers *Megaceryle alcyon* give vocal rattles in doublets during contests and as evenly spaced calls in other contexts. They are less likely to approach playbacks of doublet calls (Davis, 1988).

12.7.3 Song repertoires

Although many willow tits *Parus montanus* sing only one song (Latimer, 1977), most passerine birds have repertoires of at least two songs. The obvious explanation is that each song type has a different function, but this appears to be rare. Chestnut-sided warblers *Dendroica pensylvanica* use one of their two song types mainly to attract mates (Kroodsma *et al.*, 1989), and aquatic warblers *Acrocephalus aquaticus* use one of their three song types only when challenged by a rival male (Catchpole & Leisler, 1989). Blue-winged warblers *Vermivora pinus* use one of their two song types mainly at dawn, a pattern that they appear to learn from conspecifics (Kroodsma, 1988; see also Spector *et al.*, 1989). But it is hard to believe that the huge song repertoires of birds such as the American robin *Turdus migratorius* (Read & Weary, 1990) can be explained in this way. There is increasing evidence that song complexity *per se* improves territory defence and attraction (Baker, 1988). However, many experiments on song repertoires involve pseudoreplication (Kroodsma, 1989a), because only one version of each stimulus (e.g. a repertoire of particular size) is used. This reduces the sample size to one, regardless of the number of individuals hearing each stimulus.

Male great tits and red-winged blackbirds respond more strongly to playback of larger repertoires (Krebs, 1976; Yasukawa, 1981) and the playback of larger repertoires deters intruders from taking over vacant territories for longer (Krebs *et al.*, 1978; Yasukawa & Searcy, 1985). However, male song sparrows *Melospiza melodia*, great reed warblers *Acrocephalus arundinaceus* and sedge warblers *A. schoenbaenus* do not respond more strongly to playback of larger repertoires (Catchpole, 1983; 1989; Searcy, 1983) suggesting that song complexity is not always important in territory defence. The evidence that large repertoires increase mating success is more consistent across species. Female canaries *Serinus canaria* build nests more rapidly when they hear more complex songs (Kroodsma, 1976). Oestradiol-implanted females of several different species solicit more copulations when exposed to playback of larger repertoires (for example, Catchpole *et al.*, 1984; Searcy, 1984; Baker *et al.*, 1986; 1987; Searcy, 1988).

Repertoire size is correlated with pairing success in sedge warblers (Catchpole, 1980), great tits (McGregor & Krebs, 1982; but see Lambrechts & Dhondt, 1986) and great reed warblers (Catchpole, 1988), but not in song sparrows and red-winged blackbirds (Searcy, 1988).

Why do large repertoires have these effects? First, they appear to reduce habituation in some cases (Krebs, 1976; Yasukawa, 1981; Petrinovich, 1984), but not in others (Lemon et al., 1981; Simpson, 1984; Weary & Lemon, 1988). Second, it may be advantageous for males to match at least one song type of as many rivals as possible (Falls et al., 1982). Male great tits continue to add new songs to their repertoire throughout their lives (repertoire size does not increase with age, because others are lost). As predicted by the song-matching hypothesis these new song types resemble those of new neighbours (McGregor & Krebs, 1989). Third, large song repertoires may help to reduce physiological exhaustion, because each song type probably requires the use of different combinations of nerves and muscles (Lambrechts & Dhondt, 1988), although the evidence is contentious (Lambrechts, 1988; Weary et al., 1988). Finally, if singing more song types is costly, reportoire size may be an assessment signal used in mate choice (Catchpole, 1980) or during contests. Costs of large repertoires might also explain why repertoire size varies between individuals. Since male great tits learn more songs than they sing (McGregor & Avery, 1986), differences in repertoire size are unlikely to be due to differences in learning ability or learning opportunity alone. Great tits with larger repertoires are more likely to use song types with high frequencies or high sound content (Lambrechts & Dhondt, 1990), suggesting that large repertoires are costly because some song types are difficult to sing. If repertoire size alone is an assessment signal, it is surprising that males do not switch song types more often; male five-striped sparrows *Amphispiza quinquestriata* take hours to sing 90% of their repertoire (Groschupf & Mills, 1982). This emphasizes that repertoire size is not the only aspect of song that varies. Western meadowlarks *Sturnella neglecta* respond more strongly to short bouts of each song type (e.g. AAABBBCCC . . .) rather than to long bouts (AAAAABBBBB . . .). They are also more responsive if the song types in a repertoire are more contrasting with each other (Horn & Falls, 1988). It would be interesting to know if males sing in a way that exploits these effects.

12.8 Concluding section

What of the future? Hopefully, there will be increased communication between physiologists and ethologists. It is clear that some sensory systems have evolved soley to detect signals (Hopkins, 1983), but there is increasing evidence that signallers can exploit the tuning of receptors which evolved for other purposes (Ryan et al., 1990). Strikingly, some types of sound are better at eliciting motor responses in domestic dogs *Canis familiaris* than others, and it is possible that this is also true for other mammals and birds (McConnell, 1990). We do not know why receivers respond in this way and to what extent signallers can exploit it.

Many studies of communication tacitly assume that animals classify signals in the same way that we do. The weakness of this assumption undermines the use of Shannon information theory (section 12.3) and may lead us into exaggerating the diversity of some signals (section 12.7.1) and underestimating that of others (section 12.7.2). Recently, bird song researchers have begun to study how birds categorize songs, both in the field (Falls *et al.*, 1988) and in the laboratory (Weary, 1990). These studies, which demonstrate that our classifications can be incorrect (Weary, 1989), could be extended to other types of signal.

Evolutionarily stable strategy models of contests are increasingly incorporating the idea (Parker, 1974; 1984) that opponents gradually acquire information about each other (Enquist & Leimar, 1990). These *sequential assessment* models do not differ qualitatively from simpler models (section 12.3), but they do allow novel quantitative predictions of, for example, contest duration and probability of victory. More empirical tests of these models are needed (Englund & Olsson, 1990).

The comparative method (Clutton-Brock & Harvey, 1984) has been remarkably neglected in studies on communication (but see Weary & Lemon, 1988; Read & Harvey, 1989; Read & Weary, 1990). In particular, it would be useful to know more about the ecological correlates of interspecific variation in signal diversity (section 12.7).

References

Abrahams M.V. (1986) Patch choice under perceptual constraints: a cause for departures from an ideal free distribution. *Behav. Ecol. Sociobiol.* **19**, 409–415.
5.5.2

Abrahams M. & Dill L.M. (1989) A determination of the energetic equivalence of the risk of predation. *Ecology* **70**, 999–1007.
Introduction to Part 2, 4.3.1, 5.5.2, 6.6

Abrams J. & Eickwort G.C. (1981) Nest switching and guarding by the communal sweat bee *Agapostemon virescens* (Hymenoptera, Halictidae). *Insectes Soc.* **28**, 105–116.
11.5.1

Abrams P.A. (1986) Adaptive responses of predators to prey and prey to predators: the failure of the arms race analogy. *Evolution* **40**, 1229–1247.
Introduction to Part 2

Adams E.S. & Caldwell R.L. (1990) Deceptive communication in asymmetric fights of the stomatopod crustacean *Gonodactylus bredini. Anim. Behav.* **39**, 706–716.
12.5.2

Adams J., Greenwood P.J. & Naylor C.J. (1987) Evolutionary aspects of environmental sex determination. *Int. J. Invert. Repr. Develop.* **11**, 123–136.
8.6.2

Adams J., Rothman E.D., Kerr W.E. & Paulino Z.L. (1977) Estimation of the number of sex alleles and queen matings from diploid male frequencies in a population of *Apis mellifera. Genetics* **86**, 583–596.
11.4.2

Alatalo R.V., Carlson A., Lundberg A & Ulfstrand S. (1981) The conflict between male polygamy and female monogamy: the case of the pied flycatcher *Ficedula hypoleuca. Amer. Natur.* **117**, 738–753.
9.5.3

Alatalo R.V., Eriksson D., Gustavsson L. & Larsson K. (1987) Exploitation competition influences the use of foraging sites by tits: experimental evidence. *Ecology* **68**, 284–290.
5.6.5

Alatalo R.V., Gustafsson L. & Lundberg A. (1984) High frequency of cuckoldry in pied and collared flycatchers. *Oikos* **42**, 41–47.
9.5.2

Alatalo R.V., Lundberg A. & Glynn C. (1986) Female pied flycatchers choose territory quality and not male characteristics. *Nature* **323**, 152–153.
9.5.3

Alatalo R.V., Lundberg A. & Rätti O. (1991) Imperfect female choice in pied flycatchers: evidence for female deception. *Behav. Ecol.* (in press).
9.5.3

Alberts A.C. (1989) Ultraviolet sensitivity in desert inguanas: implications for pheromone detection. *Anim. Behav.* **38**, 129–137.
12.6.1

Albon S.D. & Clutton-Brock T.H. (1988) Climate and the population dynamics of red deer in Scotland. In: *Ecological Changes in the Uplands* (ed. M.B. Usher & D.B.A. Thompson), pp. 93–107. Blackwell Scientific Publications, Oxford.
8.2

Alcock J. (1975) Social interactions in the solitary wasp *Cerceris simplex* (Hymenoptera: Sphecidae). *Behaviour* **54**, 142–152.
11.5.1

Alcock J. (1979a) The evolution of intraspecific diversity in male reproductive strategies in some bees and wasps. In: *Sexual Selection and Reproductive competition in Insects* (ed. M.S. Blum & N.A. Blum), pp. 381–402. Academic Press, New York.
5.6.2

Alcock J. (1979b) Multiple mating in *Calopteryx maculata* (Odonata: Calopterygidae) and the advantage of non-contact guarding by males. *J. Nat. Hist.* **13**, 439–446.
9.3.4

Alcock J., Eickwort G.C. & Eickwort K.R. (1977a) The reproductive behaviour of *Anthidium maculosum* (Hymenoptera: Megachilidae) and the evolutionary significance of multiple copulations by

females. *Behav. Ecol. Sociobiol.* **2**, 385–396.
9.3.4

Alcock J., Jones C.E. & Buchmann S.L. (1977b) Male mating strategies in the bee *Centris pallida* (Hymenoptera: Anthophoridae). *Amer. Natur.* **111**, 145–155.
9.3.4

Alexander R.D. (1962) Evolutionary change in cricket acoustical communication. *Evolution* **16**, 443–467.
12.1

Alexander R.D. (1974) The evolution of social behavior. *A. Rev. Ecol. Syst.* **5**, 325–383.
8.5.1, 10.4.1, 11.2.2

Alexander R.D. (1975) Natural selection and specialized chorusing behavior in acoustical insects. In: *Insects, Science and Society* (ed. D. Pimentel), pp. 35–77. Academic Press, New York.
9.4.4

Alexander R.D. (1979) *Darwinism and Human Affairs.* University of Washington Press, Seattle.
3.1, 3.3.3, 3.4.2, 3.4.4

Alexander R.D. & Borgia G. (1979) On the origin and basis of the male–female phenomenon. In: *Sexual Selection and Reproductive Competition in Insects* (ed. M.S. Blum & N.A. Blum), pp. 417–440. Academic Press, New York.
8.1

Alexander R.D., Hoogland J.L., Howard R.D., Noonan K.M. & Sherman P.W. (1979) Sexual dimorphisms and breeding systems in pinnipeds, ungulates, primates, and humans. In: *Evolutionary Biology and Human Social Behavior: An Anthropological Perspective* (ed. N.A. Chagnon & W. Irons), pp. 402–435. Duxbury Press, North Scituate, Massachusetts.
3.3.1

Alexander R.D. & Sherman P.W. (1977) Local mate competition and parental investment in social insects. *Science* **196**, 494–500.
11.5.4

Allden W.G. (1970) The effects of nutritional deprivation on the subsequent productivity of sheep and cattle. *Nutr. Abstr. Rev.* **40**, 1167–1185.
8.2

Allen J.A. (1988) Frequency-dependent selection by predators. *Phil. Trans. Roy. Soc. B.* **319**, 485–503.
6.2.1, 6.2.2, 6.4

Allison A.C. (1954) Notes on sickle-cell polymorphism. *Ann. Hum. Genet.* **19**, 39–57.
1.2.2

Altmann S.A., Wagner S.S. & Lengington S. (1977) Two models for the evolution of polygene. *Behav. Ecol. Sociobiol.* **2**, 397–410.
8.1, 9.5.3

Andelman S. (1986) Ecological and social determinants of cercopithecine mating systems. In: *Ecological Aspects of Social Evolution* (ed. D.I. Rubenstein & R.W. Wrangham), pp. 201–216. Princeton University Press, Princeton, New Jersey.
9.2.1

Andersson M. (1982b) Female choice selects for extreme tail length in a widowbird. *Nature* **299**, 818–820.
7.3.2

Andersson M. (1976) Social behaviour and communication in the great skua. *Behaviour* **58**, 40–77.
12.3.4

Andersson M. (1980) Why are there so many threat displays? *J. Theor. Biol.* **86**, 773–781.
12.7.1

Andersson M. (1984a) Brood parasitism within species. In: *Producers and Scroungers: Strategies of Exploitation and Parasitism* (ed. C.J. Barnard), pp. 195–227. Croom Helm, London.
9.5.2

Andersson M. (1984b) The evolution of eusociality. *A. Rev. Ecol. Syst.* **15**, 165–189.
11.1

Andersson M. (1986) Evolution of condition-dependent sex ornaments and mating preferences: sexual selection based on viability differences. *Evolution* **40**, 804–816.
1.4.1, 7.3.1

Andersson M. & Eriksson M.O.G. (1982) Nest parasitism in Goldeneyes *Bucephala clangula:* some evolutionary aspects. *Amer. Natur.* **120**, 1–16.
2.5.1

Andersson M. (1982a) Sexual selection, natural selection and quality advertisement. *Biol. J. Linn. Soc.* **17**, 375–393.
7.3.1

Andrew R.J. (1963) Evolution of facial expression. *Science* **142**, 1034–1040.
12.4

Ankney C.D. (1980) Egg weight, survival and growth of lesser snow goose gos-

lings. *J. Wild. Mgmt.* **44**, 174–182.
8.2

Aoki S. (1987) Evolution of sterile soldiers in aphids. In: *Animal Societies: Theories and Facts* (ed. Y. Itô, J.L. Brown & J. Kikkawa), pp. 53–65. Japan Science Society Press, Tokyo.
11.3

Appollonio M., Festa-Bianchet M., Mari F. & Riva M. (1990) Site specific asymmetries in male copulatory success in a fallow deer lek. *Anim. Behav.* **39**, 205–212.
9.4.5

Apter D. & Vihko R. (1983) Early menarche, a risk factor for breast cancer, indicates early onset of ovulatory cycles *J. Clin. Endocrinol. Metab.* **57**, 82–88.
3.3.4

Arak A. (1983) Male–male competition and mate choice in anuran amphibians. In: *Mate Choice* (ed. P. Bateson), pp. 67–107. Cambridge University Press, Cambridge.
7.3.2

Arak A. (1988a) Female mate slection in the natterjack toad: active choice or passive attraction? *Behav. Ecol. Sociobiol.* **22**, 317–327.
7.2.3

Arak A. (1988b) Sexual dimorphism in body size: a model and a test. *Evolution* **42**, 820–825.
7.2.3, 7.3.2

Arak A. (1988c) Callers and satellites in the natterjack toad: evolutionarily stable decision rules. *Anim. Behav.* **36**, 416–432.
9.3.2, 9.4.3

Arak A., Eiriksson T. & Radesater T. (1990) The adaptive significance of acoustic spacing in male bushcrickets *Tettigonia viridissima*: a perturbation experiment. *Behav. Ecol. Sociobiol.* **26**, 1–8.
12.6.2

Arcese P. (1989) Intrasexual competition and the mating system in primarily monogamous birds: the case of the song sparrow. *Anim. Behav.* **38**, 96–111.
9.5.3

Arditi R. & d'Acorogna B. (1988) Optimal foraging on arbitrary food distributions and the definition of habitat patches. *Amer. Natur.* **131**, 837–846.
4.2.4

Arnold S.J. (1985) Quantitative genetic models of sexual selection. *Experientia* **41**, 1296–1310.
7.3.1

Armitage K.B. (1986) Marmot polygyny revisited: determinants of male and female reproductive strategies. In: *Ecological Aspects of Social Evolution* (ed. D.I. Rubenstein & R.W. Wrangham), pp. 303–331. Princeton University Press, Princeton, New Jersey.
9.2.2

Arrowood P.C. (1988) Duetting, pair-bonding and agonistic display in parakeet pairs. *Behaviour* **106**, 129–157.
12.3.2

Ashkenazie S. & Safriel U.N. (1979) Time–energy budget of the semipalmated sandpiper *Calidris pusilla* at Barrow, Alaska. *Ecology* **60**, 783–799.
9.5.5

Askenmo C. (1977) Effects of addition and removal of nestlings on nestling weight, nestling survival and female weight loss in the pied flycatcher *Ficedula hypoleuca* (Pallas). *Ornis Scand.* **8**, 1–8.
2.5.1

Askenmo C. (1979) Reproductive effort and return rate of male pied flycatchers. *Amer. Natur.* **114**, 748–753.
2.5.1

Austad S.N. & Rabenold K.N. (1985) Reproductive enhancement by helpers and an experimental examination of its mechanism in the bicolored wren: a facultatively communal breeder. *Behav. Ecol. Sociobiol.* **17**, 19–27.
10.3.1, 10.3.2

Austad S.N. & Rabenold K.N. (1986) Demography and the evolution of cooperative breeding in the bicolored wren, *Campylorhynchus griseus*. *Behaviour* **97**, 308–324.
10.3.3

Axelrod R. & Hamilton W.D. (1981) The evolution of cooperation. *Science* **211**, 1390–1396.
3.2.3, 10.6.2

Bagenal T.B. (1969) Relationships between egg size and fry survival in brown trout *Salmo trutta* L. *J. Fish Biol.* **1**, 349–353.
8.3.1

Bailey R.C. & Aunger R. (1989) Significance of the social relationships of Efe pygmy men in the Ituri forest, Zaire. *Amer. J. Phys. Anthropol.* **78**, 495–507.
3.2.2

Baker M.C. (1988) Sexual selection and the size of repertoire in song birds. In: *Acta XIX Congressus Internationalis Ornithologi* (ed. H. Ouellet), Vol. I, pp. 1358–1365. University of Ottawa Press, Ottawa.
12.7.3

Baker M.C., Bjerke T.K., Lampe H.U. & Espmark Y.O. (1987) Sexual response of

female yellowhammers to differences in regional song dialects and repertoire sizes. *Anim. Behav.* **35**, 395–401.
12.7.3

Baker M.C., Bjerke T.K., Lampe H.U. & Espmark Y.O. (1987) Sexual response of female yellowhammers to differences in regional song dialects and repertoire sizes. *Anim. Behav.* **35**, 395–401.
12.7.3

Baker R.R. (1978) *The Evolutionary Ecology of Animal Migration,* Holmes & Meier, New York.
5.6.1

Baker R.R. & Parker G.A. (1979) The evolution of bird coloration *Phil. Trans. Roy. Soc. Lond. B* **287**, 63–130.
7.3.3, 12.3.3

Baldwin J. & Krebs H.A. (1981) The evolution of metabolic cycles. *Nature* **291**, 381–382.
2.8.6

Banks M.J. & Thompson D.J. (1985) Lifetime mating success in the damselfly *Coenagrion puella. Anim. Behav.* **33**, 1175–1183.
9.3.4

Barkan C.P.L. & Withiam M.L. (1989) Profitability, rate maximization, and reward delay: a test of the simultaneous-encounter model of prey choice with *Parus atricapillus. Amer. Natur.* **134**, 254–272.
4.5.2

Barkow J.H. (1989) *Darwin, Sex, and Status: Biological Approaches to Mind and Culture.* University of Toronto Press, Toronto.
3.4.4

Barette C. & Messier F. (1980) Scent marking in free-ranging cyotes *Canis Latrans. Anim. Behav.* **28**, 814–819.
12.6.1

Bart J. & Tornes A. (1989) Importance of monogamous male birds in determining reproductive success: evidence for house wrens and a review of male-removal studies. *Behav. Ecol. Sociobiol.* **24**, 109–116.
9.5.1

Bartz S.H. (1979) Evolution of eusociality in termites. *Proc. Natl. Acad. Sci.* **76**, 5764–5768, (also see correction; **77**, 3070).
11.5.5

Bartz S.H. & Hölldobler B. (1982) Colony founding in *Myrmecocystus mimicus* Wheeler (Hymenoptera: Formicidae) and the evolution of foundress associations. *Behav. Ecol. Sociobiol.* **10**, 137–147.
11.4.2

Bateman A.J. (1948) Intra-sexual selection in *Drosophila. Heredity* **2**, 349–368.
9.1

Batra S.W.T. (1964) Behavior of the social bee *Lasioglossum zephyrum,* within the nest (Hymenoptera: Halictidae). *Insectes Soc.* **11**, 159–185.
11.4.1

Batra S.W.T (1966) The life cycle and behavior of the primitively social bee, *Lasioglossum zyphyrum* (Halictidae). *Univ. Kans. Sci. Bull.* **46**, 359–422.
11.4.1

Batra S.W.T. (1968) Behavior of some social and solitary halictine bees within their nests: a comparative study (Hymenoptera: Halictidae). *J. Kans. Ent. Soc.* **41**, 120–133.
11.4.1

Baum W.M. (1974) On two types of deviation from the matching laws: bias and under-matching. *J. Exp. Anal. Behav.* **22**, 231–242.
4.6.1

Bauwens D. & Thoen C. (1981) Escape tactics and vulnerability to predation associated with reproduction in the lizard *Lacerta vivipara. J. Anim. Ecol.* **50**, 737–743.
6.6

Bayliss J.R. (1978) Paternal behaviour in fishes: a question of investment, timing or rate? *Nature* **276**, 738.
8.4.2

Bayliss J.R. (1981) The evolution of parental care in fishes, with reference to Darwin's rule of male sexual selection. *Env. Biol. Fish.* **6**, 223–251.
8.1, 8.4, 8.4.2, 8.4.3

Bednarz J.C. (1987) Pair and group reproductive success, polyandry and cooperative breeding in Harris' Hawks. *Auk* **104**, 393–404.
12.3.1

Bednarz J.C. & Ligon J.D. (1988) A study of the ecological basis of cooperative breeding in the Harris' hawk. *Ecology* **69**, 1176–1187.
12.4.3

Beehler B.M. & Foster M.S. (1988) Hotshots, hotspots and female preferences in the organisation of mating systems. *Amer. Natur.* **131**, 203–219.
9.4.3

Beer C.G. (1970) Individual recognition of voice in the social behavior of birds. *Adv. Study Behav.* **3**, 27–74.
12.3.2

Begon M. (1985) A general theory of life-history variation. In: *Behavioural ecol-*

ogy (ed. R.M. Sibly & R.H. Smith), pp. 91–97. Blackwell Scientific Publications, Oxford.
2.7.1

Beissinger S.R. (1987) Mate desertion and reproductive effort in the snail kite. *Anim. Behav.* **35**, 1504–1519.
8.4.4

Beissinger S.R. & Snyder N.F.R. (1987) Mate desertion in the snail kite. *Anim. Behav.* **35**, 477–487.
9.5.1

Bekoff M. & Wells M.C. (1982) Behavioral ecology of coyotes: social organization, rearing patterns, space use, and resource defense. *Z. Tierpsychol.* **60**, 281–305.
10.3.1

Bell A.E. & Burris M.J. (1973) Simultaneous selection for two correlated traits in *Tribolium*. *Genet. Res.* **21**, 29–46.
2.2.2

Bell G. (1980) The costs of reproduction and their consequences. *Amer. Natur.* **116**, 45–76.
2.6.1

Bell G. (1982) *The Masterpiece of Nature.* Croom Helm, London.
11.2.1, 11.5.3

Bell G. (1984a) Measuring the cost of reproduction. I. The correlation structure of the life table of a plankton rotifer. *Evolution* **38**, 300–313.
2.2.2

Bell G. (1984b) Measuring the cost of reproduction. II. The correlation structure of the life tables of five freshwater invertebrates. *Evolution* **38**, 314–326.
2.2.2

Bell G. (1985) On the function of flowers. *Proc. Roy. Soc. Lond. B* **224**, 223–265.
7.5

Bell G. (1987) Two theories of sex and variation. In: *The Evolution of Sex and its Consequences* (ed. S.C. Stearns), pp. 117–134. Birkhauser, Basel.
2.8.6

Bell G. & Koufopanou V. (1986) The cost of reproduction. *Oxf. Surv. Evol. Biol.* **3**, 83–131.
2.1, 2.2.1, 2.2.2, 2.3.1, 2.6.1

Bellows T.S. Jr. (1982) Analytical models for laboratory populations of *Callosobruchus chinensis* and *C. maculatus* (Coleoptera, Bruchidae). *J. Anim. Ecol.* **51**, 263–287.
2.4.1

Belovsky G. (1987) Hunter–gatherer foraging: a linear programming approach. *J. anthropol. Archeol.* **6**, 29–76.
3.2.1

Benford F.A. (1978) Fisher's theory of the sex ratio applied to the social hymenoptera. *J. Theor. Biol.* **72**, 701–727.
11.5.4

Bennett P.M. & Harvey P.H. (1988) How fecundity balances mortality in birds. *Nature* **333**, 216.
2.7.2

Benson W.W. (1972) Natural selection for Mullerian mimicry in *Heliconius erato* in Costa Rica. *Science* **176**, 936–939.
6.5

Bergelson J.M. (1985) A mechanistic interpretation of prey selection by *Anax junius* larvae (Odonata: Aeschnidae). *Ecology* **66**, 1699–1705.
6.2.1

Berglund A., Rosenqvist G. & Svensson I. (1986) Reversed sex roles and parental energy investment in zygotes of two pipefish (Syngnathidae) species. *Mar. Ecol. Prog.* **29**, 209–215.
8.4.2

Bernstein C., Kacelnik A. & Krebs J.R. (1988) Individual decisions and the distribution of predators in a patchy environment. *J. Anim. Ecol.* **57**, 1007–1026.
4.7.2, 6.5.2

Bernstein C., Kacelnik A. & Krebs J.R. (in press) Individual decisions and the distribution of predators in a patchy environment. *J. Anim. Ecol.*
4.7.2

Berthold P. (1988) The control of migration in European warblers. *Acta XIX Congressus Internationalis Ornithologi* (ed. H. Ouellet), Vol 1, pp. 215–249. University of Ottawa Press. Ottawa.
5.6.1

Bertram B.C.R. (1975) Social factors influencing reproduction in wild lions. J. Zool. 177, 462–482.
10.5.2

Bertram B.C.R. (1976) Kin selection in lions and in evolution. In: *Growing Points in Ethology* (ed. P.P.G. Bateson & R.A. Hinde), pp. 281–301. Cambridge University Press, Cambridge.
1.3.2

Bertram B.C.R. (1978) Living in groups: predators and prey. In: *Behavioural Ecology: An Evolutionary Approach, 1st edn.* (ed. J.R. Krebs & N.B. Davies), pp. 64–96. Blackwell Scientific Publications, Oxford.
6.6, 10.4.1

Berven K.A. (1987) The heritable basis of variation in larval developmental patterns within populations of the wood frog (*Rana sylvatica*). *Evolution* **41**, 1088–1097.
2.2.2

Berven K.A., Gill D.E. & Smith-Gill S.J. (1979) Countergradient selection in the green frog *Rana clamitans. Evolution* **33**, 609–623.
2.8.1

Betzig L.L. (1986) *Despotism and Differential Reproduction: A Darwinian View of History.* Aldine de Gruyter, New York.
3.3.2

Betzig L.L. (1988) Mating and parenting in Darwinian perspective. In: *Human Reproductive Behaviour: A Darwinain Perspective* (ed. L. Betzig, M. Borgerhoff Mulder & P. Turke), pp. 3–20. Cambridge University Press, Cambridge.
3.3.1, 3.3.2, 3.3.4

Betzig L.L. (1989) Rethinking human ethology. *Ethol. Sociobiol.* **10**, 315–324.
3.4.4

Betzig L.L. & Turke P.W. (1986) Food sharing on Ifaluk. *Curr. Anthropol.* **27**, 397–400.
3.2.3, 3.3.3

Biebach H. (1981) Energetic costs of incubation on different clutch sizes in starlings (*Sturnus vulgaris*). *Ardea* **69**, 141–142.
2.3.1

Bierbaum T.J., Mueller L.D. & Ayala F.J. (1989) Density-dependent evolution of life-history traits in *Drosophila melanogaster. Evolution* **43**, 382–392.
2.8.6

Birkhead T.R. (1987) Sperm competition in birds. *Trends Ecol. Evol.* **2**, 268–272.
9.5.2

Birkhead T.R. (1988) Behavioural aspects of sperm competition in birds. *Adv. Study Behav.* **18**, 35–72.
9.5.2

Birkhead T.R., Pellatt J. & Hunter F.M. (1988) Extra-pair copulation and sperm competition in the zebra finch. *Nature* **334**, 60–62.
9.5.2

Birkhead T.R., Burke T., Zann R., Hunter F.M. & Krupa A.P. (1990) Extra-pair paternity and intra-specific brood parasitism in wild zebra finches *Taeniopygia guttata* revealed by DNA fingerprinting. *Behav. Ecol. Sociobiol.* **27**, 315–324.
9.5.2

Blaxter J.H.S. (1969) Development: eggs and larvae. In: *Fish Physiology* (ed. W.S. Hoar & D.J. Randall), Vol. 3, pp. 177–252. Academic Press, New York.
8.3.1

Bleckmann H. (1986) Role of the lateral line in fish behaviour. In: *The Behaviour of Teleost Fishes* (ed. T.J. Pitcher), pp. 177–202. Croom Helm, London.
12.6.2

Blumer L.S. (1986) Parental care sex differences in the brown bullhead *Ictalurus nebulosus* (Pisces, Ictaluridae). *Behav. Ecol. Sociobiol.* **19**, 97–104.
8.4.1, 8.4.3

Blurton Jones N.G. (1984) A selfish origin for human food sharing: tolerated theft. *Ethol. Sociobiol.* **5**, 1–3.
3.2.3

Blurton Jones N.G. (1986) Bushman birth spacing: a test for optimal interbirth intervals. *Ethol. Sociobiol.* **7**, 91–105.
3.3.1

Blurton Jones N.G. (1987) Bushman birth spacing: direct tests of some simple predictions. *Ethol. Sociobiol.* **8**, 183–203.
3.3.1

Boag P.T. & Noordwijk A.J. van (1987) Quantitative genetics. In: *Avian genetics* (ed. F. Cooke & P.A. Buckley), pp. 45–78. Academic Press, London.
2.8.3

Boake C.R.B. (1985) Genetic consequences of mate choice: a quantitative genetic method for testing sexual selection theory. *Science* **227**, 1061–1063.
7.3.1

Boake C.R.B. (1986) A method for testing adaptive hypotheses of mate choice. *Amer. Natur.* **127**, 654–666.
7.3.2

Boarman W.I. (1990) Avian song ecology: the relationship between habitat acoustics and vocal signal structure. *Behav. Ecol. Sociobiol.* **26**, 481–491.
12.6.2

Boomsma J.J. (1988) Empirical analysis of sex allocation in ants: from descriptive surveys to population genetics. In: *Population Genetics and Evolution* (ed. G. de Jong), pp. 42–51. Springer-Verlag, Berlin.
11.4.2, 11.5.4

Boomsma J.J. (1989) Sex investment ratios in ants: has female bias been systematically overestimated? *Amer. Natur.* **133**, 517–532.
11.5.4

Boomsma J.J. & Grafen A. (1990) Intraspecific variation in ant sex ratios and the Trivers–Hare hypothesis. *Evolution* **44**, 1026–1034.
11.5.4

Borgerhoff Mulder M. (1988a) The relevance of the polygyny threshold model

to humans. In: *Human Mating Patterns* (ed. C.G.N. Mascie-Taylor & A.J. Boyce), pp. 209–230. Cambridge University Press, Cambridge.
3.3.1

Borgerhoff Mulder M. (1988b) Reproductive success in three Kipsigis cohorts. In: *Reproductive Success* (ed. T.H. Clutton-Brock), pp. 419–435. University of Chicago Press, Chicago.
3.3.1, 3.3.2

Borgerhoff Mulder M. (1989a) Polygyny and the extent of women's contribution to subsistence. *Amer. Anthropol.* **91**, 178–180.
3.3.2

Borgerhoff Mulder M. (1989b) Reproductive consequences of sex-biased inheritance. In: *Comparative Socioecology of Mammals and Man* (ed. R. Foley & V. Standon), pp. 405–427. Basil Blackwell, London.
3.3.4

Borgerhoff Mulder M. (1989c) Early maturing Kipsigis women have higher reproductive success than later maturing women, and cost more to marry. *Behav. Ecol. Sociobiol.* **24**, 145–153.
3.3.4

Borgerhoff Mulder M. (1990) Kipsigis women prefer wealthy men: evidence for female choice in mammals? *Behav. Ecol. Sociobiol.* **27**, 255–264.
3.3.2

Borgerhoff Mulder M. (1991) Reproductive decisions in ecological context. In: *Evolutionary Ecology and Human Behavior* (ed. E.A. Smith & B. Winterhalder). Aldine de Gruyter, New York (in press).
3.3.1, 3.3.2, 3.3.4

Borgerhoff Mulder M. & Milton M. (1985) Factors affecting infant care in the Kipsigis. *J. anthropol. Res.* **41**, 231–262.
3.3.1

Borgia G. (1979) Sexual selection and the evolution of mating systems. In: *Sexual Selection and Reproductive Competition in Insects* (ed. M.S. Blum & N.A. Blum). Academic Press, New York.
8.4.3

Borgia G. (1985) Bower quality, number of decorations and mating success of male satin bowerbirds (*Ptilinorynchus violaceus*): an experimental analysis. *Anim. Behav.* **33**, 266–271.
7.3.2

Bossert W.H. & Wilson E.O. (1963) The analysis of olfactory communication among animals. *J. Theor. Biol.* **5**, 443–469.
12.6.1

Bourke A.F.G., van der Have T.M. & Franks N.R. (1988) Sex ratio determination in the slave-making ant *Harpagoxenus sublaevis*. *Behav. Ecol. Sociobiol.* **23**, 233–245.
11.4.2

Bourke A.F.G. (1988b) Worker reproduction in the higher eusocial Hymenoptera. *Q. Rev. Biol.* **63**, 291–311.
11.4.2

Bourke A.F.G. (1989) Comparative analysis of sex-investment ratios in slave-making ants. *Evolution* **43**, 913–918.
11.5.4

Bourke A.F.G., van der Have T.M. & Franks N.R. (1988) Sex ratio determination and worker reproduction in the slave-making ant *Harpagoxenus sublaevis*. *Behav. Ecol. Sociobiol.* **23**, 233–245.
11.4.2

Boyce M.S. (1984) Restitution of *r*- and *K*-selection as a model of density-dependent natural selection. *A. Rev. Ecol. Syst.* **15**, 427–447.
2.7.1

Boyce M.S. & Perrins C.M. (1987) Optimizing great tit clutch size in a fluctuating environment. *Ecology* **68**, 142–153.
2.5.1

Boyd R. & Richerson P.J. (1985) *Culture and the Evolutionary Process.* University of Chicago Press, Chicago.
3.1, 3.4.3

Boyd R. & Richerson P.J. (1988) The evolution of reciprocity in sizable groups. *J. Theor. Biol.* **132**, 337–356.
3.4.3

Boyd R. & Richerson P.J. (1989) The evolution of indirect reciprocity. *Social Networks* **11**, 213–236.
3.4.3

Bradbury J.W. (1977) Lek mating behaviour in the hammer headed bat. *Z. Tierpsychol.* **45**, 225–255.
9.4.1

Bradbury J.W. (1981) The evolution of leks. In: *Natural Selection and Social Behavior* (ed. R.D. Alexander & D.W. Tinkle), pp. 138–169. Chiron Press, New York.
7.3.1, 9.4.3

Bradbury J.W. & Andersson M.B. (1987) *Sexual Selection: Testing the Alternatives.* Wiley, Chichester.
3.4.3

Bradbury J.W. & Gibson R. (1983) Leks and

mate choice. In: *Mate Choice* (ed. P.P.G. Bateson), pp. 109–138. Cambridge University Press, Cambridge.
7.3.2, 9.4, 9.4.1

Bradbury J.W., Gibson R.M., McCarthy C.E. & Vehrencamp S.L. (1989). Dispersion of displaying male sage grouse, II. The role of female dispersion. *Behav. Ecol. Sociobiol.* **24**, 15–24.
7.3.2, 9.4.2

Bradbury J.W., Gibson R.M. & Tsai I.M. (1986) Hotspots and the evolution of leks. *Anim. Behav.* **34**, 1694–1709.
9.4.1, 9.4.2, 9.4.5

Bradbury J.W. & Vehrencamp S.L. (1977) Social organisation and foraging in emballonurid bats, III: mating systems. *Behav. Ecol. Sociobiol.* **2**, 1–17.
9.1

Bradshaw A.D. (1965) Evolutionary significance of phenotypic plasticity in plants. *Adv. Genet.* **13**, 115–155.
2.8.2, 2.8.4

Breed M.D. & Bennett B. (1987) Kin recognition in highly eusocial insects. In: *Kin Recognition in Animals* (ed. D.J.C. Fletcher & C.D. Michener), pp. 243–285. Wiley, Chichester.
11.5.6

Breed M.D., Michener C.D. & Evans H.E. (ed.) (1982) *The Biology of Social Insects.* Westview Press, Boulder, Colorado.
11.1

Breed M.D. & Page R.E. Jr (ed.) (1989) *The Genetics of Social Evolution.* Westview Press, Boulder, Colorado.
11.1, 11.4.3

Breitwisch R. (1988) Sex differences in defence of eggs and nestlings by northern mocking birds, *Mimus polyglottos.* *Anim. Behav.* **36**, 62–72.
8.4.4

Brenowitz E.A. (1982) The active space of red-winged blackbird song. *J. comp. Physiol.* **147**, 511–522.
11.4.2

Brian M.V. (1980) Social control over sex and caste in bees, wasps and ants. *Biol. Rev.* **55**, 379–415.
11.1

Brockmann H.J. (1979) Nest-site selection in the great golden digger wasp, *Sphex ichneumoneus* L. (Sphecidae). *Ecol. Entomol.* **4**, 211–224.
11.5.1

Brockmann H.J. (1984) The evolution of social behaviour in insects. In: *Behavioural Ecology: An Evolutionary Approach,* 2nd edn. (ed. J.R. Krebs &

N.B. Davies), pp. 340–361. Blackwell Scientific Publications, Oxford.
11.1, 11.4, 11.5.4

Brockmann H.J. & Dawkins R. (1979) Joint nesting in a digger wasp as an evolutionarily stable preadaptation to social life. *Behaviour* **71**, 203–245.
11.5.1

Brockmann H.J., Grafen A. & Dawkins R. (1979) Evolutionarily stable nesting strategy in a digger wasp. *J. Theor. Biol.* **77**, 473–496.
1.2.1, 11.5.1

Brodie E.D. III (1989) Genetic correlations between morphology and anti-predator behaviour in natural populations of the garter snake *Thamnophis ordinoides.* *Nature* **342**, 542–543.
6.1

Brody M.S. & Lawlor L.R. (1984) Adaptive variation in offspring size in the terrestrial isopod *Armadillium vulgare.* *Oecologia* **61**, 55–59.
8.3.1

Bronstein P.M. (1985) Predictors of dominance in male *Betta splendens. J. Comp. Psychol.* **99**, 47–55.
12.3.4

Brown C.H. (1982) Ventriloquial and locatable vocalizations in birds. *Z. Tierpsychol.* **59**, 338–350.
12.3.1

Brown C.H. (1989) The active space of blue monkey and grey-cheeked mangabey vocalisation. *Anim. Behav.* **37**, 1023–1034.
12.4.2

Brown C.H. & Schwagmeyer P.L. (1984) The vocal range of alarm calls in thirteen-lined ground squirrels. *Z. Tierpsychol.* **65**, 273–288.
12.4.2

Brown C.R. (1984) Laying eggs in a neighbor's nest: benefit and cost of colonial nesting in swallows. *Science* **224**, 518–519.
9.5.2

Brown E.D., Farabaugh S.M. & Veltman C.J. (1988) Song sharing in a group-living songbird, the Australian magpie *Gymnorhina tibicen.* I: Vocal sharing within and among social groups. *Behaviour* **104**, 1–28.
12.3.2

Brown J.L. (1964) The evolution of diversity in avian territorial systems. *Wilson Bull.* **76**, 160–169.
5.1.2

Brown J.L. (1969) The buffer effect and pro-

ductivity in tit populations. *Amer. Natur.* **103**, 347–354.
5.2

Brown J.L. (1970) A note on the sexual division of labor. *Amer. Anthropol.* **72**, 1073–1078.
3.3.1

Brown J.L. (1974) Alternate routes to sociality in jays with a theory for the evolution of altruism and communal breeding. *Amer. Zool.* **14**, 63–80.
10.4.2, 10.5.2

Brown J.L. (1975) *The Evolution of Behavior*. Norton, New York.
1.3.2, 12.1, 12.4

Brown J.L. (1978) Avian communal breeding systems. *A. Rev. Ecol. Syst.* **9**, 123–155.
10.2, 10.4.1, 10.5.3

Brown J.L. (1980) Fitness in complex avian social systems. In: *Evolution of Social Behavior* (ed. H. Markl), pp. 115–128. Verlag-Chemie, Weinheim.
10.5

Brown J.L. (1983) Cooperation: a biologist's dilemma. In: *Advances in the Study of Behavior* (ed. J.S. Rosenblatt), Vol. 13, pp. 1–37. Academic Press, New York.
10.5.3, 10.6.2

Brown J.L. (1987) *Helping and Communal Breeding in Birds*. Princeton University Press, Princeton, New Jersey.
10.1, 10.3.1, 10.4.1, 10.4.2, 10.4.4, 10.5, 10.5.1, 10.5.3, 10.6.3, 11.4

Brown J.L. & Brown E.R. (1980) Reciprocal aid-giving in a communal bird. *Z. Tierpsychol.* **53**, 313–324.
10.2

Brown J.L. & Brown E.R. (1981) Kin selection and individual fitness in babblers. In: *Natural Selection and Social Behavior* (ed. R.D. Alexander & D.W. Tinkle), pp. 244–256. Chiron Press, New York.
10.2, 10.3.2

Brown J.L. & Brown E.R. (1990) Mexican jays: uncooperative breeding. In: *Cooperative Breeding in Birds: Long-Term Studies of Ecology and Behavior* (ed. P. Stacey & W. Koenig), pp. 269–288. Cambridge University Press, Cambridge.
10.2

Brown J.L., Brown E.R., Brown S.D. & Dow D.D. (1982) Helpers: effects of experimental removal on reproductive success. *Science* **215**, 421–422.
10.3.1

Brown J.L., Dow D.D., Brown E.R. & Brown S.D. (1978) Effects of helpers on feeding of nestlings in the grey-crowned babbler (*Pomatostomus temporalis*). *Behav. Ecol. Sociobiol.* **4**, 43–59.
10.3.2

Brush J.S. & Narins P.M. (1989) Chorus dynamics of a neotropical amphibian assemblage: comparison of computer simulation and natural behaviour. *Anim. Behav.* **37**, 33–44.
12.6.2

Buckle G.R. & Greenberg L. (1981) Nestmate recognition in sweat bees (*Lasioglossum zephyrum*): does an individual recognize its own odour or only odours of its nestmates? *Anim. Behav.* **29**, 802–809.
11.5.6

Bull J.J. (1981) Sex ratio evolution when fitness varies. *Heredity* **46**, 9–26.
8.6.2

Bull J.J. (1983) *Evolution of Sex Determining Mechanisms*. Benjamin/Cummings, Menlo Park, California.
8.6.1, 11.5.5

Bull J.J. & Charnov E.L. (1988) How fundamental are Fisherian sex ratios? In: *Oxford Surveys in Evolutionary Biology* (ed. P.H. Harvey & L. Partridge), Vol 5, pp. 96–135. Oxford University Press, Oxford.
11.5.4

Bull J.J. & Charnov E.L. (1989) Enigmatic reptilian sex ratios. *Evolution* **43**, 1561–1566.
8.6.2

Bulmer M.G. (1983a) Models for the evolution of protandry in insects. *Theor. Popul. Ecol.* **23**, 314–322.
5.6.1

Bulmer M.G. (1983b) Sex ratio evolution in social hymenoptera under worker control with behavioral dominance. *Amer. Natur.* **121**, 899–902.
11.5.4

Burgess J.W. (1976) Social spiders. *Sci. Amer.* **234**, 100–106.
12.3.3

Burke T., Davies N.B., Bruford M.W. & Hatchwell B.J. (1989) Parental care and mating behaviour of polyandrous dunnocks *Prunella modularis* related to paternity by DNA finger-printing. *Nature* **338**, 249–251.
8.4.4, 9.5.2, 9.5.4

Burley N. (1980) Clutch overlap and clutch size: alternative and complementary reproductive tactics. *Amer. Natur.* **115**, 223–246.
9.5.1

Burley N. (1981) Sex ratio manipulation and selection for attractiveness. *Science* **211**, 721–722.
7.3.2

Burley N. (1983) The meaning of assortative mating. *Ethol. Sociobiol.* **4**, 191–203.
3.3.2

Burley N. (1985) Leg-band color and mortality patterns in captive breeding populations of zebra finches. *Auk* **102**, 547–651.
7.3.2

Burley N. (1986a) Sexual selection for aesthetic traits in species with biparental care. *Amer. Natur.* **127**, 415–445.
7.3.2

Burley N. (1986b) Sex-ratio manipulation in color-banded populations of zebra finches. *Evolution* **40**, 1191–1206.
7.3.2

Burley N. (1986c) Comparison of the band-color preferences of two species of estrilid finches. *Anim. Behav.* **34**, 1732–1741.
7.3.2

Burley N. (1988) Wild zebra finches have band-color preferences. *Anim. Behav.* **36**, 1235–1237.
12.1

Burns J.T., Cheng K.M. & McKinney F. (1980) Forced copulation in captive mallards. I. Fertilization of eggs. *Auk* **97**, 875–879.
9.5.2

Bush R.R. & Mosteller F. (1951) A mathematical model for simple learning. *Psychol. Rev.* **68**, 313–323.
4.6.1

Buss D.M. (1988) Sex differences in human mate preferences: evolutionary hypothesis testing in 37 cultures. *Behav. Brain Sci.* **12**, 1–49.
3.3.1

Butlin R.K., Hewitt G.M. & Webb S.F. (1985) Sexual selection for intermediate optimum in *Chorthippus brunneus* (Orthoptera: Acrididae) *Anim. Behav.* **33**, 1281–1292.
12.1

Cade W.H. (1979) The evolution of alternative male reproductive strategies in field crickets. In: *Sexual Selection and Reproductive Competition in Insects* (ed. M. Blum & N.A. Blum), pp. 343–379. Academic Press, London.
7.3.2, 12.1

Calow P. (1979) The cost of reproduction—a physiological approach. *Biol. Rev.* **54**, 23–40.
2.3

Campbell D.T. (1975) On the conflicts between biological and social evolution and between pyschology and moral tradition. *Amer. Psychol.* **30**, 1103–1126.
3.1

Capinera J.L. (1979) Qualitative variation in plants and insects: effect of propagule size on ecological plasticity. *Amer. Natur.* **114**, 350–361.
8.3.2

Capinera J.L. & Barbosa P. (1977) Influence of natural diets and larval density on gypsy moth *Lymantria dispar* (Lepidoptera: Orgyiidae) egg mass characteristics. *Can. J. Entomol.* **109**, 1313–1318.
8.3.1

Caraco T. (1987) Foraging games in a random environment. In: *Foraging Behavior* (ed. A.C. Kamil, J.R. Krebs & H.R. Pulliam). pp. 389–414. Plenum Press, New York.
5.4

Caraco T., Barkan C., Beacham J.L. *et al.* (1989) Dominance and social foraging: a laboratory study. *Anim. Behav.* **38**, 41–58.
5.4

Caraco T., Blanckenhorn W.U., Gregory G.M., Newman J.A., Recer G.M. & Zwicker S.M. (1990) Risk-sensitivity: ambient temperature affects foraging choice. *Anim. Behav.* **39**, 338–345.
4.5.1

Caraco T., Martindale S. & Whittam T.S. (1980) An empirical demonstration of risk-sensitive foraging preferences. *Anim. Behav.* **28**, 820–830.
4.5, 4.5.1

Carlin N.F. (1988) Species, kin and other forms of recognition in the brood discrimination behavior of ants. In: *Advances in Myrmecology* (ed. J.C. Trager), pp. 267–295. Brill, Leiden.
11.5.6

Carlin N.F. (1989) Discrimination between and within colonies of social insects: two null hypotheses. *Netherlands J. Zool.* **39**, 86–100.
11.5.6

Carlin N.F. & Frumhoff P.C. (1990) Scientific correspondence. *Nature* **346**, 706–707.
11.5.6

Carlin N.F., Hölldobler B. & Gladstein D.S. (1987) The kin recognition system of carpenter ants (Camponotus spp) III. Within-colony discrimination. *Behav. Ecol. Sociobiol.* **20**, 219–227.
11.5.6

Carlisle T.R. & Zahavi A. (1986) Helping at

the nest, allofeeding and social status in immature Arabian babblers. *Behav. Ecol. Sociobiol.* **18**, 339–351.
10.5.3

Caro T.M. & Borgerhoff Mulder M. (1987) The problem of adaptation in the study of human behaviour. *Ethol. Sociobiol.* **8**, 61–72.
3.4.4

Caro T.M. & Sellen D.W. (1990) The reproductive advantages of fat in women. *Ethol. Sociobiol.* **11**, 51–66.
3.4.4

Carpenter J.M. (1988) The phylogenetic system of the Stenogastrinae (Hymenoptera: Vespidae). *J. New York Entomol. Soc.* **96**, 140–175.
11.4.1

Carpenter J.M. (1989) Testing scenarios: wasp social behavior. *Cladistics* **5**, 131–144.
11.5

Cartar R.V. & Dill L.M. (1990) Why are bumble bees risk-sensitive foragers? *Behav. Ecol. Sociobiol.* **26**, 121–127.
4.5.1

Carton Y. & David J.R. (1985) Relation between the genetic variability of digging behaviour of *Drosophila* larvae and their susceptibility to a parasitic wasp. *Behav. Genet.* **15**, 143–154.
6.8

Caryl P.G. (1979) Communication by agonistic displays: what can games theory contribute to ethology? *Behaviour* **68**, 136–169.
12.3.4, 12.7.1

Caryl P.G. (1982) Animal signals: a reply to Hinde. *Anim. Behav.* **30**, 240–244.
12.3.4

Cash K.J. & Evans R.M. (1987) The occurrence, context and functional significance of aggressive begging behaviours in young American white pelicans. *Behaviour* **102**, 119–128.
12.4.2

Caswell H. (1989) Life-history strategies. In: *Ecological Concepts* (ed. J.M. Cherrett), pp. 285–307. Blackwell Scientific Publications, Oxford.
2.1.2

Catchpole C.K. (1980) Sexual selection and the evolution of complex songs among European warblers of the genus *Acrocephalus. Behaviour* **74**, 149–165.
12.7.3

Catchpole C.K. (1983) Variation in the song of the great reed warbler *Acrocephalus arundinaceus* in relation to mate attrac-

tion and territorial defence. *Anim. Behav.* **31**, 1217–1225.
12.7.3

Catchpole C.K. (1988) Sexual selection and the song of the great reed warbler. In: *Acta XIX Congressus Internationalis Ornithologici* (ed. H. Ouellet), Vol. 1, pp. 1366–1372. University of Ottawa Press, Ottawa.
12.7.3

Catchpole C.K. (1989) Responses of male sedge warblers to playback of different repertoire sizes. *Anim. Behav.* **37**, 1046–1047.
12.7.3

Catchpole C.K., Dittami J. & Leisler B. (1984) Differential responses to male song repertoire in female songbirds implanted with oestradiol. *Nature* **312**, 563–564.
12.7.3

Catchpole C.K. & Leisler B. (1989) Variation in the song of the aquatic warbler *Acrocephalus paludicola* in response to playback of different song structures. *Behaviour* **108**, 125–138.
12.7.3

Cavalli-Sforza L.L. & Feldman M.W. (1981) *Cultural Transmission and Evolution: A Quantitative Approach.* Princeton University Press, Princeton, New Jersey.
3.1

Chacko M.J. (1964) Effect of superparasitism in *Bracon gelechiae* Ashmead. *Proc. Indian Acad. Sci. B* **60**, 12–25.
2.4.1

Chagnon N.A. (1979a) Mate competition, favoring close kin, and village fissioning among the Yanomamo Indians. In: *Evolutionary Biology and Human Social Behavior: An Anthropological Perspective* (ed. N.A. Chagnon & W. Irons), pp. 86–132. Duxbury Press, North Scituate, Massachusetts.
3.3.1

Chagnon N.A. (1979b) Is reproductive success equal in egalitarian societies? In: *Evolutionary Biology and Human Social Behavior: An Anthropological Perspective* (ed. N.A. Chagnon & W. Irons), pp. 374–401. Duxbury Press, North Scituate, Massachusetts.
3.3.2

Chagnon N.A. (1988a) Life histories, blood revenge, and warfare in a tribal population. *Science* **239**, 985–992.
3.3.1

Chagnon N.A. (1988b) Male Yanomamo

manipulations of kinship classifications of female kin for reproductive advantage. In: *Human Reproductive Behaviour: A Darwinian Perspective* (ed. L. Betzig, M. Borgerhoff Mulder & P. Turke), pp. 23–48. Cambridge University Press, Cambridge.
3.3.3

Chagnon N.A. & Bugos P.E. Jr. (1979) Kin selection and conflict: an analysis of a Yanomamo ax fight. In: *Evolutionary Biology and Human Social Behavior: An Anthropological Perspective* (ed. N.A. Chagnon & W. Irons), pp. 213–238. Duxbury Press, North Scituate, Massachusetts.
3.3.3

Chagnon N.A. & Irons W.G. (ed.) (1979) *Evolutionary Biology and Human Social Behavior: An Anthropological Perspective*. Duxbury Press, North Scituate, Massachusetts.
3.1

Chapman R.F. (1971) *The Insects: Structure and Function*. English University Press, Warwick.
12.2

Charlesworth B. (1978a) Some models of the evolution of altruistic behaviour between siblings. *J. Theor. Biol.* 72, 297–319.
1.3.5

Charlesworth B. (1978b) Model for the evolution of Y chromosomes and dosage compensation. *Proc. Natl. Acad. Sci.* 75, 5618–5622.
11.5.5

Charlesworth B. (1980) *Evolution in Age-Structured Populations*. Cambridge University Press, Cambridge.
2.1.2

Charlesworth B. (1984) The cost of phenotypic evolution. *Paleobio* 10, 319–327.
7.2.1

Charlesworth B. (1987) The heritability of fitness. In: *Sexual Selection: Testing the Alternatives*. (ed. J.W. Bradbury & M. Andersson), pp. 21–40. Dahlem Workshop Report No. 39. Wiley, Chichester.
7.3.1

Charlesworth D. & Charlesworth B. (1975) Theoretical genetics of Batesian mimicry, parts I, II, and III. *J. Theor. Biol.* 55, 283–337.
6.5

Charlesworth D. & Charlesworth B. (1979) The evolution and breakdown of self-incompatibility systems. *Heredity* 43, 41–55.
7.5

Charlesworth D., Schemske D.W. & Sork V.L. (1987) The evolution of plant reproductive characters: sexual versus natural selection. In: *The Evolution of Sex and its Consequences* (ed. S.C. Stearns). Birkhäuser-Verlag, Boston.
7.5

Charnov E.L. (1976) Optimal foraging: the marginal value theorem. *Theor. Popul. Biol.* 9, 129–136.
5.6.1

Charnov E.L. (1977) An elementary treatment of the genetical theory of kin selection. *J. Theor. Biol.* 66, 541–550.
1.3.2

Charnov E.L. (1978a) Evolution of eusocial behavior: offspring choice or parental parasitism? *J. Theor. Biol.* 75, 451–465.
11.2.2

Charnov E.L. (1978b) Sex-ratio selection in eusocial hymenoptera. *Amer. Natur.* 112, 317–326.
11.5.4

Charnov E.L. (1979) The genetical evolution of patterns of sexuality: Darwinian fitness. *Amer. Natur.* 113, 465–480.
8.6.2

Charnov E.L. (1982) *The Theory of Sex Allocation*. Princeton University Press, Princeton, New Jersey.
5.6.2, 7.5, 8.6, 8.6.2

Charnov E.L. (1989a) Phenotypic evolution under Fisher's fundamental theorem of natural selection. *Heredity* 62, 113–116.
2.1.1

Charnov E.L. (1989b) Natural selection on age of maturity in shrimp. *Evol. Ecol.* 3, 236–239.
2.6.2

Charnov E.L. & Berrigan D. (1990) Dimensionless numbers and life history evolution: age of maturity versus the adult lifespan. *Evol. Ecol.* 4, 273–275.
2.7.2

Charnov E.L. & Bull J.J. (1977) When is sex environmentally determined? *Nature* 266, 828–830.
8.6.2

Charnov E.L. & Bull J.J. (1986) Sex allocation, pollinator attraction, and fruit dispersal in cosexual plants. *J. Theor. Biol.* 118, 321–326.
7.5

Charnov E.L. & Bull J.J. (1989) The primary sex ratio under environmental sex determination. *J. Theor. Biol.* 139, 431–436.
8.6.2

Charnov E.L. & Krebs J.R. (1974) On clutch-size and fitness. *Ibis* **116**, 217–219.
2.5.1

Charnov E.L., Los-den Hartogh R.L., Jones W.T. & van den Assem J. (1981) Sex ratio evolution in variable environment. *Nature* **289**, 27–33.
8.6.2

Charnov E.L. & Schaffer W.M. (1973) Life-history consequences of natural selection: Cole's result revisited. *Amer. Natur.* **107**, 791–792.
2.6.1

Charnov E.L. & Skinner S.W. (1984) Evolution of host selection and clutch size in parasitoid wasps. *Fla. Ent.* **67**, 5–21.
2.5.1

Charnov E.L. & Skinner S.W. (1985) Complementary approaches to the understanding of parasitoid oviposition decisions. *Environ. Entomol.* **14**, 383–391.
2.5.1

Chase I.D. (1980) Cooperative and non-cooperative behavior in animals. *Amer. Natur.* **115**, 827–857.
8.4.1, 8.4.4

Cheney D.L. & Seyfarth R.M. (1985) Social and non-social knowledge in vervet monkeys. *Phil. Trans. R. Soc. Lond. B* **308**, 187–201.
12.3.1

Cheverton J., Kacelnik A. & Krebs J.R. (1985) Optimal foraging: constraints and currencies. In: *Experimental Behavioural Ecology* (ed. B. Hölldobler & M. Lindauer), pp. 109–126. G. Fischer–Verlag, Stuttgart.
4.1.1, 4.2.3

Chewyreuv I. (1913) Le rôle femelles dans la determination du sexe de leur descendents dans le groupe de Ichneumonides. *C. R. Soc. Biol. (Paris)* **74**, 695–699.
8.6.2

Choe J.C. (1988) Worker reproduction and social evolution in ants (Hymenoptera: Formicidae). In: *Advances in Myrmecology* (ed. J.C. Trager), pp. 163–187. Brill, Leiden.
11.4.2

Clark A.B. (1978) Sex ratio and local resource competition in a Prosimian primate. *Science* **201**, 163–165.
3.3.4, 8.6.3

Clarke M.F. (1984) Co-operative breeding by the Australian bell miner, *Manorina melanophrys* Latham: a test of kin selection theory. *Behav. Ecol. Sociobiol.* **14**, 137–146.
10.5.2, 10.5.4, 10.6.3

Clarke M.F. (1989) The pattern of helping in the bell miner (*Manorina melanophrys*). *Ethology* **80**, 292–306.
10.3.2, 10.4.5, 10.5.2, 10.5.3, 10.5.4, 10.6.2, 10.6.3

Clutton-Brock T.H. (1983) Selection in relation to sex. In: *From Molecules to Men* (ed. D.S. Bendall), pp. 457–481. Cambridge University Press, Cambridge.
3.3.1

Clutton-Brock T.H. (1988) *Reproductive Success: Studies of Individual Variation in Contrasting Breeding Systems.* University of Chicago Press, Chicago.
7.2.2, 10.3.3, 10.5.3

Clutton-Brock T.H. (1989) Mammalian mating systems. *Proc. Roy. Soc. Lond. B* **236**, 339–372.
9.2.1, 9.4.1

Clutton-Brock T.H. (1991) *The Evolution of Parental Care.* Princeton University Press, Princeton, New Jersey.
8.1, 8.2, 8.4, 8.4.3, 8.6.1

Clutton-Brock T.H. & Albon S.D. (1982) Parental investment in male and female offspring in mammals. In: *Current Problems in Sociobiology* (ed. King's College Sociobiology Group), pp. 223–248. Cambridge University Press, Cambridge.
8.6.2

Clutton-Brock T., Albon S.D., Gibson R.M. & Guinness F.E. (1979) The logical stag: adaptive aspects of fighting in red deer (*Cervus elephos L.*). *Anim. Behav.* **27**, 211–225.
12.3.3

Clutton-Brock T.H., Albon S.D. & Guinness F.E. (1981) Parental investment in male and female offspring in polygynous mammals. *Nature* **289**, 487–489.
8.6.1, 8.6.2

Clutton-Brock T.H., Albon S.D. & Guinness F.E. (1982) *Red Deer: Behaviour and Ecology of Two Sexes.* Chicago University Press, Chicago.
7.4

Clutton-Brock T.H., Albon S.D. & Guinness F.E. (1984) Maternal dominance, breeding success, and birth sex ratios in red deer. *Nature* **308**, 358–360.
8.6.2

Clutton-Brock T.H., Albon S.D. & Guinness F.E. (1985) Parental investment and sex differences in juvenile mortality in birds and mammals. *Nature* **313**, 131–133.
8.6.1

Clutton-Brock T.H., Albon S.D. & Guinness F.E. (1986) Great expectations: maternal dominance sex ratios and offspring reproductive success in red deer. *Anim.*

Behav. **34**, 460–471.
8.6.2

Clutton-Brock T.H., Albon S.D. & Guinness F.E. (1989) Fitness costs of gestation and lactation in wild mammals. *Nature* **337**, 260–262.
8.2.

Clutton-Brock T.H., Green D., Hiraiwa-Hasegawa M. & Albon S.D. (1988) Passing the buck: resource defence, lekking and mate choice in fallow deer. *Behav. Ecol. Sociobiol.* **23**, 281–296.
9.4., 9.4.1

Clutton-Brock T.H., Guinness F.E. & Albon S.D. (1982) *Red Deer: The Behavior and Ecology of Two Sexes.* Chicago University Press, Chicago.
9.2.1

Clutton-Brock T.H. & Harvey P.H. (1977) Primate ecology and social organisation. *J. Zool. (Lond.)* **183**, 1–39.
7.2.3

Clutton-Brock T.H. & Harvey P.H. (1978) Mammals, resources and reproductive strategies. *Nature* **273**, 191–195.
9.2.3

Clutton-Brock T.H. & Harvey P.H. (1984) Comparative approaches to investigating adaptation. In: *Behavioural Ecology: An Evolutionary Approach,* 2nd edn (ed. J.R. Krebs & N.B. Davies) pp. 7–29. Blackwell Scientific Publications, Oxford.
7.4, 11.8

Clutton–Brock T.H., Hiraiwa-Hasegawa M. & Robertson A. (1989) Mate choice on fallow deer leks. *Nature* **340**, 463–465.
9.4.5

Clutton-Brock T.H. & Iason G.R. (1986) Sex ratio variation in mammals. *Q. Rev. Biol.* **61**, 339–374.
8.6.1

Coates D. (1980) Prey size intake in humbug damselfish *Dascyllus aruanus* (Pisces: Pomacentridae) living in social groups. *J. Anim. Ecol.* **49**, 335–340.
5.3.2

Cohan F.M. & Hoffman A.A. (1989) Uniform selection as a diversifying force in evolution: evidence from *Drosophila.* *Amer. Natur.* **134**, 613–637.
7.3.1

Cole B.J. (1981) Dominance hierarchies in *Leptothorax* ants. *Science* **212**, 83–84.
11.4.2, 11.4.3

Cole B.J. (1986) The social behavior of *Leptothorax allardycei:* time budgets and the evolution of worker reproduction. *Behav. Ecol. Sociobiol.* **18**, 165–173.
11.4.2, 11.4.3

Cole L.C. (1954) The population conse-

quence of life history phenomena. *Q. Rev. Biol.* **19**, 103–137.
2.1.1, 2.1.2, 2.6.1

Coleman R.M. & Whittall R.D. (1988) Clutch size and the cost of incubation in the Bengalese finch (*Lonchura striata var. domestica*). *Behav. Ecol. Sociobiol.* **23**, 367–372.
8.2

Collins N.S. (1980) Developmental responses to food limitation as indicators of environmental conditions for *Ephydra cinerea* Jones (Diptera). *Ecology* **61**, 650–661.
2.8.2

Colwell R.K. (1981) Group selection is implicated in the evolution of female-biased sex ratios. *Nature* **290**, 401–404.
8.6.3

Commons M.L., Kacelnik A. & Shettleworth S.J. (ed.) (1987) *Quantitative Analyses of Behavior. Vol. 6: Foraging.* L. Erlbaum, Hillsdale, New Jersey.
4.6

Conner D.A. (1985) Analysis of the vocal repertoire of adult pikas—ecological and evolutionary perspectives. *Anim. Behav.* **33**, 124–134.
12.3.2

Conner J. (1988) Field measurements of natural and sexual selection in the fungus beetle, *Bolitotherus cornutus.* *Evolution* **42**, 736–749.
7.2.2

Conner J. (1989) Density-dependent sexual selection in the fungus beetle, *Bolitotherus cornutus.* *Evolution* **43**, 1378–1386.
7.2.2

Conner R.C. (1986) Pseudo-reciprocity: investing in mutualism. *Anim. Behav.* **34**, 1562–1566.
10.6.2

Conover M.R. & Hunt G.L. (1984) Experimental evidence that female–female pairs in gulls result from a shortage of breeding males. *Condor* **86**, 472–476.
9.5.1

Convey P. (1989a) Influences on the choice between territorial and satellite behaviour in male *Libellula quadrimaculata* (Odonata: Libellulidae). *Behaviour* **109**, 125–141.
9.3.4

Convey P. (1989b) Post-copulatory guarding strategies in the non-territorial dragonfly *Sympetrum sangunieum* (Odonata: Libellulidae). *Anim. Behav.* **37**, 56–63.
9.3.4

Cooke F., Taylor P.D., Francis C.M. & Rockwell R.F. (1990) Directional selection and clutch size in birds. *Amer. Natur.* **136**, 261–267.
2.8.5

Cosmides L. & Tooby J. (1989) Evolutionary psychology and the generation of culture, Part II. *Ethol. Sociobiol.* **10**, 51–97.
3.4.4

Cott H.B. (1940) *Adaptive Coloration in Animals.* Methuen, London.
6.1, 6.3, 6.4.1, 6.4.2

Councilman J.J. (1977) A comparison of two populations of the grey-crowned babbler. *Bird Behav.* **1**, 43–82.
10.3.1

Cox C.R. & Le Boeuf B.J. (1977) Female incitation of male competition: a mechanism of mate selection. *Amer. Natur.* **111**, 317–335.
12.5.2

Craig J.L. (1979) Habitat variation in the social organization of a communal gallinule, the pukeko, *Porphyrio porphyrio melanotus. Anim. Behav.* **28**, 593–603.
10.4.2

Craig R. (1979) Parental manipulation, kin selection, and the evolution of altruism. *Evolution* **33**, 319–334.
11.2.2, 11.5.4

Craig R. (1980a) Sex investment ratios in social Hymenoptera. *Amer. Natur.* **116**, 311–323.
11.5.4

Craig R. (1980b) Sex ratio changes and the evolution of eusociality in the Hymenoptera: simulation and games theory studies. *J. Theor. Biol.* **87**, 55–70.
11.5.4

Craig R. (1983) Subfertility and the evolution of eusociality by kin selection. *J. Theor. Biol.* **100**, 379–397.
11.5.4

Craig R. & Crozier R.H. (1979) Relatedness in the polygynous ant *Myrmecia pilosula. Evolution* **33**, 335–341.
11.4.2

Crocker G. & Day T. (1987) An advantage to mate choice in the seaweed fly, *Coelopa frigida. Behav. Ecol. Sociobiol.* **20**, 295–301.
7.3.2

Cronk L. (1989) From hunters to herders: subsistence change as a reproductive strategy among the Mukogodo. *Curr. Anthropol.* **30**, 224–234.
3.3.5

Crook J.H. (1964) The evolution of social organization and visual communication in the weaver birds (Ploceinae). *Behaviour* **10** (Suppl.), 1–178.
9.5.1

Crook J.H. & Crook S.J. (1988) Tibetan polyandry: problems of adaptation and fitness. In: *Human Reproductive Behaviour: A Darwinian Perspective* (ed. L. Betzig, M. Borgerhoff Mulder & P. Turke), pp. 97–114. Cambridge University Press, Cambridge.
3.3.2

Crosland M.W.J. (1989a) Kin recognition in the ant *Rhytidoponera confusa* I. Environmental odour. *Anim. Behav.* **37**, 912–919.
11.5.6

Crosland M.W.J. (1989b) Kin recognition in the ant *Rhytidoponera confusa* II. Gestalt odour. *Anim. Behav.* **37**, 920–926.
11.5.6

Crozier R.H. (1989) Insect sociobiology. *Science* **245**, 313–315.
11.4.3

Crozier R.H. & Luykx P. (1985) The evolution of termite eusociality is unlikely to have been based on a male-haploid analogy. *Amer. Natur.* **126**, 867–869.
11.5.5

Crozier R.H., Pamilo P. & Crozier Y.C. (1984) Relatedness and microgeographic genetic variation in *Rhytidoponera mayri*, and Australian arid-zone ant. *Behav. Ecol. Sociobiol.* **15**, 143–150.
11.4.2

Crozier R.H., Smith B.H. & Crozier Y.C. (1987) Relatedness and population structure of the primitively eusocial bee *Lasioglossum zephyrum* (Hymenoptera: Halictidae) in Kansas *Evolution* **41**, 902–910.
11.4.1

Cruz Y.P. (1981) A sterile defender morph in a polyembryonic hymenopterous parasite. *Nature* **294**, 446–447.
11.3

Cullen J.M. (1966) Reduction of ambiguity through ritualisation. *Phil. Trans. R. Soc. B* **251**, 363–374.
12.4.2

Curio E. (1976) *The Ethology of Predation.* Springer-Verlag, Berlin.
6.1, 6.2.1, 6.2.2, 6.7, 6.8

Curry R.L. (1988a) Influence of kinship on helping behavior in Galapagos mockingbirds. *Behav. Ecol. Sociobiol.* **22**, 141–152.
10.5.4, 10.6.3

Curry R.L. (1988b) Group structure, within-group conflict and reproductive tactics in cooperatively breeding Galapagos mockingbirds, *Nesomimus*

parvulus. Anim. Behav. **36**, 1708–1728.
10.4.5, 10.6.3

Curtsinger J.W. (1986) Stay times in Scatophaga and the theory of evolutionarily stable strategies. *Amer. Natur.* **128**, 130–136.
5.6.1

Cuthill I.C. (1985) Experimental studies of optimal foraging theory. D.Phil. thesis, Oxford University, Oxford.
4.2, 4.2.3

Cuthill I.C. & Kacelnik A. (1990) Central place foraging: a re-appraisal of the 'loading effect'. *Anim. Behav.* **40**, 1087–1101.
4.2, 4.2.3

Cuthill I.C., Kacelnik A., Krebs J.R., Haccou P. & Iwasa Y. (1990) Patch use by starlings: the effect of recent experience on foraging decisions. *Anim. Behav.* **40**, 625–640.
4.2, 4.2.2, 4.2.3

Cuthill I.C. & MacDonald W.A. (1990) Experimental manipulation of the dawn and dusk chorus in the blackbird *Turdus merula. Behav. Ecol. Sociobiol.* **26**, 209–216.
12.6.2

Daan S., Dijkstra C. & Tinbergen J.M. (1990) Family planning in the kestrel (*Falco tinnunculus*): the ultimate control of covariation of laying date and clutch size. *Behaviour* **114**, 83–116.
2.5.1, 2.8.3

Daly M. & Wilson M. (1983) *Sex, Evolution and Behavior*, 2nd edn. Willard Grant Press, Boston, Massachusetts.
3.1

Daly M. & Wilson M. (1984) A sociobiological analysis of human infanticide. In: *Infanticide: Comparative and Evolutionary Perspectives* (ed. G. Hausfater & S.B. Hrdy), pp. 487–502. Aldine de Gruyter, New York.
3.3.4

Daly M. & Wilson M. (1988) *Homicide* Aldine de Gruyter, New York.
3.3.1

Danthanarayana W., Hamilton J.G. & Khoul S.P. (1982) Low-density larval crowding in the light brown apple moth, *Epiphyas postvittana* and its ecological significance. *Ent. Exp. Appl.* **31**, 353–358.
2.4.1

Darwin C. (1859) *On the Origin of Species.* Murray, London.
7.1, 11.2

Darwin C. (1871) *The Descent of Man and Selection in Relation to Sex*, 1st edn. Murray, London.
7.1, 7.3, 7.3.2, 7.3.3

Darwin C. (1872) *The Expression of Emotions in Man and Animals.* Murray, London.
12.4.1

Darwin C. (1874) *The Descent of Man and Selection in Relation to Sex*, 2nd edn. Murray, London.
7.1, 7.3.2

Davies N.B. (1978a) Ecological questions about territorial behaviour. In: *Behavioural Ecology: An Evolutionary Approach* (ed. J.R. Krebs & N.B. Davies), pp. 317–350. Blackwell Scientific Publications, Oxford.
5.6.3

Davies N.B. (1978b) Territorial defence in the speckled wood butterfly (*Pararge aegeria*): the resident always wins. *Anim. Behav.* **26**, 138–147.
12.3.3

Davies N.B. (1982) Behaviour and competition for scarce resources. In: *Current Problems in Sociobiology* (ed. King's College Sociobiology Group), pp. 363–380. Cambridge University Press, Cambridge.
5.6.2

Davies N.B. (1985) Cooperation and conflict among dunnocks, *Prunella modularis* in a variable mating system. *Anim. Behav.* **33**, 628–648.
8.4.4, 9.5.4

Davies N.B. (1986) Reproductive success of dunnocks *Prunella modularis* in a variable mating system. I: Factors influencing provisioning rate, nestling weight and fledging success. *J. Anim. Ecol.* **55**, 123–138.
9.5.3

Davies N.B. (1989) Sexual conflict and the polygamy threshold. *Anim. Behav.* **38**, 226–234.
9.5.3, 9.5.4

Davies N.B., Bourke A.F.G. & Brooke M. de L. (1989) Cuckoos and parasitic ants: interspecific brood parasitism as an evolutionary arms race. *Trends Ecol. Evol.* **4**, 274–278.
6.8

Davies N.B. & Halliday T.R. (1979) Competitive mate searching in common toads *Bufo bufo. Anim. Behav.* **27**, 1253–1267.
9.3.2

Davies N.B. & Houston A.I. (1984) Territory economics. In: *Behavioural Ecology: An Evolutionary Approach*, 2nd edn (ed.

J.R. Krebs & N.B. Davies), pp. 148–169. Blackwell Scientific Publications, Oxford.
5.1.2, 5.6.3

Davies N.B. & Houston A.I. (1986) Reproductive success of dunnocks *Prunella modularis* in a variable mating system. II: Conflicts of interests among breeding adults. *J. Anim. Ecol.* **55**, 139–154.
3.4.1, 9.5.3, 9.5.4

Davies N.B. & Lundberg A. (1984) Food distribution and a variable mating system in the dunnock *Prunella modularis*. *J. Anim. Ecol.* **53**, 895–912.
9.5.4

Davis J.W.F. & O'Donald P. (1976) Sexual selection for a handicap: a critical analysis of Zahavi's model. *J. Theor. Biol.* **57**, 345–354.
7.3.1

Davis M.S. (1987) Acoustically mediated neighbor recognition in the North American bullfrog, *Rana catesbeiana*. *Behav. Ecol. Sociobiol.* **21**, 185–190.
12.3.2

Davis W.M.J. (1988) Acoustic communication in the belted kingfisher: an example of temporal coding. *Behaviour* **106**, 1–24.
12.7.2

Dawkins R. (1976) *The Selfish Gene*. Oxford University Press, Oxford.
3.1, 8.4.4, 8.5.1, 10.6.2, 12.3.3

Dawkins R. (1982) *The Extended Phenotype* W.H. Freeman, Oxford.
1.2, 6.4.1

Dawkins R. (1989) *The Selfish Gene*, 2nd edn. Oxford University Press, Oxford.
1.4.1

Dawkins R. & Carlisle T.R. (1976) Parental investment, mate desertion and a fallacy. *Nature* **262**, 131–133.
8.4.2, 8.4.3

Dawkins R. & Krebs J.R. (1978) Animal signals: information or manipulation. In: *Behavioural Ecology: An Evolutionary Approach*, 1st edn. (ed. J.R. Krebs & N.B. Davies), pp. 282–309. Blackwell Scientific Publications, Oxford.
1.4.1, 12.1, 12.3.3, 12.3.4, 12.3.5

Dawkins R. & Krebs J.R. (1979) Arms races between and within species. *Proc. Roy. Soc. B* **205**, 489–511.
6.8

Dewsbury D.A. (1984) Sperm competition in muroid rodents. In: *Sperm Competition and the Evolution of Animal Mating Systems* (ed. R.L. Smith), pp. 547–571. Academic Press, London.
9.2.2

Dewsbury D.A. (1988) Sperm competition in deer mice *Peromyscus maniculatus bairdi*. *Behav. Ecol. Sociobiol.* **22**, 251–256.
9.2.2

Diamond J. (1988) Experimental study of bower decoration by the bowerbird *Amblyornis inornatus* using colored poker chips. *Amer. Natur.* **131**, 631–653.
12.1

Dickemann M. (1979a) Female infanticide, reproductive strategies and social stratification: a preliminary model. In: *Evolutionary Biology and Human Social Behavior: An Anthropological Perspective* (ed. N.A. Chagnon & W. Irons), pp. 321–367. Duxbury Press, North Scituate, Massachusetts.
3.3.2

Dickemann M. (1979b) The ecology of mating systems in hypergynous dowry systems. *Soc. Sci. Inform.* **18**, 163–195.
3.3.2

Dijkstra C., Bult A., Bijlsma S., Daan S., Meijer T. & Zijlstra M. (1990) Brood size manipulations in the kestrel (*Falco tinnunculus*): effects on offspring and adult survival. *J. Anim. Ecol.* **59**, 269–285.
2.3.1, 2.4.1, 2.5.1

Dill L.M. (1987) Animal decision making and its ecological consequences: the future of aquatic ecology and behavior. *Can. J. Zool.* **65**, 803–811.
4.3.1

Dill L.M. & Fraser A.H.G. (1984) Risk of predation and feeding behaviour of juvenile coho salmon (*Onchorynchus kisutch*). *Behav. Ecol. Sociobiol.* **16**, 65–71.
4.3.1

Dingle H. & Hegmann J.P. (1982) *Evolution and Genetics of Life Histories*. Springer-Verlag, New York.
2.1.1

Dixson A.F. (1987) Observations on the evolution of the genitalia and copulatory behaviour in male primates. *J. Zool.* **213**, 423–443.
7.4

Dobzhansky T. (1951) *Genetics and the Origin of Species*, 3rd edn. Columbia University Press, New York.
2.8.1

Dodson S. (1989) Predator-induced reaction norms. *Bioscience* **39**, 447–452.
2.8

Dominey W.J. (1981) Anti-predator function of bluegill sunfish nesting colonies. *Nature* **290**, 586–588.
8.2

Douwes P., Sivusaari L., Niklasson M. & Stille B. (1987) Relatedness among queens in polygynous nests of the ant *Leptothorax acervorum*. *Genetica* **75**, 23–29.
11.4.2

Drent R.H. & Daan S. (1980) The prudent parent: energetic adjustments in avian breeding. *Ardea* **68**, 225–252.
8.2

Dunbar R.I.M. (1982) Intraspecific variations in mating strategy. In: *Perspectives in Ethology* (ed. P.P.G. Bateson & P.H. Klopfer), Vol. 5, pp. 385–431. Plenum Press, New York.
5.6.2

Dunbar R.I.M. (1984) *Reproductive Decisions: An Economic Analysis of Gelada Baboon Social Strategies.* Princeton University Press, Princeton, New Jersey.
9.2.1

Dunbar R.I.M. (1988a) Darwinizing man: a commentary. In: *Human Reproductive Behaviour: A Darwinian Perspective* (ed. L. Betzig, M. Borgerhoff Mulder & P. Turke), pp. 161–169. Cambridge University Press, Cambridge.
3.4.2

Dunbar R.I.M. (1988b) *Primate Social Systems.* Chapman & Hall, London.
3.4.2

Dunham D.W. (1966) Agonistic behaviour in captive rose-breasted grosbeaks, *Pheucticus ludovicianus* (L.). *Behaviour* **27**, 160–173.
12.3.4

Dupré J. (ed.) (1987) *The Latest on the Best. Essays on Evolution and Optimality.* MIT Press, Cambridge, Massachusetts.
4.7.3

Dyer M. & Fry C.H. (1980) The origin and role of helpers in bee-eaters. *Proc. Int. Ornithol. Congr.* **17**, 862–868.
10.3.1

Dyson M.L. & Passmore N.I. (1988a) Two choice phonotaxis in *Hyperolius marmoratus* (Anura: Hyperoliidae): the effect of temporal variation in presented stimuli. *Anim. Behav.* **36**, 648–652.
12.6.2

Dyson M.L. & Passmore N.I. (1988b) The combined effect of intensity and the temporal relationship of stimuli on phonotaxis in female painted reed frogs *Hyperolius marmoratus*. *Anim. Behav.* **36**, 1555–1556.
7.3.2

Eberhard W.G. (1975) The ecology and behaviour of a subsocial pentatomid bug and two scelionid wasps: strategy and counterstrategy in a host and its parasites. *Smithson. Contr. Zool.* **205**, 1–39.
8.2

Eberhard W.G. (1977) Aggressive chemical mimicry by a bolas spider. *Science* **198**, 1173–1175.
11.1

Eberhard W.G. (1985) *Sexual Selection and Animal Genitalia.* Harvard University Press, Cambridge, Massachusetts.
7.4

Eden S.F. (1987) When do helpers help? Food availability and helping in the moorhen, *Gallinula chloropus*. *Behav. Ecol. Sociobiol.* **21**, 191–195.
10.3.2

Edmunds M. (1974) *Defense in Animals: A survey of Anti-Predator Defenses.* Longmans, London.
6.1, 6.3, 6.4, 6.4.1, 6.8

Eibl Eibesfeldt I. (1989) *Human Ethology.* Aldine de Gruyter, New York.
3.4.1

Eickwort G.C. (1981) Presocial insects. In: *Social Insects* (ed. H.R. Hermann), vol. 2, pp. 199–280. Academic Press, New York.
11.3

Eickwort G.C. (1982) Introduction to symposium V, presocial behavior. In: *The Biology of Social Insects* (ed. M.D. Breed, C.D. Michener & H.E. Evans), pp. 151–153. Westview Press, Boulder, Colorado.
11.3

Ekman P., Friesen W.V. & Scherer K.R. (1976) Body movement and voice pitch in deceptive interaction. *Semiotica* **16**, 23–27.
12.5

Ekman P. & Johansson-Allende A. (1990) Egg size investment of tits: does number conflict with size? In: *Population Biology of Passerine Birds. An Integrated Approach* (ed. J. Blondel, A. Gosler, J.D. Lebreton & R.H. McCleery), pp. 247–256. NATO ASI series, Springer-Verlag, Berlin.
2.7.2

El-Sawaf S.K. (1956) Some factors affecting the longevity, oviposition, and rate of development in the southern cowpea weevil, *Callosobruchus maculatus* F. (Coleoptera: Bruchidae). (*Bull. Soc. Entomol. Egypte* **40**, 29–95.
2.2.2, 2.3.1

Ellison P.T. (1991) Reproductive ecology and human fertility. In: *Applications of Biological Anthropology to Human Affairs* (ed. G.W. Lasker & C.G.N.

Mascie-Taylor). Cambridge University Press, Cambridge (in press).
3.3.4

Emlen S.T. (1981) Altruism, kinship, and reciprocity in the white-fronted bee-eater. In: *Natural Selection and Social Behavior: Recent Research and New Theory* (ed. R.D. Alexander & D.W. Tinkle), pp. 245–281. Chiron Press, New York.
10.2, 10.6.2

Emlen S.T. (1982a) The evolution of helping. I: An ecological constraints model. *Amer. Natur.* **119**, 29–39.
10.4.1, 10.4.2, 10.4.5, 10.4.6

Emlen S.T. (1982b) The evolution of helping. II. The role of behavioral conflict. *Amer. Natur.* **119**, 40–53.
10.4.1, 10.4.5

Emlen S.T. (1984) Cooperative breeding in birds and mammals. In: *Behavioral Ecology: An Evolutionary Approach.* 1st edn (ed. J.R. Krebs & N.B. Davies), pp. 305–339. Sinauer, Sunderland, Massachusetts.
10.4.6, 10.6.2

Emlen S.T. (1990) The white-fronted bee-eater: helping in a colonially nesting species. In: *Cooperative Breeding in Birds: Long-Term Studies of Ecology and Behavior* (ed. P. Stacey & W. Koenig), pp. 489–526. Cambridge University Press, Cambridge.
10.2

Emlen S.T., Emlen J.M. & Levin S.A. (1986) Sex-ratio selection in species with Helpers-at-the-nest. *Amer. Natur.* **127**, 1–8.
8.6.3, 10.4.5

Emlen S.T. & Oring L.W. (1977) Ecology, sexual selection, and the evolution of mating systems. *Science* **197**, 215–223.
3.3.2, 9.1, 9.4.1

Emlen S.T., Ratnieks F.L.W., Reeve H.K., Shellman-Reeve J., Sherman P.W. & Wrege P.H. (1991) Adaptive versus non-adaptive explanations of behavior: the case of alloparental helping. *Amer. Natur.* (in press).
10.5

Emlen S.T. & Wrege P.H. (1988) The role of kinship in helping decisions among white-fronted bee-eaters. *Behav. Ecol. Sociobiol.* **23**, 305–315.
10.2, 10.5.4, 10.6.3

Emlen S.T. & Wrege P.H. (1989) A test of alternate hypotheses for helping behavior in white-fronted bee-eaters of Kenya. *Behav. Ecol. Sociobiol.* **25**, 303–320.
10.3.1, 10.5, 10.5.3, 10.5.4, 10.6.2, 10.6.3

Emlen S.T. & Wrege P.H. (1991) Breeding biology of white-fronted bee-eaters at Nakuru. Part. I: The influence of helpers on breeder fitness. *J. Anim. Ecol.* (in press).
10.2, 10.3.1

Emsley A., Dickerson G.E. & Kashyap T.S. (1977) Genetic parameters in progeny-test selection for field performance of strain-cross layers. *Poult. Sci.* **56**, 121-146.
2.4.1

Endler J.A. (1978) A predator's view of animal coloration. *Evol. Biol.* **11**, 319–393.
6.1, 6.2.2, 6.3, 6.6

Endler J.A. (1981) An overview of the relationships between mimicry and crypsis. *Biol. J. Linn. Soc. Lond.* **16**, 25–31.
6.1

Endler J.A. (1983) Natural and sexual selection on color patterns in poeciliid fishes. *Envir. Biol. Fishes* **9**, 173–190.
6.3

Endler J.A. (1984) Progressive background matching in moths and a quantitative measure of crypsis. *Biol. J. Linn. Soc. Lond.* **22**, 187–231.
6.1, 6.3, 6.4.3

Endler J.A. (1986a) Defense against predators. In: *Predator–Prey Relationships: Perspectives and Approaches from the Study of Lower Vertebrates* (ed. M.E. Feder & G.V. Lauder), pp. 109–134. University of Chicago Press, Chicago.
6.1, 6.2.1, 6.3, 6.4.1, 6.6, 6.7

Endler J.A. (1986b) *Natural Selection in the Wild.* Princeton University Press, Princeton, New Jersey.
6.2.1, 6.8.

Endler J.A. (1987) Predation, light intensity, and courtship behaviour in *Poecilia reticulata* (Pisces: Poeciliidae). *Anim. Behav.* **35**, 1376–1385.
6.3

Endler J.A. (1988) Frequency-dependent predation, crypsis and aposematic coloration. *Phil. Trans. Roy. Soc. B* **319**, 505–523.
6.2.1, 6.2.2, 6.3, 6.4, 6.4.1, 6.4.2, 6.4.3, 6.5, 6.7

Endler J.A. (1990) On the measurement and classification of colour in studies of animal colour patterns. *Biol. J. Linn. Soc. Lond.* **41** (in press).
6.3

Endler J.A. (1991) Variation in the appearance of guppy color patterns to guppies and their predators under different visual conditions. *Vision Res.* **31** (in press).
6.3

Endler J.A. & Lyles A.M. (1989) Bright ideas about parasites. *Trends Ecol. Evol.* **4**, 246–248.
6.3

Endler J.A. & McLellan T. (1988) The processes of evolution: towards a newer synthesis. *Ann. Rev. Ecol. Syst.* **19**, 395–421.
6.3

Englund G. & Olsson T.I. (1990) Fighting and assessment in the net-spinning caddis larva *Arctopsyche ladogensis*: a test of the sequential assessment game. *Anim. Behav.* **39**, 55–62.
12.8

Enquist M. (1985) Communication during aggressive interactions with particular reference to variation in choice of behaviour. *Anim. Behav.* **33**, 1152–1161.
1.4.1

Enquist M. & Leimar O. (1983) Evolution of fighting behaviour: decision rules and assessment of relative strength. *J. Theor. Biol.* **102**, 387–410.
12.3.3

Enquist M. & Leimar O. (1990) The evolution of fatal fighting. *Anim. Behav.* **39**, 1–9.
12.3.3, 12.8

Enquist M., Plane E. & Roed B.J. (1985) Aggressive communication in fulmars (*Fulmaris glacialis*) competing for food. *Anim. Behav.* **33**, 1107–1120.
12.7.1

Erckmann W.J. (1983) The evolution of polyandry in shorebirds: an evaluation of hypotheses. In: *Social Behavior of Female Vertebrates* (ed. S.K. Wasser), pp. 113–168. Academic Press, New York.
8.4.4, 9.5.5

Escalante G. & Rabinovich J.E. (1979) Population dynamics of *Telenomus fariai* (Hymenoptera: Scelionidae), a parasite of Chagas' disease vectors. IX. Larval competition and population size regulation under laboratory conditions. *Res. Popul. Ecol.* **20**, 235–246.
2.4.1

Evans H.E. (1958) The evolution of social life in wasps. *Proc. Tenth Internatl. Cong. Entomology* **2**, 449–457.
11.4

Evans H.E. (1977) Extrinsic versus intrinsic factors in the evolution of insect sociality. *Bioscience* **27**, 613–617.
11.2.2., 11.5.1

Evans H.E. & Hook A.W. (1982a) Communal nesting in Australian *Cerceris* digger wasps. In: *The Biology of Social Insects* (ed. M.D. Breed, C.D. Michener & H.E. Evans), pp. 159–163. Westview Press, Boulder, Colorado.
11.5.1

Evans H.E. & Hook A.W. (1982b) Communal nesting in the digger wasp *Cerceris australis* (Hymenoptera: Sphecidae). *Aust. J. Zool.* **30**, 557–568.
11.5.1

Evans H.E. & Hook A.W. (1986) Nesting behavior of Australian *Cerceris* digger wasps, with special reference to nest reutilization and nest sharing (Hymenoptera, Sphecidae). *Sociobiology* **11**, 275–302.
11.5.1

Evers C.A. & Seeley T.D. (1986) Kin discrimination and aggression in honey bee colonies with laying workers. *Anim. Behav.* **34**, 924–925.
11.5.6

Ewald P.W. & Rohwer S. (1982) Effects of supplemental feeding on timing of breeding, clutch size and polygamy in red-winged blackbirds, *Agelaius phoeniceus*. *J. Anim. Ecol.* **51**, 429–450.
9.5.3

Ezaki Y. (1988) Mate desertion by male great reed warblers *Acrocephalus arundinacus* at the end of the breeding season. *Ibis* **130**, 427–437.
8.4.4

Ezaki Y. (1990) Female choice and the causes and adaptiveness of polygyny in great reed warblers. *J. Anim. Ecol.* **59**, 103–119.
9.5.3

Fagen R. (1987) A generalized habitat matching rule. *Evol. Ecol.* **1**, 5–10.
5.6.3

Fagerstrom T & Wiklund C. (1982) Why do males emerge before females? Protandry as a mating strategy in male and female butterflies. *Oecologia* **52**, 164–166.
5.6.1

Fairbanks L.A. & McGuire M.T. (1986) Age, reproductive value and dominance related behaviour in vervet monkey females: cross-generational social influences on social relationships and reproduction. *Anim. Behav.* **34**: 1718–1721.
8.2

Falconer D.S. (1981) *An Introduction to Quantitative Genetics*, 2nd edn. Longman, London.
6.8

Falconer D.S. (1989) *Introduction to Quantitative Genetics*, 3rd edn. Longman, Harlow, Essex.
2.2.2

Falls J.B., Horn A.G. & Dickinson T.E. (1988) How western meadowlarks classify their songs: evidence from song matching. *Anim. Behav.* **36**, 579–585.
12.8

Falls J.B., Krebs J.R. & McGregor P.K. (1982) Song matching in the great tit *Parus major*: the effect of similarity and familiarity. *Anim. Behav.* **30**, 997–1009.
12.7.3

Fantino E. (1969) Choice and rate of inforcement. *J. Exp Anal. Behav.* **12**, 723–730.
4.5.2

Fantino E. & Abarca N. (1985) Choice, optimal foraging and the delay-reduction hypothesis. *Behav. Brain Sci.* **8**, 315–330.
4.6

Farabaugh S.M. (1982) The ecological and social significance of duetting. In: *Acoustic Communication in Birds* (ed. D.E. Kroodsma & E.H. Miller), Vol. 2, pp. 85–124. Academic Press, New York.
12.3.2.

Finke M.A., Milinkovich D.J. & Thompson C.F. (1987) The evolution of clutch size: an experimental test in the house wren (*Troglodytes aedon*). *J. Anim. Ecol.* **56**, 99–114.
2.5.1

Fisher R.A. (1915) The evolution of sexual preference. *Eugen. Rev.* **7**, 184–192.
7.3.1

Fisher R.A. (1930) *The Genetical Theory of Natural Selection*. Clarendon Press, Oxford.
2.1.2, 5.6.2, 6.4.1, 6.3.1, 8.6, 8.6.1, 8.6.3, 11.2.1

Fitzgibbon C.D. & Fanshawe J.H. (1988). Stotting in Thompson's gazelles: an honest signal of condition. *Behav. Ecol. Sociobiol.* **23**, 69–74.
6.1, 12.3.3

Fitzpatrick J.W. & Woolfenden G.E. (1988) Components of lifetime reproductive success in the Florida scrub jay. In: *Reproductive Success: Studies of Individual Variation in Contrasting Breeding Systems.* (ed. T. Clutton-Brock), pp. 305–320. University of Chicago Press, Chicago.
10.2

Fitzpatrick S.M. & McNeil J.N. (1989) Lifetime mating potential and reproductive success in males of the true armyworm, *Pseudaletia unipuncta* (Haw.) (Lepidoptera: Noctuidae). *Funct. Ecol.* **3**, 37–44.
2.3.1

Fjeldså (1983) Ecological character displacement and character release in grebes, Podicepididae. *Ibis* **125**, 463–481.
5.6.5

Flesness N.R. (1978) Kinship asymmetry in diploids. *Nature* **276**, 495–496.
12.5.5

Fletcher D.J.C. & Michener C.D. (ed.) (1987) *Kin Recognition in Animals*. Wiley, New York.
11.1, 12.3.2

Fletcher D.J.C. & Ross K.G. (1985) Regulation of reproduction in eusocial Hymenoptera. *A. Rev. Entomol.* **30**, 319–343.
11.1, 11.4.1, 11.4.2

Flinn M.V. (1981) Uterine versus agnatic kinship variability and associated cousin marriage preferences: an evolutionary biological analysis. In: *Natural Selection and Social Behavior: Recent Research and New Theory* (ed. R.D. Alexander & D.W. Tinkle), pp. 439–475. Chiron Press, New York.
3.3.3

Flinn M.V. & Low B.S. (1986) Resource distribution. social competition. and mating patterns in human societies. In: *Ecological Aspects of Social Evolution* (ed. D.I. Rubenstein & R.W. Wrangham), pp. 217–243. Princeton University Press, Princeton. New Jersey.
3.3.2

Foley R. (1987) *Another Unique Species*. Longman, Harlow.
3.4.4

Foley R. (1988) Hominids, humans and hunter–gatherers: an evolutionary perspective. In: *Hunters and Gatherers. 1: History, Evolution and Social Change* (ed. T. Ingold, D. Riches & J. Woodburn), pp. 207–221. Berg, Oxford.
3.4.4

Foley R. & Lee P.C. (1989) Finite social space, evolutionary pathways, and reconstructing hominid behavior. *Science* **243**, 901–906.
3.4.2

Foltz D.W. (1981) Genetic evidence for long-term monogamy in a small rodent *Peromyscus polionotus*. *Amer. Natur.* **117**, 665–675.
9.2.2

Foltz D.W. & Hoogland J.L. (1981) Analysis of the mating system in the black-tailed prairie dog *Cynomys ludovicianus* by likelihood of paternity. *J. Mamm.* **62**, 706–712.
9.2.2

Foltz D.W. & Schwagmeyer P.L. (1988)

Sperm competition in the thirteen-lined ground squirrel: differential fertilisation success under field conditions. *Amer. Natur.* **133**, 257–265.
9.2.2

Ford E.B. (1975) *Ecological Genetics*, 4th edn. Chapman & Hall, London.
6.1, 6.2.1, 6.2.2

Forester D.C. (1979) The adaptiveness of parental care in *Desmognathus ochrophaeus* (Urodela: Plethodontidae). *Copeia* **1979**, 332–341.
8.2

Forester D.C. & Harrison W.K. (1987) The significance of antiphonal vocalisation by the spring peeper *Pseudacris crucifer* (Amphibia: Anura) *Behaviour* **103**, 1–15.
12.6.2

Formanowicz D.R., Jr (1984) Foraging tactics of one aquatic insect: partial consumption of prey. *Anim. Behav.* **32**, 774–781.
4.1.1

Fowler K. & Partridge L. (1989) A cost of mating in female fruitflies. *Nature* **338**, 760–761.
2.3.1

Frame L.H., Malcolm J.R., Frame G.W. & van Lawick H. (1979) Social organization of African wild dogs (*Lycaon pictus*) on the Serengeti Plains, Tanzania 1967–1978. *Z. Tierpsychol.* **50**, 225–249.
10.4.5

Franck D. & Ribowski A. (1989) Escalating fights for rank-order position between male swordtails (*Xiphophorus helleri*): effects of prior rank-order experience and information transfer. *Behav. Ecol. Sociobiol.* **24**, 133–144.
12.3.4

Frank S.A. (1985) Hierarchical selection theory and sex ratios. II: On applying the theory, and a test with fig wasps. *Evolution* **39**, 949–964.
8.6.3

Frank S.A. (1986) Hierarchical selection theory and sex ratios. I: General solutions for structured populations. *Theor. Pop. Biol.* **29**, 312–342.
8.6.3

Frank S.A. (1987a) Individual and population sex allocation patterns. *Theor. Pop. Biol.* **31**, 47–74.
8.6.2

Frank S.A. (1987b) Variable sex ratio among colonies of ants. *Behav. Ecol. Sociobiol.* **20**, 195–201.
11.5.4

Frank S.A. & Crespi B.J. (1989) Synergism between sib-rearing and sex ratio in

Hymenoptera. *Behav. Ecol. Sociobiol.* **24**, 155–162.
11.5.4

Frank S.A. & Swingland I.R. (1988) Sex ratio under conditional sex expression. *J. Theor. Biol.* **135**, 415–418.
8.6.2

Franks N.R. & Scovell E. (1983) Dominance and reproductive success among slave-making worker ants. *Nature* **304**, 724–725.
11.4.2, 11.4.3

Fraser D.F. (1980) On the environmental control of oocyte maturation in a plethodontid salamander. *Oecologia (Berlin)* **46**, 302–307.
8.3.1

Free C.A., Beddington J.R. & Lawton J.H. (1977) On the inadequacy of simple models of mutual interference for parasitism and predation. *J. Anim. Ecol.* **46**, 543–554.
5.1.1

Fretwell S.D. (1972) *Populations in a Seasonal Environment*. Princeton University Press, Princeton, New Jersey.
5.2, 5.6.3

Fretwell S.D. & Lucas H.J. Jr. (1970) On territorial behavior and other factors influencing habitat distribution in birds. *Acta Biotheor.* **19**, 16–36.
5.2

Fricke H. (1974) Oko-ethologie des monogamen Anemonefisches *Amphiprion bicinctus* (Freiswasser-untersuchung aus dem Rotem meer). *Z. Tierpsychol.* **39**, 429–512.
8.4.3

Fricke H. (1980) Mating systems, maternal and biparental care in triggerfish Balistidae. *Z. Tierpsychol.* **53**, 105–122.
8.4.3

Friedmann S.A. (1960) The parasitic weaverbirds. *Bull. U.S. Nat. Mus.* **223**, 1–196.
12.5.1

Frisch R.E. (1984) Body fat, puberty, and fertility. *Science* **199**, 22–30.
3.3.4

Frumhoff P.C. & Baker J. (1988) A genetic component to division of labour within honey bee colonies. *Nature* **333**, 358–361.
11.4.3

Frumhoff P.C. & Schneider S. (1987) The social consequences of honey bee polyandry: the effects of kinship on worker interactions within colonies. *Anim. Behav.* **35**, 255–262.
11.5.6

Fryer G. & Iles T.D. (1972) *The Cichlid Fishes of the Great Lakes of Africa.* T.F.H. Publishing, Neptune City, New Jersey.
8.4.3

Fullick T.G. & Greenwood J.J.D (1979) Frequency dependent food selection in relation to two models. *Amer. Natur.* **113**, 762–765.
6.2.1

Gadagkar R. (1985) Kin recognition in social insects and other animals—a review of recent findings and a consideration of their relevance for the theory of kin selection. *Proc. Indian Acad. Sci. (Anim. Sci.)* **94**, 587–621.
11.1, 11.5.6

Gadgil M. (1972) Male dimorphism as a consequence of sexual selection. *Amer. Natur.* **106**, 574–580.
7.2.1

Gadgil M. & Bossert W.H. (1970) Life historical consequences of natural selection. *Amer. Natur.* **104**, 1–24.
2.1.2, 2.6.1

Galbraith H. (1988) Effects of egg size and composition on the size, quality and survival of lapwing *Vanellus vanellus*. *J. Zool. Lond.* **21**, 383–298.
8.2

Galdikas B.M.F. (1979) Orangutan adaptation at Tarijung Puting Reserve: mating and ecology. In: *The Great Apes* (ed. D.A. Hamburg & E.R. McCoun), pp. 194–233. Benjamin Cummings, Menlo Park, California.
12.1, 12.4.1

Gale W.F. & Deutsch W.G. (1985) Fecundity and spawning frequency of captive tessellated darters—fractional spawners. *Trans. Amer. Fish. Soc.* **114**, 220–229.
8.4.3

Gall G.A.E. (1975) Genetics of reproduction in domesticated rainbow trout. *J. Anim. Sci.* **40**, 19–28.
2.4.1

Gamboa G.J. (1988) Sister, aunt–niece, and cousin recognition by social wasps. *Behav. Genet.* **18**, 409–423.
11.5.6

Gamboa G.J., Reeve H.K. & Pfennig D.W. (1986) The evolution and ontogeny of nestmate recognition in social wasps. *A. Rev. Entomol.* **31**, 431–454.
11.1, 11.5.6

Gardner R. (1990) Mating calls. *Nature* **344**, 495.
7.3.1

Garnett V. (1978) Sibling aggression in the Black Eagle in the Matapos, Rhodesia. *Ostrich* **49**, 57–63.
8.5.2

Garson P.J. & Hunter M.L. (1979) Effects of temperature and time of year on the singing behaviour of wrens (*Troglodytes troglodytes*) and great tits (*Parus major*). *Ibis* **121**, 481–487.
4.4.2, 7.3.2

Gaston A.J. (1978a) Demography of the jungle babbler, *Turdoides striatus*. *J. Anim. Ecol.* **47**, 845–897.
10.4.2, 10.5.2, 10.5.3

Gaston A.J. (1978b) The evolution of group territorial behavior and cooperative breeding. *Amer. Natur.* **112**, 1091–1100.
10.5.1, 10.5.2

Gaulin S.J.C. & Boster J. (1985) Cross-cultural differences in sexual dimorphism: is there any variance to be explained? *Ethol. Sociobiol.* **6**, 219–225.
3.3.1

Gebhardt M.D. & Stearns S.C. (1988) Reaction norms for developmental time and weight at eclosion in *Drosophila mercatorum*. *J. Evol. Biol.* **1**, 335–354.
2.2.2

Gendron R.P. (1986) Searching for cryptic prey: evidence for optimal search rates and the formation of search images in quail. *Anim. Behav.* **34**, 898–912.
6.2.1

Gendron R.P. & Staddon J.E.R. (1983) Searching for cryptic prey: the effect of search rate. *Amer. Natur.* **121**, 172–186.
4.1.1, 6.2.1

Gerhardt H.C. (1982) Sound pattern recognition in some North American treefrogs (Anura: Hylidae): implications. for mate choice. *Amer. Zool.* **22**, 581–595.
7.3.2, 12.1

Gerhardt H.C. (1983) Communication and the environment. In: *Communication* (ed. T.R. Halliday & P.J.B. Slater), pp. 82–113. Blackwell Scientific Publications, Oxford.
12.2, 12.6.1

Gerhardt H.C. (1988) Acoustic properties used in call recognition by frogs and toads. In: *The Evolution of the Amphibian Auditory System. B.* (ed. B. Fritzsch, T. Hethington, M. Ryan, W. Wilczynski & W. Walkowiak), pp. 455–483. Wiley, New York.
7.3.2

Gerhardt H.C. & Doherty J.A. (1988) Acoustic communications in the gray treefrog, *Hyla versicolor*: evolutionary and

neurobiological implications. *J. comp. Physiol.* **162**, 261–278.
7.3.2

Getz W.M. & Smith K.B. (1986) Honey bee kin recognition: learning self and nestmate phenotypes. *Anim. Behav.* **34**, 1617–1626.
11.5.6

Ghiselin M.T. (1969) The evolution of hermaphroditism among animals. *Q. Rev. Biol.* **44**, 189–208.
5.6.2

Gibbon J., Church R.M., Fairhurst S. & Kacelnik A.(1988) Scalar expectancy theory and choice between delayed rewards. *Psychol. Rev.* **95**, 102–114.
4.5.2, 4.6.1, 4.6.2

Gibbons D.W. (1987) Juvenile helping in the moorhen, *Gallinula chloropus*. *Anim. Behav.* **35**, 170–181.
10.3.1, 10.3.2

Gibson R.M. (1989) Field playback of male display attracts females in lek breeding sage grouse. *Behav. Ecol. Sociobiol.* **24**, 439–443.
9.4.5

Gibson R.M. & Bradbury J. (1985) Sexual selection in lekking sage grouse: phenotypic correlates of male mating success. *Behav. Ecol. Sociobiol.* **18**, 117–123.
7.3.2, 9.4.5, 12.4.2

Giesel J.T. (1979) Genetic covariation of survivorship and other fitness indices in *Drosophila melanogaster*. *Exp. Geront.* **14**, 323–328.
2.2.2

Giesel J.T. (1986) Genetic correlation structure of life history variables in outbred. wild *Drosophila melanogaster*: effects of photoperiod regime. *Amer. Natur.* **128**, 593–603.
2.2.2

Giesel J.T.. Murphy P.A. & Manlove M.N. (1982) The influence of temperature on genetic interrelationships of life history traits in a population of *Drosophila melanogaster*: what tangled data sets we weave. *Amer. Natur.* **119**, 464–479.
2.2.2

Giesel J.T. & Zettler E.E. (1980) Genetic correlations of life historical parameters and certain fitness indices in *Drosophila melanogaster*: r_m. r_s. diet breadth. *Oecologia* **47**, 299–302.
2.2.2

Gill F.B. & Wolf L.L. (1975) Economics of feeding territoriality in the golden-winged sunbird. *Ecology* **56**, 333–345.
5.1.2

Gillespie R.G. & Caraco T. (1987) Risk-sensitive foraging strategies of two spider populations. *Ecology* **68**, 887–899.
4.5.1

Gilliam J.F. & Fraser D.F. (1987) Habitat selection when foraging under predation hazard: a model and a test with stream-dwelling minnows. *Ecology* **68**, 1227–1253.
4.3.1

Gilliard E.T. (1969) *Birds of Paradise*. Natural History Press, New York.
7.3.3

Gittleman J.L. & Harvey P.H. (1980) Why are distasteful prey not cryptic? *Nature* **286**, 149–150.
6.4.2

Glass C.W. & Huntingford F.A. (1988) Initiation and resolution of fights between swimming crabs *Liocarcinus depurator*. *Ethology* **77**, 237–249.
12.3.4

Godfray H.C.J. (1987a) The evolution of clutch size in invertebrates. *Oxf. Surv. Evol. Biol.* **4**, 117–154.
2.4, 2.5.1

Godfray H.C.J. (1987b) The evolution of clutch size in parasitic wasps. *Amer. Natur.* **129**, 221–233.
8.5.2

Godfray H.C.J. & Grafen A. (1988) Unmatedness and the evolution of eusociality. *Amer. Natur.* **131**, 303–305.
11.5.4

Godfray H.C.J. & Harper A.B. (1990) The evolution of brood reduction by siblicide in birds. *J. Theor. Biol.* (in press).
1.3.5, 8.5.2

Godfray H.C.J. & Ives A.R. (1988) Stochasticity in invertebrate clutch-size models. *Theor. Popul. Biol.* **33**, 79–101.
2.5.1

Godfray H.C.J. & Waage J.K. (1990) The evolution of highly skewed sex ratios in aphelinid wasps. *Amer. Natur.* (in press).
8.6.1

Godin J.G.J. & Keenleyside M.H.A. (1984) Foraging on patchily distributed prey by a cichlid fish (Teleostei, Cichlidae): a test of the ideal free distribution theory. *Anim. Behav.* **32**, 120–131.
5.2.2

Goldizen A.W. (1987) Facultative polyandry and the role of infant-carrying in wild saddle-backed tamarins. (*Sanguinus fuscicollis*). *Behav. Ecol. Sociobiol.* **20**, 99–109.
9.5.4, 10.2, 10.4.4

Goldizen A.W. (1989) Social relationships

in a cooperatively polyandrous group of tamarins *Saguinus fuscicollis. Behav. Ecol. Sociobiol.* **24**, 79–89.
9.5.4

Goldizen A.W. & Terborgh J. (1989) Demography and dispersal patterns of a tamarin population: possible causes of delayed breeding. *Amer. Natur.* **134**, 208–224.
10.2, 10.4.2

Goodenough U. (1984) *Genetics.* Saunders College Publishing, Philadelphia.
12.1

Goodman D. (1982) Optimal life histories, optimal notation, and the value of reproductive value. *Amer. Natur.* **119**, 803–823.
2.1.2

Gordon D.M. (1989) Ants distinguish neighbors from strangers. *Oecologia* **81**, 198–200.
11.5.6

Gosling L.M. (1986) The evolution of mating strategies in male antelope. In: *Ecological Aspects of Social Evolution* (ed. D.I. Rubenstein & R.W. Wrangham), pp. 244–281. Princeton University Press, Princeton, New Jersey.
9.4.2

Gosling M. (1987) Scent marking in an antelope lek territory. *Anim. Behav.* **35**, 620–622.
12.6.1

Goss-Custard J.D. (1977) Optimal foraging and the size- selection of worms by redshank *Tringa totanus* in the field. *Anim. Behav.* **25**, 10–29.
4.1.1

Goss-Custard J.D., Clarke R.T. & Durell S.E.A. Le V. dit (1984) Rates of food intake and aggression of oystercatchers *Haematopus ostralegus* on the most and least preferred mussel *Mytilus edulis* beds of the Exe Estuary. *J. Anim. Ecol.* **53**, 233–235.
5.5.2

Gottlander K. (1987) Variation in the song rate of the male pied flycatcher *Ficedula hypoleuca*: causes and consequences. *Anim. Behav.* **35**, 1037–1043.
7.3.2

Gould S.J. & Lewontin R.C. (1979) The spandrels of San Marco and the Panglossian paradigm. *Proc. R. Soc. Lond. B* **205**, 581–598.
3.4.2, 4.1.1, 4.7.3

Gouzoules H. & Gouzoules S. (1989) Design features and developmental modification of pigtail macaque, *Macaca nemes-trina* agonistic screams. *Anim. Behav.* **37**, 383–401.
12.7.2

Gouzoules S., Gouzoules H. & Marler P. (1984) Rhesus monkey (*Macaca mulatta*) screams: representational signalling in the recruitment of agonistic aid. *Anim. Behav.* **32**, 182–193.
12.7.2

Gowaty P.A. (1981) An extension of the Orians–Verner–Willson model to account for mating systems besides polygyny. *Amer. Natur.* **118**, 851–859.
9.5.4

Gowaty P.A. (1985) Multiple parentage and apparent monogamy in birds. *Ornithol. Monogr.* **37**, 11–21.
9.5.2

Gowaty P.A. & Lennartz M.R. (1985) Sex ratios of nestling and fledgling red-cockaded woodpeckers (*Picoides borealis*) favor males. *Amer. Natur.* **126**, 347–353.
3.3.4, 8.6.3

Grafen A. (1984) Natural selection, kin selection and group selection. In: *Behavioural Ecology, An Evolutionary Approach*, 2nd edn (ed. J.R. Krebs & N.B. Davies), pp. 62–84. Blackwell Scientific Publications, Oxford.
1.3.3, 2.1.1, 3.3.3, 8.6.3

Grafen A. (1985) A geometric view of relatedness. In: *Oxford Surveys in Evolutionary Biology* (ed. R. Dawkins & M. Ridley), Vol. 2, pp. 28–89. Oxford University Press, Oxford.
1.3.6, 11.2.1

Grafen A. (1986) Split sex ratios and the evolutionary origins of eusociality. *J. Theor. Biol.* **122**, 95–121.
11.5.4

Grafen A. (1990a) Biological signals as handicaps. *J. Theor. Biol.* **144**, 517–546.
1.4, 1.4.1, 1.4.2, Introduction to Part 3

Grafen A. (1990b) Sexual selection unhandicapped by the Fisher process. *J. Theor. Biol.* **144**, 473–516.
1.4, 1.4.1, 1.4.2, Introduction to Part 3

Grafen A. (1990c) Do animals really recognize kin? *Anim. Behav.* **39**, 42–54.
12.3.2

Gray P.J. (1985) *Primate Sociobiology.* HRAF Press, New Haven, Connecticut.
3.3.3, 3.4.1

Gray P.J. & Wolfe L.D. (1980) Height and sexual dimorphism of stature among human societies. *Am. J. Phys. Anthropol.* **53**, 441–456.
4.3.1

Gray R.D. (1987) Faith and foraging: a critique of the 'paradigm argument from design'. In: *Foraging Behavior* (ed. A.C. Kamil, J.R. Krebs & H.R. Pulliam), pp. 69–140. Plenum Press, New York.
4.1.1, 4.2.5, 4.7.3

Green S. (1975) Communication by a graded vocal system in Japanese monkeys. In: *Primate Behaviour* (ed. L.A. Rosenblum), Vol. 4, pp. 61–69. Academic Press, New York.
12.7.2

Greenberg L. (1988) Kin recognition in the sweat bee, *Lasioglossum zephyrum*. *Behav. Genet.* **18**, 425–437.
11.5.6

Greene H.W. (1988) Anti-predator mechanisms in reptiles. In *Biology of the Reptilia, Vol. 16 Ecology B* (ed. C. Gans & R.B. Huey, pp. 1–152. Alan Liss, New York.
6.1

Greene H.W. (1989) Defense behavior and feeding biology of the Asian mock viper, *Psammodynastes pulverulentus* (Colubridae), a specialized predator on Scincid lizards. *Chinese Herpetol. Res.* **2**, 21–32.
6.8

Greenfield M.B. (1988) Interspecific acoustic interactions among katydids *Neoconocephalus:* inhibition-induced shifts in diel periodicity. *Anim. Behav.* **36**, 684–695.
12.6.2

Greenlaw J.S. & Post W. (1985) Evolution of monogamy in seaside sparrows *Ammodramus maritimus*: tests of hypotheses. *Anim. Behav.* **33**, 373–383.
9.5.1

Greenslade P.J.M. (1983) Adversity selection and the habitat templet. *Amer. Natur.* **122**, 352–365.
2.7.1

Greenslade P.J.M. (1983) Adversity selection and the habitat templet. *Amer. Natur.* **122**, 352–365.
2.7.1

Greenwood J.J.D. (1984) The functional basis of frequency-dependent selection. *Biol. J. Linn. Soc. Lond.* **23**, 177–199.
6.2.1, 6.2.2, 6.4

Greenwood P.J. (1980) Mating systems, philopatry and dispersal in birds and mammals. *Anim. Behav.* **28**, 1140–1162.
3.3.3

Griffiths D. (1980) Foraging costs and relative prey size. *Amer. Natur.* **116**, 743–752.
6.7, 6.8

Griffiths N.T. & Godfray H.C.J. (1988) Local mate competition, sex ratio and clutch size in bethylid wasps. *Behav. Ecol. Sociobiol.* **22**, 211–217.
8.6.3

Grime J.P. (1977) Evidence for the existence of three primary strategies in plants and its relevance to ecological and evolutionary theory. *Amer. Natur.* **111**, 1169–1194.
2.7.1

Grimes L.G. (1980) Observations of group behaviour and breeding biology of the yellow-billed shrike, *Corvinella corvina*. *Ibis* **122**, 166–192.
10.5.2

Groeters F.R. & Dingle H. (1987) Genetic and maternal influences on life history plasticity in response to photoperiod by milkweed bugs (*Oncopeltus fasciatus*). *Amer. Natur.* **129**, 332–346.
2.2.2, 2.8.6

Grosberg R.K. & Quinn J.F. (1986) The genetic control and consequences of kin recognition by the larvae of a colonial marine invertebrate. *Nature* **322**, 457–459.
12.3.2

Groschupf K. & Mills G.S. (1982) Singing behavior of the five-striped sparrow. *Condor* **84**, 226–236.
12.7.3

Gross M.R. & Sargent R.C. (1985) The evolution of male and female parental care in fishes. *Amer. Zool.* **25**, 807–822.
8.2, 8.4.3

Gross M.R. & Shine R. (1981) Parental care and mode of fertilization in ectothermic vertebrates. *Evolution* **35**, 775–793.
8.4, 8.4.3

Guilford T. (1985) Is kin selection involved in the evolution of warning coloration? *Oikos* **45**, 31–36.
6.4.1, 6.4.2, 6.4.3

Guilford T. (1989) Studying warning signals in the laboratory. In: *Ethoexperimental Approaches to the Study of Behavior* (ed. R.J. Blanchard, P.F. Brain, D.C. Blanchard & S. Parmigiani), pp. 87–103. Kluwer Academic Publishers, Dordrecht.
6.1, 6.4

Guilford T. (1990) The evolution of aposematism. In: *Insect Defense: Adaptive Mechanisms and Strategies of Prey and Predators* (ed. D.L. Evans & J.O. Schmidt), pp. 23–61. State University of New York Press, New York.
6.4. 6.4.1, 6.4,2

Guilford T. & Dawkins M.S. (1987) Search

images not proven. a reappraisal of recent evidence. *Anim. Behav.* 35, 1838–1845.
6.2.1

Guilford T., Nicol C., Rothschild M. & Moore B.P. (1987) The biological roles of pyrazines: evidence for a warning odour function. *Biol. J. Linn. Soc.* 31, 113–128.
6.4

Guinness F.E., Clutton-Brock T.H. & Albon S.D. (1978) Factors affecting calf mortality in red deer. *J. Anim. Ecol.* 47, 817–832.
8.2

Gupta A.P. & Lewontin R.C. (1982) A study of reaction norms in natural populations of *Drosophila pseudobscura*. *Evolution* 36, 934–948.
2.8.6

Gustafsson L. & Sutherland W.J. (1988) The costs of reproduction in the collared flycatcher *Ficedula albicollis*. *Nature* 335, 813–815.
2.3.1, 2.5.1, 2.7.2, 2.8.3

Gwynne D.T. (1980) Female defence polygyny in the bumblebee wolf *Philanthus bicinctus* (Hymenoptera: Specidae). *Behav. Ecol. Sociobiol.* 7, 213–225.
9.4.1

Gwynne D.T. (1984) Male mating effort, confidence of paternity, and insect sperm composition. In: *Sperm Competition and the Evolution of Mating Systems* (ed. R.L. Smith), pp. 117–149. Academic Press, New York.
8.1

Gyllensten U.B., Jakobsson S. & Temrin H. (1990) No evidence for illegitimate young in monogamous and polygynous warblers. *Nature* 343, 168–170.
9.5.2

Ha J.C., Lehner P.N. & Farley S.D. (1990) Risk-prone foraging behaviour in captive grey jays *Perisoreus canadensis*. *Anim. Behav.* 39, 91–96.
4.5.1

Haartman L. von (1954) Der Trauerfliegenschäpper. III. Die Nahrungsbiologie. *Acta Zoologia Fennica 83*, 1–196.
2.5.1

Haftorn S. & Reinertsen R.E. (1985) The effect of temperature and clutch size on the energetic cost of incubation in a free-living blue tit (*Parus caeruleus*). *Auk* 102, 470–478.
2.3.1

Hagedorn M. & Zelick R. (1989) Relative dominance among males is expressed in the electric organ discharge characteristics of a weakly electric fish. *Anim. Behav.* 38, 520–525.
12.6.3

Hagen R.H., Smith D.R. & Rissing S.W. (1988) Genetic relatedness among cofoundresses of two desert ants, *Veromessor pergandei* and *Acromyrmex versicolor* (Hymenoptera: Formicidae). *Psyche* 95, 191–201.
11.4.2

Hahn P.D. & Stuart A.M. (1987) Sibling interactions in two species of termites: a test of the haplodiploid analogy (Isoptera: Kalotermitidae: Rhinotermitidae). *Sociobiology* 13, 83–92.
11.5.5

Haigh J. & Rose M.R. (1980) Evolutionary game auctions. *J. Theor. Biol.* 85, 381–397.
7.2.1

Hairston N.G. (1983) Growth, survival and reproduction of *Plethodon jordani*: trade-offs between selection pressures. *Copeia* 1983, 1024–1035.
8.2

Haldane J.B.S (1932) *The Causes of Evolution*. Longman, New York.
11.2.1

Halliday T. (1983) Information and communication. In: *Animal Behaviour, 2: Communication* (ed. T.R. Halliday & P.J.B. Slater), pp. 43–81. Blackwell Scientific Publications, Oxford.
12.3.2

Halliday T. & Arnold S.J. (1987) Multiple mating by females: a perspective from quantitative genetics. *Anim. Behav.* 35, 939–941.
10.5

Hames R.B. (1979) A comparison of the efficiencies of the shotgun and the bow in neotropical forest hunting. *Human Ecol.* 7, 219–252.
3.2.1

Hames R.B. (1987) Garden labor exchange among the Ye'kwana. *Ethol. Sociobiol.* 8, 259–284.
3.3.3

Hames R.B. (1988a) Game conservation or efficient hunting? In: *The Question of the Commons: The Culture and Ecology of Communal Resources* (ed. B.J. McCay & J.M. Acheson), pp. 92–107. University of Arizona Press, Tucson.
3.2.4

Hames R.B. (1988b) The allocation of parental care among the Ye'kwana. In:

Human Reproductive Behaviour: A Darwinian Perspective (ed. L. Betzig, M. Borgerhoff Mulder & P. Turke), pp. 237–251. Cambridge University Press, Cambridge.
3.3.1

Hames R.B. (1990) Sharing among the Yanomamo: Part I, The effects of risk. In: *Risk and Uncertainty in Tribal and Peasant Economies* (ed. E.A. Cashdan), pp. 89–105. Westview Press, Boulder, Colorado.
3.2.3

Hames R.B. & Vickers W.T. (1982) Optimal diet breadth theory as a model to explain variability in Amazonian hunting. *Amer. Ethnol.* **9**, 358–378.
3.2.1

Hamilton W.D. (1963) The evolution of altruistic behavior. *Amer. Natur.* **97**, 354–356.
11.2.1

Hamilton W.D. (1964) The genetical evolution of social behaviour. *J. Theor. Biol.* **7**, 1–52.
1.1, 1.3, 1.3.2, 1.3.3, 8.5.1, 10.1, 10.6.1, 10.6.3, 11.2.1

Hamilton W.D. (1967) Extraordinary sex ratios. *Science* **156**, 477–488.
1.1, 1.3.2, 8.6.3, 11.5.4

Hamilton W.D. (1970) Selfish and spiteful behaviour in an evolutionary model. *Nature,* **228**, 1218–1220.
1.1, 1.3, 1.3.2

Hamilton W.D. (1972) Altruism and related phenomena, mainly in social insects. *A. Rev. Ecol. Syst.* **3**, 193–232.
1.3.2, 11.2, 11.2.1, 11.5.4, 11.5.5

Hamilton W.D. (1975) Innate social aptitudes of man: an approach from evolutionary genetics. In: *Biosocial Anthropology* (ed. R. Fox), pp. 133–155. John Wiley, New York.
1.3.2

Hamilton W.D. (1979) Wingless and fighting males in fig wasps and other insects. In: *Sexual Selection and Reproductive Competiton in Insects* (ed. M.S. Blum & N.A. Blum), pp. 167–220. Academic Press, New York.
8.6.3, 11.3

Hamilton W.D. & Zuk M. (1982) Heritable true fitness and bright birds: a role for parasites. *Science* **218**, 384–387.
3.3.2, Introduction to Part 3, 7.3.2, 7.3.3

Hamilton W.D. & Zuk M. (1989) Parasites and sexual selection: reply. *Nature* **341**, 289–290.
7.3.3

Hamilton W.J. & Busse C. (1982) Social dominance and predatory behaviour of chacma baboons. *J. Human Evol.* **11**, 567–573.
12.3.3

Hammerstein P. (1981) The role of asymmetries in animal contests. *Anim. Behav.* **29**, 193–205.
5.1.2, 11.3.3

Hammerstein P. & Reichert S.E. (1988) Payoffs and strategies in territorial contests: ESS analyses of two ecotypes of the spider *Agelenopsis aperta. Evol. Ecol.* **2**, 115–138.
12.3.3

Hanken J. & Sherman P.W. (1981) Multiple paternity in Belding's ground squirrel litters. *Science* **212**, 351–353.
9.2.2

Hannon S.J. (1984) Factors limiting polygyny in willow ptarmigon. *Anim. Behav.* **32**, 153–161.
9.5.1

Hannon S.J., Mumme R.L., Koenig W.D. & Pitelka F.A. (1985) Replacement of breeders and within-group conflict in the cooperatively breeding acorn woodpecker. *Behav. Ecol. Sociobiol.* **4**, 303–312.
10.4.6, 10.5.2

Hansen A.J. & Rohwer (1986) Coverable badges and resource defence in birds. *Anim. Behav.* **34**, 69–76.
12.1, 12.6.3

Hanson L.A. (1970) *Rapan Lifeways: Society and History on a Polynesian Island.* Little, Brown, Boston, Massachusetts.
3.3.3

Harcourt A.H., Harvey P.H., Larson S.G. & Short R.V. (1981) Testis weight, body weight and breeding system in primates. *Nature* **293**, 55–57.
7.4

Harcourt A.H. & Stewart K.J. (1987) The influence of help in contests on dominance rank in primates: hints from gorillas. *Anim. Behav.* **35**, 182–190.
8.2

Harder L.D. & Real L.A. (1987) Why are bumble bees risk averse? *Ecology* **68**, 1104–1108.
4.5.1

Hardy J.W. (1961) Studies in behavior and phylogeny of certain New World jays (*Garrulinae*). *Univ. Kansas Sci. Bull.* **42**, 13–149.
10.4.2

Hardy J.W., Webber T.A. & Raitt R.J. (1981) Communal social biology of the South-

ern San Blas jay. *Bull. Fla. St. Mus.* **26**, 203–264.
10.4.2

Harper A.B. (1986) The evolution of begging: sibling competiton and parent–offspring conflict. *Amer. Natur.* **128**, 99–114.
8.5.1

Harper D.G.C. (1982) Competitive foraging in mallards: 'ideal free' ducks. *Anim. Behav.* **30**, 575–584.
5.2.2, 5.6.2

Hartshorne C. (1973) *Born to Sing.* Indiana University Press, Bloominton.
7.3.3

Hartung J. (1985) Matrilineal inheritance: new theory and analysis. *Behav. Brain Sci.* **8**, 661–688.
3.3.3, 3.3.4

Harvell C.D. (1984) Predator-induced defense in a marine bryozoan. *Science* **224**, 1357–1359.
2.8.

Harvell C.D. & Grosberg R.K. (1988) The timing of sexual maturity in clonal animals. *Ecology* **69**, 1855–1864.
2.8, 2.8.4

Harvey P.H. & Arnold S.J. (1982) Female mate choice and runaway sexual selection. *Nature* **297**, 533–534.
7.3.1

Harvey P.H. & Bennett P.M. (1985) Sexual dimorphism and reproductive strategies. In: *Human Sexual Dimorphism* (ed. J. Ghesquiere, R.D. Martin & F. Newcombe), pp. 43–59. Taylor & Francis, London.
7.4

Harvey P.H. & Clutton-Brock T.H. (1985) Life history variation in primates. *Evolution* **39**, 559–581.
2.7.2

Harvey P.H. & Harcourt A.H. (1984) Sperm competition, testes size and breeding systems in primates. In: *Sperm Competition and the Evolution of Animal Mating Systems* (ed. R.L. Smith), pp. 589–600. Academic Press, New York.
7.4

Harvey P.H., Kavanagh M. & Clutton-Brock T.H. (1978) Sexual dimorphism in primate teeth. *J. Zool.* **186**, 475–486.
7.2.3, 7.4

Harvey P.H. & May R.M. (1989) Out for the sperm count. *Nature* **337**, 508–509.
7.4

Harvey P.H. & Partridge L. (1982) Bird coloration and parasites—a task for the future. *Nature* **300**, 480–481.
7.3.3

Harvey P.H., Read A.F., John J.L., Gregory R.D. & Keymer A.E. (1991) An evolutionary perspective. In: *Parasitism: Coexistence or Conflict. Ecological, Physiological and Immunological Aspects* (ed. C.A. Toft & A. Aeschlimann), pp. 344–355. Oxford University Press, Oxford.
7.3.3

Hassell M.P. & May R.M. (1985) From individual behaviour to population dynamics. In: *Behavioural Ecology* (ed. R.M. Sibley & R.H. Smith), pp. 3–32. Blackwell Scientific Publications, Oxford.
4.7.2

Hassell M.P. & Varley G.C. (1969) New inductive population model for insect parasites and its bearing on biological control. *Nature* **223**, 1133–1136.
5.3.1

Hatchwell B.J. & Davies N.B. (1990) Provisioning of nestlings by dunnocks *Prunella modularis* in pairs and trios: compensation reactions by males and females. *Behav. Ecol. Sociobiol.* **27**, 199–209.
8.4.4, 9.5.1

Hauser M.D. (1988) How infant vervet monkeys learn to recognise starling alarm calls: the role of experience. *Behaviour* **105**, 187–201.
12.3.1

Hawkes K. (1990) Why do men hunt? Benefits for risky choices. In: *Risk and Uncertainty in Tribal and Peasant Economies* (ed. E.A. Cashdan), pp. 145–166. Westview Press, Boulder, Colorado.
3.2.3, 3.3.1

Hawkes K., Hill K. & O'Connell J.F. (1982) Why hunters gather: optimal foraging and the Ache of eastern Paraguay. *Amer. Ethnol.* **9**, 379–398.
3.2.1

Hawkes K., Kaplan H., Hill K. & Huratado A. M. (1987) Ache at the settlement: contrasts between farming and foraging. *Human Ecol.* **15**, 133–161.
3.2.1

Hawkes K., O'Connell J.F. & Blurton Jones N.G. (1989) Hardworking Hadza grandmothers. In: *Comparative Socioecology* (ed. V. Standen & R.A. Foley), pp. 341–366. Basil Blackwell, London.
3.3.1

Hawkins A.D. (1986) Underwater sound and fish behaviour. In: *The Behaviour of Teleost Fishes* (ed. T.J. Pitcher), pp. 144–155. Croom Helm, London.
12.6.2

Heatwole H. (1968) Relationship of escape behavior and camouflage in anoline lizards. *Copeia* **1968**, 109–113.
6.6

Hegmann J.P. & Dingle H. (1982) Phenotypic and genetic covariance structure in milkweed bug life history traits. In: *Evolution and Genetics of Life Histories* (ed. H. Dingle & J.P. Hegmann), pp. 177–185. Springer-Verlag, New York.
2.2.2

Hegner R.E. & Emlen S.T. (1987) Territorial organization of white-fronted bee-eater in Kenya. *Ethology* **76**, 189–222.
10.2, 10.4.3, 10.5.1

Heisler I.L. (1984) A quantitative genetic model for the origin of mating preferences. *Evolution* **38**, 1283–1295.
7.3.1

Heisler I.L. (1985) Quantitative genetic models of female choice based on 'arbitrary' male characters. *Heredity* **55**, 187–198.
7.3.1

Henwood K. & Fabrik A. (1979) A quantitative analysis of the dawn chorus: temporal selection for communicatory optimization. *Amer. Natur.* **114**, 260–274.
4.4.2

Hepper P.G. (1986) Kin recognition: functions and mechanisms. A review. *Biol. Rev.* **61**, 63–93.
12.3.2

Hepper P.G. (1991) *Kin Recognition*. Cambridge University Press, Cambridge.
11.1

Herbers J.M. (1984) Queen–worker conflict and eusocial evolution in polygynous ant species. *Evolution* **38**, 631–643.
11.5.4

Hermann H.R. (ed.) (1979) *Social Insects, Vol. 1*. Academic Press, New York.
11.1

Hermann H.R. (ed.) (1981) *Social Insects, Vol. 2*. Academic Press, New York.
11.1

Hermann H.R. (ed.) (1982a) *Social Insects, Vol. 3*. Academic Press, New York.
11.1

Hermann H.R. (ed.) (1982b) *Social Insects, Vol. 4*. Academic Press, New York.
11.1

Herre E.A. (1985) Sex ratio adjustment in fig wasps. *Science* **228**, 896–898.
8.6.3

Herre E.A. (1987) Optimality, plasticity and selective regime in fig wasp sex ratios. *Nature* **329**, 627–629.
8.6.3, 11.2.1

Herrnstein R.J. (1961) Relative and absolute response as a function of reinforcement. *J. Exp. Anal. Behav.* **4**, 267–272.
4.6.1

Herrnstein R.J. (1964) Aperiodicity as a factor in choice. *J. Exp. Analy. Behav.* **7**, 178–182.
4.5.2

Herrnstein R.J. & Vaughan W. (1980) Melioration and behavioural allocation. In: *Limits to Action* (ed. J.E.R. Staddon), pp. 143–176. Academic Press, New York.
4.6.1

Heyman G.M. & Luce R.D. (1979) Operant matching is not a logical consequence of maximizing reinforcement rate. *Anim. Learning Behav.* **7**, 133–140.
4.6.1

Hewlett B.S. (1988) Sexual selection and paternal investment among Aka pygmies. In: *Human Reproductive Behaviour: A Darwinian Perspective* (ed. L. Betzig, M. Borgerhoff Mulder & P Turke), pp. 263–276. Cambridge University Press, Cambridge.
3.3.1

Hewlett B.S. (1991) *The Father's Role: Cultural and Evolutionary Perspectives*. Aldine de Gruyter, New York.
3.3.1

Hildrew A.G. & Townsend C.R. (1987) Organization in freshwater benthic communities. In: *Organization of Communities Past and Present* (ed. J.H.R. Gee & P.S. Giller), pp. 347–372. Blackwell Scientific Publications, Oxford.
2.7.1

Hill K. (1988) Macronutrient modifications of optimal foraging theory: an approach using indifference curves applied to some modern foragers. *Human Ecol.* **16**, 157–197.
3.2.1

Hill K. & Hawkes K. (1983) Neotropical hunting among the Ache of eastern Paraguay. In: *Adaptive Responses of Native Amazonians* (ed. R.B. Hames & W.T. Vickers), pp. 139–188. Academic Press, New York.
3.2.1, 3.2.2

Hill K. & Kaplan H. (1988) Tradeoffs in male and female reproductive strategies among the Ache: Parts 1 & 2. In: *Human Reproductive Behaviour: A Darwinain*

Perspective (ed. L. Betzig, M. Bogerhoff Mulder & P. Turke), pp. 277–305. Cambridge University Press, Cambridge.
3.3.1, 3.3.2

Hill K., Kaplan H., Hawkes K. & Hurtado A.M. (1987) Foraging decisions among Ache hunter–gatherers: new data and implications for optimal foraging models. *Ethol. Sociobiol.* **8**, 1–36.
3.2.1

Hills S. (1980) Incubation capacity as a limiting factor of shorebird clutch size. *Amer. Zool.* **20**, 774.
9.5.5

Hinde R.A. (1970) *Animal Behaviour: A Synthesis of Ethology and Comparative Psychology.* McGraw-Hill, Kogakusha.
12.4

Hinde R.A. (1981) Animal signals: ethological and games-theory approaches are not incompatible. *Anim. Behav.* **29**, 535–542.
12.3.4

Hinde R.A. (1987) *Individuals, Relationships and Culture.* Cambridge University Press, Cambridge.
3.4.1, 3.4.2

Hines W.G.S. (1987) Evolutionarily stable strategies: a review of basic theory. *Theor. Pop. Biol.* **31**, 195–272.
1.2.2

Hingston R.W.G. (1933) *The Meaning of Animal Coloration and Adornment.* Edward Arnold, London.
7.3.3

Hinson J.M. & Staddon J.E.R. (1983) Hill-climbing by pigeons. *J. Exp. Anal. Behav.* **39**, 25–47.
4.6.1

Hiraizumi Y. (1961) Negative correlation between rate of development and female fertility in *Drosophila melanogaster*. *Genetics* **46**, 615–624.
2.2.2

Hixon M.A. (1987) Territory area as a determinant of mating system. *Amer. Zool.* **27**, 229–247.
9.2.3

Hoffman S.G. (1983) Sex-related foraging behaviour in sequentially hermaphroditic hogfishes (Bodianus spp). *Ecology* **64**, 798–808.
9.1

Hogendoorn K. & Velthuis H.H.W (1988) Influence of multiple mating on kin recognition by worker honeybees. *Naturwissenshaften* **75**, 412–413.
11.5.6

Höglund J. (1989a) *Sexual selection and the evolution of leks in the great snipe, Gallinago media.* PhD thesis. Uppsala University, Sweden.
7.3.2

Höglund J. (1989b) Size and plumage dimorphism in lek-breeding birds: a comparative analysis. *Amer. Natur.* **134**, 72–87.
7.3.2, 7.3.3

Höglund J. (1989c) Pairing and spawning patterns in the comman toad *Bufo bufo*: The effects of sex ratios and the time available for male–male competition. *Anim. Behav.* **38**, 423–429.
7.3.2, 9.3.2

Hogsett M.L. & Nordskog A.W. (1958) Genetic economic value in selecting for egg production rate, body weight and egg weight. *Poult. Sci.* **37**, 1404–1417.
2.4.1

Högstedt G. (1980) Evolution of clutch size in birds: adaptive variation in relation to territory quality. *Science* **210**, 1148–1150.
2.2.1, 2.8.3, 2.8.5, 3.3.4

Högstedt G. (1981) Should there be a positive or negative correlation between survival of adults in a bird population and their clutch size? *Amer. Natur.* **118**, 568–571.
2.2.1, 2.3.1

Hölldobler B. & Wilson E.O. (1977) The number of queens: an important trait in ant evolution. *Naturwissenshaften* **64**, 8–15.
11.4.2

Hölldobler B. & Wilson E.O. (1990) *The Ants.* Harvard University Press, Cambridge, Massachusetts.
11.1, 11.4.2, 11.5.1

Holling C.S. (1965) The functional response of predators to prey density and its role in mimicry and population regulation. *Mem. Entomol. Soc. Can.* **45**, 1–60.
6.2.1

Holling C.S. (1966) The functional response of invertebrate predators to prey density. *Mem. Entomol. Soc. Can.* **47**, 3–86.
6.2.1

Holm E. (1988) Environmental restraints and life-strategies: a habitat templet matrix. *Oecologia* **75**, 141–145.
2.7.1

Hoogland J.L. (1981) Nepotism and cooperative breeding in the black-tailed prairie dog (Sciuridae: *Cynomys ludovicianus*). In: *Natural Selection and*

Social Behavior (ed. R.D. Alexander & D.W. Tinkle), pp. 283–310. Chiron Press, New York.
9.2.2

Hoogland J.L. & Sherman P.W. (1976) Advantages and disadvantages of bank swallow (*Riparia riparia*) coloniality. *Ecol. Monogr.* **46**, 33–58.
10.4.1

Hook A.W. (1987) Nesting behavior of Texas *Cerceris* digger wasps with emphasis on nest reutilization and nest sharing (Hymenoptera: Sphecidae). *Sociobiology* **13**, 93–118.
11.5.1

Hopkins C.D. (1977) Electric communication. In: *How Animals Communicate* (ed. T.A. Sebeok), pp. 263–289. Indiana University Press, Bloomington.
12.6.3

Hopkins C.D. (1983) Sensory mechanisms in animal communication. In: *Communication* (ed. T.R. Halliday & P.J.B. Slater), pp. 114–155. Blackwell Scientific Publications, Oxford.
12.1, 12.6.2, 12.6.3, 12.8

Horn A.G. & Falls J.B. (1988) Responses of western meadowlark *Sturnella neglecta* to song repetition and contrast. *Anim. Behav.* **36**, 291–293.
12.7.3

Horn H.S. (1978) Optimal tectics of reproduction and life-history. In: *Behavioural Ecology, An Evolutionary Approach*, 1st edn (ed. J.R. Krebs & N.B. Davies), pp. 411–429. Blackwell Scientific Publications, Oxford.
2.1.2, 2.6.1, 2.7.1

Horn H.S. & Rubenstein D.I. (1984) Behavioural adaptations and life history. In: *Behavioural Ecology, An Evolutionary Approach*, 2nd edn (ed. J.R. Krebs & N.B. Davies), pp. 279–298. Blackwell Scientific Publications, Oxford.
2.6.1, 2.7.1

Houde A. (1987) Mate choice based upon naturally occurring color-pattern variation in a guppy population. *Evolution* **41**, 1–10.
6.3

Houde A. (1988) Genetic differentiation in female choice between two guppy populations. *Anim. Behav.* **36**, 511–516.
6.3

Houston A.I. (1983) Optimality theory and matching. *Behav. Anal. Lett.* **3**, 1–15.
4.6.1

Houston A.I. (1987) Optimal foraging by birds feeding dependent young. *J. Theor. Biol.* **124**, 251–274.
4.2.1

Houston A.I., Clarke C.W., McNamara J.M. & Mangel M. (1988) Dynamic models in behavioural and evolutionary ecology. *Nature* **332**, 29–34.
4.4, 4.4.2

Houston A.I. & Davies N.B. (1985) The evolution of cooperation and life history in the dunnock *Prunella modularis*. In: *Behavioural Ecology* (ed. R.M. Sibley & R.H. Smith), pp. 471–487. Blackwell Scientific Publications, Oxford.
3.4.4, 9.5.1

Houston A.I. & McNamara J.M. (1982) A sequential approach to risk-taking. *Anim. Behav.* **30**, 1260–1261.
4.5

Houston A.I. & McNamara J.M. (1985) The choice of two prey types that minimises the probability starvation. *Behav. Ecol. Sociobiol.* **17**, 135–141.
4.1.1, 4.4.3, 4.5

Houston A.I. & McNamara J.M. (1987) Switching between resources and the ideal free distribution. *Anim. Behav.* **35**, 301–302.
5.5.2

Houston A.I. & McNamara J.M. (1988) The ideal free distribution when competitive abilities differ: an approach based on statistical mechanics. *Anim. Behav.* **36**, 166–174.
5.2.3

Houtman A.N. (1991) Sexual selection in the zebra finch (*Poephila guttata*). PhD thesis, University of Oxford.
7.3.1

Howard R.D. (1978) The evolution of mating strategies in bullfrogs *Rana caterbeiana*. *Evolution* **32**, 850–871.
9.3.2

Howard R.D. (1980) Mating behaviour and mating success in woodfrogs *Rana sylvatica*. *Anim. Behav.* **28**, 705–716.
9.3.2

Hrdy S.B. (1986) Empathy, polyandry, and the myth of the coy female. In: *Feminist Approaches to Science* (ed. R. Bleier), pp. 119–146. Pergamon Press, New York.
3.3.1

Hrdy S.B. (1987) Sex-biased parental investment among primates and other mammals: a critical evaluation of the Trivers–Willard hypothesis. In: *Child Abuse and Neglect: Biosocial Dimen-*

sions (ed. R. Gelles & J. Lancaster), pp. 97–147. Aldine de Gruyter, New York.
3.3.1, 3.3.4

Huck U.W., Labov J.B. & Lisk R.D. (1986) Food restricting young hamsters (*Mesocricetus auratus*) affects sex ratio and growth of subsequent offspring. *Biol. Reprod.* **35**, 592–598.
8.2

Huck U.W., Labov J.B, & Lisk R.D. (1987) Food-restricting first generation juvenile female hamsters (*Mesocricetus auratus*) affects sex ratio and growth of third generation offspring. *Biol. Reprod.* **37**, 612–617.
8.2

Huck U.W., Tonias B.A. & Lisk R.D. (1989) The effectiveness of competitive male inseminations in golden hamsters *Mesocricetus auratus* depends on an interaction of mating order, time delay between males and the timing of mating relative to ovulation. *Anim. Behav.* **37**, 674–680.
9.2.2

Hughes A.L. (1988) *Evolution and Human Kinship.* Oxford University Press, Oxford.
3.3.3

Hultsch H. & Toat D. (1989) Memorization and reproduction of songs in nightingales (*Luscinia megarhynches*): evidence for package formation. *J. Comp. Physiol. A* **165**, 167–203.
12.0

Humphrey N.K. (1976) The social function of intellect. In: *Growing Points in Ethology* (ed. P.P.G. Bateson & R.A. Hinde), pp. 303–317. Cambridge University Press, Cambridge.
12.5

Hunter J. (1837) An account of an extraordinary pheasant. In: *Observations on Certain Parts of the Animal Oeconomy* with notes by Richard Owen. Longmans, London.
7.1

Hunter J. (1861) Observations on generation. In: *Essays and Observations on Natural History, Anatomy, Physiology, Psychology and Geology. Vol. 1.* John van Voorst, London.
7.1

Hunter L.A. (1985) The effects of helpers in cooperatively breeding purple gallinules. *Behav. Ecol. Sociobiol.* **18**, 147–153.
10.3.1

Hunter M.S. (1989) Sex allocation and egg distribution of an autoparasitoid *Encasia pergandiella* (Hymenoptera: Aphelinidae). *Ecol. Ent.* **14**, 57–67.
8.6.1

Hurtado A.M., Hawkes K., Hill K. & Kaplan H. (1985) Female subsistence strategies among Ache hunter–gatherers of eastern Paraguay. *Human Ecol.* **13**, 1–28.
3.3.1

Hussell D.J.T. (1972) Factors affecting clutch size in arctic passerines. *Ecol. Monogr.* **42**, 317–364.
2.8.4

Ikawa T. & Okabe H. (1985) Regulation of egg number per host to maximize the reproductive success in the gregarious parasitoid, *Apanteles glomeratus* L. (Hymenoptera: Braconidae). *Appl. Ent. Zool.* **20**, 331–339.
2.4.1

Inman A.I., Lefebvre L. & Giraldeau L.A. (1987) Individual diet differences in feral pigeons: evidence for resource partitioning. *Anim. Behav.* **35**, 1902–1903.
5.6.4

Inman A.J. (1990) Studies of group foraging. D.Phil. thesis, Oxford University, Oxford.
4.6.2

Ims R.A. (1987) Responses in spatial organisation and behaviour to manipulations of the food resource in the vole *Clethrionomys rufocanus. J. Anim. Ecol.* **56**, 585–596.
9.2.3

Ims R.A. (1988) Spatial clumping of sexually receptive females induces space sharing among male voles. *Nature* **335**, 541–543.
9.2.3, 9.3.3

Irons W. (1979a) Natural selection, adaptation and human social behaviour. In: *Evolutionary Biology and Human Social Behavior: An Anthropological Perspective* (ed. N.A. Chagnon & W. Irons), pp. 4–39. Duxbury Press, North Scituate, Massachusetts.
3.1, 3.4,1

Irons W. (1979b) Investment and primary social dyads. In: *Evolutionary Biology and Human Social Behavior: An Anthropological Perspective* (ed. N.A. Chagnon & W. Irons), pp. 181–213. Duxbury Press, North Scituate, Massachusetts.
3.3.3

Irons W. (1983) Human female reproductive strategies. In: *Social Behavior of Female Vertebrates* (ed, S.K. Wasser),

pp. 169–213. Academic Press, New York.
3.3.2

Itô Y. (1980) *Comparative Ecology.* Cambridge University Press, Cambridge.
8.3.1

Itô Y. (1987) Role of pleometrosis in the evolution of eusociality in wasps. In: *Animal Societies: Theories and Facts* (ed. Y. Itô, J.L. Brown & J. Kikkawa), pp. 17–34. Japan Science Society Press, Tokyo.
11.4.1

Itô Y. (1989) The evolutionary biology of sterile soldiers in aphids. *Trends Ecol. Evol.* **4**, 69–73.
11.1, 11.3

Itô Y., Brown J.L. & Kikkawa J. (eds) (1987) *Animal Societies: Theories and Facts.* Japan Science Society Press, Tokyo.
11.1

Ives A.R. (1989) The optimal clutch size of insects when many females oviposit per patch. *Amer. Natur.* **133**, 671–687.
2.5.1

Iwasa Y. (1981) Role of sex ratio in the evolution of eusociality in haplodiploid social insects. *J. Theor. Biol.* **93**, 125–142.
11.5.4

Iwasa Y. & Obara Y. (1989) A game model for daily activity schedule of male butterfly. *J. Insect. Behav.* **2**, 589–608.
5.6.1

Iwasa Y., Odendaal F.J., Murphy D.D., Ehrlich P.R. & Launer A.E. (1983) Emergence patterns of male butterflies: a hypothesis and a test. *Theor. Popul. Biol.* **23**, 363–379.
5.6.1

Iwasa Y., Pomiankowski A. & Nee S. Quantitative genetic models for the evolution of costly mate preference. *Evolution* (submitted).
7.3.1

Jackson J.A., Lennartz M.R. & Hooper R.G. (1979) Tree age and cavity initiation by red-cockaded woodpeckers. *J. Forestry* **77**, 102–103.
10.4.6

Jaffe W.P. (1966) Egg production, body weight and egg quality characters; their heritability and the correlations between them. *Poult. Sci.* **7**, 91–98.
2.4.1

Jakobsen P.J. & Johnsen G.H. (1987) Behavioural response of the water flea *Daphnia pulex* to a gradient in food concentration. *Anim. Behav.* **35**, 1891–1895.
5.2.1

Jakobsen P.J., Johnsen G.H. & Holm J.C. (1987) Increased growth rate in Atlantic salmon parr (*Salmo salar*) by using a two-coloured diet. *Can. J. Fish Aquat. Sci.* **44**, 1079–1081.
5.6.4

Jakobsson S., Radesater T. & Jarvi T. (1979) On the fighting behaviour of *Nannacara anomala* (Pisces, Cichlidae). *Z. Tierpsychol.* **49**, 210–220.
12.3.4

Jamieson I.G. (1986) The functional approach to behavior: is it useful? *Amer. Natur.* **127**, 195–208.
10.3, 10.5

Jamieson I.G. (1989) Behavioral heterochrony and the evolution of birds' helping at the nest: an unselected consequence of communal breeding? *Amer. Natur.* **133**, 394–406.
10.3, 10.5

Jamieson I.G. & Craig J.L. (1987) Critique of helping behaviour in birds: a departure from functional explanations. In: *Perspectives in Ethology* (ed. P. Bateson & P. Klopfer), Vol. 7, pp. 79–98. Plenum Press, New York.
10.3, 10.5

Janzen D.H. (1977) Promising direction of study in tropical plant–animal interactions. *Ann. Mo. Bot. Gard.* **61**, 706–736.
8.3.2

Janzen D.H. (1980) When is it coevolution? *Evolution* **34**, 611–612.
6.8

Jarman P.J. (1974) The social organisation of antelope in relation to their ecology. *Behaviour* **48**, 215–267.
9.2.3

Jarman P.J. (1988) On being thick-skinned: dermal shields in large mammalian herbivores. *Biol. J. Linn. Soc.* **36**, 169–191.
7.2.3

Jarvi T. & Bakken M. (1984) The function of the variation in the breast-stripe of the great tit (*Parus major*). *Anim. Behav.* **327**, 590–596.
12.3.3

Jarvis J.U.M. (1981) Eusociality in a mammal: cooperative breeding in naked mole-rat colonies. *Science* **212**, 571–573.
10.2, 11.4

Jeanne R.L. (1980) The evolution of social behavior in the Vespidae. *A. Rev. Entomol.* **25**, 317–396.
11.4.1

Jeanne R.L. (ed.) (1988) *Interindividual Behavioral Variability in Social Insects.* Westview Press, Boulder, Colorado.
11.1, 11.4.3

Jenni D.A. & Collier G. (1972) Polyandry in the American jacana *Jacana spinosa*. *Auk* **89**, 743–765.
9.5.6

Jerome F.M., Henderson C.R. & King S.C. (1956) Heritability, gene interactions and correlations with certain traits in the domestic fowl. *Poult. Sci.* **35**, 995–1013.
2.4.1

Johansson S.R. (1984). Deferred infanticide: excess female mortality during childhood. In: *Infanticide* (ed. G.Hausfater & S.B. Hrdy), pp. 463–485. Aldine, New York.
3.3.4

Johnson C.N. (1985) *Ecology, social behaviour and reproductive success in a population of red-necked wallabies*. PhD thesis, University of New England, Armidale, New South Wales, Australia.
8.2

Johnson C.N. (1987) Relationship between mother and infant red-necked wallabies (*Macropus rufogriseus banksianus*). *Ethology* **74**, 1–20.
8.2

Johnson C.N. (1988) Dispersal and the sex ratio at birth in primates. *Nature* **322**, 726–728.
8.6.3

Jones W.T. (1982) Sex ratio and host size in a parasitoid wasp. *Behav. Ecol. Sociobiol.* **10**, 207–210.
8.6.2

Jong G. de (1990) Quantitative genetics of reaction norms. *J. Evol. Biol.* **3**, 447–468.
2.1.1

Joste N.E., Ligon J.D. & Stacey P.B. (1985) Shared paternity in the acorn woodpecker. *Behav. Ecol. Sociobiol.* **17**, 39–41.
10.2

Kacelnik A. (1984) Central place foraging in starlings (*Sturnus vulgaris*). I. Patch residence time. *J. Anim. Ecol.* **53**, 283–299.
4.2, 4.2.1

Kacelnik A. (1988) Short-term adjustments of parental effort in starlings. In: *Acta XIX Congressus Internationalis Ornithologici*, (ed. H. Ouellet), Vol. II, pp. 1843–1856. University of Ottawa Press, Ottawa.
4.2

Kacelnik A., Bruner D. & Gibbon J. (1990) Timing mechanisms in optimal foraging: some applications of scalar expectancy theory. In: *Behavioural Mechanisms of Food Selection* (ed. R.N. Hughes), pp. 61–82. NATO ASI Series G. Ecological

Sciences, Vol. 20. Springer-Verlag, Berlin.
4.2, 4.6

Kacelnik A. & Cuthill I.C. (1987) Starlings and optimal foraging theory: modelling in a fractal world. In: *Foraging Behaviour* (ed. A.C. Kamil, J.R. Krebs & H.R. Pulliam), pp. 303–333. Plenum Press, New York.
4.1.1, 4.2, 4.2.3

Kacelnik A. & Cuthill I.C. (1990) Central place foraging in starlings (*Sturnus vulgaris*) II. Food allocation to chicks. *J. Anim. Ecol.* **59**, 655–674.
4.2

Kacelnik A., Houston A.I. & Schmid-Hempel P. (1986) Central-place foraging in honey bees: the effect of travel time and nectar flow on crop filling. *Behav. Ecol. Sociobiol.* **19**, 19–24.
4.2.3

Kacelnik A. & Krebs J.R. (1983) The dawn chorus of the great tit: proximate and ultimate causes. *Behaviour* **83**, 287–308.
4.4.2

Kagel J.H., Green L. & Caraco T. (1986a) When foragers discount the future: constraint or adaptation? *Anim. Behav.* **34**, 271–283.
4.5.2

Kagel J.H., MacDonald D.N., Battalio R.C., White S. & Green L. (1986b) Risk aversion in rats *Rattus norvegicus* under varying levels of resource availability. *J. Comp. Psychol.* **100**, 95–100.
4.5.2

Kaiser H. (1985) Availabilty of receptive females at the mating place and mating chances of males in the dragonfly *Aeschna cyanea*. *Behav. Ecol. Sociobiol.* **18**, 1–7.
9.3.4

Kamil A.C. (1989) Studies of learning and memory in natural contexts: integrating functional and mechanistic approaches to behavior. In: *Ethoexperimental Approaches to the Study of Behavior* (ed. R.J. Blanchard, P. Brain, D.C. Blanchard & S. Parmigiani), pp. 30–50. Kluwer Academic Publishers, Dordrecht.
6.2.1

Kamil A.C. & Roitblat H.L. (1985) The ecology of foraging behavior: implications for animal learning and memory. *A. Rev. Psychol.* **36**, 141–169.
4.6

Kaplan H. & Cooper W.S. (1984) The evolution of developmental plasticity in reproductive characteristics: an application of the 'adaptive coin flipping' principle.

Amer. Natur. **123**, 393–410.
8.3.2

Kaplan H. & Dove H. (1987) Infant development among the Ache of eastern Paraguay. *Dev. Psch.* **23**, 190–198.
3.3.1

Kaplan H. & Hill K. (1985a) Hunting ability and reproductive success among male Ache foragers. *Curr. Anthropol.* **26**, 131–133.
3.2.1

Kaplan H. & Hill K. (1985b) Food sharing among Ache foragers: tests of explanatory hypotheses. *Curr. Anthropol.* **26**, 223–246.
3.2.3

Kaplan H. & Hill K. (1991) The evolutionary ecology of food acquisition. In: *Evolutionary Ecology and Human Behavior* (ed. E.A. Smith & B. Winterhalder). Aldine de Gruyter, New York (in press).
3.2.1

Kaplan H., Hill K. & Hurtado A.M. (1990) Risk, foraging and food sharing among the Ache. In: *Risk and Uncertainty in Tribal and Peasant Economies* (ed. E.A. Cashdan) pp. 107–143. Westview Press, Boulder, Colorado.
3.2.1, 3.2.2, 3.2.3

Karban R. (1989) Fine-scale adaptation of herbivorous thrips to individual host plants. *Nature* **340**, 60–61.
2.8.4

Keeney R.C. & Raiffa H. (1976) *Decisions with Multiple Objectives: Preferences and Value Trade-offs.* Wiley, New York.
4.5

Keenleyside M. (1983) Mate desertion in relation to adult sex ratio in biparental cichlid fish *Herotilapia multispinosa. Anim. Behav.* **31**, 683–688.
9.5.1

Kenagy G.J. & Tromulak S.C. (1986) Size and function of mammalian testes in relation to body size. *J. Mammal.* **67**, 1–22.
7.4

Kennedy J.S. & Marsh D. (1974) Pheromone-regulated anemotaxis in flying moths. *Science* **184**, 999–1001.
12.6.1

Kent E.B. (1981) Life-history responses to resource variation in a sessile predator, the ciliate protozoan *Tokophyra lemnarum* Stein. *Ecology* **62**, 296–302.
2.2.2

King J.R. (1972) Adaptive periodic fat storage by birds. In: *Proceedings of the XVth International Ornithological Congress* (ed. K.H. Voous), pp. 200–217. Brill, Leiden.
4.4.2

King P.E. (1973) *Pycnogonids.* Hutchinson, London.
8.2

King B.H. (1987) Offspring sex ratios in parasitoid wasps. *Q. Rev. Biol.* **62**, 367–396.
8.6.2

Kirkpatrick M. (1982) Sexual selection and the evolution of female choice. *Evolution* **36**, 1–12.
7.3.1

Kirkpatrick M. (1985) Evolution of female choice and parental investment in polygynous species: the demise of the 'sexy son'. *Amer. Natur.* **125**, 788–810.
7.3.1

Kirkpatrick M. (1986) The handicap mechanism of sexual selection does not work. *Amer. Natur.* **127**, 222–240.
7.3.1

Kirkpatrick M. (1987a) Sexual selection by female choice in polygynous animals. *A. Rev. Ecol. Syst.* **18**, 43–70.
7.3.1

Kirkpatrick M. (1987b) The evolutionary forces acting on female mating preferences in polygynous animals. In: *Sexual Selection: Testing the Alternatives* (ed. J. Bradbury & M. Andersson), pp. 67–82. Wiley, Chichester.
Introduction to Part 3, 7.3.1

Kirkpatrick M., Price T. & Arnold S.J. (1990) The Darwin–Fisher theory of sexual selection in monogamous birds. *Evolution* (in press).
7.3.1

Kitcher P. (1985) *Vaulting Ambition.* MIT Press, Boston, Massachusetts.
3.4.1, 3.4.4

Klemperer H.G. (1983) The evolution of parental care in Scarabeinae (Coleoptera, Scarabeidae): an experimental approach. *Ecol. Entomol.* **8**, 49–59.
2.4

Klomp H. & Teerink B.J. (1962) Host selection and number of eggs per oviposition in the egg-parasite *Trichogramma embryophagum* Htg. *Nature* **195**, 1020–1021.
2.4.1

Klump G.M. & Gerhardt H.C. (1987) Use of non-arbitary acoustic criteria in mate choice by female tree frogs. *Nature* **326**, 286–288.
7.3.2

Kolding S. & Fenchel T.M. (1981) Patterns of reproduction in different populations of five species of the amphipod genus *Gammarus. Oikos* **37**, 167–172.
8.3.1

Knerer G. (1983) The biology and social behaviour of *Evylaeus linearis* (Schenck) (Apoidea; Halictinae). *Zool. Anz.* **211**, 177–186.
11.4.1

Koenig W.D. (1981) Space competition in the acorn woodpecker: power struggles in a cooperative breeder. *Anim. Behav.* **29**, 396–427.
10.4.6, 10.5.2, 10.5.3

Koenig W.D. (1987) Reciprocal altruism in birds: a critical review. *Ethol. Sociobiol.* **9**, 73–84.
10.6.2

Koenig W.D. & Albano S.S. (1985) Patterns of territoriality and mating success in the white-tailed skimmer *Plathemis lydia* (Odonata: Anisoptera). *Amer. Midl. Natur.* **114**, 1–12.
9.3.4

Koenig W.D. & Mumme R.L. (1987) *Population Ecology of the Cooperatively Breeding Acorn Woodpecker.* Princeton University Press, Princeton, New Jersey.
10.2, 10.3.1, 10.3.3, 10.4.2, 10.4.4, 10.4.6, 10.5.2, 10.5.3, 10.6.3

Koenig W.D. & Mumme R.L. (1990) Levels of analysis, and the functional significance of helping behavior. In: *Interpretation and Explanation in the Study of Animal Behavior* (ed. M. Bekoff & D. Jamieson), pp. 268–303. Westview Press, Boulder, Colorado.
10.3.1, 10.5

Koenig W.D., Mumme R.L. & Pitelka F.A. (1983) Female roles in cooperatively breeding acorn woodpeckers. In: *Social Behavior of Female Vertebrates* (ed. S.K. Wasser), pp. 235–261. Academic Press, New York.
10.2

Koenig W.D. & Pitelka F.A. (1981) Reproductive success, group size and the evolution of cooperative breeding in the acorn woodpecker. *Amer. Natur.* **117**, 421–443.
10.4.2, 10.4.4

Koenig W.D. & Stacey P.B. (1990) Acorn woodpeckers: group-living and food storage under contrast ecological conditions. In: *Cooperative Breeding in Birds: Long-Term Studies of Ecology and Behavior* (ed. P. Stacey & W. Koenig), pp. 415–453. Cambridge University Press, Cambridge.
10.2

Kornoa R. (1989) Ideal free distribution of unequal competitiors can be determined by the form of competition. *J. Theor. Biol.* **138**, 347–352.
5.2.3

Korpimäki E. (1988) Costs of reproduction and success of manipulated broods under varying conditions in Tengmalm's owl. *J. Anim. Ecol.* **57**, 1027–1039.
2.3.1, 2.5.1

Kosuda K. (1985) The aging effect on male mating activity in *Drosophila melanogaster. Behav. Genet.* **15**, 297–303.
2.2.2

Krebs J.R. (1976) Habituation and song repertoires in the great tit. *Behav. Ecol. Sociobiol.* **1**, 215–227.
12.7.3

Krebs J.R. (1977) Song and territory in the great tit. In: *Evolutionary Ecology* (ed. B. Stonehouse & C.M. Perrins), pp. 47–62. Macmillan, London.
12.3.3

Krebs J.R. (1979) Bird Colours. *Nature* **282**, 14.
7.3.3

Krebs J.R. (1980) Optimal foraging, predation risk and territory defence. *Ardea* **68**, 83–90.
4.1.1, 4.3.1

Krebs J.R. (1982) Territorial defense in great tits: do residents always win? *Behav. Ecol. Sociobiol.* **4**, 185–194.
12.3.3

Krebs J.R., Ashcroft R. & Webber M.I. (1978) Song repertoires and territory defence in the great tit (*Parus major*). *Nature* **271**, 539–542.
12.7.3

Krebs J.R. & Davies N.B. (ed.) (1978) *Behavioural Ecology: An Evolutionary Approach,* 1st edn. Blackwell Scientific Publications, Oxford.
1.4

Krebs J.R. & Davies N.B. (ed.) (1984) *Behavioural Ecology: An Evolutionary Approach,* 2nd edn. Blackwell Scientific Publications, Oxford.
1.4

Krebs J.R. & Davies N.B. (1987) *An Introduction to Behavioural Ecology.* Blackwell Scientific Publications, Oxford.
5.2.1, 12.4.2

Krebs J.R. & Dawkins R. (1984) Animal signals: mind-reading and manipulation. In: *Behavioural Ecology: An Evolutionary Approach,* 2nd edn. (ed. J.R. Krebs & N.B. Davies), pp. 380–402. Blackwell Scientific Publications, Oxford.
1.4.1, 12.1, 12.3, 12.3.3, 12.3.5, 12.4.2

Krebs J.R. & Houston A.I. (1989) Optimization in ecology. In: *Ecological Concepts* (ed. J.M. Cherrett), pp. 309–338.

Blackwell Scientific Publications. Oxford.
4.7.3

Krebs J.R. & McCleery R.H. (1984) Optimisation in behavioural ecology. In: *Behavioural Ecology: An Evolutionary Approach,* 2nd edn. (ed. J.R. Krebs & N.B. Davies), pp. 91–121. Blackwell Scientific Publications, Oxford.
4.1.1

Kroodsma D.E. (1976) Reproductive development in a female song bird: differential stimulation by quality of male song. *Science* 192, 574–575.
12.7.3

Kroodsma D.E. (1988) Song types and their use: developmental flexibility of the male blue-winged warbler. *Ethology* 79, 235–247.
12.7.3

Kroodsma D.E. (1989a) Suggested experimental designs for song playbacks. *Anim. Behav.* 37, 600–609.
12.7.3

Kroodsma D.E. (1989b) Inappropriate experimental designs inpede progress in bioacoustic research: a reply. *Anim. Behav.* 38, 717–719.
12.7.3

Kroodsma D.E., Bereson R.C., Byres B.E. & Minear E. (1989) Use of song type by the chestnut-sided warbler: evidence for both intra- and inter-sexual functions. *Can. J. Zool.* 67, 447–456.
12.7.3

Kruijt J.P. & de Vos G.J. (1988) Individual variation in reproductive success in male black grouse, *Tetrao tetrix* L. In: *Reproductive Success: Studies in Individual Variation in Contrasting Breeding Systems* (ed. T.H. Clutton-Brock), pp. 279–290. University of Chicago Press, Chicago.
7.3.2

Kruijt J.P. & Hogan J.A. (1967) Social behaviour on the lek in black grouse. *Ardea* 55, 203–240.
7.3.2

Kukuk P.F. (1989) Evolutionary genetics of a primitively eusocial halictine bee, *Dialictus zephyrus.* In: *The Genetics of Social Evolution* (ed. M.D. Breed & R.E. Page Jr), pp. 183–202. Westview Press, Boulder, Colorado.
11.4.1

Kukuk P.F., Eickwort G.C., Raveret-Richter M. *et al.,* (1989) Importance of the sting in the evolution of sociality in the Hymenoptera. *A. Ent. Soc. Amer.* 82, 1–5.
11.5.2

Kukuk P.F. & Schwartz M. (1988) Macrocephalic male bees as functional reproductives and probable guards. *Pan-Pacific Entomol.* 64, 131–137.
11.3

Lacey E.A. & Sherman P.W. (1990) Social organization of naked mole-rat colonies: evidence for division of labor. In: *The Biology of the Naked Mole Rat* (ed. P.W. Sherman, J.U.M. Jarvis & R.D. Alexander), pp. 275–336. Princeton University Press, Princeton, New Jersey.
10.2

Lack D. (1947) The significance of clutch size. *Ibis* 89, 302–352.
2.1, 2.1.1, 2.5.1

Lack D. (1954) *The Natural Regulation of Animal Numbers.* Clarendon Press, Oxford.
2.5.1, 2.8.4, 8.3.2

Lack D. (1966) *Population Studies of Birds.* Clarendon Press, Oxford.
2.8.4, 8.5.2

Lack D. (1968) *Ecological Adaptations for Breeding in Birds.* Chapman & Hall, London.
2.7.2, 9.5.1, 9.5.6

Lacy R.C. (1980) The evolution of eusociality in termites: a haplodiploid analogy? *Amer. Natur.* 116, 449–451.
11.5.5

Lacy R.C. (1984) The evolution of termite eusociality: reply to Leinaas. *Amer. Natur.* 123, 876–878.
11.5.5

Lai C. & MacKay T. (1990) Hybrid dysgeneies induced quantitative variation on the *X* chromosome of *Drosophila melanogaster. Genetics* (in press).
7.3.1

Lam P.K. & Calow P. (1989) Intraspecific life-history variation in *Lymnaea peregra* (Gastropda: pulmonata). II. Environmental or genetic variance? *J. Anim. Ecol.* 58, 589–602.
2.2.2

Lambrechts M. (1988) Great tit song output is determined both by motivation and by constraints in singing ability: a reply to Weary *et al. Anim. Behav.* 36, 1244–1246.
12.7.3

Lambrechts M. & Dhondt A.A. (1986) Male quality, reproduction and survival in the great tit (*Parus major*). *Behav. Ecol. Sociobiol.* 19, 57–63.
12.7.3

Lambrechts M. & Dhondt A.A. (1988) The anti-exhaustion hypothesis: a new hypothesis to explain song performance and song switching in the great tit. *Anim. Behav.* **36**, 327–334.
12.7.3

Lambrechts M.M. & Dhondt A.A. (1990) A relationship between the composition and size of great tit song repertoires. *Anim. Behav.* **39**, 213–218.
12.7.3

Lancaster J.B. & Lancaster C.S. (1987) The watershed: change in parental-investment and family-formation in the course of human evolution. In: *Parenting Across the Life Span: Biosocial Dimensions* (ed. J.B. Lancaster *et al.*). pp. 187–205. Aldine de Gruyter, New York.
3.3.1, 3.3.4

Lande R. (1980) Sexual dimorphism, sexual selection and adaption in polygenic characters. *Evolution* **34**, 292–305.
7.3.1

Lande R. (1981) Models of speciation by sexual selection on polygenic characters. *Proc. Nat. Acad. Sci. USA* **78**, 3721–3725.
7.3.1

Lande R. (1982a) Elements of a quantitative genetic model of life-history evolution. In: *Evolution and Genetics of Life Histories.* (ed. H. Dingle & J.P. Hegmann), pp. 21–29. Springer-Verlag, New York.
2.1.1, 2.2.2

Lande R. (1982b) A quantitative genetic theory of life history evolution. *Ecology* **63**, 607–615.
2.1.1, 2.1.2

Lande R. (1987) Genetic correlations between the sexes in the evolution of sexual dimorphism and mating preferences. In: *Sexual Selection: Testing the Alternatives* (ed. J. Bradbury & M. Andersson), pp. 83–94. Wiley, Chichester.
7.3.1

Lande R. & Arnold S.J. (1985) Evolution of mating preference and sexual dimorphism. *J. Theor. Biol.* **117**, 651–664.
7.3.1

Lande R. & Kirkpatrick M. (1988) Ecological speciation by sexual selection. *J. Theor. Biol.* **133**, 85–98.
7.3.1

Lank D.B., Oring L.W. & Maxson S.J. (1985) Mate and nutrient limitation of egg laying in a polyandrous shorebird. *Ecology* **66**, 1513–1524.
9.5.6

Latimer W.A. (1977) A comparative study of the songs and alarm calls of some *Parus* species. *Z. Tierpsychol.* **45**, 414–433.
12.7.3

Lawton J.H., Beddington J. & Bonser R. (1974) Switching in invertebrate predators. In: *Ecological Stability* (ed. M.B. Usher & M.H. Williamson), pp. 141–158. Chapman & Hall, London.
6.2.1

Lazarus J. (1990) The logic of mate desertion. *Anim. Behav.* **39**, 672–684.
8.4.4

Lazarus J. & Inglis I.R. (1978) The breeding behaviour of the pink-footed goose: parental care and vigilant behaviour during the fledging period. *Behaviour* **65**, 62–88.
8.1

Lazarus J. & Inglis I.R. (1986) Shared and unshared parental investment, parent–offspring conflict and brood size. *Anim. Behav.* **34**, 1791–1804.
2.4, 8.1, 8.5.1

Le Boeuf B.J. (1974) Male–male competition and reproductive success in elephant seals. *Amer. Zool.* **14**, 163–176.
9.2.1

Le Boeuf B.J. (1978) Social behaviour in some marine and terrestrial carnivores. In: *Contrasts in Behavior* (ed. E.S. Reese & F.J. Lighter), pp. 251–279. Wiley, New York.
9.2.1

Le Mesurier A.D. (1987) A comparative study of the relationship between host size and brood size in *Apanteles* spp. (Hymenoptera: Bracoridae). *Ecol. Ent.* **12**, 383–393.
8.5.2

Lee R.D. (1979) *The !Kung San: Men, Women and Work in a Foraging Society.* Cambridge University Press, Cambridge.
3.3.2

Leimar O., Enquist M. & Sillén-Tullberg B. (1986) Evolutionary stability of aposematic coloration and prey unprofitability: a theoretical analysis. *Amer. Natur.* **128**, 469–490.
6.4.3, 6.7

Leinaas H.P. (1983) A haplodiploid analogy in the evolution of termite eusociality? Reply to Lacy. *Amer. Natur.* **121**, 302–304.
11.5.5

Lemon R.E., Fieldes M.A. & Struger J. (1981) Testing the monotony threshold hypothesis of bird song. *Z. Tierpsychol.* **56**, 359–379.
12.7.3

Lenington S. (1984) The evolution of polyandry in shorebirds. In: *Shorebirds: Breeding Behaviour and Populations* (ed. J. Burger & B.L. Olla), pp. 149–167. Plenum Press, New York.
9.5.5

Lennartz M.R., Hooper R.G. & Harlow R.F. (1987) Sociality and cooperative breeding of red-cockaded woodpeckers, *Picoides borealis*. *Behav. Ecol. Sociobiol.* **20**, 77–88.
10.3.1

Lenington S. (1983) Social preferences for partners carrying 'good genes' in wild house mice. *Anim. Behav.* **31**, 325–333.
7.3.2

Leonard M.L., Horn A.G. & Eden S.F. (1989) Does juvenile helping enhance breeder reproductive success? A removal experiment on moorhens. *Behav. Ecol. Sociobiol.* **25**, 357–361.
10.3.1

Leonard M.L. & Picman J. (1987) Female settlement in marsh wrens: is it affected by other females? *Behav. Ecol. Sociobiol.* **21**, 135–140.
9.5.3

Lessells C.M. (1986) Brood size in Canada Geese: a manipulation experiment. *J. Anim. Ecol.* **55**, 669–689.
2.3.1, 2.4.1, 2.5.1

Lessells C.M. & Avery M.I. (1987) Sex-ratio selection in species with helpers at the next: some extensions of the repayment model. *Amer. Natur.* **129**, 610–620.
8.6.3

Lessells C.M. & Avery M.I. (1989) Hatching asynchrony in European bee-eaters *Merops apiaster*. *J. Anim. Ecol.* **58**, 815–836.
8.3.2

Lessells C.M., Cooke F. & Rockwell R.F. (1989) Is there a trade-off between egg weight and clutch size in wild lesser snow geese (*Anser c. caerulescens*)? *J. Evol. Biol.* **2**, 457–472.
2.2.2, 2.4.1, 2.8.5

Levins R. (1968) *Evolution in Changing Environments*. Princeton University Press, Princeton.
2.1

Lewis D.M. (1981) Determinants of reproductive success of the white-browed sparrow weaver, *Plocepasser mahali*. *Behav. Ecol. Sociobiol.* **9**, 83–93.
10.3.2, 10.4.6

Lewis D.M. (1982) Dispersal in a population of white-browed sparrow weavers. *Condor* **84**, 306-312.
10.4.6, 10.5.2

Lewontin R.C. (1979) Fitness, survival and optimality. In: *Analysis of Ecological Systems* (ed. D.J. Horn, G.R. Stairs & R.D. Mitchell), pp. 3–21. Ohio State University Press, Columbus, Ohio.
2.1.1, 4.7.3

Lewontin R.C. (1987) The shape of optimality. In: *The Latest on the Best* (ed. J. Dupré), pp. 151–159. MIT Press, Cambridge, Massachusetts.
4.1.1

Lewontin R.C., Rose S. & Kamin L.J. (1985) *Not in Our Genes: Biology, Ideology and Human Nature.* Pantheon Book, New York.
3.4.1

Lightbody J.P. & Weatherhead P.J. (1987) Interactions among females in polygynous yellow-headed blackbirds. *Behav. Ecol. Sociobiol.* **21**, 23–30.
9.5.3

Lightbody J.P. & Weatherhead P.J. (1988) Female settling patterns and polygyny: tests of a neutral-mate-choice hypothesis. *Amer. Natur.* **132**, 20–33.
9.5.3

Ligon J.D. (1981) Demographic patterns and communal breeding in the green woodhoopoe, *Phoeniculus purpureus*. In: *Natural Selection and Social Behavior: Recent Research and New Theory* (ed. R.D. Alexander & D.W. Tinkle), pp. 231–243. Chiron Press, New York.
10.5.2, 10.5.3

Ligon J.D. (1983) Cooperation and reciprocity in avian social systems. *Amer. Natur.* **121**, 366–384.
10.6.2

Ligon J.D. & Ligon S.H. (1978a) The communal social system of the green woodhoopoe in Kenya. *Living Bird* **17**, 159–198.
10.3.2, 10.4.2, 10.5.2, 10.5.3

Ligon J.D. & Ligon S.H. (1978b) Communal breeding in green woodhoopoes as a case for reciprocity. *Nature* **276**, 496–498.
10.5.2, 10.5.3, 10.6.2

Ligon J.D. & Ligon S.H. (1988) Territory quality: key determinant of fitness in the group-living green woodhoopoe. In: *The Ecology of Social Behaviour* (ed. C.N. Slobodchifoff), pp. 229–254. Academic Press, New York.
10.4.4

Lill A. (1974) Social organisation and space utilisation in the lek-forming white-bearded manakin, *M. manacus trinitatis*.

Z. Tierpsychol. **36**, 513–530.
9.4.5

Lill A. (1976) Sexual behavior of the lek-forming white-bearded manakin, *Manacus manacus trinitatis*. *Z. Tierpsychol.* **45**, 225–255.
7.3.2

Lima S.L. (1986) Predation risk and unpredictable feeding conditions: determinants of body mass in birds. *Ecology* **67**, 377–385.
4.4.2

Lin N. (1964) Increased parasitic pressure as a major factor in the evolution of social behavior in halictine bees. *Insectes Sociaux* **11**, 187–192.
11.5.1

Lin N. & Michener C.D. (1972) Evolution of sociality in insects. *Q. Rev. Biol.* **47**, 131–159.
11.4, 11.5.1

Lindén M. (1988) Reproductive trade-off between first and second clutches in the great tit *Parus major*: an experimental study. *Oikos* **51**, 285–290.
2.3.1

Lints F.A. & Hoste C. (1974) The Lansing effect revisited. I. Life span. *Exp. Gerontol.* **9**, 51–69.
2.2.2

Lints F.A. & Hoste C. (1977) The Lansing effect revisited. II. Cumulative and spontaneously reversible parental age effects on fecundity in *Drosophila melanogaster*. *Evolution* **31**, 387–404.
2.2.2

Lints F.A., Stoll J., Gruwez G. & Lints C.V. (1979) An attempt to select for increased longevity in *Drosophila melanogaster*. *Gerontology* **25**, 192–204.
2.2.2

Lively C.M. (1986) Canalization versus developmental conversion in a spatially variable environment. *Amer. Natur.* **128**, 561–569.
2.8.4

Lloyd D.G. (1977) Genetic and phenotypic models of natural selection. *J. Theor. Biol.* **69**, 543–560.
1.2

Lloyd D.G. (1987) Selection of offspring size at independence and other size-versus-number strategies. *Amer. Natur.* **129**, 800–817.
8.3.1

Lloyd J.E. (1971) Bioluminescent communication insects. *A. Rev. Entomol.* **16**, 97–122.
12.6.3

Loehle C. (1988) Problems with the triangular model for representing plant strategies. *Ecology* **69**, 284–286.
2.7.1

Logue A.W. (1988) Research on self-control: an integrating framework. *Behav. Brain Sci.* **11**, 665–709.
4.5.2

Loiselle P.V. & Barlow G.W. (1978) Do fishes lek like birds? In: *Contrasts in Behavior* (ed. E.S. Reese & F.J. Lighter), pp. 31–76. Wiley, New York.
8.4.3

Lopez P.T., Narins P.M., Lewis E.R. & Moore S.W. (1988) Acoustically induced call modification in the white-lipped frog, *Leptodactylus albilabris*. *Anim. Behav.* **36**, 1295–1308.
7.3.2

Low B.S. (1978) Environmental uncertainty and the parental strategies of marsupials and placentals. *Amer. Natur.* **112**, 197–213.
8.1

Low B.S. (1988) Pathogen stress and polygyny in humans. In: *Human Reproductive Behaviour: A Darwinian Perspective* (ed. L. Betzig, M. Borgerhoff Mulder & P. Turke), pp. 115–127. Cambridge University Press, Cambridge.
3.3.2

Low B.S. (1989) Cross-cultural patterns in the training of children. *J. Comp. Psychol.* **103**, 311–319.
3.3.1

Low B.S. (1990) Human responses to environmental extremeness and uncertainty: a cross-cultural perspective. In: *Risk and Uncertainty in Tribal and Peasant Economies* (ed. E.A. Cashdan), pp. 229–255. Westview Press, Boulder, Colorado.
3.3.2

Lucas J.R. (1990) Time scale and diet choice decisions. In: *Behavioural Mechanisms of Food Selection* (ed. R.N. Hughes), pp. 165–185. NATO ASI Series G. Ecological Sciences, Vol. 20. Springer-Verlag, Berlin.
4.4.2

Luckinbull L.S. (1978) *r* and *K* selection in experimental populations of *Escherichia coli*. *Science* **202**, 1201–1203.
2.2.2

Luckinbull L.S. (1979) Selection and the *r/K* continuum in experimental populations of protozoa. *Amer. Natur.* **113**, 427–437.
2.2.2

Luckinbull L.S., Arking R., Clare M.J., Cirocco W.C. & Buck S.A. (1984) Selection for delayed senescence in *Drosophila melanogaster. Evolution* **38**, 996–1003.
2.2.2

Luckinbull L.S. & Clare M.J. (1985) Selection for life span in *Drosophila melanogaster. Heredity* **55**, 9–18.
2.2.2

Luckinbull L.S., Clare M.J., Krell W.L., Cirocco W.C. & Richards P.A. (1987) Estimating the number of genetic elements that defer senescence in *Drosophila. Evol. Ecol.* **1**, 37–46.
2.2.2

Lundberg C.A. & Vaisanen R.A. (1979) Selective correlation of egg size with chick mortality in the black headed gull (*Larus ridibundus*). *Condor* **81**, 141–156.
8.2

Lundberg P. (1988) The evolution of partial migration in birds. *Trends Ecol. Evol.* **3**, 172–175.
5.6.1

Luykx P., Michel J. & Luykx J. (1986) The spatial distribution of the sexes in colonies of the termite *Incisitermes schwarzi* Banks (Isoptera: Kalotermitidae). *Insectes Sociaux* **33**, 406–421.
11.5.5

Luykx P. & Syren R.M. (1979) The cytogenetics of *Incistermes schwarzi* and other Florida termites. *Sociobiology* **4**, 191–209.
11.5.5

Lukyx P. & Syren R.M. (1981) Multiple sex-linked reciprocal translocations in a termite from Jamaica. *Experientia* **37**, 819–820.
11.5.5

Lynch M. (1984) The limits of life history evolution in *Daphnia. Evolution* **38**, 465–483.
2.2.2

Lyon B., Montgomerie R.D. & Hamilton L.D. (1987) Male parental care and monogamy in snow buntings. *Behav. Ecol. Sociobiol.* **20**, 377–382.
9.5.1

Lythgoe J.N. (1979) *The Ecology of Vision.* Oxford University Press, Oxford.
6.1, 6.3

MacArthur R.H. & Pianka E. (1966) On the optimal use of a patchy environment. *Amer. Natur.* **100**, 603–609.
3.2.1, Introduction to Part 2

MacArthur R.H. & Wilson E.O. (1967) *The Theory of Island Biogeography.* Princeton University Press. Princeton, New Jersey.
2.7.1

McConnell P.B. (1990) Acoustic structure and receiver response in domestic dogs, *Canis familiaris. Anim. Behav.* **39**, 897–904.
12.8

McCorquodale D.B. (1988) Relatedness among nestmates in a primitively social wasp, *Cerceris antipodes* (Hymenoptera: Sphecidae). *Behav. Ecol. Sociobiol.* **23**, 401–406.
11.5.1

McCorquodale D.B. & Naumann D.I. (1988) A new Australian species of communal ground nesting wasp, in the genus *Spilomena* Shuckard (Hymenoptera: Sphecidae: Pemphredoninae). *J. Aust. Ent. Soc.* **27**, 221–231.
11.4.1

McCorquodale D.B. & Thomson C.E. (1988) A nest shared by the solitary wasps, *Cerceris antipodes* Smith and *C. australis* Saussure (Hymenoptera: Sphecidae). *J. Aust. Ent. Soc.* **27**, 9–10.
11.5.1

MacDonald D.W. (1979) Helpers in fox society. *Nature* **282**, 69–71.
10.3.1, 10.5.2

MacDonald D.W. & Moehlman P.D. (1982) Cooperation, altruism, and restraint in the reproduction of carnivores. In: *Perspectives in Ethology* (ed. P. Klopfer & P. Bateson), Vol. 5, pp. 433–466. Plenum Press, New York.
10.4.2

Mace R. (1987) The dawn chorus in the great tit *Parus major* is directly related to female fertility. *Nature* **330**, 745–746.
12.6.2

McFarland D.J. (1977) Decision making in animals. *Nature* **269**, 15–21.
4.3.1, 4.4

McGinley M.A., Temme D.H. & Geber M.A. (1987) Parental investment in offspring in variable environments: theoretical and empirical considerations. *Amer. Natur.* **130**, 370–398.
8.3.1, 8.3.2

McGowan K.J. & Woolfenden G.E. (1989) A sentinel system in the Florida scrub jay. *Anim. Behav.* **37**, 1000–1006.
10.2

McGregor P.K. & Avery M.I. (1986) The unsung songs of great tits (*Parus major*): learning neighbour's songs for discrimination. *Behav. Ecol. Sociobiol,* **18**, 311–316.
12.7.3

McGregor P.K. & Krebs J.R. (1982) Mating and song types in the great tit. *Nature* **297**, 60–61.
12.7.3

McGregor P.K. & Krebs J.R. (1989) Song learning in adult great tits (*Parus major*): effects of neighbours. *Behaviour* **103**, 139–159.
12.7.3

McGregor P.K., Krebs J.R. & Perrins C.M. (1981) Song repertoires and life-time reproductive success in the great tit (*Parus major*). *Amer. Natur.* **118**, 149–159.
1.2.1

McLean E.B. & Godin J.-G. (1989) Distance to cover and fleeing from predators in fish with different amounts of defensive armour. *Oikos* **55**, 281–290.
6.6

MacLean G.L. (1972) Clutch size and evolution in the Charadrii. *Auk* **89**, 299–324.
9.5.5

McMahon T.A. & Bonner J.T. (1983) *On Size and Life.* Scientific American Library, New York.
7.1

Macnair M.R. (1978) An ESS for the sex ratio in animals, with particular reference to the social Hymenoptera. *J. Theor. Biol.* **70**, 449–459.
11.5.4

Macnair M.R. & Parker G.A. (1978) Models of parent–offspring conflict. II. Promiscuity. *Anim. Behav.* **26**, 111–122.
8.5.1

Macnair M.R. & Parker G.A. (1979) Models of parent–offspring conflict. III. Intrabrood conflict. *Anim. Behav.* **27**, 1202–1209.
8.5.1

McNamara J.M. (1985) An optimal sequential policy for controlling a Markov neural process. *J. Appl. Prob.* **22**, 324–335.
4.2.2

McNamara J.M. & Houston A.I. (1985) Optimal foraging and learning. *J. Theor. Biol.* **117**, 231–249.
4.2.2

McNamara J.M. & Houston A.I. (1987a) Partial preferences and foraging. *Anim. Behav.* **35**, 1084–1099.
4.1.1, 4.7.3

McNamara J.M. & Houston A.I. (1987b) Starvation and predation as factors limiting population size. *Ecology* **68**, 1515–1519.
4.7.2

McNamara J.M. & Houston A.I. (1987c) A general framework for understanding the effects of variability and interruptions on foraging behaviour. *Acta Biotheor* **36**, 3–22.
4.5.2

McNamara J.M. & Houston A.I. (1990a) Starvation and predation in a patchy environment. In: *Living in a Patchy Environment* (ed. I. Swingland, N.C. Stenseth & B. Shorrocks). Oxford University Press, Oxford.
4.4.1

McNamara J.M. & Houston A.I. (1990b) The value of fat reserves and the tradeoff between starvation and predation. *Acta Biotheor.* **38**, 37–61.
Preface, 4.4.2

McNamara J.M. & Houston A.I. (1990c) State-dependent ideal free distributions. *Evol. Ecol.* **4**, 298–311.
5.5.2

McNamara J.M. & Houston A.I. & Krebs J.R. (1990) Why hoard? The economics of food storing in tits. *Behav. Ecol.* (in press).
4.4.1, 4.4.2

McNamara J.M., Mace R.H. & Houston A.I. (1987) Optimal daily routines of singing and foraging in a bird singing to attract a mate. *Behav. Ecol. Sociobiol.* **20**, 399–405.
4.4.2

McVey M.E. (1988) The opportunity for sexual selection in a territorial dragonfly *Erythemis siaplicicollis*. In: *Reproductive Success* (ed. T.H. Clutton-Brock). Chicago University Press, Chicago.
9.3.4

Magrath R.D. (1989) Hatching asynchrony and reproductive success in the blackbird. *Nature* **339**, 536–538.
2.8.4, 8.3.2

Magrath R.D. (1990) Hatching asynchrony in altricial birds. *Biol. Rev.* **65**, 587–622.
2.8.4

Magurran A.E. (1986) Individual differences in fish behaviour. In: *The Behaviour of Teleost Fishes* (ed. T.J. Pitcher), pp. 338–365. Croom Helm, London
5.5.2

Magurran A.E. (1989) Population differences in minnow anti-predation behavior. In: *Ethoexperimental Approaches to the Study of Behavior* (ed. R.J. Blanchard, P. Brain, D.C. Blanchard & S. Parmigiani), pp. 192–199. Kluwer Academic Publishers, Dordrecht.
6.6

Malcolm J.R. & Marten K. (1982) Natural selection and the communal rearing of pups in African wild dogs (*Lycaon*

pictus). *Behav. Ecol. Sociobiol.* **10**, 1–13.
8.6.3

Mallet J. (1986) Hybrid zones of *Heliconius* butterflies and the stability of movement of warning colour clines. *Heredity* **56**, 191–202.
6.5

Mallet J. & Barton N.H. (1989) Strong natural selection in a warning-color hybrid zone. *Evolution* **43**, 421–431.
6.5

Mallet J. & Singer M. (1987) Individual selection, kin selection, and the shifting balance in the evolution of warning colors: the evidence from butterflies. *Biol. J. Linn. Soc. Lond.* **32**, 337–350.
6.4.3

Mangel M. & Clark C.W. (1988) *Dynamic Modeling in Behavioral Ecology.* Princeton University Press, Princeton, New Jersey.
4.4, 4.4.2, 4.4.3

Manzur M.I. & Fuentes E.R. (1979) Polygyny and agonistic behaviour in the tree-dwelling lizard *Liolaemus tenius* (Iguanidae). *Behav. Ecol. Sociobiol.* **6**, 23–28.
9.3.1

Markl H. (1983) Vibrational communication. In: *Neuroethology and Behavioural Phsyiology* (ed. F. Huber & H. Markl), pp. 332–353. Springer-Verlag, Berlin.
12.6.2

Marler P. (1955) Characteristics of some alarm calls. *Nature* **176**, 6–8.
12.3.1

Marler P. & Mundinger P.C. (1975) Vocalisations, social organisation and breeding biology of twite *Acanthis flavirostris. Ibis* **117**, 1–17.
12.3.2

Marshall N.B. (1979) *Developments in Deep-Sea Biology.* Blandford, Poole.
12.6.3

Martin K., Cooch F.G., Rockwell R.F. & Cooke F. (1985) Reproductive performance in lesser snow geese: are two parents essential? *Behav. Ecol. Sociobiol.* **17**, 257–263.
9.5.1

Martin K. & Cooke F. (1987) Bi-parental care in willow ptarmigan: a luxury? *Anim. Behav.* **35**, 369–379.
8.4.4, 9.5.1

Martin S.G. (1974) Adaptations for polygynous breeding in the bobolink, *Dolichonyx oryzivorous. Anim. Zool.* **14**, 109–119.
8.4.4

Marzluff J.M. & Balda R.P. (1990) Pinyon jays: making the best of a bad situation by helping. In: *Cooperative Breeding in Birds: Long-term studies of Ecology and Behavior* (ed. P. Stacey & W. Koenig), pp. 199–238. Cambridge University Press, Cambridge.
10.3.1, 10.4.5

Masataka M. (1989) Motivational references of contact calls in Japanese monkeys. *Ethology* **80**, 265–273.
12.7.2

Matthews R.W. (1968a) *Microstigmus comes:* sociality in a sphecid wasp. *Science* **160**, 787–788.
11.4.1

Matthews R.W. (1968b) Nesting biology of the social wasp *Microstigmus comes* (Hymenoptera: Sphecidae, Pemphredoninae). *Psyche* **75**, 23–45.
11.4.1

Matthews R.W. (1991) Evolution of social behavior in sphecid wasps. In: *The Social Biology of Wasps* (ed. K.G. Ross & R.W. Matthews). Cornell University Press, Ithaca, New York.
11.4.1

Matthews R.W. & Naumann I.D. (1988) Nesting biology and taxonomy of *Arpactophilus mimi,* a new species of social sphecid (Hymenoptera: Sphecidae) from northern Australia. *Aust. J. Zool.* **36**, 585–597.
11.4.1

Maynard Smith J. (1964) Group selection and kin selection. *Nature* **201**, 1145–1147.
11.2.1

Maynard Smith J. (1974) The theory of games and the evolution of animal conflicts. *J. Theor. Biol.* **47**, 209–221.
12.3.3

Maynard Smith J. (1976) Sexual selection and the handicap principle. *J. Theor. Biol.* **57**, 239–242.
7.3.1

Maynard Smith J. (1977) Parental investment—a prospective analysis. *Anim. Behav.* **25**, 1–9.
3.3.1, Introduction to Part 3, 8.4.1, 8.4.2, 9.1

Maynard Smith J. (1978a) Optimization theory in evolution. *A. Rev. Ecol. Syst.* **9**, 13–56.
2.1.1

Maynard Smith J. (1978b) The handicap principle—a comment. *J. Theor. Biol.* **70**, 251–252.
7.3.1

Maynard Smith J. (1978c) *The Evolution of*

Sex. Cambridge University Press, Cambridge.
8.4.1

Maynard Smith J. (1980) A new theory of sexual investment. *Behav. Ecol. Sociobiol.* **7**, 247–251.
8.6.2

Maynard Smith J. (1981) Will a sexual population evolve to an ESS? *Amer. Natur.* **117**, 1015–1018.
1.2.2

Maynard Smith J. (1982) *Evolution and the Theory of Games.* Cambridge University Press, Cambridge.
1.2.2, 1.4, 5.1.2, 8.4.1, 11.3.4

Maynard Smith J. (1985a) Biology and the behaviour of man. *Nature* **318**, 121–122.
3.4.1

Maynard Smith J. (1985b) Sexual selection, handicaps and true fitness. *J. Theor. Biol.* **115**, 1–8.
7.3.1

Maynard Smith J. (1985c) Appendix: adumbrations. In: On Being the Right Size and Other Essays, *by J.B.S. Haldane* (ed. J. Maynard Smith), pp. 178–187. Oxford University Press, Oxford.
11.2.1

Maynard Smith J. (1989) *Evolutionary Genetics.* Oxford University Press, Oxford.
6.4.1

Maynard Smith J. & Brown R.L.W. (1986) Competition and body size. *Theor. Popul. Biol.* **30**, 166–179.
7.2.1

Maynard Smith J. & Harper D.G.C. (1988) The evolution of aggression—can selection generate variability? *Phil. Trans. R. Soc. B* **319**, 557–570.
12.3.3, 12.3.4

Maynard Smith J. & Parker G. (1976) The logic of asymmetric contests. *Anim. Behav.* **24**, 159–175.
3.2.3, 5.1.2, 12.3.3

Maynard Smith J. & Price G.R. (1973) The logic of animal conflict. *Nature* **246**, 15–18.
1.2.1, 5.1.2

Maynard Smith J. & Ridpath M.G. (1972) Wife sharing in the Tasmanian native hen, *Tribonyx mortierii*: a case of kin selection? *Amer. Natur.* **106**, 447–452.
10.4.5

Mayr E. (1963) *Animal Species and Evolution.* Harvard University Press, Cambridge, Massachusetts.
2.8.1

Mazur J.E. (1984) Tests of an equivalence rule for fixed and variable reinforcer delays. *J. Exp. Psychol. Anim. Behav. Proc.* **10**, 426–436.
4.5.2

Mazur J.E., Snyderman M. & Coe D. (1985) Influences of delay and rate of reinforcement on discrete-trial choice. *J. Exp. Psychol. Anim. Behav. Proc.* **11**, 565–575.
4.5.2

Mead P.S. & Morton P.L. (1985) Hatching asynchrony in the mountain white-crowned sparrow (*Zonotrichia leucuphrys oriantha*): a selected or incidental trait? *Auk* **102**, 781–792.
2.8.6

Mech L.D. (1970) *The wolf: The Ecology and Behaviour of an Endangered Species.* Natural History Press, New York.
10.5.2

Medawar P.B. (1946) Old age and natural death. *Modern Quarterly* **1**, 30–56.
2.1.1

Medawar P.B. (1952) *An Unsolved Problem in Biology.* H.K. Lewis, London.
2.1.1

Mertz D.B. (1975) Senescent decline in flour beetle strains selected for early adult fitness. *Physiol. Zool.* **48**, 1–23.
2.2.2

Metcalf R.A. & Whitt G.S. (1977) Relative inclusive fitness in the social wasp *Polistes metricus.* A genetic analysis. *Behav. Ecol. Sociobiol.* **2**, 353–360.
1.3.2

Meyburg B.-U. (1974) Sibling aggression and mortality among nestling eagles. *Ibis* **116**, 224–228.
8.5.2

Michelsen A., Fink F., Gogala M. & Traue D. (1982) Plants as transmission channels for insect vibrational songs. *Behav. Ecol. Sociobiol.* **11**, 269–281.
12.6.2

Michelsen A. & Larson O.N. (1983) Strategies for acoustic communication in complex environments. In: *Neuroethology and Behavioural Physiology.* (ed. F. Huber & H. Markle), pp. 321–331. Springer-Verlag, Berlin.
12.6.2

Michener C.D. (1969) Comparative social behavior of bees. *A. Rev. Entomol.* **144**, 299–342.
11.2.3, 11.4

Michener C.D. (1982) Early stages in insect social evolution: individual and family odor differences and their functions. *Bull. Entomol. Soc. Amer.* **28**, 7–11.
11.4.1

Michener C.D. (1985) From solitary to eusocial: need there be a series of intervening species? *Fortsch. Zoologie* **31**, 293–305.
11.4, 11.4.1

Michener C.D. & Bennett F.D. (1977) Geographic variation in nesting biology and social organization of *Halictus ligatus*. *Univ. Kansas Sci. Bull.* **51**, 233–260.
11.4.1

Michener C.D. & Brothers D.J. (1974) Were workers of eusocial Hymenoptera initially altruistic or oppressed? *Proc. Natl. Acad. Sci.* **71**, 671–674.
11.2.2, 11.4.1

Michener C.D. & Smith B.H. Kin recognition in primitively eusocial insects. In: *Kin Recognition in Animals* (ed. D.J.C. Fletcher & C.D. Michener), pp. 209–242. Wiley, Chichester.
11.5.6

Milinski M. (1979) An evolutionary stable feeding strategy in sticklebacks. *Z. Tierpyschol.* **51**, 36–40.
5.2.2, 5.5.1

Milinski M. (1982) Optimal foraging: the influence of intraspecific competition on diet selection. *Behav. Ecol. Sociobiol.* **11**, 109–115.
5.6.4

Milinski M. (1984a) A predator's cost of overcoming the confusion effect of swarming prey. *Anim. Behav.* **32**, 233–242.
4.3.1

Milinski M. (1984b) Competitive resource sharing: an experimental test of a learning rule for ESSs. *Anim. Behav.* **32**, 233–242.
5.2.2, 5.2.3, 5.5.1

Milinski M. (1986) A review of competitive resource sharing under constraints in sticklebacks. *J. Fish. Biol.* **29** (Suppl. A), 1–14.
5.2.3, 5.5.1

Milinski M. (1987) Competition for non-depleting resources: the ideal free distribution in sticklebacks. In: *Foraging Behavior* (ed. A.C. Kamil, J.R. Krebs & H.R. Pulliam), pp. 363–388. Plenum Press, New York.
5.5.1

Milinski M. (1988) Games fish play: making decisions as a social forager. *Trends Ecol. Evol.* **3**, 325–330.
5.2.2, 5.2.3

Milinski M. & Bakker T.C.M. (1990) Female sticklebacks use male coloration in mate choice and hence avoid parasitized males. *Nature* **344**, 330–333.
7.3.2

Milinski M. & Heller R. (1978) Influence of a predator on the optimal foraging behaviour of sticklebacks (*Gasterosteus aculeatus*). *Nature* **275**, 642–644.
4.3.1, 4.4.1

Minchella D.J. & Loverde P.T. (1981) A cost of increased early reproductive effort in the snail *Biomphalaria glabrata*. *Amer. Natur.* **118**, 876–881.
2.8

Mobley J.R., Herman L.M. & Frankel A.S. (1988) Responses of wintering humpback whales (*Megaptera novaeangliae*) to playback of recordings of winter and summer vocalisations and of synthetic sound. *Behav. Ecol. Sociobiol.* **23**, 211–223.
12.6.2

Mock D.W. (1983) On the study of avian mating systems. In: *Perspectives in Ornithology* (ed. A.M. Brush & G.A. Clark), pp. 55–85. Cambridge University Press, Cambridge.
9.5.2

Mock D.W. (1984) Siblicidal aggression and resource monopolization in birds. *Science* **225**, 731–733.
8.3.2, 8.5.2

Moehlman P.D. (1979) Jackal helpers and pup survival. *Nature* **277**, 382–383.
10.2, 10.3.1, 10.5.2

Moehlman P.D. (1983) Socioecology of silverbacked and golden jackals, *Canis mesolmelas* and *C. Aureus*. In: *Recent Advances in the Study of Mammalian Behavior* (ed. J.F. Eisenberg & D.G. Kleiman), pp. 423–453. Special Publication No 7. The American Society of Mammalogists.
10.2, 10.3.1

Moehlman P.D. (1986) Ecology of cooperation in canids. In: *Ecological Aspects of Social Evolution: Birds and Mammals* (ed. D.I. Rubenstein & R.W. Wrangham), pp. 64–86. Princeton University Press, Princeton, New Jersey.
10.2

Møller A.P. (1986) Mating systems among European passerines: a review. *Ibis* **128**, 234–250.
9.5.1

Møller A.P. (1987a) Behavioural aspects of sperm competition in swallows *Hirundo rustica*. *Behaviour* **100**, 92–104.
9.5.2

Møller A.P. (1987b) Intraspecific nest parasitism and anti-parasite behaviour in swallows, *Hirundo rustica*. *Anim. Behav.* **35**, 247–254.
9.5.2

Møller A.P. (1987c) Variation badge size in male house sparrows *Passer domesticus*—evidence for status signalling *Anim. Behav.* **35**, 1637–1644.
12.3.3

Møller A.P. (1987d) Social control of deception among status signalling house sparrows *Passer domesticus*. *Behav. Ecol. Sociobiol.* **20**, 307–311.
12.3.3, 12.3.4

Møller A.P. (1988a) Female mate choice selects for male sexual tail ornament in the monogamous swallow. *Nature* **332**, 640–642.
7.3.1, 7.3.2, 9.5.2

Møller A.P. (1988b) Testes size, ejaculate quality and sperm competition in birds. *Biol. J. Linn. Soc.* **33**, 273–283.
7.4.

Møller A.P. (1988c) Ejaculate quality, testes size and sperm control in primates. *J. Hum. Evol.* **17**, 479–488.
7.4

Møller A.P. (1988d) Paternity and paternal care in the swallow, *Hirundo rustica*. *Anim. Behav.* **36**, 996–1005.
9.5.2

Møller A.P. (1988e) Badge size in the house sparrow *Passer domesticus*: effects of intrasexual selection. *Behav. Ecol. Sociobiol.* **22**, 373–378.
12.3.3

Møller A.P. (1988f) False alarm calls as a means of resource usurpation in the Great Tit *Parus major*. *Ethology* **79**, 25–30.
12.5.1

Møller A.P. (1989a) Viability costs of male tail ornaments in a swallow. *Nature* **339**, 132–135.
5.1.1, 7.3.2

Møller A.P. (1989b) Ejaculate quality, testes size and sperm production in mammals. *Funct. Ecol.* **3**, 91–96.
7.4

Møller H., Smith R.H. & Sibly R.M. (1989) Evolutionary demography of a bruchid beetle. I. Quantitative genetical analysis of the female life history. *Funct. Ecol.* **3**, 673–681.
2.2.2

Monaghan P. (1980) Dominance and dispersal between feeding sites in the herring gull (*Larus argentatus*). *Anim. Behav.* **28**, 521–527.
5.5.2

Monaghan P. & Metcalfe N.B. (1985) Group foraging in wild brown hares: effects of resource distribution and social status. *Anim. Behav.* **33**, 993–999.
5.1.1

Moore F.E. & Simm P.A. (1986) Risk-sensitive foraging by a migratory bird *Dendroica coronata*. *Experientia* **42**, 1054–1056.
4.5.1

Moreno J., Gustafsson L., Carlson A. & Part T. (1991) The cost of incubation in relation to clutch size in the collared flycatcher *Ficedula albicollis Ibis* **133**, (in press).
2.3.1

Morris D.J. (1957) 'Typical intensity' and its relation to the problem of ritualisation. *Behaviour* **11**, 1–12.
12.4.2

Morton E.S. (1987) Variation in mate guarding intensity by male purple martins. *Behaviour* **101**, 211–224.
9.5.2

Mountford M.D. (1968) The significance of litter-size. *J. Anim. Ecol.* **37**, 363–367.
2.5.1

Mueller L.D. & Ayala F.J. (1981) Trade-off between *r*-selection and *K*-selection in *Drosophila* populations. *Proc. Natl. Acad. Sci.* **78**, 1303–1305.
2.2.2

Mukai T. & Yamazaki T. (1971) The genetic structure of natural populations of *Drosophila melanogaster*. X. Developmental time and viability. *Genetics* **69**, 385–398.
2.2.2

Muma K.E. & Weatherhead P.J. (1989) Male traits expressed in females: direct or indirect sexual selection? *Behav. Ecol. Sociobiol.* **25**, 23–32.
12.1

Mumme R.L. (in press) Helping behaviour in the Florida scrub jay: nonaptation, exaptation, or adaptation? *Proceedings XX International Ornithological Congress.*
10.3.1

Mumme R.L., Koenig W.D. & Pitelka F.A. (1983) Reproductive competition in the communal acorn woodpecker: sisters destroy each other's eggs. *Nature* **306**, 583–584.
10.2

Mumme R.L., Koenig W.D. & Ratnieks F.L.W. (1989) Helping behaviour, reproductive value, and the future component of indirect fitness. *Anim. Behav.* **38**, 331–343.
10.3.3, 10.5.4

Mumme R.L., Koenig W.D., Zink R.M. & Martin J.A. (1985) Genetic variation and parentage in a California population of acorn woodpeckers. *Auk* **102**, 312–320.
10.2

Munn C.A. (1986) Birds that 'cry wolf'. *Nature* 319, 143–145.
12.5.1

Murdoch W.W. (1969) Switching in general predators: experiments on predator specificity and stability of prey populations. *Ecol. Mongr.* 39, 335–354.
6.2.1, 6.2.2

Murdoch W.W. & Stewart Oaten A. (1989) Aggregation by parasitoids and predators: effects on equilibrium and stability. *Amer. Natur.* 134(2), 288–310.
4.7.2

Murdock G.P. (1967) *Ethnographic Atlas.* Pittsburgh University Press, Pittsburgh.
3.3.2, 3.3.4

Murphy G.I. (1968) Pattern in life history and the environment. *Amer. Natur.* 102, 391–403.
2.6.1

Murton R.K., Westwood N.J. & Isaacson A.J. (1974) Factors affecting egg-weight, body-weight and moult of the wood-pigeon *Columba palumbus. Ibis* 116, 52–73.
2.5.1

Myers J.P. (1981) Cross-seasonal interactions in the evolution of sandpiper social systems *Behav. Ecol. Sociobiol.* 8, 195–202.
9.5.6

Myers J.P., Connors P.G. & Pitelka F.A. (1981) Optimal territory size and the sanderling: compromise in a variable environment. In: *Foraging Behavior: Ecological, Ethological and Psychological Approaches* (ed. A.C. Kamil & T.D. Sargent), pp. 135–158. Garland STPM Press, New York.
5.6.3

Myerson J. & Miezin F.M. (1980) The kinetics of choice: an operant systems analysis. *Psychol. Rev.* 87, 160–174.
4.6.1

Myles T.G. & Nutting W.L. (1988) Termite eusocial evolution: a re-examination of Bartz's hypothesis and assumptions. *Q. Rev. Biol.* 63, 1–23.
11.1, 11.4, 11.5.5

Myrberg A.A. & Riggio R.J. (1985) Acoustically mediated individual recognition by a coral-reef fish (*Pomacentrus partitus*). *Anim. Behav.* 33, 411–416.
12.3.2

Narayanan E.S. & Subba Rao B.R. (1955) Studies in insect parasitism I–III. The effect of different hosts on the physiology, on the development and behaviour and on the sex ratio of *Microbra-con gelechiae. Beitr. Entomol.* 5, 36–60.
2.4.1

Neal E. (1970) The banded mongoose, *Mungos mungo. East Afr. Wildl. J.* 8, 63–71.
10.2

Nei M. (1970) Accumulation of nonfunctional genes on sheltered chromosomes. *Amer. Natur.* 104, 311–322.
11.5.5

Newman J.A. & Caraco T. (1987) Foraging, predation hazard, and patch use in grey squirrels. *Anim. Behav.* 35, 1804–1813.
4.3.1

Newman J.A., Recer G.M., Zwicker S.M. & Caraco T. (1988) Effects of predation hazard on foraging 'constraints': patch-use strategies in grey squirrels. *Oikos* 53, 93–97.
4.1.1, 4.3.1

Newman R.A. (1989) Developmental plasticity of *Scaphiopus couchii* tadpoles in an unpredictable environment. *Ecology* 70, 1775–1787.
2.8, 2.8.2

Nias R.C. (1986) Nest-site characteristics and reproductive success in the superb fairy-wren. *Emu* 86, 139–144.
10.3.1, 10.3.2

Noirot C. & Pasteels J.M. (1987) Ontogenetic development and evolution of the worker caste in termites. *Experientia* 43, 851–860.
11.4

Nonacs P. (1986) Ant reproductive strategies and sex allocation theory. *Q. Rev. Biol.* 61, 1–21.
11.1, 11.5.4

Noonan K.N. (1981) Individual strategies of inclusive fitness maximizing in *Polistes fuscatus* foundresses. In: *Natural Selection and Social Behavior* (ed. R.D. Alexander & D.W. Tinkle), pp. 18–44. Chiron Press, New York.
11.4.1

Noordwijk A.J. van (1990) The effects of forest damage on caterpillars and their effect on the breeding biology of the great tit: an overview. In: *Population Biology of Passerine Birds. An Integrated Approach* (ed. J. Blondel, A.G. Gosler, J.D. Lebreton & R.H. McCleery), pp. 215–222. NATO ASI Series, Springer-Verlag, Berlin.
2.8.4

Noordwijk A.J. van & Jong, G. de (1986) Acquisition and allocation of resources: their influence on variation in life history tactics. *Amer. Natur.* 128, 137–142.
2.2.1

Nordskog A.W. (1977) Success and failure of quantitative genetic theory in poultry. In: *Proceedings of the International Conference on Quantitative Genetics* (ed. E. Pollacks, O. Kempthorne & T.H. Bailey), pp. 568–569. Iowa State University, Columbus.
8.2

Nordskog A.W., Tolman H.S., Casey D.W. & Lin C.Y. (1974) Selection in small populations of chickens. *Poultry Sci.* **53**, 1188–1219.
8.2

Norton S. (1988) Role of gastropod shell and operculum in inhibiting predation by fish. *Science* **241**, 92–94.
6.1, 6.8

Nowicki S. (1989) Vocal plasticity in captive black-capped chickadees: the acoustic basis and rate of call convergence. *Anim. Behav.* **37**, 64–73.
12.3.1

Nunney L. (1985) Female-biased sex ratios: Individual or group selection. *Evolution* **39**, 349–361.
8.6.3

Nur N. (1984a) Feeding frequencies of nestling blue tits (*Parus caeruleus*): costs, benefits and a model of optimal feeding frequency. *Oecologia* **65**, 125–137.
2.3.1

Nur N. (1984b) The consequences of brood size for breeding blue tits. II. Nestling weight, offspring survival and optimal brood size. *J. Anim. Ecol.* **53**, 497–517.
2.5.1

Nur N. (1988) The consequences of brood size for breeding blue tits. III. Measuring the cost of reproduction: survival, future fecundity, and differential dispersal. *Evolution* **42**, 351–362.
2.3.1

Nur N. & Hasson O. (1984) Phenotypic plasticity and the handicap principle. *J. Theor. Biol.* **110**, 275–297.
1.4.1

Nussbaum R.A. (1985) The evolution of parental care in salamanders. *Misc. Publ. Mus. Zool. Univ. Michigan* **169**, 1–50.
8.4

O'Connell J.F. & Hawkes K. (1984) Food choice and foraging sites among the Alyawara. *J. Anthropol. Res.* **40**, 504–535.
3.2.1

O'Connor R.J. (1978) Brood reduction in birds: selection for fratricide, infanticide and suicide. *Anim. Behav.* **26**, 79–96.
1.3.5, 8.3.2, 8.5.2

O'Donald P. (1962) The theory of sexual selection. *Heredity* **17**, 541–552.
7.3.1

O'Donald P. (1967) A general model of sexual and natural selection. *Heredity* **22**, 499–518.
7.3.1

O'Donald P. (1980) *Genetic Model of Sexual Selection.* Cambridge University Press, Cambridge.
7.3.1

O'Donald P. (1983) Sexual selection by female choice. In: *Mate Choice* (ed. P. Bateson) pp. 53–66. Cambridge University Press, Cambridge.
7.3.1

Oftedal O.T. (1985) Pregnancy and lactation. In: *Bioenergetics of Wild Herbivores* (ed. R.J. Hudson & R.G. White), pp. 215–238. CRC Press, Florida.
8.2

Ollason J.G. (1980) Learning to forage—optimality? *Theor. Popul. Biol.* **18**, 44–56.
4.7.3

Orell M. (1990) Effects of brood size manipulations on adult and juvenile survival and future fecundity in the willow tit *Parus montanus*. In: *Population Biology of Passerine Birds. An Integrated Approach* (ed. J. Blondel, A.G. Gosler, J.D. Lebreton & R.H. McCleery), pp. 297–306. NATO ASI Series. Springer-Verlag, Berlin.
2.3.1

Orell M. & Koivula K. (1988) Cost of reproduction: parental survival and production of recruits in the willow tit *Parus montanus*. *Oecologia* **77**, 423–432.
2.3.1

Orians G.H. (1969) On the evolution of mating systems in birds and mammals. *Amer. Natur.* **103**, 589–603.
3.3.2, 5.2, 9.5.3

Orians G.H. (1980) *Some Adaptations of Marsh-Nesting Blackbirds.* Princeton University Press, Princeton, New Jersey.
9.5.3

Orians G.H. & Pearson N.E. (1979) On the theory of central place foraging. In: *Analysis of Ecological Systems* (ed. D.J. Horn, R.D. Mitchell & C.R. Stairs), pp. 154–177. Ohio State University Press, Columbia.
4.2

Oring L.W. (1982) Avian mating systems. In: *Avian Biology* (ed. D.S. Fayner, J.R. King & K.C. Parkes), Vol. 6, pp. 1–92. Academic Press, New York.
9.4, 9.5.1

Oring L.W. & Lank D.B. (1984) Breeding area fidelity, natal philopatry and the social system of sandpipers. In: *Shorebirds: Breeding Behaviour and Populations* (ed. J. Burger & B.L. Olla), pp. 125–147. Plenum Press, New York. 9.5.5

Orzack S.H. (1986) Sex ratio control in a parasitic wasp, *Nasonia vitripennis*. II. Experimental analysis of an optimal sex ratio model. *Evolution* **40**, 341–356. 8.6.3

Oster G.F., Eshel I. & Cohen D. (1977) Worker–queen conflict and the evolution of social insects. *Theor. Pop. Biol.* **12**, 49–85. 11.5.4

Oster G.F. & Wilson E.O. (1978) *Caste and Ecology in the Social Insects*. Princeton University Press, Princeton, New Jersey. 11.2.4

Ostfeld R.S. (1986) Territoriality and mating systems of California voles. *J. Anim. Ecol.* **55**, 691–706. 9.2.3

Owens D.D. & Owens M.J. (1984) Helping behaviour in brown hyenas. *Nature* **308**, 843–845. 10.3.1, 10.5.4, 10.6.3

Pacala S.W., Hassell M.P. & May R.M. (1990) Host-parasitoid associations in patchy environments. *Nature* **374**, 150–153. 4.7.2

Packer C. (1979) Male dominance and reproductive activity in *Papio anubis*. *Anim. Behav.* **27**, 37–45. 12.3.3

Packer C. (1980) Male care and exploitation in *Papio anubis*. *Anim. Behav.* **28**, 512–520. 12.3.3

Packer C. (1983) Sexual dimorphism: the horns of African antelopes. *Science* **221**, 1191–1193. 7.2.3

Packer C., Herbst L., Pusey A.E. *et al.* (1988) Reproductive success in lions. In: *Reproductive Success: Studies of Individual Variation in Contrasting Breeding Systems* (ed. T.H. Clutton-Brock), pp. 363–383. University of Chicago Press, Chicago. 10.5.2

Packer L. (1986a) The social organisation of *Halictus ligatus* (Hymenoptera; Halictidae) in southern Ontario. *Can. J. Zool.* **64**, 2317–2324. 11.4.1

Packer L. (1986b) Multiple-foundress associations in a temperate population of *Halictus ligatus* (Hymenoptera; Halictidae). *Can. J. Zool.* **64**, 2325–2332. 11.4.1, 11.6

Packer L. (1986c) The biology of a subtropical population of *Halictus ligatus*. IV. A cuckoo-like caste. *J. New York Ent. Soc.* **94**, 458–466. 11.4.1

Packer L. & Knerer G. (1986) The biology of a subtropical population of *Halictus ligatus* Say (Hymenoptera: Halictidae). I. Phenology and social organisation. *Behav. Ecol. Sociobiol.* **18**, 363–375. 11.4.1

Packer L. & Knerer G. (1987) The biology of a subtropical population of *Halictus ligatus* Say (Hymenoptera; Halictidae). III. The transition between annual and continuously brooded colony cycles. *J. Kansas. Ent. Soc.* **60**, 510–516. 11.4.1

Page R.E. Jr (1986) Sperm utilization in social insects. *A. Rev. Entomol.* **31**, 297–320. 11.4.2

Page R.E. Jr & Breed M.D. (1987) Kin recognition in social bees. *Trends Ecol. Evol.* **2**, 272–275. 11.1, 11.5.6

Page R.E. Jr & Erickson E.J. (1986) Kin recognition during emergency queen rearing in honeybees (Hymenoptera: Apidae). *A. Ent. Soc. Amer.* **79**, 460–467. 11.5.6

Page R.E. Jr, Robinson G.E. & Fondrk M.K. (1989) Genetic specialists, kin recognition and nepotism in honey-bee colonies. *Nature* **338**, 576–579. 11.5.6

Paley W. (1828) *Natural Theology*, 2nd edn. J. Vincent, Oxford. 4.1.1

Palmer J.O. & Dingle H. (1986) Direct and correlated responses to selection among life-history traits in milkweed bugs (*Oncopeltus fasciatus*). *Evolution* **40**, 767–777. 2.2.2

Pamilo P. (1981) Genetic organization of *Formica sanguinea* populations. *Behav. Ecol. Sociobiol.* **9**, 45–50. 11.4.2

Pamilo P. (1982a) Genetic evolution of sex ratios in eusocial hymenoptera: allele frequency simulations. *Amer. Natur.* **119**, 638–656. 11.5.4

Pamilo P. (1982b) Genetic population structure in polygynous *Formica* ants. *Heredity* **48**, 95–106.
11.4.2

Pamilo P. (1984a) Genotypic correlation and regression in social groups: multiple alleles, multiple loci and subdivided populations. *Genetics* **107**, 307–320.
11.4.1

Pamilo P. (1984b) Genetic relatedness and evolution of insect sociality. *Behav. Ecol. Sociobiol.* **15**, 241–248.
11.5.5

Pamilo P. & Rosengren R. (1983) Sex ratio strategies in *Formica* ants. *Oikos* **40**, 24–35.
11.5.4

Pamilo P. & Varvio-Aho S.-L. (1979) Genetic structure of nests in the ant *Formica sanguinea*. *Behav. Ecol. Sociobiol.* **6**, 91–98.
11.4.2

Pani S.N. & Lasley J.F. (1972) *Genotype Environment Interactions in Animals*. Research Bulletin No. 992, University of Missouri—Columbia College of Agricultural Experiment Station, Columbia, Missouri.
2.8.6

Papageorgis C. (1975) Mimicry in neotropical butterflies. *Amer. Scient.* **63**, 522–532.
6.3

Parker G.A. (1970a) The reproductive behaviour and the nature of sexual selection in *Scatophaga stercoraria* L. II. The fertilization rate and the spatial and temporal relationships of each sex around the site of mating and oviposition. *J. Anim. Ecol.* **39**, 205–228.
5.2, 5.6.1

Parker G. (1970b) Sperm competition and its evolutionary effect on copula duration in the fly *Scatophaga stercoraria*. *J. Insect Physiol.* **16**, 1301–1328.
7.4, 9.3.4

Parker G.A. (1974a) The reproductive behavior and the nature of sexual selection in *Scatophaga stercoraria* L. IX. Spatial distribution of fertilization rates and evolution of male search strategy within the reproductive area. *Evolution* **28**, 93–108.
5.2

Parker G.A. (1974b) Assessment strategy and the evolution of fighting behaviour. *J. Theor. Biol.* **47**, 223–243.
12.3.3, 12.8

Parker G.A. (1978a) Searching for mates. In: *Behavioural Ecology: An Evolutionary Approach*, 1st edn (ed. J.R. Krebs & N.B. Davies), pp. 214–244. Blackwell Scientific Publications, Oxford.
5.2.2

Parker G.A. (1978b) Evolution of competitive mate searching. *A. Rev. Ent.* **23**, 173–196.
9.4.4

Parker G.A. (1982) Phenotype-limited evolutionarily stable strategies. In: *Current Problems in Sociobiology* (ed. King's College Sociobiology Group, Cambridge), pp. 173–201. Cambridge University Press, Cambridge.
5.6.2, 7.3.2

Parker G.A. (1983) Arms race in evolution: an ESS to the opponent-independent costs game. *J. Theor. Biol.* **101**, 619–648.
7.2.1, 7.2.3

Parker G.A. (1984a) The producer/scrounger model and its relevance to sexuality. In: *Producers and Scroungers: Strategies of Exploitation and Parasitism* (ed. C.J. Barnard), pp. 127–153. Croom Helm, London.
5.6.1, 5.6.2

Parker G.A. (1984b) Evolutionarily stable strategies. In: *Behavioural Ecology: An Evolutionary Approach*, 2nd edn (ed. J.R. Krebs & N.B. Davies), pp. 30–61. Blackwell Scientific Publications, Oxford.
12.3.3, 12.5.2, 12.8

Parker G.A. (1985a) Population consequences of evolutionarily stable strategies. In: *Behavioural Ecology* (ed. R.M. Sibly & R.H. Smith), pp. 33–58. Blackwell Scientific Publications, Oxford.
5.6.1

Parker G.A. (1985b) Models of parent offspring conflict. V. Effects of the behaviour of the two parents. *Anim. Behav.* **33**, 519–533.
8.5.1

Parker G.A. & Begon M. (1986) Optimal egg size and clutch size: effects of environment and maternal phenotype. *Amer. Natur.* **128**, 573–592.
8.3.1

Parker G.A. & Courtney S.P. (1983) Seasonal incidence: adaptive variation in the timing of life history stages. *J. Theor. Biol.* **105**, 147–155.
5.6.1

Parker G.A. & Courtney S.P. (1984) Models of clutch size in insect oviposition. *Theor. Popul. Biol.* **26**, 27–48.
2.5.1

Parker G.A. & Knowlton N. (1980) The evolution of territory size—some ESS models. *J. Theor. Biol.* **84**, 445–476.
5.6.3

Parker G.A. & Macnair M.R. (1978) Models of parent–offspring conflict. I. Monogamy. *Anim. Behav.* **26**, 97–110.
8.3.1, 8.5.1

Parker G.A. & Macnair M.R. (1979) Models of parent–offspring conflict. IV. Suppression: evolutionary retaliation of the parent. *Anim. Behav.* **27**, 1210–1235.
8.5.1

Parker G.A. & Maynard Smith J. (1987) The distribution of stay times in *Scatophaga:* reply to Curtsinger. *Amer. Natur.* **129**, 621–628.
5.6.1

Parker G.A. & Rubenstein D.I. (1981) Role assessment, reserve strategy, and acquisition of information in asymmetric animal conflicts. *Anim. Behav.* **29**, 221–240.
12.3.3

Parker G.A. & Stuart R.A. (1976) Animal behavior as a strategy optimizer: evolution of resource assessment strategies and optimal emigration threshholds. *Amer. Natur.* **110**, 1055–1076.
5.6.1

Parker G.A. & Sutherland W.J. (1986) Ideal free distributions when individuals differ in competitive ability: phenotype-limited ideal free models. *Anim. Behav.* **34**, 1222–1242.
5.2.2, 5.2.3, 5.3.2, 5.4, 5.6.5

Parsons J. (1970) Relationship between egg size and post-hatching chick mortality in the herring gull (*Larus argentatus*). *Nature* **228**, 1221–1222.
8.2

Parsons J. (1976) Factors determining the number and size of eggs laid in the herring gull. *Condor* **78**, 481–492.
8.3.2

Parsons P.A. (1977) Genotype–environment interaction for longevity in natural populations of *Drosophila simulans*. *Exp. Gerontol.* **12**, 241–244.
2.8.6

Partridge L. (1980) Mate choice increases a component of offspring fitness in fruit flies. *Nature* **283**, 290–291.
7.3.2

Partridge L. (1989) Lifetime reproductive success and life-history evolution. In: *Lifetime Reproduction in Birds* (ed. I. Newton), pp. 421–440. Academic Press, London.
2.2.1, 2.3.1, 2.6.1

Partridge L. & Harvey P.H. (1988) The ecological context of life history evolution. *Science* **241**, 1449–1455.
2.1.2, 2.2.1, 2.2.2, 2.6.1

Pasteels J.M. & Deneubourg J.-L. (ed.) (1987) *From Individual to Collective Behavior in Social Insects.* Birkhäuser-Verlag, Basel.
11.1

Paton D. (1986) Communication by agonistic displays. II. Perceived information and the definition of agonistic displays. *Behaviour* **99**, 157–174.
12.3.4, 12.7.1

Paton D. & Caryl P.G. (1986) Communication by agonistic displays. I. Variation in information content between samples. *Behaviour* **18**, 213–239.
12.3.4, 12.7.1

Payne R.B.(1984) Sexual selection, lek and arena behaviour, and sexual size dimorphism in birds. *Ornithol. Monogr.* **33**, 52.
7.3.3

Pearson B. (1983) Intra-colonial relatedness amongst workers in a population of nests of the polygynous ant, *Myrmica rubra* Latreille. *Behav. Ecol. Sociobiol.* **12**, 1–4.
11.4.2

Pease C.M. & Bull, J.J. (1988) A critique of methods for measuring life history trade-offs. *J. Evol. Biol.* **1**, 293–303.
2.2.1

Peckarsky B.L. (1985) *Factors affecting the evolution of stonefly–mayfly interactions.* Abstract, Third International Congress of Systematic and Evolutionary Biology (ICSEB III), Brighton, Sussex, England, 4–10 July 1985.
6.8

Peeters C. & Crozier R.H. (1988) Caste and reproduction in ants: not all mated egg-layers are 'queens'. *Psyche* **95**, 283–288.
11.4.3

Pennington R. & Harpending H. (1988) Fitness and fertility among Kalahri !Kung. *Am. J. Phys. Anthropol.* **77**, 303–319.
3.3.1

Perrins C.M. & Moss D. (1975) Reproductive tactics in the great tit. *J. Anim. Ecol.* **44**, 695–706.
2.8.3

Perrone M. & Zaret T.M. (1979) Parental care patterns of fishes. *Amer. Natur.* **113**, 351–361.
8.4.3

Persson O. & Öhrström P. (1989) A new avian mating system: ambisexual

polygamy in the penduline tit *Remiz pendulinus*. *Ornis. Scand.* **20**, 105–111.
9.5.1

Petrie M., Halliday T.R. & Sanders C. (1990) Peahens prefer peacocks with elaborate trains. *Anim. Behav.* **40** (in press).
7.3

Petrinovich L. (1984) A two-factor dual process theory of habitation and sensitization. In: *Habituation, Sensitization and Behavior* (ed. H.V.S. Peeke & L. Petrinovich), pp. 17–55. Academic Press, Orlando.
12.7.3

Pettifor R.A. (1989) Individual variation in clutch-size in tits. D. Phil. thesis, University of Oxford, Oxford.
2.3.1

Pettifor R.A., Perrins C.M. & McCleery R.H. (1988) Individual optimization of clutch size in great tits. *Nature* **336**, 160–162.
2.5.1, 2.8.3

Pianka E.R. (1970) On *r*- and *K*-selection. *Amer. Natur.* **104**, 592–597.
2.7.1

Pianka E.R. & Parker W.S. (1975) Age-specific reproductive tactics. *Amer. Natur.* **109**, 453–464.
2.6.1

Picman J., Leonard M. & Horn A. (1988) Antipredation role of clumped nesting by marsh-nesting red-winged blackbirds. *Behav. Ecol. Sociobiol.* **22**, 9–15.
9.5.3

Pierce G.J. & Ollason J.G. (1987) Eight reasons why optimal foraging theory is a complete waste of time. *Oikos* **49**, 111–118.
4.7.3

Piercy J.E. & Embelton T.F.W. (1977) Review of noise propagation in the atmosphere. *J. Acoust. Soc. Amer.* **61**, 1403–1418.
12.6.2

Pitcher T.J., Lang S.H. & Turner J.R. (1988) A risk-balancing trade-off between foraging rewards and predation risk in shoaling fish. *Behav. Ecol. Sociobiol.* **22**, 225–228.
5.5.2

Pleszczynska W.K. (1978) Microgeographic prediction of polygyny in the lark bunting. *Science* **201**, 935–937.
9.5.3

Plowright R.C. & Laverty T.M. (1984) The ecology and sociobiology of bumblebees. *A. Rev. Entomol.* **29**, 175–199.
11.4.2

Poethke H.J. & Kaiser H. (1987) The territoriality threshold: a model for mutual avoidance in dragonfly mating systems. *Behav. Ecol. Sociobiol.* **20**, 11–19.
9.3.4

Pomiankowski A. (1987a) The 'handicap principle' does work—sometimes. *Proc. Roy. Soc. B* **127**, 123–145.
7.3.1

Pomiankowski A. (1987b) The costs of choice in sexual selection. *J. Theor. Biol.* **128**, 195–218.
Introduction to Part 7, 7.3.1, 7.3.2

Pomiankowski A. (1988) The evolution of female mate preferences for male genetic quality. In: *Oxford Surveys in Evolutionary Biology* (ed. P.H. Harvey & L. Partridge), Vol. 5, pp. 136–184.
7.3.1, 7.3.2

Pomiankowski A. & Guilford T. (1990) Mating calls. *Nature* **344**, 495–496.
7.3.1

Poole J.H. (1989) Announcing intent: aggressive state of musth in African elephants. *Anim. Behav.* **37**, 140–152.
12.3.4

Poole J.H., Payne K., Longbauer W.R. & Moss C.J. (1988) The social contexts of some very low-frequency calls of African elephants. *Behav. Ecol. Sociobiol.* **22**, 385–392.
12.6.2

Popper A.N. & Coombs S. (1980) Auditory mechanisms in teleost fishes. *Amer. Sci.* **68**, 429–440.
12.6.2

Poran N.S., Coss R.G. & Benjamini E. (1987) Resistance to california ground squirrels (*Spermophilus beecheyi*) to the venom of the nothern pacific rattlesnake (*Crotalus viridis oreganus*): a study of adaptive variation. *Toxicon* **25**, 767–777.
6.1, 6.6.

Possingham H.P., Houston A.I. & McNamara J.M. (1990) Risk-averse foraging in bees: a comment on the model of Harder and Real. *Ecology* **71**, 1622–1624.
4.5

Poulton E.B. (1898) Natural selection the cause of mimetic resemblance and common warning colours. *Zool. J. Linn. Soc. Lond.* **26**, 557–612+5 plates.
6.5

Powell R.A. (1982) Evolution of black-tipped tails in weasels: predator confusion. *Amer. Natur.* **119**, 126–131.
6.1

Pravosudov W. (1985) Search for and storage of food by *Parus cinctus japponicus*

and *P. montanus borealis* (Paridae). *Zool. Zh.* **64**, 1031–1034.
4.4.2

Prelec D. (1982) Matching, maximising, and the hyperbolic reinforcement feedback function. *Psychol. Rev.* **89**, 189–230.
4.6.1

Price T.D. (1984a) Sexual selection on body size, territory and plumage variables in a population of Darwin's finches. *Evolution* **38**, 327–343.
7.2.2

Price T.D. (1984b) The evolution of sexual size dimorphism in Darwin's finches. *Amer. Natur.* **123**, 500–518.
7.2.2

Price T.D., Kirkpatrick M. & Arnold S.J. (1988) Directional selection and evolution of breeding date in birds. *Science* **240**, 798–799.
2.8.5

Pruett-Jones S.G. (1985) The evolution of lek mating behavior in Lawe's parotia, *Aves: Patotia lawesii*. PhD thesis, University of California, Berkeley.
7.3.2

Pruett-Jones S.G. (1988) Lekking versus solitary display: temporal variations in dispersion in the buff-breasted sandpiper. *Anim. Behav.* **36**, 1740–1752.
9.4.1

Pruett-Jones S.G. & Lewis M.J. (1990) Habitat limitation and sex ratio promote delayed dispersal in Superb fairy-wrens. *Nature* **348**, 541–542.
10.4.6

Pruett-Jones S.G. & Pruett-Jones M.A. (1990) Sexual selection through female choice in Lawes' Parotia, a lek mating bird of paradise. *Evolution* (in press).
9.4.2

Pulliam H.R. & Caraco T. (1984) Living in groups: is there an optimal group size? In: *Behavioural Ecology: An Evolutionary Approach*, 2nd edn (ed. J.R. Krebs & N.B. Davies), pp. 122–147. Blackwell Scientific Publications, Oxford.
3.2.2, 5.2.2, 5.4

Punnett R.C. (1915) *Mimicry in Butterflies*. Cambridge University Press, Cambridge.
6.5

Pyke H.G., Pulliam H.R. & Charnov E.L. (1977) Optimal foraging theory: a selective review of theory and tests. *Q Rev. Biol.* **52**, 137–154.
3.2.1

Qasim S.Z. (1956) Time and duration of the spawning season in some marine teleosts

in relation to their distribution. *J. Conseil* **21**, 144–155.
8.4.3

Queller D.C. (1987) Sexual selection in flowering plants. In: *Sexual Selection: Testing the Alternatives* (ed. J.W. Bradbury & M.B. Andersson), pp. 165–179. Wiley, New York.
7.5

Queller D.C. (1989) The evolution of eusociality: reproductive head starts of workers. *Proc. Natl. Acad. Sci.* **86**, 3224–3226.
11.5.1

Queller D.C. & Strassman J.E. (1988) Reproductive success and group nesting in the paper wasp, *Polistes annularis*. In: *Reproductive Success* (ed. T.H. Clutton-Brock), pp. 76–96. University of Chicago Press, Chicago.
11.4.1

Queller D.C. & Strassman J.E. (1989) Measuring inclusive fitness in social wasps. In: *The Genetics of Social Evolution* (ed. M.D. Breed & R.E. Page Jr), pp. 103–122. Westview Press, Boulder, Colorado.
11.4.1

Queller D.C., Strassman J.E. & Hughes C.R. (1988) Genetic relatedness in colonies of tropical wasps with multiple queens. *Science* **242**, 1155–1157.
11.4.2

Quiring D.T. & McNeil J.N. (1984a) Exploitation and interference intraspecific larval competition in the dipteran leaf miner, *Agromyza frontella* (Rondani). *Can. J. Zool.* **62**, 421–427.
2.4.1

Quiring D.T. & McNeil J.N. (1984b) Influence of intra-specific larval competition and mating on the longevity and reproductive performance of females of the leaf miner *Agromyza frontella* (Rondani) (Diptera: Agromyzidae). *Can. J. Zool.* **62**, 2197–2200.
2.4.1

Rabenold K.N. (1984) Cooperative enhancement of reproductive success in tropical wren societies. *Ecology* **65**, 871–885.
10.3.1, 10.3.2, 10.4.3, 10.4.4, 10.4.6

Rabenold K.N. (1985) Cooperation in breeding by nonreproductive wrens: kinship, reciprocity, and demography. *Behav. Ecol. Sociobiol.* **17**, 1–17.
10.5.3, 10.6.2

Rachlin H., Battalio R., Kagel J. & Green L. (1981) Maximization theory and behavior. *Behav. Brain Sci.* **4**, 371–388.
4.6.1

Radabaugh D.C. (1989) Seasonal colour changes and shifting antipredator tactics in darters. *J. Fish. Biol.* **34**, 679–685.
6.1, 6.6

Raitt R.J., Winterstein S.R. & Hardy J.W. (1984) Structure and dynamics of communal groups in the Beechey jay. *Wilson Bull.* **96**, 206–227.
10.3.2

Randall J.A. (1989) Individual footdrumming signature in banner-tailed kangaroo rats *Dipodomys spectabilis. Anim. Behav.* **38**, 620–630.
12.3.2

Rasa O.A.E. (1973) Marking behaviour and its social significance in the African dwarf mongoose, *Helogale undulata rufula. Z. Tierpsychol.* **32**, 293–318.
12.3.2

Ratnieks F.L.W. (1988) Reproductive harmony via mutual policing by workers in eusocial Hymenoptera. *Amer. Natur.* **132**, 217–236.
11.4.2

Ratnieks F.L.W. & Visscher P.K. (1989) Worker policing in the honeybee. *Nature* **342**, 796–797.
11.4.2

Read A.F. (1988) Sexual selection and the role of parasites. *Trends Ecol. Evol.* **3**, 97–102.
7.3.2

Read A.F. (1990) Parasites and the evolution of host sexual behaviour. In: *Parasitism and Host Behaviour* (ed. C. Barnard & J.M. Belinke), pp. 117–157. Taylor & Francis, London.
7.3.2

Read A.F. & Harvey P.H. (1989a) Life history differences among the eutherian radiations. *J. Zool. Lond.* **219**, 329–353.
2.7.2, 3.4.2

Read A.F. & Harvey P.H. (1989b) Reassessment of comparative evidence for Hamilton and Zuk theory on the evolution of secondary sexual characters. *Nature* **339**, 618–620.
7.3.3, 12.8

Read A.F. & Harvey P.H. (1989c) Reply to Zuk. *Nature* **340**, 104–105.
7.3.3

Read A.F. & Weary D.M. (1990) Sexual selection and the evolution of bird song: a test of the Hamilton–Zuk hypothesis. *Behav. Ecol. Sociobiol.* **26**, 47–56.
7.3.3, 11.7.3, 12.8

Real L.A. (1981) Uncertainty and pollinator–plant interactions: the foraging behavior of bees and wasps on artificial flowers. *Ecology* **62**, 20–26.
4.5.1

Real L.A. & Caraco T. (1986) Risk and foraging in stochastic environments: theory and evidence. *A. Rev. Ecol. Syst.* **17**, 371–390.
4.5.1

Real L.A., Ott J. & Silverfine E. (1982) On the trade-off between the mean and the variance in foraging: effect of spatial distribution and color preference. *Ecology* **63**, 1617–1623.
4.5.1, 5.5.2

Recer G.M. & Caraco T. (1989) Sequential-encounter prey choice and effects of spatial resource variability. *J. Theor. Biol.* **139**, 239–249.
4.7.3

Rechten C., Avery M.I. & Stevens J.A. (1983) Optimal prey selection: why do great tits show partial preferences? *Anim. Behav.* **31**, 576–584.
4.1.1, 4.4.3, 4.7.3

Reeve H.K., Westneat D.F., Noon W.A., Sherman P.W. & Aquadro C.F. (1990) DNA 'fingerprinting' reveals high levels of inbreeding in the eusocial naked mole-rat. *Proc. Nat. Acad. Sci.* **87**, 2496–2500.
10.2

Regelmann K. (1984) Competitive resource sharing: a simulation model. *Anim. Behav.* **32**, 226–232.
5.5.2

Regelmann K. & Curio E. (1986) How do great tit (*Parus major*) pair-mates cooperate in brood defence? *Behaviour* **97**, 10–36.
8.4.4

Reid M.L. (1987) Costliness and reliability in the singing vigour of Ipswich sparrows. *Anim. Behav.* **35**, 1735–1744.
4.4.2, 7.3.2

Reid W.V. (1987) The cost of reproduction in the glaucous-winged gull. *Oecologia* **74**, 458–467.
2.3.1

Reyer H.-U. (1980) Flexible helper structure as an ecological adaptation in the pied kingfisher *(Ceryle rudis). Behav. Ecol. Sociobiol.* **6**, 219–227.
10.3.1, 10.3.2, 10.4.5, 10.5.2, 10.6.3

Reyer H.-U. (1984) Investment and relatedness: a cost/benefit analysis of breeding and helping in the pied kingfisher (*Ceryle rudis*). *Anim. Behav.* **32**, 1163–1178.
10.3.3, 10.4.5, 10.5.2, 10.6.3

Reyer H.-U. (1986) Breeder-helper-inter-

actions in the pied kingfisher reflect the costs and benefits of cooperative breeding. *Behaviour* **96**, 277–303.
10.5.2, 10.6.3

Reynolds J.D. (1987) Mating system and nesting biology of the red-necked phalarope *Phalaropus lobatus*: what constrains polyandry? *Ibis* **129**, 225–242.
9.5.6

Reznick D. (1985) Costs of reproduction: an evaluation of the empirical evidence. *Oikos* **44**, 257–267.
2.2.1, 2.2.2

Richards D.G. (1981) Alerting and message components in songs of nifous-sided towhees. *Behaviour* **76**, 223–249.
12.4.1

Richards O.W. (1978) The Australian social wasps (Hymenoptera: Vespidae). *Aust. J. Zool. Suppl. Ser.* **61**, 1–132.
11.4.1

Richerson P.J. & Boyd R. (1991) Cultural inheritance and evolutionary ecology. In: *Ecology, Evolution, and Human Behavior* (ed. E.A. Smith & B. Winterhalder). Aldine de Gruyter, New York (in press).
3.4.3

Ricklefs R.E. (1974) Energetics of reproduction in birds. In: *Avian Energetics* (ed. R.A. Paynter), pp. 152–297. Nuttall Ornithological Club, Cambridge, Massachusetts.
8.2

Ridpath M.G. (1972) The Tasmanian native hen, *Tribonyx mortierrii*. *CSIRO Wildl. Res.* **17**, 1–118.
10.3.1, 10.4.2, 10.5.2, 10.5.3

Riechert S.E. (1978) Games spiders play: behavioral variability in territorial disputes. *Behav. Ecol. Sociobiol.* **3**, 135–162.
12.3.3

Riedman M.L. (1982) The evolution of alloparental care and adoption in mammals and birds. *Q. Rev. Biol.* **57**, 405–435.
10.1

Rippin A.B. & Boag D.A. (1974) Spatial organisation among male sharp-tailed grouse on arenas. *Can. J. Zool.* **52**, 591–597.
9.4.5

Rissing S.W. & Pollock G.B. (1988) Pleometrosis and polygyny in ants. In; *Interindividual Behavioral Variability in Social Insects* (ed. R.L. Jeanne), pp. 179–222. Westview Press, Boulder, Colorado.
11.4.2

Rissing S.W., Pollock G.B., Higgins M.R.,

Hagen R.H. & Smith D.R. (1989) Foraging specialization without relatedness or dominance among co-founding ant queens. *Nature* **338**, 420–422.
11.4.2

Robbins C.T. (1983) *Wildlife Feeding and Nutrition*. Academic Press, New York.
8.2

Robertson D.R. & Hoffman S.G. (1977) The roles of female mate choice and predation in the mating systems of some tropical labroid fishes. *Z. Tierpsychol.* **45**, 298–320.
9.3.3

Robertson J.G.M. (1986) Female choice, male strategies and the role of vocalization in the Australian frog *Uperoleia rugosa*. *Anim. Behav.* **34**, 773–784.
7.3.2

Robinson G.E. & Page R.E. Jr (1988) Genetic determination of guarding and undertaking in honey-bee colonies. *Nature* **333**, 356–358.
11.4.3

Robinson S.K. (1986) The evolution of social behaviour and mating systems in the blackbirds (Icterinae). In: *Ecological Aspects of Social Evolution* (ed. D.I. Rubenstein & R.W. Wrangham), pp. 175–200. Princeton University Press, Princeton, New Jersey.
9.5.3

Rockwell R.F., Findlay C.S. & Cooke F. (1987) Is there an optimal clutch size in lesser snow geese? *Amer. Natur.* **130**, 839–863.
2.8.5

Roelofs W. (1979) Production and perception of Lepidopterous pheromone blends. In: *Chemical Ecology: Odour Communication in Animals* (ed. F.J. Ritter), pp. 159–168. Elsevier, Amsterdam.
12.6.1

Roff D.A. (1986) Predicting body size with life history models. *Bioscience* **36**, 316–323.
2.6.2

Rogers A. (1990) The evolutionary economics of reproduction. *Ethol. Sociobiol.* (in press).
3.3.4

Rohwer F.C. (1985) The adaptive significance of clutch size in prairie ducks. *Auk* **102**, 354–361.
2.4.1, 2.5.1

Rohwer F.C. (1988) Inter- and intraspecific relationships between egg size and clutch size in waterfowl. *Auk* **105**, 161–176.
2.7.2

Rood J.P. (1974) Banded mongoose males guard young. *Nature* **248**, 176.
10.2

Rood J.P. (1978) Dwarf mongoose helpers at the den. *Z.Tierpsychol.* **48**, 277–287.
10.2

Rood J.P. (1980) Mating relationships and breeding suppression in dwarf mongoose. *Anim. Behav.* **28**, 143–150.
10.2

Rood J.P. (1990) Group size, survival, reproduction, and routes to breeding in dwarf mongooses. *Anim. Behav.* **39**, 566–572.
10.3.1, 10.3.3, 10.5.2

Roper T.J. & Redston S. (1987) Conspicuousness of distasteful prey affects the strength and durability of one-trial avoidance learning. *Anim. Behav.* **35**, 739–747.
6.4.2

Roper T.J. & Wistow R. (1986) Aposematic coloration and avoidance learning in chicks. *Q. J. Exp. Psychol.* **38B**, 141–149.
6.4.2

Rose M.R. (1983) Theories of life history evolution. *Amer. Zool.* **23**, 15–23.
2.1

Rose M.R. (1984a) Laboratory evolution of postponed senescence in *Drosophila melanogaster*. *Evolution* **38**, 1004–1010.
2.2.2

Rose M.R. (1984b) Genetic covariation in *Drosophila* life history: untangling the data. *Amer. Natur.* **123**, 565–569.
2.2.2

Rose M.R. & Charlesworth B. (1980) A test of evolutionary theories of senescence. *Nature* **287**, 141–142.
2.2.2

Rose M.R. & Charlesworth B. (1981a) Genetics of life history in *Drosophila melanogaster*. I. Sib analysis of adult females. *Genetics* **97**, 173–186.
2.2.2

Rose M.R. & Charlesworth B. (1981b) Genetics of life history in *Drosophila melanogaster*. II. Exploratory selection experiments. *Genetics* **97**, 187–196.
2.2.2

Rose M.R., Service P.M. & Hutchinson E.W. (1987) Three approaches to trade-offs in life-history evolution. In: *Genetic Constraints on Adaptive Evolution* (ed. V. Loeschke), pp. 91–105. Springer-Verlag, Berlin.
2.1.1

Rosenberg G. (1989) Aposematism evolves by individual selection: evidence from marine gastropods with pelagic larvae. *Evolution* **43**, 1811–1813.
6.4.1

Rosenzweig M.L. (1981) A theory of habitat selection *Ecology* **62**, 327–335.
Introduction to Part 2, 5.6.5

Rosenzweig M.L. (1985) Some theoretical aspects of habitat selection. In: *Habitat Selection in Birds* (ed. M.L. Cody), pp. 517–540. Academic Press, New York.
Introduction to Part 2, 5.6.5

Ross K.G. & Fletcher D.J.C. (1985) Comparative study of genetic and social structure in two forms of the fire ant *Solenopsis invicta* (Hymenoptera: Formicidae). *Behav. Ecol. Sociobiol.* **17**, 349–356.
11.4.2

Ross K.G. & Matthews R.W. (1989a) New evidence for eusociality in the sphecid wasp *Microstigmus comes*. *Anim. Behav.* **38**, 613–619
11.4.1

Ross K.G. & Matthews R.W. (1989b) Population structure and social evolution in the sphecid wasp *Microstigmus comes*. *Amer. Natur.* **134**, 574–598.
11.4.1

Ross K.G. & Matthews R.W. (ed.) (1991) *The Social Biology of Wasps.* Cornell University Press, Ithaca, New York.
11.1

Rosser A.M. (1987) Resource defence in an African antelope, the puku *Kobus vardoni*. PhD thesis, University of Cambridge, Cambridge.
9.4.1

Roubik D.W. (1989) *Ecology and Natural History of Tropical Bees.* Cambridge University Press, Cambridge.
11.1

Rowland J.J. (1989) Mate choice and the supernormality effect in female sticklebacks *Gasterosteus aculeatus*. *Behav. Ecol. Sociobiol.* **24**, 433–438.
12.1

Rowley I. (1965) The life history of the superb blue wren, *Malurus cyaneus*. *Emu* **64**, 251–297.
10.3.2, 10.4.5, 10.4.6

Rowley I. (1981) The communal way of life in the splendid wren, *Malurus splendens*. *Z. Tierpsychol.* **55**, 228–267.
10.3.2, 10.4.5, 10.6.3

Rowley I. & Russell E. (1990) Splendid fairy-wrens: demonstrating the importance of longevity. In: *Cooperative Breeding in Birds: Long-Term Studies of Ecology and Behavior* (ed. P.B. Stacey &

W.D. Koenig), pp. 3–30. Cambridge University Press, Cambridge.
10.3, 10.4.5, 10.5.3

Royama T. (1966) Factors governing feeding rate, food requirement and brood size of nestling great tits *Parus major*. *Ibis* **108**, 313–347.
2.4

Rubenstein D.I. (1981) Population density, resource patterning, and territoriality in the everglades pygmy sunfish. *Anim. Behav.* **29**, 155–172.
5.3.2

Rubenstein D.I. (1986) Ecology and sociality in horses and zebras. In: *Ecological Aspects of Social Evolution* (ed. D.I. Rubenstein & R.W. Wrangham), pp. 282–302. Princeton University Press, Princeton, New Jersey.
9.2.1

Rubenstein D.I. & Wrangham R.W. (ed.) (1986) *Ecological Aspects of Social Evolution*. Princeton University Press, Princeton, New Jersey.
9.2.1

Rushton J.P. (1988) Race differences in behaviour: a review and evolutionary analysis. *Person. Indiv. Diff.* **9**, 1009–1024.
3.4.1

Russell E.M. & Rowley I.C.R. (1988) Helper contributions to reproductive success in the splendid fairy-wren (*Malurus splendens*). *Behav. Ecol. Sociobiol.* **22**, 131–140.
10.3.2, 10.3.3, 10.6.3

Ryan M.J. (1985) *The Túngara Frog: A Study in Sexual Selection and Communication*. University of Chicago Press, Chicago.
7.3.2, 12.1

Ryan M.J. (1988) Energy, calling and selection. *Amer. Zool.* **28**, 885–898.
7.3.2, 12.1, 12.6.2

Ryan M.J., Fox J.H., Wilczynski W. & Rand A.S. (1990) Sexual selection for sensory exploitation in the frog *Physalaemus pustulosus*. *Nature* **343**, 66–68.
7.3.1, 12.8

Ryan M.J., Tuttle M.D. & Rand A.S. (1982) Bat predation and sexual advertisment in a neotropical anuran. *Amer. Natur.* **119**, 136–139.
7.3.2

Ryan M.J., Tuttle M.D. & Taft L.K. (1981) The costs and benefits of frog chorusing behavior. *Behav. Ecol. Sociobiol.* **8**, 273–278.
9.4.5

Safriel U.N. (1975) On the significance of clutch size in nidifugous birds. *Ecology* **56**, 703–708.
2.4.1, 2.5.1

Sakagami S.F. & Maeta Y. (1982) Further experiments on the artificial induction of multifemale associations in the principally solitary bee genus *Ceratina*. In: *The Biology of Social Insects* (ed. M.D. Breed, C.D. Michener & H.E. Evans), pp. 171–174. Westview Press, Boulder, Colorado.
11.4.1

Sakagami S.F. & Maeta Y. (1984) Multifemale nests and rudimentary castes in the normally solitary bee *Ceratina japonica* (Hymenoptera: Xylocopinae). *J. Kans. Ent. Soc.* **57**, 639–656.
11.4.1

Sakagami S.F. & Maeta Y. (1987) Sociality, induced and/or natural, in the basically solitary small carpenter bees (*Ceratina*). In: *Animal Societies: Theories and Facts* (ed. Y. Itô, J.L. Brown & J. Kikkawa), pp. 1–16. Japan Science Society Press, Tokyo.
11.4.1

Sakagami S.F. & Munakata M. (1972) Distribution and bionomics of a trans-palaearctic eusocial halictine bee, *Lasioglossum (Evylaeus) calceatum* in northern Japan with notes on its solitary life cycle at high altitude. *J. Fac. Sci. Hokkaido Univ. Zool.* **18**, 411–439.
11.4.1

Sakaluk S.K. & Belwood J.J. (1984) Gecko phontoaxis to cricket calling song—a case of satellite predation. *Anim. Behav.* **32**, 659–662.
12.1

Salt G. (1940) Experimental studies in insect parasitism. VII. The effects of different hosts on the parasite *Trichogramma evanescens* Westw. (Hym. Chalcidoidea). *Proc. R. Ent. Soc. Lond. A* **15**, 81–95.
2.4.1

Salthe S.N. (1969) Reproductive modes and number and size of ova in the Urodeles. *Amer. Midl. Natur.* **81**, 467–490.
8.4

Sargent T.D. (1969) Behavioral adaptations of cryptic moths III. Resting attitude of two bark-like species. *Anim. Behav.* **17**, 670–672.
6.3

Sasvari L. (1986) Reproductive effort of widowed birds. *J. Anim. Ecol.* **55**, 553–564.
9.5.1

Schaeffer S.W., Brown C.J. & Anderson W.W. (1984) Does mate choice affect fitness. *Genetics* **107**, 94.
7.3.2

Schaller G.B. (1972) *The Serengeti Lion.* University of Chicago Press, Chicago.
10.5.2

Schantz T. von (1981) Female cooperation, male competition, and dispersal in the red fox, *Vulpes vulpes. Oikos* **37**, 63–68.
10.5.2

Schantz T. von, Goransson G., Andersson G. *et al.* (1989) Female choice selects for a viability-based male trait in pheasants. *Nature* **337**, 166–169.
7.1

Scheiner S.M., Caplan R.L. & Lyman R.F. (1989) A search for trade-offs among life history traits in *Drosophila melanogaster. Evol. Ecol.* **3**, 51–63.
2.2.2

Schindler M. & Lamprecht J. (1987) Increase of parental investment with brood size in a nidifugous bird. *Auk* **104**, 688–693.
2.4.1

Schmid-Hempel P., Kacelnik A. & Houston A.I. (1985) Honeybees maximise efficiency by not filling their crop. *Behav. Ecol. Sociobiol.* **17**, 61–66.
4.2.3

Schnebel E.M. & Grossfield J. (1988) Antagonistic pleiotropy: an interspecific *Drosophila* comparison. *Evolution* **42**, 306–311.
2.7.2

Schneider D. (1974) The sex-attractant receptor in moths. *Sci. Amer.* **231**, 28–35.
12.4.2

Schoener T.W. (1974) Resource partitioning in ecological communities. *Science* **185**, 27–39.
5.1.1, 5.6.5

Schoener T.W. (1986) Mechanistic approaches to community ecology: a new reductionism. *Amer. Zool.* **26**, 81–106.
4.7.2

Schoener T.W. (1987) A brief history of optimal foraging theory. In: *Foraging Behavior* (ed. A.C. Kamil, J.R. Krebs & H.R. Pulliam), pp. 5–67. Plenum Press, New York.
4.2.5

Schoener T.W. & Schoener A. (1982) Intraspecific variation in home range size in some *Anolis* lizards. *Ecology* **63**, 809–823.
9.3.1

Schultz T. & Bernard G.D. (1989) Pointillistic mixing of interference colours in cryptic tiger beetles. *Nature* **337**, 72–73.
6.3

Schwagmeyer P.L. (1988) Scramble competition polygyny in an asocial mammal: male mobility and mating success. *Amer. Natur.* **131**, 885–892.
9.2.2

Schwartz J.J. (1987) The function of call alternation in anuran amphibians: a test of three hypotheses. *Evolution* **41**, 461–471.
12.6.2

Schwartz O.A. & Armitage K.B. (1980) Genetic variation in social mammals: the marmot model. *Science* **207**, 665–667.
9.2.2

Schwarz M.P. (1987) Intra-colony relatedness and sociality in the allodapine bee *Exoneura bicolor. Behav. Ecol. Sociobiol.* **21**, 387–392.
11.4.1

Schwinning S. & Rosenzweig M.L. (1990) Periodic oscillations in an ideal-free predator–prey distribution. *Oikos* **59**, 85–91.
5.4

Scott D.K. (1988) Reproductive success in Bewick's swans. In: *Reproductive Success* (ed. T.H. Clutton-Brock), pp. 220–236. Chicago University Press, Chicago.
9.5.1

Searcy W.A. (1979) Female choice of mates: a general model for birds and its application to red-winged blackbirds. *Amer. Natur.* **114**, 77–100.
9.5.3

Searcy W.A. (1983) Response to multiple song types in male song sparrows and field sparrows. *Anim. Behav.* **31**, 948–949.
12.7.3

Searcy W.A. (1984) Song repertoire size and female preferences in song sparrows. *Behav. Ecol. Sociobiol.* **14**, 281–286.
12.7.3

Searcy W.A. (1988a) Do female red-winged balckbirds limit their own breeding densities? *Ecology* **69**, 85–95.
9.5.3

Searcy W.A. (1988b) Dual intersexual and intrasexual functions of song in red-winged blackbirds. In: *Acta XIX Congressus Internationalis Ornithologic* (ed. H. Ouellet), Vol. 1, pp. 1373–1381. University of Ottawa Press, Ottawa.
12.7.3

Searcy W.A. & Yakusawa K. (1989) Altern-

ative models of territorial polygyny in birds. *Amer. Natur.* **134**, 323–343.
9.5.3

Seeley T.D. (1985) *Honey Bee Ecology: A Study of Adaptation in Social Life.* Princeton University Press, Princeton, New Jersey.
12.6.1

Seger J. (1983) Partial bivoltinism may cause sex-ratio biases that favour eusociality. *Nature* **301**, 59–62.
11.5.4

Seger J. (1985) Unifying genetic models for the evolution of female choice. *Evolution* **39**, 1185–1193.
7.3.1

Seger J. & Brockman H.J. (1987) What is bet-hedging? In: *Oxford Surveys of Evolutionary Biology* (ed. P. Harvey & L. Partridge), 182–211. Oxford University Press, Oxford.
8.3.2

Seigel R.A., Huggins M.M. & Ford N.B. (1987) Reduction in locomotor activity as a cost of reproduction in gravid snakes. *Oecologia* **73**, 481–485.
2.3

Selander R.K. (1964) Speciation in wrens of the genus *Campylorhynchus. Univ. Calif. Pub. Zool.* **74**, 1–224.
10.4.2

Selous E. (1927) *Realities of Bird Life.* Constable, London.
7.3.2

Service P.M. (1989) The effect of mating status on lifespan, egg laying and starvation resistance in *Drosophila melanogaster* in relation to selection on longevity. *J. Insect Physiol.* **35**, 447–452.
2.3.1

Service P.M. & Rose M.R. (1985) Genetic covariation among life-history components: the effect of novel environments. *Evolution* **39**, 943–945.
2.2.2

Seyfarth R.M. & Cheney D.L. (1986) Vocal development in vervet monkeys. *Anim. Behav.* **34**, 1640–1658.
12.3.1

Seyfarth R.M., Cheney D.L. & Marler P. (1980) Vervet monkey alarm calls: semantic communication in a free-ranging primate. *Anim. Behav.* **28**, 1070–1094.
12.3.1

Shalter M.D. (1978) Localisation of passerine seet and mobbing calls by goshawks and pygmy owls. *Z. Tierpsychol.* **46**, 260–267.
12.3.1

Shannon C.E. & Weaver W. (1949) *The Mathematical Theory of Communication.* University of Illinois Press, Urbana.
12.3

Sharrock J.T.R. (1976) *The Atlas of Breeding Birds in Britain and Ireland.* Poyser, Berkhamstead.
12.5.2

Shelly T.E., Greenfield M.D. & Downum K.R. (1987) Variation in host plant quality: influences on the mating system of a desert grasshopper. *Anim. Behav.* **35**, 1200–1209.
9.3.4

Sherman K.J. (1983) The adaptive significance of post-copulatory mate guarding in a dragonfly *Pachydiplax longipennis. Anim. Behav.* **31**, 1107–1115.
9.3.4

Sherman P.W. (1989) Mate guarding as paternity insurance in Idaho ground squirrels. *Nature* **338**, 418–420.
9.2.2

Sherman P.W., Jarvis J.U.M. & Alexander R.D. (ed.) (1990) *The Biology of the Naked Mole Rat.* Princeton University Press, Princeton, New Jersey.
10.2, 11.4

Sherman P.W. & Morton M.L. (1984) Demography of Belding's ground squirrels. *Ecology* **65**, 1617–1628.
9.2.2

Sherman P.W. & Morton M.L. (1988) Extra-pair fertilizations in mountain white-crowned sparrows. *Behav. Ecol. Sociobiol.* **22**, 413–420.
9.5.2

Shettleworth S.J. (1988) Foraging as operant behavior and operant behavior as foraging: what have we learned? In: *The Psychology of Learning & Motivation: Advances in Research and Theory.* (ed. G. Bower), Vol. 22, pp. 1–49. Academic Press, New York.
4.6

Shettleworth S.J., Krebs J.R., Stephens D. & Gibbon J. (1988) Tracking a changing environment: a study of sampling. *Anim. Behav.* **36**, 87–105.
5.6.2

Shine R. (1988) The evolution of large body size in females: a critique of Darwin's 'fecundity' advantage model. *Amer. Natur.* **131**, 124–131.
8.4.3

Shorey H.H. (1976) *Animal Communication by Pheromones.* Academic Press, New York.
12.6.1

Short R.V. (1977) Sexual selection and the descent of man. In: *Proceedings of the Canberra Symposium on Reproduction and Evolution*, pp. 3–19. Australian Academy of Sciences, Canberra.
7.4

Short R.V. (1979) Sexual selection and its component parts, somatic and genital selection, as illustrated by man and the great apes. *Adv. Study Behav.* **9**, 131–158.
7.4

Short R.V. (1981) Sexual selection in man and the great apes. In: *Reproductive Biology of the Great Apes* (ed. C.E. Graham), pp. 319–341. London, Academic Press.
7.4

Shumway C.A. & Zelick R.D. (1988) Sex recognition and neuronal coding of electric organ discharge wave-form in the pulse-type weakly electric fish, *Hypopomus occidentalis*. *J. Comp. Physiol.* **163A**, 465–478.
12.6.3

Shykoff J.A. & Schmid-Hempel P. (1991) Parasites and the advantage of genetic variability within social insect colonies. *Proc. Roy. Soc. B* (in press)
Introduction to Part 4

Sibly R.M. & Calow P. (1985) Classification of habitats by selection pressures: a synthesis of life-cycle and *r*/*K* theory. In: *Behavioural Ecology* (ed. R.M. Sibly & R.H. Smith), pp. 75–90. Blackwell Scientific Publications, Oxford.
2.7.1

Sieff D.F. (1990) Explaining biased sex ratios in human populations. *Curr. Anthropol.* **31**, 25–48.
3.3.4

Sigurjònsdòttir H. & Parker G.A. (1981) Dung fly struggles: evidence for assessment strategy. *Behav. Ecol. Sociobiol.* **8**, 219–230.
5.6.2

Sih, A. (1987) Predators and prey lifestyles: an evolutionary and ecological overview. In: *Predation: Direct and Indirect Impacts on Aquatic Communities.* (ed. W.C. Kerfoot & A. Sih), pp. 203–224. University of New England Press, Hanover, New Hampshire.
4.3.1

Sillén-Tullberg B. (1988) Evolution of gregariousness in aposematic butterfly larvae: a phylogenetic analysis. *Evolution* **42**, 293–305.
6.4.1

Silk J.B. (1983) Local resource competition and facultative adjustment of sex ratios in relation to competitive activities. *Amer. Natur.* **12**, 56–66.
8.2

Silk J.B. (1990) Human adoption in evolutionary perspective. *Human Nature* **1**, 25–52.
3.3.3

Simmons L.W. (1986) Inter-male competition and mating success in the field cricket, *Gryllus bimaculatus* (de Geer). *Anim. Behav.* **34**, 567–579.
7.3.2

Simmons L.W. (1987) Female choice contributes to offspring fitness in the field cricket *Gryllus Bimaculatus* (de Geer). *Behav. Ecol. Sociobiol.* **21**, 313–321.
7.3.2

Simmons M.J., Preston C.R. & Engels W.R. (1980) Pleiotropic effects on fitness of mutations affecting viability in *Drosophila melanogaster*. *Genetics* **94**, 467–475.
2.2.2

Simon M.P. (1983) The ecology of parental care in a terrestrial breeding frog from New Guinea. *Behav. Ecol. Sociobiol.* **14**, 61–67.
8.2

Simpson B.S. (1984) Test of habituation to song repertoires by Carolina Wrens. *Auk* **101**, 244–254.
12.7.3

Simpson M.J.A. (1968) The display of Siamese fighting fish, *Betta splendens*. *Anim. Behav. Monogr.* **1**, 1–73.
12.3.4

Sinclair A.R.E. (1989) Population regulation in animals. In: *Ecological Concepts* (ed. J.M. Cherrett), pp. 197–241. Blackwell Scientific Publications, Oxford.
4.7.2

Skinner S.W. (1985) Clutch size as an optimal foraging problem for insects. *Behav. Ecol. Sociobiol.* **17**, 231–238.
2.5.1

Skutch A.F. (1935) Helpers at the nest. *Auk* **52**, 257–273.
10.1, 10.2

Skutch A.F. (1976) *Parent Birds and their Young*. University of Texas Press, Austin.
12.5.1

Skutch A.F. (1983) *Birds of Tropical America*. University of Texas Press, Austin, Texas.
7.3.1

Slagsvold T. (1990) Fisher's sex ratio

theory may explain hatching patterns in birds. *Evolution* **44**, 1009–1017.
2.8.4

Slagsvold T. & Lifjeld J.T. (1988) Ultimate adjustment of clutch size to parental feeding capacity in a passerine bird. *Ecology* **69**, 1918–1922.
2.8.3, 8.4.4

Slagsvold T. & Lifjeld J.T. (1989a) Hatching asynchrony in birds: the hypothesis of sexual conflict over parental investment. *Amer. Natur.* **134**, 239–253.
2.8.4

Slagsvold T. & Lifjeld J.T. (1989b) Constraints on hatching asynchrony and egg size in pied flycatchers. *J. Anim. Ecol.* **58**, 837–850.
8.3.2

Slagsvold T., Sandvik J., Rofstad G., Lorentsen Ø. & Husby M. (1984) On the adaptive value of intra-clutch egg-size variation in birds. *Auk.* **101**, 685–697.
8.3.2

Smith A.P. & Alcock J. (1980) A comparative study of the mating systems of Australian eumenid wasps (Hymenoptera). *Z. Tierpsychol.* **53**, 41–60.
9.3.4

Smith B.H. (1987) Effects of genealogical relationship and colony age on the dominance hierarchy in the primitively eusocial bee *Lasioglossum zephyrum*. *Anim. Behav.* **35**, 211–217.
11.4.1, 11.5.6

Smith B.H., Carlson R.G. & Frazier J. (1985) Identification and bioassay of macrocyclic lactone sex pheromone of the halictine bee *Lasioglossum zephyrum*. *J. Chem. Ecol.* **11**, 1447–1456.
11.5.6

Smith B.H. & Wenzel J.W. (1988) Pheromonal covariation and kinship in social bee *Lasioglossum zephyrum* (Hymenoptera: Halictidae). *J. Chem. Ecol.* **14**, 87–94.
11.5.6

Smith C.C. & Fretwell S.D. (1974) The optimal balance between the size and number of offspring. *Amer. Natur.* **108**, 499–506.
2.5.1, 8.3.1

Smith E.A. (1983) Anthropological applications of optimal foraging theory: a critical review. *Curr. Anthropol.* **24**, 625–651.
3.2.1

Smith E.A. (1984) Anthropology, evolutionary ecology, and the explanatory limitations of the ecosystem concept. In: *The Ecosystem Concept in Anthropology* (ed. E.F. Moran), pp. 51–85. Westview Press, Boulder, Colorado.
3.1

Smith E.A. (1985) Inuit foraging groups: some simple models incorporating conflicts of interest, relatedness, and central-place sharing. *Ethol. Sociobiol.* **6**, 27–47.
3.2.2

Smith E.A. & Boyd R. (1990) Risk and reciprocity: hunter–gatherer socioecology and the problem of collective action. In: *Risk and Uncertainty in Tribal and Peasant Economies* (ed. E.A. Cashdan), pp. 167–191. Westview Press, Boulder, Colorado.
3.4.2

Smith H.G., Källander H., Fontell K. & Ljungström M. (1988) Feeding frequency and parental division of labour in the double-brooded great tit *Parus major*. *Behav. Ecol. Sociobiol.* **22**, 447–453.
2.3.1

Smith H.G., Källander H. & Nilsson J.-Å. (1987) Effect of experimentally altered brood size on frequency and timing of second clutches in the great tit. *Auk* **104**, 700–706.
2.3.1

Smith H.G,, Källander H. & Nilsson J.-Å. (1989) The trade-off between offspring number and quality in the great tit *Parus major*. *J. Anim. Ecol.* **58**, 383–401.
2.5.1

Smith J.L.D., McDougal C. & Miquelle D. (1989) Scent marking in free-ranging tigers *Panthera tigris*. *Anim. Behav.* **37**, 1–10.
12.6.1

Smith J.M.N., Yom-Tov Y. & Moses R. (1982) Polygyny, male parental care and sex ratio in song sparrows: an experimental study. *Auk* **99**, 555–564.
8.4.4, 9.5.1

Smith R.H. & Lessells C.M. (1985) Oviposition, ovicide and larval competition in granivorous insects. In: *Behavioural Ecology* (ed. R.M. Sibly & R.H. Smith), pp. 423–448. Blackwell Scientific Publications, Oxford.
2.5.1

Smith S.M. (1978) The underworld in a territorial sparrow: adaptive strategy for floaters. *Amer. Natur.* **112**, 571–582.
10.4.4

Smith S.M. (1988) Extra-pair copulations in black-capped chickadees: the role of the female. *Behaviour* **107**, 15–23.
9.5.2

Smith-Gill S.J. (1983) Developmental plasticity: developmental conversion *versus* phenotypic modulation. *Amer. Zool.* **23**, 47–55.
2.8.1

Smythe N. (1970) On the existence of 'pursuit invitation' signals in mammals. *Amer. Natur.* **104**, 491–494.
12.3.3

Snow D.W. (1958) *A Study of Blackbirds.* Unwin, London.
12.5.2

Snyderman M. (1983) Optimal prey selection: the effects of food deprivation *Behav. Anal. Lett.* **3**, 359–369.
4.4.3

Sokal R.R. (1970) Senescence and genetic load: evidence from *Tribolium. Science* **167**, 1733–1734.
2.2.2

Sokolowski M.B. (1985) Genetics and ecology of *Drosophila melanogaster* larval foraging and pupation behaviour. *J. Insect Physiol.* **31**, 857–864.
6.8

Soliman M.H. (1982) Directional and stabilizing selection for developmental time and correlated response in reproductive fitness in *Tribolium castaneum. Theor. Appl. Genet.* **63**, 111–116.
2.2.2

Sorjonen J. (1986) Factors affecting the structure of song and the singing behaviour of some Northern European passerine birds. *Behaviour* **98**, 286–304.
12.6.2

Southwood T.R.E. (1977) Habitat, the templet for ecological strategies. *J. Anim. Ecol.* **46**, 337–365.
2.7.1

Southwood T.R.E. (1988) Tactics, strategies and templets. *Oikos* **52**, 3–18.
2.7.1

Spector D.A., McKin L.K. & Kroodsma D.E. (1989) Yellow warblers are able to learn songs and situations in which to use them. *Anim. Behav.* **38**, 723–725.
12.7.3

Stacey P.B. (1979a) Habitat saturation and communal breeding in the acorn woodpecker. *Anim. Behav.* **27**, 1153–1166.
10.4.2

Stacey P.B. (1979b) Kinship, promiscuity, and communal breeding in the acorn woodpecker. *Behav. Ecol. Sociobiol.* **6**, 53–66.
10.2

Stacey P.B. (1982) Female promiscuity and male reproductive success in social birds and mammals. *Amer. Natur.* **120**, 51–64.
10.4.1

Stacey P.B. & Koenig W.D. (ed.) (1990) *Cooperative Breeding in Birds: Long-Term Studies of Ecology and Behavior.* Cambridge University Press, Cambridge.
10.1, 10.4.6

Stacey P.B. & Ligon J.D. (1987) Territory quality and dispersal options in the acorn woodpecker, and a challenge to the habitat-saturation model of cooperative beeding. *Amer. Natur.* **130**, 654–676.
10.4.1, 10.4.3, 10.4.4, 10.4.6, 10.5.1, 10.6.3

Staddon J.E.R. (1983) *Adaptive Behavior and Learning.* Cambridge University Press, Cambridge.
3.4.3

Staddon J.E.R. & Horner J.M. (1989) Stochastic choice models: a comparison between Bush–Mosteller and a source-independent reward-following model. *J. Exp. Anal. Behav.* **52**, 57–64.
4.6.1

Stallcup J.A. & Woolfenden G.E. (1978) Family status and contribution to breeding by Florida scrub jays. *Anim. Behav.* **26**, 1144–1156.
10.3.2

Stamps J.A. (1983) Sexual selection, sexual dimorphism and territoriality. In: *Lizard Ecology* (ed. R.B. Huey, E.R. Pianka & T.W. Schoener), pp. 169–204. Harvard University Press, Cambridge, Massachusetts.
9.3.1

Stamps J.A., Clark A., Arrowood P. & Kus B. (1985) Parent–offspring conflict in budgerigars. *Behaviour* **94**, 1–39.
8.5.1

Starr C.K. (1985) Enabling mechanisms in the origin of sociality in the Hymenoptera—the sting's the thing. *A. Ent. Soc. Amer.* **78**, 837–840.
11.5.2

Starr C.K. (1989) In reply, is the sting the thing? *A. Ent. Soc. Amer.* **82**, 6–8.
11.5.2

Stearns S.C. (1976) Life history tactics: a review of the ideas. *Q. Rev. Biol.* **51**, 3–47.
2.6.1

Stearns S.C. (1982) The role of development in the evolution of life histories. In: *Evolution and Development, Dahlem Konferenzen 1982* (ed. J.T. Bonner), pp. 237–258. Springer-Verlag, Berlin.
2.8.1

Stearns S.C. (1983a) The genetic basis of differences in life-history traits among

six populations of mosquitofish (*Gambusia affinis*) that shared ancestors in 1905. *Evolution* **37**, 618–627.
2.2.2

Stearns S.C. (1983b) The influence of size and phylogeny on patterns of covariation among life-history traits in the mammals. *Oikos* **41**, 173–187.
2.7.2

Stearns S.C. (1989) Trade-offs in life-history evolution. *Funct. Ecol.* **3**, 259–268.
2.1, 2.2.1, 2.8.6

Stearns S.C. & Koella J.C. (1986) The evolution of phenotypic plasticity in life-history traits: predictions of reaction norms for age and size at maturity, *Evolution* **40**, 893–913.
2.1.2, 2.6.2, 2.8

Stearns S.C. & Schmid-Hempel P. (1987) Evolutionary insights should not be wasted. *Oikos* **49**, 118–125.
4.7.3

Steger R. & Caldwell R.L. (1983) Intra-specific deception by bluffing: a defense strategy of newly moulted stomatopods. *Science* **21**, 558–560.
12.5.2

Stemberger R.S. & Gilbert J.J. (1987) Multiple-species induction of morphological defenses in the rotifer *Keratella testudo. Ecology* **68**, 370–378.
2.8

Stenmark G., Slagsvold T. & Lifjeld J.T. (1988) Polygyny in the pied flycatcher *Ficedula hypoleuce*: a test of the deception hypothesis. *Anim. Behav.* **36**, 1646–1657.
9.5.3

Stephens D.W. (1981) The logic of risk-sensitive foraging preferences. *Anim. Behav.* **29**, 628–629.
5.5

Stephens D.W. (1985) How important are partial preferences? *Anim. Behav.* **33**, 667–669.
4.7.3

Stephens D.W. (1987) On economically tracking a variable environment. *Theor. Pop. Biol.* **32**, 15–25.
4.6.2

Stephens D.W. (in press) Risk and uncertainty in behavioral ecology. In: *Risk and Uncertainty in Tribal and Peasant Economies* (ed. E. Cashdan). Westview Press, Boulder, Colorado.
4.5.1

Stephens D.W. & Krebs J.R. (1986) *Foraging*

Theory. Princeton University Press, Princeton, New Jersey.
2.1.2, 3.2.1, Introduction to Part 2, 4.1.1, 4.2.2, 4.2.5, 4.3.1, 4.7.1, 4.7.3, 5.1.1, 6.2.1, 6.7

Stephens D.W. & Paton S.R. (1986) How constant is the constant of risk-aversion? *Anim. Behav.* **34**, 1659–1667.
4.5.1

Stephenson A.G. & Bertin R.I. (1983) Male competition, female choice and sexual selection in plants. In: *Pollination Biology* (ed. L. Real), pp. 109–149. Academic Press, Orlando, Florida.
7.5

Steven D. de (1980) Clutch size, breeding success and parental survival in the tree swallow (*Iridoprocne bicolor*). *Evolution* **34**, 278–291.
2.5.1

Stevens, T.A. & Krebs J.R. (1986) Retrieval of stored seeds by marsh tits in the field. *Ibis* **128**, 513–525.
4.4.2

Stinson C.H. (1979) On the selective advantage of fratricide in raptors. *Evolution* **33**, 1219–1225.
1.3.5, 8.5.2

Stokes A.W. (1962) Agonistic behaviour among blue tits at a winter feeding station. *Behaviour* **19**, 118–138.
12.3.4, 12.7.1

Strahl S.D. & Schmitz A. (1990) Hoatzins: cooperative breeding in a folivorous neotropical bird. In: *Cooperative Breeding in Birds: Long-Term Studies of Ecology and Behavior* (ed. P. Stacey & W. Koenig), pp. 133–155. Cambridge University Press, Cambridge.
10.3.1, 10.3.2, 10.4.2, 10.5.2

Strassmann J.E. (1989) Altruism and relatedness at colony foundation in social insects. *Trends Ecol. Evol.* **4**, 371–374.
11.1, 11.4.2

Strassmann J.E., Hughes C.R., Queller D.C. *et al.* (1989) Genetic relatedness in primitively eusocial wasps. *Nature* **342**, 268–269.
11.4.1

Strassmann J.E. & Queller D.C. (1989) Ecological determinants of social evolution. In: *The Genetics of Social Evolution* (ed. M.D. Breed & R.E. Page Jr), pp. 81–101. Westview Press, Boulder, Colorado.
11.4.1

Strassmann J.E., Queller D.C. & Hughes C.R. (1988) Predation and the evolution of sociality in the paper wasp *Polistes*

bellicosus. Ecology **69**, 1497–1505.
11.4.1

Streeter L.A., Krauss R.M., Geller V., Olson C. & Apple W. (1977) Pitch changes during attempted deception. *J. Pers. Sound Psychol.* **35**, 345–350.
12.5

Stuart R.J. (1988) Collective cues as a basis for nestmate recognition in polygynous leptothoracine ants. *Proc. Natl. Acad. Sci. USA* **85**, 4572–4575.
11.5.6

Stuart R.J., Francoeur A. & Loiselle R. (1987) Lethal fighting among dimorphic males of the ant, *Cardiocondyla wroughtonii. Naturwissenschaften* **74**, 548–549.
11.3

Stubblefield J.W. & Charnov E.L. (1986) Some conceptual issues in the origin of eusociality. *Heredity* **57**, 181–187.
11.2.2, 11.5.4

Sudd J.H. & Franks N.R. (1987) *The Behavioural Ecology of Ants.* Blackie, Glasgow.
11.1

Sullivan B.K. (1983) Sexual selection in Woodhouse's toad (*Bufo woodhousei*). II. Female choice. *Anim. Behav.* **31**, 1011–1017.
7.3.2

Sussman R.W. & Garber P.A. (1987) A new interpretation of the social organization and the mating system of the Callitrichidae. *Int. J. Primatol.* **8**, 73–92.
10.2

Sutherland W.J. (1983) Aggregation and the 'ideal free' distribution. *J. Anim. Ecol.* **52**, 821–828.
5.3.1

Sutherland W.J., Grafen A. & Harvey P.H. (1986) Life history correlations and demography. *Nature* **320**, 88.
2.7.2

Sutherland W.J. & Parker G.A. (1985) Distribution of unequal competitors. In: *Behavioural Ecology* (ed. R.M. Sibly & R.H. Smith), pp. 255–274. Blackwell Scientific Publications, Oxford.
5.2.2, 5.2.3, 5.3.2

Sutherland W.J., Townsend C.R. & Patmore J.M. (1988) A test of the ideal free distribution with unequal competitors. *Behav. Ecol. Sociobiol.* **23**, 51–53.
5.2.3

Sutherland W.S. (1986) Patterns of fruit set; what controls fruit–flower ratios in plants? *Evolution* **40**, 117–128.
7.5

Sydeman W.J. (1989) Effects of helpers on nestling care and breeder survival in pygmy nuthatches. *Condor* **91**, 147–155.
10.3.1

Symons D. (1979) *The Evolution of Human Sexuality.* Oxford University Press, New York.
3.1, 3.3.1, 3.4.4

Symons D. (1987) If we're all Darwinians what's the fuss about? In: *Sociobiology and Psychology* (ed. C.B. Crawford, M.F. Smith & D.L. Krebs), pp. 121–146. Lawrence Erlbaum Associates, Hillsdale, New Jersey.
3.4.4

Symons D. (1989) A critique of Darwinian anthropology. *Ethol. Sociobiol.* **10**, 131–144.
3.4.4

Syren R.M. & Luykx P. (1977) Permanent segmental interchange complex in the termite *Incisitermes schwarzi. Nature* **266**, 167–168.
11.5.5

Syren R.M. & Luykx P. (1981) Geographic variation of sex-linked translocation heterozygosity in the termite *Kalotermes approximatus* Snyder (Insecta: Isoptera). *Chromosoma* **82**, 65–88.
11.5.5

Taborsky M. (1984) Brood care helpers in the cichlid fish *Lamprologus brichardi*: their costs and benefits. *Anim. Behav.* **32**, 1236–1252.
10.7

Tagaki M. (1985) The reproductive strategy of the gregarious parasitoid, *Pteromalus puparum* (Hymenoptera: Pteromalidae). *Oecologia* **68**, 1–6.
2.4.1

Taigen T.L. & Wells K.D. (1985) Energetics of vocalization by an anuran amphibian, *Hyla versicolor. J. Comp. Physiol.* **155**, 163–170.
7.3.2

Tallamy D.W. & Denno R.F. (1982) Life history trade-offs in *Gargaphia solani* (Hemiptera: Tingidae): the cost of reproduction. *Ecology* **63**, 616–620.
2.3.1

Tallamy D.W. & Wood T.K. (1986) Convergence patterns in subsocial insects. *A. Rev. Entomol.* **31**, 369–390.
11.3

Tamm S. (1987) Tracking varying environments: sampling by hummingbirds. *Anim. Behav.* **35**, 1725–1734.
4.6.2

Tantawy A.O. & El-Helw M.R. (1966)

Studies on natural populations of Drosophila. V. Correlated response to selection in *Drosophila melanogaster*. *Genetics* **53**, 97–110.
2.2.2

Tantawy A.O. & Rakha F.A. (1964) Studies on natural populations of *Drosophila*. IV. Genetic variances of and correlations between four characteristics in *D. melanogaster* and *D. simulans*. *Genetics* **50**, 1349–1355.
2.2.2

Taylor A.D. (1988) Host effects on larval competition in the gregarious parasitoid *Bracon hebetor*. *J. Anim. Ecol.* **57**, 163–172.
2.4.1

Taylor C.E. & Condra C. (1980) *r*- and *K*-selection in *Drosophila pseudobscura*. *Evolution* **34**, 1183–1193.
2.2.2

Taylor C.E. & McGuire M.T. (1987) Reciprocal altruism: 15 years later. *Ethol. Sociobiol.* **9**, 67–72.
10.6.2

Taylor C.E., Pereda A.D. & Ferrari J.A. (1987) On the correlation between mating success and offspring quality in *Drosophila melanogaster*. *Amer. Natur.* **129**, 721–729.
7.3.2

Taylor H.M., Gourley R.S., Lawrence C.E. & Kaplan R.S. (1974) Natural selection of life history attributes: an analytical approach. *Theor. Popul. Biol.* **5**, 104–122.
2.1.2

Taylor P.D. (1981) Intra-sex and inter-sex sibling interactions as sex ratio determinants. *Nature* **291**, 64–66.
8.6.3

Temin R.G. (1966) Homozygous viability and fertility loads in *Drosophila melanogaster*. *Genetics* **53**, 27–46.
2.2.2

Temrin H. (1989) Female pairing options in polyterritorial wood warblers *Phylloscopus sibilatrix*: are females deceived? *Anim. Behav.* **37**, 579–586.
9.5.3

Tepedino V.J. & Parker F.D. (1988) Alternation of sex ratio in a partially bivoltine bee. *Megachile rotundata* (Hymenoptera: Megachilidae). *A. Ent. Soc. Amer.* **81**, 467–476.
11.5.4

Terborgh J. & Goldizen A.W. (1985) On the mating system of the cooperatively breeding saddle-backed tamarin (*San-guinus fuscicollis*). *Behav. Ecol. Sociobiol.* **16**, 293–299.
9.5.4, 10.2

Thaler R. (1980) Toward a positive theory of consumer choice. *J. Econom. Behav. Org.* **1**, 39–60.
4.5

Thomas B. (1985a) On evolutionarily stable sets. *J. Math. Biol.* **22**, 105–115.
1.2.2

Thomas B. (1985b) Genetical ESS-models I. Concepts and basic model. *Theor. Pop. Biol.* **28**, 18–32.
1.2.2

Thomas B. (1985c) Genetical ESS-models II. Multi-strategy models and multiple alleles. *Theor. Pop. Biol.* **28**, 33–49.
1.2.2

Thomas B. (1985d) Evolutionarily stable sets in mixed-strategist models. *Theor. Pop. Biol.* **28**, 332–341.
1.2.2

Thorne B.L. (1982) Multiple primary queens in termites: phyletic distribution, ecological context, and a comparison to polygyny in Hymenoptera. In: *The Genetics of Social Evolution* (ed. M.D. Breed & R.E. Page Jr), pp. 206–211. Westview Press, Boulder, Colorado.
11.4

Thorne E.T., Dean R.E. & Hepworth W.G. (1976) Nutrition during gestation in relation to successful reproduction in elk. *J. Wildl. Mgmt.* **40**, 330–335.
8.2

Thornhill R. (1980) Sexual selection within mating swarms of the lovebug *Plecia nearctica*. (Diptera: Bibionidae). *Anim. Behav.* **28**, 405–412.
9.3.4

Thornhill R. (1981) *Panorpa* (Mecoptera: Panorpidae) scorpionflies: systems for understanding resource-defence polygyny and alternative male reproductive efforts. *A. Rev. Ecol. Syst.* **12**, 355–386.
9.3.4

Thornhill R. & Alcock J. (1983) *The Evolution of Insect Mating Systems*. Harvard University Press, Cambridge, Massachusetts.
9.3.4, 9.4, 9.4.4

Thornhill R. & Gwynne D.T. (1986) The evolution of sex differences in insects. *Amer. Sci.* **74**, 382–389.
8.1

Thorson G. (1950) Reproductive and larval ecology of marine bottom invertebrates. *Biol. Rev.* **25**, 1–45.
8.3.1

Thresher R.E. (1984) *Reproduction in Reef Fishes.* T.F.H. Publications, Neptune City, New Jersey.
8.4.3

Tinbergen J. (1981) Foraging decisions in starlings. *Ardea* **69**, 1–67.
4.2

Tinbergen J.M. & Daan S. (1990) Family planning in the great tit (*Parus major*): optimal clutch size as integration of parent and offspring fitness. *Behaviour* **114**, 161–190.
2.5.1, 2.8.3

Tinbergen N. (1959) Comparative studies of the behaviour of gulls (Laridae): a progress report. *Behaviour* **15**, 1–70.
12.7.1

Tinbergen N. (1963) On aims and methods of ethology. *Z. Tierpsychol.* **20**, 410–433.
Preface, 3.4.2

Tomlinson I. & O'Donald P. (1989) The coevolution of multiple mating preferences and preferred male characters: the 'gene-for-gene' hypothesis of sexual selection. *J. Theor. Biol.* **139**, 219–238.
7.3.1, 7.3.2

Tooby J. & Cosmides L. (1989) Evolutionary psychology and the generation of culture, Part I. *Ethol. Sociobiol.* **10**, 29–49.
3.4.4

Torok J. & Toth L. (1990) Costs and benefits of reproduction of the collared flycatcher, *Ficedula albicollis*. In: *Population Biology of Passerine Birds. An Integrated Approach* (ed. J. Blondel, A.G. Gosler, J.D. Lebreton & R.H. McCleery), pp. 307–322. NATO ASI Series, Springer-Verlag, Berlin.
2.3.1

Towers S.R. & Coss R.G. (1990) Confronting snakes in the burrow: snake-species discrimination and antisnake tactics of two california ground squirrel populations. *Ethology*, **84**, 177–192.
6.1, 6.6

Townsend D.S. (1986) The costs of male parental care and its evolution in a neotropical frog. *Behav. Ecol. Sociobiol.* **19**, 187–195.
8.2

Trager J.C. (ed.) (1988) *Advances in Myrmecology.* Brill, Leiden.
11.1

Trail P.W. & Adams E.S. (1989) Active mate choice at cock-of-the-rock leks: tactics of sampling and comparison. *Behav. Ecol. Sociobiol.* **25**, 283–292.
7.3.2

Travis J., Emerson S.B. & Blouin M. (1987) A quantitative-genetic analysis of larval life-history traits in *Hyla crucifer*. *Evolution* **41**, 145–156.
2.2.2

Trivers R.L. (1971) The evolution of reciprocal altruism. *Q. Rev. Biol.* **46**, 35–57.
10.6.2

Trivers R.L. (1972) Parental investment and sexual selection. In: *Sexual Selection and the Descent of Man* (ed. B. Campbell). pp. 136–179. Aldine, Chicago.
8.1, 8.4.2, 8.4.3, 9.1, 9.5.1, 9.5.2

Trivers R.L. (1974) Parent-offspring conflict. *Amer. Zool.* **14**, 249–264.
8.5.1, 10.1, 11.2, 12.4.2

Trivers R.L. (1976) Sexual selection and resource accruing abilities in *Anolis garmani*. *Evolution* **30**, 253–269.
9.3.1

Trivers R.L. (1985) *Social Evolution.* Benjamin-Cummings, Menlo Park, California.
10.6.2, 11.1, 11.4

Trivers R.L. & Hare H. (1976) Haplodiploidy and the evolution of the social insects. *Science* **191**, 249–263.
1.3.2, Introduction to Part 4, 11.2, 11.2.1, 11.5.4

Trivers R.L. & Willard D.E. (1973) Natural selection of parental ability to vary the sex ratio of offspring. *Science* **179**, 90–92.
3.3.4, 8.6.2

Tsuji K. (1988) Obligate parthenogenesis and reproductive division of labor in the Japanese queenless ant *Pristomyrmex pungens*: comparison of intranidial and extranidial workers. *Behav. Ecol. Sociobiol.* **23**, 247–255.
11.6

Tsuji K. (1990) Reproductive division of labour related to age in the Japanese queenless ant, *Pristomyrmex pungens*. *Anim. Behav.* **39**, 843–849.
11.6

Tucić N., Cvetović D. & Milanović D. (1988) The genetic variation and covariation among fitness components in *Drosophila melanogaster* females and males. *Heredity* **60**, 55–60.
2.2.2

Tucić N., Milosević M., Gliksman I., Milanović D. & Aleksic I. (1991) The effects of larval density on genetic variation and covariation among life history traits in bean weevil. *Funct. Ecol.* (in press).
2.2.2

Turchin P. & Kareiva P. (1989) Aggregation in *Aphis varians*: an effective strategy for reducing predation risk. *Ecology* **70**, 1008–1016.
6.1

Turelli M. (1988) Phenotypic evolution, constant covariances, and the maintenance of additive variance. *Evolution* **42**, 1342–1347.
7.3.1

Turke P.W. (1988) Helpers at the nest: childcare networks on Ifaluk. In: *Human Reproductive Behaviour: A Darwinian Perspective* (ed. L. Betzig, M. Borgerhoff Mulder & P. Turke), pp. 173–188. Cambridge University Press, Cambridge.
3.3.1, 3.3.4

Turner G. & Huntingford F. (1986) A problem for games theory analysis: assessment and intention in male moth brooder contests. *Anim. Behav.* **34**, 961–970.
12.3.4

Turner J.R.G. (1984) Mimicry: the palatability spectrum and its consequences. In: *The Biology of Butterflies* (ed. R.I. Vane-Wright & P.R. Ackery), pp. 141–161. Academic Press, New York.
6.5

Turner J.R.G., Kearney E.P. & Exton L.S. (1984) Mimicry and the monte carlo predator: the palatability spectrum and the origins of mimicry. *Biol. J. Linn. Soc. Lond.* **23**, 247–268.
6.5

Tyson J.J. (1984) Evolution of eusociality in diploid species. *Theor. Pop. Biol.* **26**, 283–295.
11.5.5

Ubukata H. (1987) Mating system of the dragonfly *Cordulia aenea amurensis* Selys, and a model of mate searching and territorial behaviour in Odonata. In: *Animal Societies: Theories and Facts* (ed. Y. Ito, J.L. Brown & J. Kikkawa), pp. 213–228. Japan Science Society Press, Tokyo.
9.3.4

Unger L.M. & Sargent R.C. (1988) Allopaternal care in the fathead minnow, *Pimephales promelas*: females prefer males with eggs. *Behav. Ecol. Sociobiol.* **23**, 27–32.
8.4.3

Uyenoyama M.K. & Bengtsson B.O. (1981) Towards a genetic theory for the evolution of the sex ratio. II. Haplodiploid and diploid models with sibling and parental control of the brood sex ratio and brood size. *Theor. Pop. Biol.* **20**, 57–79.
11.5.4

van Alphen J.J.M. & Visser M.E. (1990) Superparasitism as an adaptive strategy for insect parasitoids. *A. Rev. Entomol.* **35**, 59–79.
4.7.2

van der Have T.M., Boomsma J.J. & Menken S.B.J. (1988) Sex-investment ratios and relatedness in the monogynous ant *Lasius niger* (L.). *Evolution* **42**, 160–172.
11.4.2, 11.5.4

van Schaik C.P. (1983) Why are diurnal primates living in groups: *Behaviour* **87**, 120–144.
8.2

Vance R.R. (1973a) On reproductive strategy in marine benthic invertebrates. *Amer. Natur.* **107**, 339–352.
8.3.1

Vance R.R. (1973b) More on reproductive strategies in marine benthic invertebrates. *Amer. Natur.* **107**, 339–352.
8.3.1

Vaughan W. Jr & Herrnstein R.J. (1987) Stability, amelioration and natural selection. In: *Advances in Behavioral Economics* (ed. L. Green & J.H. Kagel), pp. 185–215. Ablex Publishing Co., Norwood, New Jersey.
4.6.1

Veen J. (1987) Ambivalence in the structure of display vocalisations of gulls and terns: new evidence in favour of Tinbergen's conflict hypothesis? *Behaviour* **100**, 33–49.
12.7.2

Vehrencamp S.L. (1979) The roles of individual, kin, and group selection in the evolution of sociality. In: *Handbook of Behavioral Neurobiology* (ed. P. Marler & J.G. Vandenbergh), Vol. 3, pp. 351–394. Plenum Press, New York.
10.4.1, 10.6.3

Vehrencamp S.L. (1983) A model for the evolution of despotic versus egalitarian societies. *Anim. Behav.* **31**, 667–682.
3.3.2

Vehrencamp S.L. & Bradbury J.W. (1984) Mating systems and ecology. In: *Behavioural Ecology: An Evolutionary Approach*, 2nd edn. (ed. J.R. Krebs & N.B. Davies), pp. 251–278. Blackwell Scientific Publications, Oxford.
8.4.1, 9.1

Vehrencamp S.L., Bradbury J.B. & Gibson R.M. (1989) The energetic cost of display in male sage grouse. *Anim. Behav.* **38**, 885–896.
7.3.2

Vermeij G.I. (1987) *Evolution and Escalation; An Ecological History of Life.* Princeton University Press, Princeton, New Jersey.
6.1, 6.6, 6.7, 6.8

Verner J. (1964) The evolution of polygamy in the long-billed marsh wren. *Evolution* **18**, 252–261.
3.3.2

Verner J. & Willson M.F. (1966) The influence of habitats on mating systems of North American passerine birds. *Ecology* **47**, 143–147.
3.3.2, 9.5.3

Via S. (1984) The quantitative genetics of polyphagy in an insect herbivore. II. Genetic correlations in larval performance within and among host plants. *Evolution* **38**, 896–905.
2.2.2

Via S. & Lande R. (1985) Genotype–environment interaction and the evolution of phenotypic plasticity. *Evolution* **39**, 505–522.
2.1.1, 2.8.1

Visscher P.K. (1989) A quantitative study of worker reproduction in honey bee colonies. *Behav. Ecol. Sociobiol.* **25**, 247–254.
11.4.2

Voland E. (1988) Differential infant and child mortality in evolutionary perspective: data from late 17th to 19th century Ostfriesland (Germany). In: *Human Reproductive Behaviour: A Darwinian Perspective* (ed. L. Betzig, M Borgerhoff Mulder & P. Turke), pp. 253–261. Cambridge University Press, Cambridge.
3.3.1

Voland E. & Engel C. (1990) Female choice in humans: a conditional mate selection strategy of the Krummhörn women (Germany, 1720–1874). *Ethology* **84**, 144–154.
3.3.2

Waage J.K. (1984) Sperm competition and the evolution of Odonate mating systems. In: *Sperm Competition and the Evolution of Animal Mating Systems* (ed. R.L. Smith), pp. 251–290. Academic Press, London.
9.3.4

Waage J.K. (1986) Family planning in parasitoids: adaptive patterns of progeny and sex allocation. In: *Insect Parasitoids* (ed. J.K. Waage & D. Greathead), pp. 63–96. Academic Press, London.
8.6.3

Waage J.K. & Godfray H.C.J. (1985) Reproductive strategies and population ecology of insect parasitoids. In: *Behavioural Ecology* (ed. R.M. Sibly & R.H. Smith), pp. 449–470. Blackwell Scientific Publications, Oxford.
2.4.1, 2.5.1

Waage J.K. & Lane J.A. (1984) The reproductive strategy of a parasitic wasp. II. Sex allocation and local mate competition in *Trichogramma evanescens. J. Anim. Ecol.* **53**, 417–426.
8.6.3

Waage J.K. & Ng Sook Ming (1984) The reproductive strategy of a parasitic wasp. I. Optimal progeny and sex allocation in *Trichogramma evanescens. J. Anim. Ecol.* **53**, 401–415.
2.4.1

Waldman B., Frunhoff P.C. & Sherman P.W. (1988) Problems of kin recognition. *Trends Ecol. Evol.* **3**, 8–13.
12.3.2

Waller D.M. & Green D. (1981) Implications of sex for the analysis of life histories. *Amer. Natur.* **114**, 179–196.
2.6.1

Walters J.R. (1990) Red-cockaded woodpeckers: a 'primitive' cooperative breeder. In: *Cooperative Breeding in Birds: Long-term Studies of Ecology and Behavior* (ed. P.B. Stacey & W.D. Koenig), pp. 69–101. Cambridge University Press, Cambridge.
10.3.1

Walters J.R., Doerr P.D. & Carter J.H. III (1988) The cooperative breeding system of the red-cockaded woodpecker. *Ethology* **78**, 275–305.
10.3.1, 10.4.2

Walters J.R., Doerr P.D. & Carter J.H. III (in press) Cooperative breeding as a life history strategy in the red-cockaded woodpecker: test of a demographic model. *Amer. Natur.*
10.4.1, 10.4.3, 10.4.4, 10.4.6

Waltz E.C. (1981) Reciprocal altruism and spite in gulls: a comment. *Amer. Natur.* **118**, 588–592.
10.6.2

Wanless S., Harris M.P. & Morris J.A. (1988) The effect of radio transmitters on the behavior of common murres and razorbills during chick rearing. *Condor* **90**, 816–824.
8.4.4

Ward P.S. (1983a) Genetic relatedness and colony organization in a species complex

of ponerine ants I. Phenotypic and genotypic composition of colonies. *Behav. Ecol. Sociobiol.* **12**, 285–299.
11.4.2

Ward P.S. (1983b) Genetic relatedness and colony organization in a species complex of ponerine ants II. Patterns of sex ratio investment. *Behav. Ecol. Sociobiol.* **12**, 301–307.
11.5.4

Warner R.R. (1987) Female choice of sites versus mates in a coral reef fish, *Thalassoma bifasciatum. Anim. Behav.* **35**, 1470–1478.
9.3.3

Warner R.R. (1988) Traditionality of mating-site preferences in a coral reef fish. *Nature* **335**, 719–721.
9.3.3

Warner R.R. (1990) Male versus female influences on mating site determination in a coral reef fish. *Anim. Behav.* **39**, 540–548.
9.3.3

Waser P.M. (1988) Resources, philopatry, and social interactions among mammals. In: *The Ecology of Social Behavior* (ed. C.N. Slobodchikoff), pp. 109–130. Academic Press, New York.
10.4.4

Watson J.A.L., Okot-Kotber B.M. & Noirot C. (ed.) (1985) *Caste Differentiation in Social Insects.* Pergamon Press, Oxford.
11.1

Wattiaux J.M. (1968a) Parental age effects in *Drosophila pseudobscura. Exp. Gerontol.* **3**, 55–61.
2.2.2

Wattiaux J.M. (1968b) Cumulative parental age effects in *Drosophila subobscura. Evolution* **22**, 406–421.
2.2.2

Wcislo W.T. (1987) The roles of seasonality, host synchrony, and behaviour in the evolutions and distributions of nest parasites in Hymenoptera, with special reference to bees (Apoidea). *Biol. Rev.* **62**, 515–543.
11.4.1

Weary D.M. (1989) Categorical perception of bird song: how do great tits (*Parus major*) perceive temporal variation in their own song? *J. Comp. Psychol.* **103**, 320–325.
12.8

Weary D.M. (1990) Categorization of song notes in great tits: which acoustic features are used and why? *Anim. Behav.* **39**, 450–457.
12.8

Weary D.M., Krebs J.R., Eddyshaw R., McGregor P.K. & Horn A. (1988) Decline in song output by great tits: exhaustion or motivation? *Anim. Behav.* **36**, 1242–1244.
12.7.3

Weary D.M. & Lemon R.E. (1988) Evidence against the continuity–versatility relationship in bird song. *Anim. Behav.* **36**, 1379–1383.
12.7.3, 12.8

Weary D.M., Lemon R.E. & Date E.M. (1987) Neighbor–stranger discrimination by song in the veery, a species with song repertoires. *Can. J. Zool.* **65**, 1206–1209.
12.3.2

Weatherhead P.J. (1979) Ecological correlates of monogamy in tundra-breeding savannah sparrows. *Auk* **96**, 391–401.
8.4.4

Weatherhead P.J. & Robertson R.J. (1979) Offspring quality and the polygyny threshold: 'the sexy son hypothesis'. *Amer. Natur.* **113**, 201–208.
9.5.3

Weber N.A. (1979) Fungus-culturing by ants. In: *Insect–Fungus Symbiosis: Mutualism and Commensalism* (ed. L.R. Batra), pp. 77–116. Allanheld & Osmun, Montclair, New Jersey.
11.1

Wegge P. & Rolstad J. (1986) Size and spacing of capercaillie leks in relation to social behaviour and habitat. *Behav. Ecol. Sociobiol.* **19**, 401–408.
9.4.4

Wells K.D. (1977) The social behaviour of anuran amphibians. *Anim. Behav.* **25**, 666–693.
9.3.2, 9.4

Wells K.D. (1981) Parental behavior of male and female frogs. In: *Natural Selection and Social Behavior* (ed. R.D. Alexander & D.W. Tinkle), pp. 184–198. Chiron Press, New York.
8.4

Wells K.D. & Taigen T.L. (1986) The effect of social interactions on calling energetics in the gray treefrog (*Hyla versicolor*). *Behav. Ecol. Sociobiol.* **19**, 9–18.
7.3.2

Wells K.D. & Taigen T.L. (1989) Calling energetics of a neotropical treefrog, *Hyla microcephala. Behav. Ecol. Sociobiol.* **25**, 13–22.
7.3.2, 12.6.2

Wemmer C., Collins L.R., Beck B.B. & Rettberg B. (1983) The ethogram. In: *The*

Biology and Management of an Extinct Species: Pere David's Deer (ed. B.B. Beck & C. Wemmer), pp. 91–121. Noyes Publications, New Jersey.
12.3.4

Wenzel J.W. (1987) *Ropalidia formosa*, a nearly solitary paper wasp from Madagascar (Hymenoptera: Vespidae). *J. Kansas Ent. Soc.* **60**, 549–556.
11.4.1

Werner E.E. (1986) Amphibian metamorphosis: growth rate, predation risk and the optimal time to transform. *Amer. Natur.* **128**, 319–341.
5.6.2

Werner E.E. & Gilliam J.F. (1984) The ontogenetic niche and species interactions in sizestructured populations. *A. Rev. Ecol. Syst.* **15**, 393–425.
4.3.1, 5.6.2

Werner E.E., Gilliam J.F., Hall D.J. & Mittelbach G.G. (1983) An experimental test of the effects of predation risk on habitat use in fish. *Ecology* **64**, 1540–1548.
Introduction to Part 2, 5.6.2

Werren J.H. (1980) Sex ratio adaptations to local mate competition in a parasitic wasp. *Science* **208**, 1157–1159.
8.6.3

Werren J.H. (1983) Sex ratio evolution under local mate competition in a parasitic wasp. *Evolution* **37**, 116–124.
8.6.3

Werren J.H. (1987) Labile sex ratios in wasps and bees. *Bioscience* **37**, 498–506.
11.2.1

Werren J.H. & Charnov E.L. (1978) Facultative sex ratios and population dynamics. *Nature* **272**, 349–350.
11.5.4

West Eberhard M.J. (1969) The social biology of polistine wasps. *Miscellaneous Publications, Museum of Zoology* **140**, 1–101 (University of Michigan, Ann Arbor, Michigan).
11.4.1

West Eberhard M.J. (1975) The evolution of social behavior by kin selection. *Q. Rev. Biol.* **50**, 1–33.
10.6.2

West Eberhard M.J. (1978a) Polygyny and the evolution of social behavior in wasps. *J. Kansas Ent. Soc.* **51**, 832–856.
11.2.3

West Eberhard M.J. (1978b) Temporary queens in *Metapolybia* wasps: nonreproductive helpers without altruism? *Science* **200**, 441–443
11.2.3

West Eberhard M.J. (1979) Sexual selection, social competition and evolution. *Proc. Amer. Philos. Soc.* **123**, 222–234.
7.3.1

West Eberhard M.J. (1987) Flexible strategy and social evolution. In: *Animal Societies: Theories and Facts* (ed. Y. Itô, J.L. Brown & J. Kikkawa), pp. 35–51. Japan Science Society Press, Tokyo.
11.2.4, 11.4.3

Westneat D.F. (1987) Extra-pair fertilizations in a predominantly monogamous bird: genetic evidence. *Anim. Behav.* **35**, 877–886.
9.5.2

Westneat D.F. (1988) Male parental care and extrapair copulations in the indigo bunting. *Auk* **105**, 149–160.
9.5.2

Wetterer M. (1989) Central place foraging theory: when load size affects travel time. *Theor. Popul. Biol.* **36**, 267–280.
4.2.3

Wheeler D.E. (1986) Developmental and physiological determinants of caste in social Hymenoptera: evolutionary implications. *Amer. Natur.* **128**, 13–34.
11.2.4

Wheeler W.M. (1910) *Ants: Their Structure, Development and Behavior.* Columbia University Press, New York.
11.4.2

Wheeler W.M. (1923) *Social Life Among the Insects.* Harcourt Brace, New York.
11.4

White D.R. & Burton M.L. (1988) Cause of polygyny: ecology, economy, kinship and warfare. *Amer. Anthropol.* **90**, 871–887.
3.3.2

Whitfield D.P. (1986) Plumage variability and territoriality in breeding turnstone *Arenaria interpres*: status signalling or individual recognition? *Anim. Behav.* **34**, 1471–1482.
12.3.2

Whitfield D.P. (1987) Plumage variability, status signalling and individual recognition in avian flocks. *Trends Ecol. Evol.* **2**, 13–18.
12.3.3

Whitfield D.P. (1988) The social significance of plumage variability in wintering turnstone *Arenaria interpres*. *Anim. Behav.* **36**, 408–417.
12.3.2

Whitham T.G. (1980) The theory of habitat selection: examined and extended using

Pemphigus aphids. *Amer. Natur.* **115**, 449–466.
5.5.2

Whitney C.L. & Krebs J.R. (1975) Spacing and calling in Pacific tree frogs *Hyla regilla. Can. J. Zool.* **53**, 1519–1527.
12.6.2

Whittingham L.A. (1989) An experimental study of paternal behavior in red-winged blackbirds. *Behav. Ecol. Sociobiol.* **251**, 73–80.
8.4.4

Wickler W. (1968) *Mimicry in Plants and Animals.* World University Library, New York.
6.1, 6.5

Wickler W. (1980) Vocal duetting and the pairbond I. Coyness and partner commitment, a hypothesis. *Z. Tierpsychol.* **52**, 201–209.
12.3.2

Wicklund C. & Jarvi T. (1982) Survival of distasteful insects after being attacked by naive birds: a reappraisal of the theory of aposematic coloration evolving through individual selection. *Evolution* **36**, 998–1002.
6.4.1

Wilbur H.M. (1980) Complex life cycles. *A. Rev. Ecol. Syst.* **11**, 67–93.
5.6.2

Wilcox R.S. (1979) Sex discrimination in *Gerris remigis*: role of a surface wave signal. *Science* **206**, 1325–1327.
12.6.2

Wiley R.H. (1973) Territoriality and non-random mating in sage grouse *Centrocercus urophasianus. Anim. Behav. Monogr.* **6**, 85–169.
9.4.5

Wiley R.H. (1983) The evolution of communication: information and manipulation. In: *Communication* (ed. T.R. Halliday & P.J.B. Slater), pp. 82–113. Blackwell Scientific Publications, Oxford.
12.2, 12.4.1, 12.4.2

Wiley R.H. & Rabenold K.N. (1984) The evolution of cooperative breeding by delayed reciprocity and queuing for favorable social position. *Evolution* **38**, 609–621.
10.5.2, 10.5.3

Wiley R.H. & Richards D.G. (1978) Physical constraints on acoustic communication in the atmosphere: implications for the evolution of animal vocalisations. *Behav. Ecol. Sociobiol.* **3**, 69–94.
12.6.2

Wilkes A. (1963) Environmental causes of variation in the sex ratio of an arrhenotokous insect, *Dahlbominus fuliginosus* (Nees) (Hymenoptera: Eulophidae). *Can. Ent.* **95**, 183–202.
2.4.1

Wilkinson G.S. (1987) Equilibrium analysis of sexual selection in *Drosophila melanogaster. Evolution* **41**, 11–21.
7.2.2

Wilkinson G.S., Fowler K. & Partridge L. (1990) Resistance of genetic correlation structure to directional selection in *Drosophila melanogaster. Evolution.*
7.3.1

Williams G.C. (1966a) Natural selection, the costs of reproduction and a refinement of Lack's principle. *Amer. Natur.* **100**, 687–690.
2.1, 2.1.2, 2.5.1, 2.6.1

Williams G.C. (1966b) *Adaptation and Natural Selection.* Princeton University Press, Princeton, New Jersey.
8.1

Williams G.C. (1975) *Sex and Evolution.* Princeton University Press, Princeton, New Jersey.
8.4.3

Williams G.C. (1985) In defense of reductionism in evolutionary biology. In: *Oxford Surveys in Evolutionary Biology* (ed. R. Dawkins & M. Ridley), Vol. 2, pp. 1–27.
3.4.4

Williams G.C. & Williams D.C. (1957) Natural selection of individually harmful social adaptations among sibs with special reference to social insects. *Evolution* **11**, 32–39.
11.2.1

Willson M.F. & Burley N. (1983) *Mate Choice in Plants: Tactics, Mechanisms and Consequences.* Princeton University Press, Princeton, New Jersey.
7.5

Wilson D.S. & Colwell R.K. (1981) Evolution of sex ratio in structured demes. *Evolution* **35**, 882–897.
8.6.3

Wilson D.S. & Sober E. (1989) Reviving the superorganism. *J. Theor. Biol.* **136**, 337–356.
11.1

Wilson E.O. (1971) *The Insect Societies.* Harvard University Press, Cambridge, Massachusetts.
8.4, 11.3, 11.4, 11.4.1

Wilson E.O. (1975) *Sociobiology.* Harvard

University Press, Boston, Massachusetts.
3.4.1, 11.4

Wilson K. (1988) Egg laying decisions by the bean weevil *Callosobruchus maculatus*. *Ecol. Entomol.* **13**, 107–118.
2.8.6

Wilson K. (1989) The evolution of oviposition behaviour in the bruchid *Callosobruchus maculatus*. PhD thesis, University of Sheffield.
2.2.1, 2.2.2, 2.3.1, 2.4.1, 2.5.1

Winkler D.W. (1987) A general model for parental care. *Amer. Natur.* **130**, 526–543.
8.4.4

Winkler D.W. & Wallin K. (1987) Offspring size and number: a life history model linking effort per offspring and total effort. *Amer. Natur.* **129**, 708–720.
8.3.1

Winterhalder B.P. (1981) Foraging strategies in the boreal forest: an analysis of Cree hunting and gathering. In: *Hunter–Gatherer Foraging Strategies: Ethnographic and Archeological Analyses* (ed. B. Winterhalder & E.A. Smith), pp. 66–98, University of Chicago Press, Chicago.
3.2.1

Winterhalder B.P. (1986) Diet choice, risk, and food sharing in a stochastic environment. *J. Anthropol. Archaeol.* **5**, 369–392.
3.2.3

Winterhalder B.P. & Smith E.A. (ed.) (1981) *Hunter-Gatherer Foraging Strategies.* University of Chicago Press, Chicago.
3.1

Wittenberger J.F. (1976) The ecological factors selecting for polygyny in altricial birds. *Amer. Natur.* **110**, 779–799.
9.5.3

Wittenberger J.F. (1979) The evolution of mating systems in birds and mammals. In: *Handbook of Behavioral Neurobiology: Social Behavior and Communication* (ed. P. Master & J. Vandenburgh), pp. 271–349. Plenum Press, New York.
8.1

Wittenberger J.F. (1980) Group size and polygamy in social mammals. *Amer. Natur.* **115**, 197–222.
9.2.1

Wolf L., Ketterson E.D. & Nolan V. Jr. (1988) Paternal influence on growth and survival of dark-eyed junco young: do parental males benefit? *Anim. Behav.* **36**, 1601–1618.
8.2, 8.4.4, 9.5.1

Wolf L., Ketterson E.D. & Nolan V. Jr. (1989) Behavioural response of female dark-eyed juncos to experimental removal of their mates: implications for the evolution of parental care. *Anim. Behav.* **39**, 125–134.
8.4.4

Woltereck R. (1909) Weitere experimentelle Untersuchungen über Artveränderung, speziell über das Wesen quantitativer Artunderschiede bei Daphniden. *Verh. D. Tsch. Zool. Ges.* **1909**, 110–172.
2.8

Woolfenden G.E. (1975) Florida scrub jay helpers at the nest. *Auk* **92**, 1–15.
10.2, 10.3.1, 10.4.2

Woolfenden G.E. & Fitzpatrick J.W. (1978) The inheritance of territory in group-breeding birds. *Bioscience* **28**, 104–108.
10.5.2

Woolfenden G.E. & Fitzpatrick J.W. (1984) The *Florida Scrub Jay: Demography of a Cooperative Breeding Bird.* Princeton University Press, Princeton, New Jersey.
10.2, 10.3.1, 10.3.3, 10.4.1, 10.4.2, 10.4.4, 10.4.6, 10.5.2, 10.5.3, 10.6.3

Woolfenden G.E. & Fitzpatrick J.W. (1986) Sexual asymmetries in the life history of the Florida scrub jay. In: *Ecological Aspects of Social Evolution: Birds and Mammals* (ed. D. Rubenstein & R.W. Wrangham), pp. 87–107. Princeton University Press, Princeton, New Jersey.
10.5.2

Wootton R.J. (1979) Energy costs of egg production and environmental determinants of fecundity in teleost fishes. *Symp. Zool. Soc. Lond.* **44**, 133–159.
8.3.1

Wootton R.J. (1990) *Ecology of Teleost Fishes.* Chapman & Hall, London.
12.1

Wourms M.K. & Wasserman F.W. (1985) Butterfly wing markings are more advantageous during handling than during the initial strike of an avian predator. *Evolution* **39**, 845–851.
6.1

Wrangham R.W. (1980a) An ecological model of female-bonded primate groups. *Behaviour* **75**, 262–300.
9.2.3

Wrangham R.M. (1980b) Female choice of least costly males: a possible factor in the evolution of leks. *Z. Tierpsychol.* **54**, 357–367.
9.4.4

Wright J. & Cuthill I. (1989) Manipulation of sex differences in parental care. *Behav. Ecol. Sociobiol.* **25**, 171–181.
8.4.4, 9.5.1

Wright S. (1931) Evolution in Mendelian populations. *Genetics* **16**, 97–159.
2.8.1

Wright S. (1945) Tempo and mode in evolution: a critical review. *Ecology* **26**, 415–419.
11.2.1

Wright S. (1969) *Evolution and the Genetics of Populations. Vol. 2. The Theory of Gene Frequencies.* University of Chicago Press, Chicago.
1.3.2, 7.5

Wrigley E.A. (1978) Fertility strategy for the individual and the group. In: *Historical Studies of Changing Fertility* (ed. C. Tilly), pp. 135–154. Princeton University Press, Princeton, New Jersey.
3.3.2

Wunderle J.M., Santa Castro M. & Fetcher N. (1987) Risk-averse foraging by bananaquits on negative energy budgets. *Behav. Ecol. Sociobiol.* **21**, 249–255.
4.5.1

Wyatt A.J. (1954) Genetic variation and covariation in egg production and other economic traits in chickens. *Poult. Sci.* **33**, 1266–1274.
2.4.1

Wylie H.G. (1965) Effects of super-parasitism on *Nasonia vitripennis* (Walk.) (Hymenoptera: Pteromalidae). *Can. Ent.* **97**, 326–331.
2.4.1

Yanega D. (1988) Social plasticity and early-diapausing females in a primitively social bee. *Proc. Natl. Acad. Sci. USA* **85**, 4374–4377.
11.4.1

Yanega D. (1989) Caste determination and differential diapause within the first brood of *Halictus rubicundus* in New York (Hymenoptera: Halictidae). *Behav. Ecol. Sociobiol.* **24**, 97–107.
11.4.1, 11.5.4

Yasukawa K. (1981) Song repertoires in the red-winged blackbird *Agelaius phoeniceus*: a test of the Beau Geste hypothesis. *Anim. Behav.* **29**, 114–125.
12.7.3

Yasukawa K. & Searcy W.A. (1985) Song repertoires and density assessment in red-winged blackbirds: further tests of the Beau Geste hypothesis. *Behav. Ecol. Sociobiol.* **16**, 171–176.
12.7.3

Ydenberg R.C., Giraldeau L.A. & Falls J.B. (1988) Neighbours, strangers, and the asymmetric war of attrition. *Anim. Behav.* **36**, 343–347.
12.3.2

Yodzis P. (1981) Concerning the sense in which maximizing fitness is equivalent to maximizing reproductive value. *Ecology* **62**, 1681–1682.
2.1.2

Yokel D.A. (1989) Payoff asymmetries in contests among male brown-headed cowbirds. *Behav. Ecol. Sociobiol.* **24**, 209–216.
12.3.3

Zack S. & Ligon J.D. (1985a) Cooperative breeding in Lanius shrikes. I. Habitat and demography of two sympatric species. *Auk* **102**, 766–773.
10.3.1

Zack S. & Ligon J.D. (1985b) Cooperative breeding in Lanius shrikes. II. Maintenance of group-living in a nonsaturated habitat. *Auk* **102**, 766–773.
10.3.1, 10.4.3

Zack S. & Rabenold K.N. (1989) Assessment, age and proximity in dispersal contests among cooperative wrens: field experiments. *Anim. Behav.* **38**, 235–247.
10.4.3, 10.4.6

Zahavi A. (1974) Communal nesting by the Arabian babbler: a case of individual selection. *Ibis* **113**, 203–211.
10.4.2, 10.5.2

Zahavi A. (1975) Mate selection—a selection for a handicap. *J. Theor. Biol.* **53**, 205–214.
1.1, 1.4, 1.4.1, 7.3.1, 11.4.2

Zahavi A. (1977a) The cost of honesty (further remarks on the handicap principle). *J. Theor. Biol.* **67**, 603–605.
1.1, 1.4, 1.4.1, 7.3.1, 11.4.2

Zahavi A. (1977b) Reliability in communication systems and the evolution of altruism. In: *Evolutionary Ecology* (ed. B. Stonehouse & C.M. Perrins). pp. 253–259. Macmillan, London.
12.4.2

Zahavi A. (1978) Decorative patterns and the evolution of art. *New Scientist* **19**, 182–184.

Zahavi A. (1979) Ritualisation and the evolution of movement signals. *Behaviour* **72**, 77–81.
12.4.2

Zavhavi A. (1986) Reliability in signalling motivation. *Behav. Brain* **9**, 741–772.
12.4.2

Zahavi A. (1987)The theory of signal selection and some of its implications. In: *International Symposium of Biological Evolution* (ed. V.P. Delfino) Adriatica Editrice, Bari.
1.4.2

Zammuto R.M. (1986) Life histories of birds: clutch size, longevity and body mass among North American game birds. *Can. J. Zool.* **64**, 2739–2749.
2.7.2

Zeh D.W. & Smith R.L. (1985) Paternal investment in terrestrial arthropods. *Amer. Zool.* **25**, 785–805.
8.1

Zeh D.W. & Zeh J.A. (1988) Condition-dependent sex ornaments and field tests of sexual selection theory. *Amer. Natur.* **132**, 454–459.
7.3.1

Zentall T.R. & Galef B.G. Jr (1988) *Social Learning: Psychological and Biological Perspectives.* Lawrence Erlbaum Associates, Hillsdale, New Jersey.
3.4.3

Zuk M. (1989) Validity of sexual selection in birds. *Nature* **340**, 104.
7.3.3

Index